CONTEMPORARY ROBOTICS
- CHALLENGES AND SOLUTIONS

Edited by
A. D. RODIĆ

CONTEMPORARY ROBOTICS - Challenges and Solutions
Edited by A. D. Rodić

Republished by InTech
Janeza Trdine 9, 51000 Rijeka, Croatia

Technical Editor Teodora Smiljanic

Additional hard copies can be obtained from orders@intechopen.com

CONTEMPORARY ROBOTICS - Challenges and Solutions, Edited by A. D. Rodić
p. cm.
ISBN 978-953-307-038-4

For EU product safety concerns:
IN TECH d.o.o., Prolaz Marije Krucifikse Kozulić 3, 51000 Rijeka, Croatia,
info@intechopen.com or visit our website at intechopen.com.

Preface

According to the *Oxford English Dictionary*, the word *robotics* was first used in print by Isaac Asimov, in his science fiction short story "Liar!", published in May 1941 in *Astounding Science Fiction*. Asimov was unaware that he was coining the term; since the science and technology of electrical devices is *electronics*, he assumed *robotics* already referred to the science and technology of robots. However, in some of Asimov's other works, he states that the first use of the word *robotics* was in his short story *Runaround* (Astounding Science Fiction, March 1942). The word *robotics* was derived from the word *robot*, which was introduced to the public by Czech writer Karel Čapek in his play R.U.R. (*Rossum's Universal Robots*), which premiered in 1921.

The field of robotics was born in the middle of the last century when emerging computers were altering every field of science and engineering. Stories of artificial helpers and companions and attempts to create them have a long history, but fully autonomous machines only appeared in the 20th century. The first digitally operated and programmable robot, the Unimate, was installed in 1961 to lift hot pieces of metal from a die casting machine and stack them. Today, commercial and industrial robots are in widespread use performing jobs more cheaply or more accurately and reliably than humans. They are also employed in jobs which are too dirty, dangerous, or dull to be suitable for humans. Robots are widely used in manufacturing, assembly, and packing; transport; earth and space exploration; surgery; weaponry; laboratory research; safety; and mass production of consumer and industrial goods.

In actuality any machines, including familiar household appliances, which have microprocessors directing their actions can be considered as robots. In addition to vacuum cleaners, there are washing machines, refrigerators, and dishwashers that could be easily marketed as robotic devices. There are of course a wide range of possibilities, including those machines that have sensory environmental feedback and decision-making capabilities. In actual practice, in devices considered to be robotic, the amount of sensory and decision making capability may vary from a great deal to none.

In recent decades the study of robotics has expanded from a discipline centered on the study of mechatronic devices to a much broader interdisciplinary subject. An example of this is the area called human-centered robotics. Here one deals with the interactions between humans and intelligent machines. This is a growing area where the study of the interactions between robots and humans has enlisted expertise from outside the classical robotics domain. Concepts such as emotions in both robots and people are being studied, and older areas such as human physiology and biology are being incorporated into the mainstream of robotics research. These activities enrich the field of robotics, as they introduce new engineering and science dimensions into the research discourse.

The field of autonomous robots, a widely recognized test-bed, has recently benefited from salient contributions in robot planning using the results of algorithmic geometry as well as of a stochastic framework approach applied both to environmental modeling and robot localization problems (SLAM, simultaneous localization and maping), and further from the development of decisional procedures via Bayesian estimation and decision approaches.

For the last decade of the millennium, robotics largely dealt with the intelligent robot paradigm, blending together robots and machine-intelligence generic research within themes covering advanced sensing and perception, task reasoning and planning, operational and decisional autonomy, functional integration architectures, intelligent human–machine interfaces, safety, and dependability.

The evolution levels for robotics stress the role of theoretical aspects, moving from application domains to the technical and scientific area. The organization of this thematic book illustrates these different levels. The edited book is a collection of 18 chapters written by internationally recognized experts and well-known professionals of the field. Chapters contribute to diverse facets of contemporary robotics and autonomous systems. The volume is organized in four thematic parts according to the main subjects, regarding the recent advances in the contemporary robotics.

The first thematic topics of the book are devoted to the theoretical issues. This includes development of the algorithms for automatic trajectory generation using redundancy resolution scheme, intelligent algorithms for robotic grasping, modeling approach for reactive mode handling of flexible manufacturing and design of an advanced controller for robot manipulators.

The second part of the book concerns with different aspects of robot calibration and sensing. This includes a geometric and threshold calibration of a multiple robotic line-vision system, robot-based inline 2D/3D quality monitoring using picture-giving and laser triangulation, and a study on prospective polymer composite materials for flexible tactile sensors.

The third part addresses issues of mobile robots and multi-agent systems, including SLAM of mobile robots based on fusion of odometry and visual data, configuration of a localization system by a team of mobile robots, development of generic real-time motion controller for differential mobile robots, control of fuel cells of mobile robots, modeling of omni-directional wheeled-based robots, building of hunter-hybrid tracking environment, as well as design of a cooperative control in distributed population-based multi-agent approach.

The fourth part presents recent approaches and results in humanoid and bioinspirative robotics. That concerns with design of adaptive control of anthropomorphic biped gait, building of dynamic-based simulation for humanoid robot walking, building controller for perceptual motor control dynamics of humans and biomimetic approach to control mechatronic structure using smart materials.

The content of this thematic book admirably reflects the complementary aspects of theory and practice which have taken place in the last years. Certainly, the content of this book will serve as a valuable handbook to those who work in research and development of advanced robotic devices.

The editors are greatfull to the authors for their excellent work and interesting contributions. Thanks are also due to the renomeus publisher for their editorial assistance and excellent technical arrangement of the book.

December, 2009

A. D. Rodić

Contents

1

Automatic Trajectory Generation using Redundancy Resolution Scheme Based on Virtual Mechanism

Bojan Nemec and Leon Žlajpah
Robotics Laboratory, Jožef Stefan Institute, Jamova 39, 1001 Ljubljana Slovenia

1. Introduction

In recent times, while markets are reaching their saturation limits and customers are becoming more demanding, a paradigm shift has been taking place from mass production to customized mass production. The concept of customization focuses on satisfying a customer's unique needs with the help of new technologies. Typically, the products are similar but they differ in some parameters which make manual teaching and manual preparation of the manufacturing programs not acceptable. The customized mass production requires that all production phases are prepared in advance during the design phase of the specific product. This requires that standard production procedures are modified and prepared for each specific product. It is also required that the adaptation is done automatically without any human intervention. In modern production systems, CAD models of the product are used to generate specific machining programs. In the case of industrial robots, automatically generated programs have to consider various limitations, such as joint limits, wrist singularity and possible collisions of the robot with the environment. Although the off-line programming enables detection of such situations during the program preparation, it does not solve the basic goal - the automatic generation of feasible collision free trajectories. One of the most promising approaches to solving these problems is based on redundancy resolution control schemes, where the primary task is assigned to the trajectory tracking while the secondary task optimizes robot trajectories using various optimization goals, such as obstacle and singularity avoidance, staying within the available join limits, etc..

The basic definition of the kinematic redundancy is that the robot has more degrees of freedom than needed to accomplish the specific task. In the past, many control schemes were presented which use kinematic redundancy for the optimization of secondary tasks, such as obstacle avoidance, torque optimization, singularity avoidance, etc. All these schemes rely on a non-square Jacobian, which maps the joint velocities to the task space, which is in most cases described by the Cartesian coordinates. If the redundancy of the task can be easily described in Cartesian coordinates, i.e. the task is redundant in one of the Cartesian coordinates, then the solution is trivial an we can directly apply one of the existing control schemes. But there are tasks, such as brushing, polishing, grinding, sawing, etc. where the kinematic redundancy is hidden. It reveals when the circular shape of the tool is considered. Note that all six Cartesian coordinates are still needed to describe and to accomplish the given task.

Many authors have noticed this type of redundancy, but an efficient way how to solve it was not yet proposed (Kapoor et al., 2004; Nemec & Zlajpah, 2008; Sevier et al., 2006). As a solution to this problem we propose a virtual mechanism approach, where the tool is described as a virtual kinematic chain. In this case, the task space preserves its dimension, while the dimension of the joint space is increased by the dimension of the virtual mechanism. This approach has two major advantages. First, we can use existing robot Jacobian, which is assumed to be known. Second, the augmented part of the Jacobian, which describes the virtual mechanism, has a very simple structure in most cases. Using this formalism, we can directly apply any control and optimization algorithms developed for the kinematically redundant manipulators.

Additionally, we present some optimization procedures for secondary motion optimization. It is shown how to generate collision and wrist singularity free trajectories and how to avoid joint limits. We discuss also the case, where the given task redundancy does not allow to meet all optimization goals simultaneously. In such a case, we propose an on-line adaptation method, which reassigns a part of the primary tasks to the secondary task. The proposed approach is validated with illustrative experiments and examples - automated cell for finishing operations in the shoe production, shoe grinding cell and object tracking with the humanoid robot equipped with humanoid vision.

2. Task redundancy resolution

Robotic systems under study are n degrees of freedom (DOF) serial manipulators. We consider redundant systems, which have more DOF than needed to accomplish the task, i.e. the dimension of the joint space n exceeds the dimension of the task space m, $n > m$ and $r = n - m$ denote the degree of the redundancy. Let the configuration of the manipulator be represented by the a vector $\boldsymbol{q}_r$ of n joint positions, and the end-effector position (and orientation) by m-dimensional vector $\boldsymbol{x}_r$ of the robot tool center point positions (and orientations). The relation between the joints and the task velocities is given by the following well known expression

$$\dot{\boldsymbol{x}}_r = \mathbf{J}_r \dot{\boldsymbol{q}}_r \tag{1}$$

where $\mathbf{J}_r$ is the $m \times n$ manipulator Jacobian matrix. The solution of the above equation for $\dot{\boldsymbol{q}}_r$ can be given as a sum of particular and homogeneous solution

$$\dot{\boldsymbol{q}}_r = \bar{\mathbf{J}}_r \dot{\boldsymbol{x}}_r + \mathbf{N}_r \xi \tag{2}$$

where

$$\bar{\mathbf{J}}_r = \mathbf{W}^{-1} \mathbf{J}_r^T (\mathbf{J}_r \mathbf{W}^{-1} \mathbf{J}_r^T)^{-1}. \tag{3}$$

Here, $\bar{\mathbf{J}}_r$ is the weighted generalized-inverse of $\mathbf{J}_r$, $\mathbf{W}$ is the weighting matrix, $\mathbf{N}_r = (\mathbf{I} - \bar{\mathbf{J}}_r \mathbf{J}_r)$ is a $n \times n$ matrix representing the projection into the null space of $\mathbf{J}_r$, and ξ is an arbitrary n dimensional vector. We will denote this solution as the generalized inverse based redundancy resolution at the velocity level (Nenchev, 1989). The homogenous part of the solution belongs to the null-space of the Jacobian. Therefore, we will denote it as $\dot{\boldsymbol{q}}_n$, $\dot{\boldsymbol{q}}_n = \mathbf{N}_r \xi$.

Now consider the case where the robot Jacobian matrix $\mathbf{J}_w$ is defined in Cartesian (world) coordinate system and the dimension of the Jacobian is $6 \times n$, but the task is described in the another coordinate system, denoted with $\mathbf{p}$. The relation between the task velocities and cartesian velocities can be described as

$$\dot{\mathbf{p}} = \mathbf{J}_t \dot{\boldsymbol{x}}, \tag{4}$$

where $\mathbf{J}_t$ represents the task Jacobian. Let consider the case where the dimension of the task space m is less than the dimension of the Cartesian space, which is 6. It follows

$$\dot{x} = \bar{\mathbf{J}}_t \dot{\mathbf{p}} + \mathbf{N}_t \mu. \tag{5}$$

Here, N_t is the 6×6 task null space matrix and μ an arbitrary 6 dimensional vector. The redundancy resolution for such case can be expressed as

$$\dot{q}_r = \bar{\mathbf{J}}_w(\bar{\mathbf{J}}_t \dot{\mathbf{p}} + \mathbf{N}_t \mu) + \mathbf{N}_r \xi, \tag{6}$$

where vectors ξ and μ can be used for the secondary task accomplishment. The problem with the above approach is that the task Jacobian $\mathbf{J}_t$ becomes very complex even for a simple relation between then task and the cartesian coordinates. In many cases the analytical solution might even not exist.

We will demonstrate this with the shoe bottom roughing task. In the shoe bottom roughing process, we have to press the shoe bottom against a rotary grindstone and control the contact force and the contact position between the shoe bottom and the grindstone. Robot holds the shoe, while the position/orientation of the tool, grindstone in this case, is fixed. In general, the contact position on the grindstone can be freely chosen. Let define a polar task coordinates system, which describes rotary brush, as shown in the figure 1 The cartesian

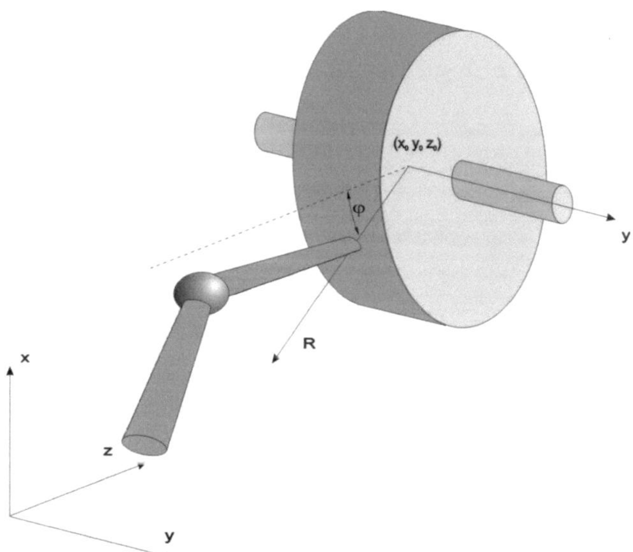

Fig. 1. Polar coordinate system of the rotary tool

coordinates are described with the $\boldsymbol{x}_r = (x, y, z, \phi, \theta, \psi)^T$ and the polar coordinates with the $\mathbf{p} = (R, y, \varphi, \phi, \nu, \psi)^T$, where ϕ, θ, ψ are the roll, pitch and yaw angles respectively, R is the radius of the polar coordinates, φ is the angle of the polar coordinate and $\nu = \theta + \varphi$. Coordinates x_0, y_0 and z_0 denote displacement of the center of the grindstone from the robot base. Obviously, coordinate φ can be freely chosen, because it does not matter which part of the rotary brush is used for the grinding. In general, also the coordinate y could be freely chosen, but since the grindstone is narrow in most cases, we will treat it as a restricted coordinate. The resulting task Jacobian, where the third line is canceled due to the task redundancy, has the form

$$\mathbf{J}_t = \begin{bmatrix} -\dfrac{(x-x_0)}{(-z_0+z)\sqrt{\frac{\eta}{(-z_0+z)^2}}} & 0 & -\dfrac{1}{\sqrt{\frac{\eta}{(-z_0+z)^2}}} & 0 & 0 & 0 \\ 0 & 1 & 0 & 0 & 0 & 0 \\ 0 & 0 & 0 & 1 & 0 & 0 \\ \dfrac{-z_0+z}{\eta} & 0 & -\dfrac{x-x_0}{\eta} & 0 & 1 & 0 \\ 0 & 0 & 0 & 0 & 0 & 1 \end{bmatrix},$$

where we used the substitute

$$\eta = (z_0 - z)^2 + (x - x_0)^2.$$

We can notice that even for the simplest case, the task Jacobian $\mathbf{J}_t$ becomes rather complex. In the case for the toroidal shaped brush we were not able to find the analytical solution of the task Jacobian using the Matlab symbolical computation toolbox.
As an alternative approach we propose to model the tool as a serial kinematic link. Let consider more general case where the robot holds the object to be machined and the work tool is fixed, as illustrated in Fig. 2. In such a case, we can define direct kinematic transformation as

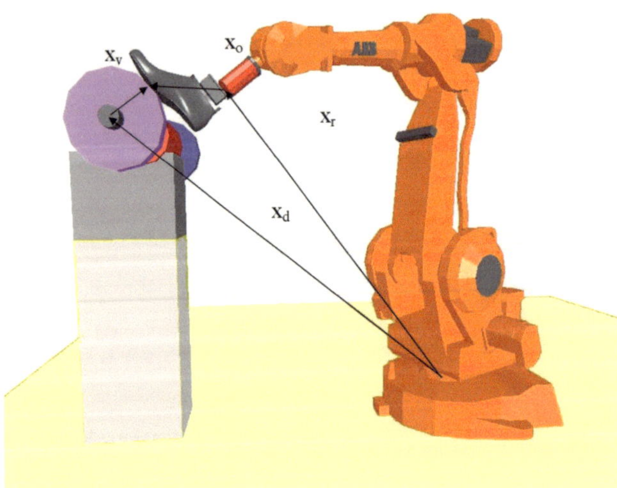

Fig. 2. The case when the robot holds an object and the work toll is fixed

$$\boldsymbol{x}_r + \left[\mathbf{R} \,\vdots\, \mathbf{I}\right] \boldsymbol{x}_o = \boldsymbol{x}_d + \boldsymbol{x}_v \tag{7}$$

where $\boldsymbol{x}_r$ is the robot Cartesian position and orientations, $\mathbf{R}$ is the robot tool rotation 3×3 dimensional matrix, $\boldsymbol{x}_o$ is the 6 dimensional vector of the object position and orientation, $\boldsymbol{x}_v$ is the 6 dimensional vector of position and orientation of the top of the virtual mechanism and

6 dimensional vector $\boldsymbol{x}_d$ describes the distance and orientation between the base coordinates system and the work tool coordinate system. Let consider robot and virtual mechanism as one mechanism with $n + n_v$ degrees of freedom, where n_v is the degree of freedom of the virtual mechanism. The configuration of the virtual mechanism can be described with the n_v dimensional vector $\boldsymbol{q}_v$. The new Cartesian position is

$$\boldsymbol{x} = \boldsymbol{x}_r - \boldsymbol{x}_v \tag{8}$$

The Jacobian of this new mechanism can be expressed as

$$\mathbf{J} = [\mathbf{J}_r \vdots - \mathbf{J}_v] \tag{9}$$

where $\mathbf{J}_r$ is the Jacobian of the robot and $\mathbf{J}_v$ Jacobian of the virtual mechanism is defined as

$$\mathbf{J}_v = \frac{\partial \boldsymbol{x}_v}{\partial \boldsymbol{q}_v}. \tag{10}$$

Note that we assume that the work tool rotation remains fixed during the execution of the task. Vector $\boldsymbol{q}_v$ of dimension n_v corresponds to the joints of the virtual mechanism.
As we can see, the task space preserves its dimension, while the joint space is increased with the dimension of the virtual mechanism. This approach has two major advantages. First, we can use existing robot Jacobian, which is assumed to be known. Second, the augmented part of the Jacobian has very simple structure in most cases. Another benefit of the proposed approach compared to the approach described with the equation 6 is that we have one larger null-space instead of two small null-spaces, which is more convenient for the trajectory optimization using the null-space motion.
As an example we present the direct kinematic transformation for the work cell shown in Fig. 5. The surface of the grinding disc is naturally described with outer surface of the torus, where R and r are the corresponding radii of the brush, as shown in the Fig. 3 and $\boldsymbol{x}$ is the task (Cartesian) coordinate of the whole system. Assuming that the robot tool position and

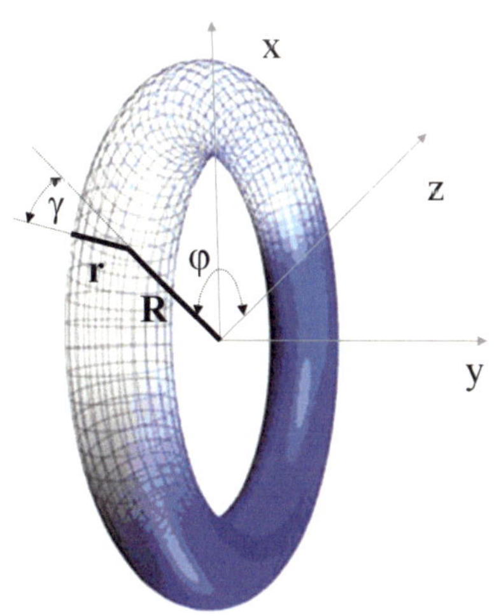

Fig. 3. Rotary brush presented as torus

robot Jacobian is known, the forward kinematics can be easily expressed as

$$\boldsymbol{x} = \boldsymbol{x}_r + \begin{bmatrix} s_\varphi (R + r\, c_\gamma) \\ r\, s_\gamma \\ c_\varphi (R + r\, c_\gamma) \\ 0 \\ -\varphi \\ \gamma \end{bmatrix}, \tag{11}$$

and the corresponding Jacobian is

$$\mathbf{J} = \begin{bmatrix} \mathbf{J}_r & \begin{matrix} c_\varphi (R + r\, c_\gamma) & -s_\varphi r\, s_\gamma \\ 0 & r\, c_\gamma \\ -s_\varphi (R + r\, c_\gamma) & -c_\varphi r\, s_\gamma \\ 0 & 0 \\ -1 & \\ 0 & 1 \end{matrix} \end{bmatrix}. \tag{12}$$

Here, we used the abbreviation $c_\varphi = \cos(\varphi)$, $c_\gamma = \cos(\gamma)$, $s_\varphi = \sin(\varphi)$ and $s_\gamma = \sin(\gamma)$. Note that Eq. 7 does not handle orientations correctly, since orientation vectors can not be simply added in general case. If orientations are important, we can use equation 7 for the calculation of positions only, while the orientations have to be calculated using rotation matrices as follows.

$$\mathbf{R}_o = \mathbf{R}_v{}^T \mathbf{R} \tag{13}$$

Here, $\mathbf{R}_o$ and $\mathbf{R}_v$ are 3×3 rotation matrices describing object rotation against virtual mechanism and virtual mechanism rotation expressed in the robot base coordinate system. The corresponding orientation vector can be than obtained using the transformation of the rotation matrix to the orientation vector described with euler or roll pitch yaw notation. On the other hand, orientation angles are additive for small angles. Let denote roll, pitch and yaw angles with ϕ,θ and ψ respectively. If angles are small, it holds

$$\begin{bmatrix} \frac{\Delta\phi_{rv}}{\Delta t} \\ \frac{\Delta\varphi_{rv}}{\Delta t} \\ \frac{\Delta\psi_{rv}}{\Delta t} \end{bmatrix} \simeq \begin{bmatrix} \frac{\Delta\phi_{v}}{\Delta t} \\ \frac{\Delta\varphi_{v}}{\Delta t} \\ \frac{\Delta\psi_{v}}{\Delta t} \end{bmatrix} + \begin{bmatrix} \frac{\Delta\phi_{v}}{\Delta t} \\ \frac{\Delta\varphi_{v}}{\Delta t} \\ \frac{\Delta\psi_{v}}{\Delta t} \end{bmatrix} = \mathbf{J}_r^{rot} \boldsymbol{q}_r + \mathbf{J}_v^{rot} \boldsymbol{q}_v. \tag{14}$$

Here, subscript r and v denotes the robot and the virtual mechanism respectively and *rot* denotes rotational part of the corresponding Jacobian. Equation 14 shows that even if rotation angles are not additive, velocities and thus Jacoians of the robot and virtual mechanism are additive and equation 9 hold also for the orientations. Therefore, we can directly apply any control algorithm based on Jacobian matrices and thus control and optimization algorithms developed for the kinematically redundant manipulators.

3. Control

As we mentioned in the previous section, we can directly apply any control algorithm for the kinematically redundant robot. Here we will briefly present a control law, based on the generalized inverse redundancy resolution at the velocity level in the extended operational space. Redundancy resolution at the velocity level is favorable because it enables direct implementation of the gradient optimization scheme for the secondary task. Although the control law using generalized inverse-based redundancy resolution at velocity level can not completely decouple the task and the null space (Nemec et al., 2007; Oh et al., 1998; Park et al., 2002), it enables good performance in real implementation. The joint space control law is

$$\begin{aligned}\tau_c = \; & \mathbf{H}\bar{\mathbf{J}}^T(\ddot{\boldsymbol{x}}_d + \mathbf{K}_v\dot{\boldsymbol{e}}_x + \mathbf{K}_p\boldsymbol{e}_x + \mathbf{K}_f\boldsymbol{e}_f - \dot{\mathbf{J}}\dot{\boldsymbol{q}}) + \\ & \mathbf{H}\mathbf{N}(\ddot{\boldsymbol{q}}_{nd} + \mathbf{K}_n\dot{\boldsymbol{e}}_n - \dot{\mathbf{N}}\dot{\boldsymbol{q}}) + \boldsymbol{h} + \mathbf{J}^T\boldsymbol{f}\end{aligned} \tag{15}$$

where $\bar{\mathbf{J}}$ is the inertia weighted pseudo-inverse of the Jacobian matrix $\mathbf{J}$, $\mathbf{H}$ is $n \times n$ the inertia matrix, $\boldsymbol{h}$ is n-dimensional vector of the centrifugal, coriolis and gravity forces, $\boldsymbol{F}$ is n-dimensional vector of the external forces acting on the manipulator's end effector and $\mathbf{K}_p$, $\mathbf{K}_v$, $\mathbf{K}_f$ and $\mathbf{K}_n$ are the corresponding $n \times n$ diagonal matrices with the positional, velocity, force and the null-space feedback gains. The first term of the control law corresponds to the task-space control τ_x, the second to the null-space control τ_n and the third and the fourth correspond to the compensation of the non-linear system dynamics and the external force, respectively. Here, $\boldsymbol{e}_x = \boldsymbol{x}_d - \boldsymbol{x}$ is the task-space tracking error, $\boldsymbol{e}_f = \boldsymbol{f}_d - \boldsymbol{f}$ and $\dot{\boldsymbol{e}}_n = \dot{\boldsymbol{q}}_{nd} - \dot{\boldsymbol{q}}_n$ are the force and the null-space tracking error. $\boldsymbol{x}_d$ and $\dot{\boldsymbol{q}}_{nd}$ are the desired task coordinates and the null space velocity, respectively. The details of the control law derivation can be found in (Nemec et al., 2007).

An attention should be paid on the selection of the inertia of the virtual link. The inertia matrix $\mathbf{H}$ has the form

$$\mathbf{H} = \begin{bmatrix} \mathbf{H}_m & 0 \\ 0 & \mathbf{H}_v \end{bmatrix} \tag{16}$$

where $\mathbf{H}_m$ is the manipulator inertia matrix and $\mathbf{H}_v$ is the diagonal matrix describing the virtual mechanism inertia. Clearly, $\mathbf{H}_v$ can not be zero, but arbitrary small values can be chosen describing the lightweight virtual mechanism. Selection of the inertia matrix of the virtual mechanism affects only the null space behavior of the whole system. Heavy virtual links with high inertia will slow down the movements of the virtual links. Therefore, low inertia of the virtual links makes suitable choice. On contrary, we can assume that the virtual links have no gravity and no coriolis and centrifugal forces and the corresponding terms in the vector $\boldsymbol{h}$ can be set to zero. Control law 15 assumes the feedback from all joints, including non-existing virtual joints. There are multiple choices how to provide the joint coordinates and the joint velocities of the virtual link. A suitable method is to build a simple model composed of a double integrator

$$\begin{aligned}\dot{\boldsymbol{q}}_v &= \int \mathbf{H}_v^{-1}\tau_{cv} \\ \boldsymbol{q}_v &= \int \dot{\boldsymbol{q}}_v\end{aligned} \tag{17}$$

where τ_{cv} is the part of the control signals corresponding to the virtual link.

4. Null space motion determination trough optimization

As we mentioned previously, one of the main problems in automatic trajectory generation is the inability to assure that the generated trajectory is feasible using a particular robot, either because of possible collisions with the environment or because of the limited workspace of the particular robot. Limitations in the workspace are usually not subjected to the tool position, but rather to the tool orientation. Another sever problem are wrist singularities, which can not be predicted in the trajectory design phase on a CAD system. A widely used solution in such cases is off-line programming with graphical simulation, where such situation can be detected in the design phase of the trajectory. Unfortunately this is a tedious and time consuming process and therefore not applicable in customized production, where almost each work piece can vary from the previous one (Dulio & Boer, 2004; Nemec & Zlajpah, 2008). The problem can be efficiently solved using the kinematic redundancy and null space motion, which changes the robot configuration, but does not affect the task space motion. The force and the position tracking are of the highest priority for a force controlled robot and are therefore considered as the primary task. The secondary task can be defined as a result of the local optimization of a given cost function. Here we will use the gradient projection technique, which has been widely implemented for the calculation of the null space velocity that optimizes the given criteria. The reason for this is that a variety of performance criteria can be easily expressed as gradient function of joint coordinates.

Let be the desired cost function, which has to be maximized or minimized. Then the velocities

$$\dot{\boldsymbol{q}}_n = \mathbf{N}\mathbf{H}^{-1}\frac{\partial p}{\partial \boldsymbol{q}}k, \tag{18}$$

maximize cost function for any $k > 0$ and minimize cost function for any $k < 0$ (Asada & Slotine, 1986), where k is an arbitrary scalar which defines the optimization step. In our case we have selected a compound p which maximizes the distances between obstacle and the robot links or robot work object, maximizes the distance to the singular configuration of the robot and maximizes the distance in joint coordinates between current joint angle and joint angle limit.

For the obstacle avoidance we use approach based on the potential field pointing away from the obstacle as illustrated in the figure 4 (Khatib, 1986; 1987).

$$p_a = \frac{1}{2}\boldsymbol{d}_i^T\boldsymbol{d}_i \tag{19}$$

where $\boldsymbol{d}_i$ is the shortest distance between the obstacle and the robot body. In our case the desired objective is fulfilled if the imaginary force is applied only on the robot joints and we can obtain the cost function gradient in a simple form as

$$\frac{\partial p_a}{\partial \boldsymbol{q}} = \boldsymbol{d}_1^T\mathbf{J}_{0,1}^{pos} + \boldsymbol{d}_2^T\mathbf{J}_{0,2}^{pos} + ... + \boldsymbol{d}_{n-1}^T\mathbf{J}_{0,n-1}^{pos}, \tag{20}$$

where $\mathbf{J}_{0,i}^{pos}$ denotes Jacobian matrices between base (the first index in the subscript) and i-th joint (the second index in the subscript) regarding the robot positions only.

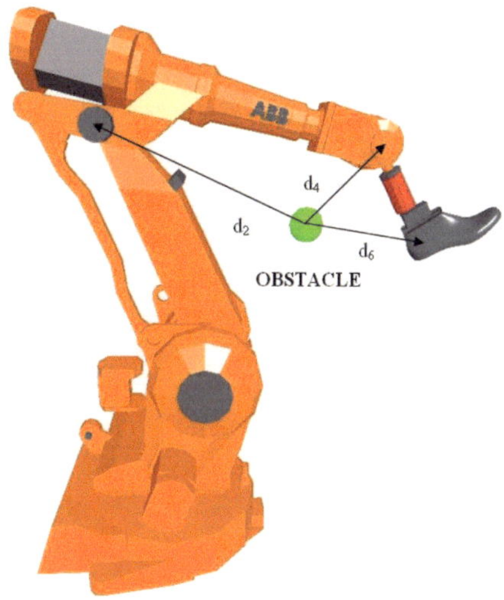

Fig. 4. Obstacle avoidance using potential field approach

The cost function for the joint limits avoidance is defined as (Chaumette & Marchand, 2001; Nemec & Zlajpah, 2008)

$$p_l = \frac{1}{2} \left\{ \begin{array}{c} (q_{max} - q)^2, |q_{max} - q| < \epsilon \\ 0 \\ (q_{min} - q)^2, |q_{min} - q| < \epsilon \end{array} \right\} \tag{21}$$

where ϵ is a positive constant defining the neighborhood of joint limits. Cost function gradient for obstacle avoidance is thus

$$\frac{\partial p_l}{\partial \boldsymbol{q}} = \boldsymbol{q}_{lim} - \boldsymbol{q}, \tag{22}$$

where $\boldsymbol{q}_{lim}$ denotes closed joint limits, that can be either $\boldsymbol{q}_{max}$ or $\boldsymbol{q}_{min}$.
For the singularity avoidance we use the manipulability index defined as (Asada & Slotine, 1986)

$$p_s = \sqrt{|\mathbf{J}\mathbf{J}^T|}, \tag{23}$$

where the gradient can be expressed as (Park et al., 1999)

$$\frac{\partial p_s}{\partial \boldsymbol{q}} = p_s[trace(\frac{\partial p_s}{\partial q_1}\bar{\mathbf{J}}) \quad trace(\frac{\partial p_s}{\partial q_2}\bar{\mathbf{J}}) \quad ...... \quad trace(\frac{\partial p_s}{\partial q_n}\bar{\mathbf{J}})]. \tag{24}$$

Note that in most cases the singularity is due to the spherical wrist of the robot and the equation 24 can be reduced by taking into the consideration only the last three joints which correspond to the wrist movement. Unfortunately, the partial derivative $\frac{\partial p_s}{\partial \boldsymbol{q}}$ is not easy to calculate. However, we can use the numerical derivative of the manipulability measure instead.
In practice we deal only with one optimization goal at the time, depending on which secondary task is the most critical one. Note that simple addition of different optimization gradients my not be appropriate, although it is often used in practice. Namely, it could

happen that two optimizations gradients act in opposite, e.g. obstacle avoidance might push joints toward the physical limits.
It might happen that there is no enough redundancy to fulfil the required secondary task. It the secondary task is critical, like obstacle avoidance or joint limit avoidance, the task execution has to be interrupted. The only possibility in such a case is to modify the primary task. For some tasks like polishing it is not necessary to assure the strict orientations of the tool. Therefore we can define orientation tracking as a secondary task. Using this approach we obtain additional degrees of redundancy, which can be successful used for the accomplishment of the critical secondary task. In such a case we have to assure that the tool orientation follows the desired orientation whenever it is possible. To do this, we define orientation tracking as an optimization procedure, where we minimize the difference between the desired orientation vector $\boldsymbol{x}_{od}$ and actual orientation vector $\boldsymbol{x}_o$.

$$p_o = \frac{1}{2}(\boldsymbol{x}_{od} - \boldsymbol{x}_o)^T(\boldsymbol{x}_{od} - \boldsymbol{x}_o) \tag{25}$$

$$\frac{\partial p_o}{\partial \boldsymbol{q}} = -(\boldsymbol{x}_{od} - \boldsymbol{x}_o)^T \mathbf{J}^{rot} \tag{26}$$

The dimension of the vector $\boldsymbol{x}_o$ depends from the specific application depending on how much redundancy we need in order to accomplish the most critical secondary task. $\mathbf{J}^{rot}$ denotes the corresponding rotational part of the Jacobian which relates to the selected components of $\boldsymbol{x}_o$.

5. Shoe grinding example

In the shoe assembly process, in order to attach the upper with the corresponding sole, it is necessary to remove a thin layer of the material off the upper surface so that the glue can penetrate the leather. To do this, the robot has to press the shoe against the grinding disc with the desired force while executing the desired trajectory. In the past, there were several approaches how to automate this operation. For mass production, there are special NC machines available. Their main drawback is relatively complicated setup and are therefore not suitable for the custom made shoes. Required flexibility is offered by the robot based grinding cell. In the EUROShoeE project (Dulio & Boer, 2004), a special force controlled grinding head has been designed. The robot manipulated with the grinding head while the shoe remained fixed on the conveyor belt (Jatta et al., 2004) The main drawback of this approach is relatively heavy and expensive grinding head. Additionally, force control can be applied only in one direction. In our approach, the robot holds the shoe and presses it against the grinding disc of a standard grinding machine as used in the shoe production industry. The impedance force control was accomplished by the robot using universal force-torque sensor mounted between the robot wrist and the gripper which holds the shoe last. It is well known that the kinematic redundancy enables greater flexibility in execution of complex trajectories. For example, also humanoid hand dexterity is subjected by its kinematical redundancy. We used Mitsubishi Pa10 robot with 7 D.O.F in our roughing cell, which has one degree of redundancy. Additional two degrees of redundancy were obtained by treating the grinding disc as a virtual mechanism. The surface of the grinding disc can be naturally described with the outer surface of the torus, where R and r are the corresponding radius of the grinding disc, as shown in the Fig. 3. Thus we have 9 degrees of freedom, 6 of them are required to describe the grinding task, while the remaining three degrees of freedom are used for the obstacle avoidance, joint limits avoidance and singularity avoidance. The prototype of the

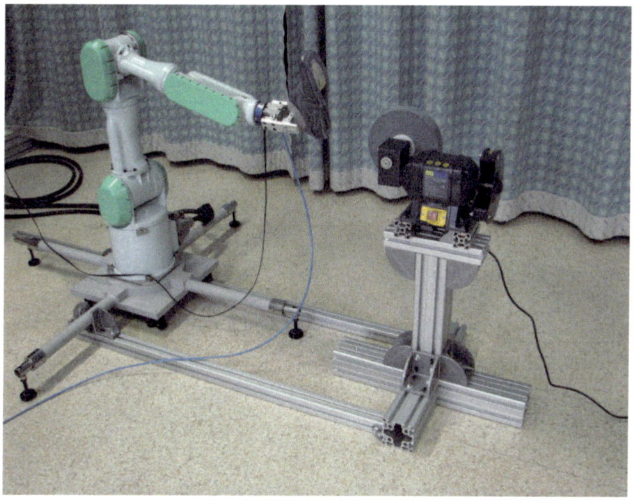

Fig. 5. Experimental cell for shoe bottom roughing

cell is shown in figure 5. It consists of the Mitsubishi Pa10 robot with a force/torque sensor Jr3 mounted in the robot wrist, a grinding machine, a Pa10 robot controller and a cell control computer, which coordinates the task and calculates the required robot torques. The control computer is connected to the robot controller using ArcNet. The frequency rate of the control algorithm (Eq. 15) and the motion optimization algorithm (Eq. 18) is 700 Hz. The grinding path is obtained from CAD model of the shoe. For this purpose, the control computer is connected to the shoe database computer using Ethernet. Unfortunately, CAD model itself can not supply all necessary data for the grinding process. CAD models are usually available for the reference shoe size, therefore, non-linear grading of the shoe shape is necessary for the given size. Additionally, some technological parameters such as material characteristics and shoe sole gluing technology have to be taken into account during the grinding trajectory preparation. For this purpose, we have developed a special CAD expert program, which enables the operator to define additional technological and material parameters. The program then automatically generates the grinding trajectory. In order to show the efficiency of the proposed algorithm, we defined the shoe grinding trajectory as seen in the Fig 4. Note that without using trajectory optimization is is very hard to execute the given task without splitting the desired trajectory in two or more fragments. Fig 5 shows how the system rotated joints of virtual mechanism in order to avoid the joint limits and to minimize joint velocities of the robot and virtual mechanism.

6. Automation of finishing operations in shoe assembly

Finishing operations in shoe manufacturing process comprises operations such as application of polishing wax, polishing cream and spray solvents, and brushing in order to achieve high gloss. These operations require skilled worker and are generally difficult to automate due to the complex motion trajectories. The finishing cell consists of the shoe polishing machine, machine for application of polishing creme, spray cabin for application of the polishing solvents and an industrial robot, as seen in Fig 4. The 6. d.o.f robot is a commercially available product from ABB, rest of the cell components were not available and had to be developed

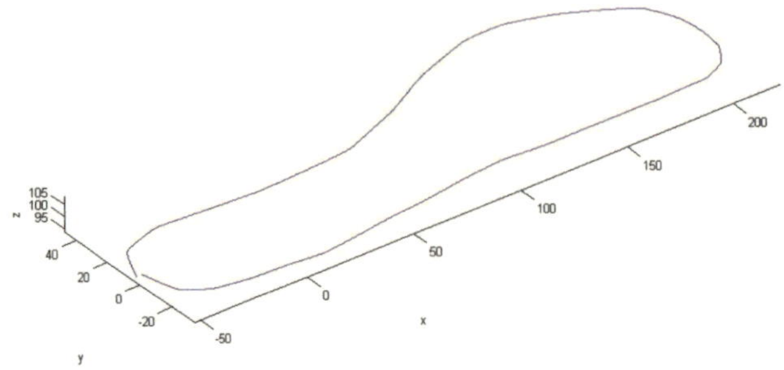

Fig. 6. Shoe grinding trajectory

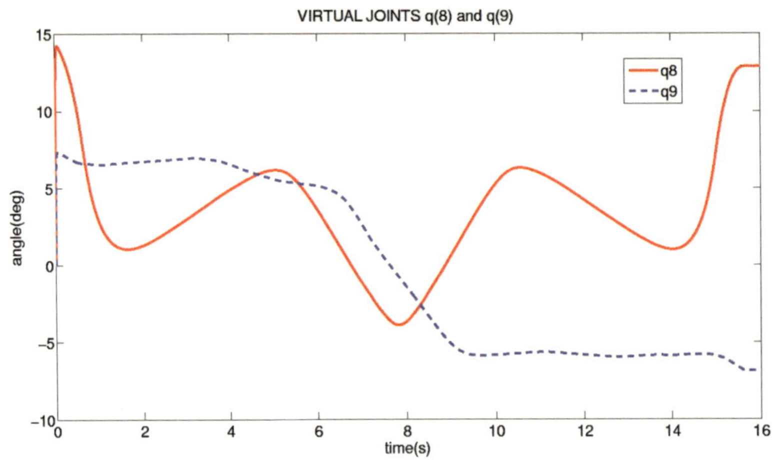

Fig. 7. Virtual mechanism angles q8 and q9

especially for this purpose. Customized mass production differs from the mass production because virtually any product item can differ from the previous one. Therefore, manual teaching and manual preparation of the manufacturing programs is not acceptable. The customized mass production requires that all production phases are prepared in advance during the design phase of the specific shoe model. Modification of the part programs for the specific shoe model, required for the customization, has to be done automatically without any human intervention. Therefore, new CAD tools for finishing operations had to be developed. One of the main problems in automatic trajectory generation is the inability to assure that the generated trajectory is feasible using a particular robot, either because of the possible collisions with the environment or because of the limited workspace of the particular robot. Limitations in the workspace are usually not subjected to the tool position, but rather to the tool orientation. Another sever problem are wrist singularities, which can not be predicted in the trajectory design phase on a CAD system. A widely used solution is such case is off-line programming with graphical simulation, where such situation can be detected in the design phase of the trajectory. Unfortunately this is a tedious and time consuming process and therefore not applicable in customized production, where almost each work piece can vary from the previous one (Dulio & Boer, 2004). The problem was efficiently solved using the trajectory optimization based on kinematic redundancy of the manipulator (Nemec &

Fig. 8. Finishing cell

Zlajpah, 2008). For a given task, the obstacle avoidance can be accomplished only if the robot is redundant. Note that the degree of redundancy depends on the task the robot is performing. For example, a 6 D.O.F robot is kinematicaly redundant for spraying and creaming operations in shoe production. Due to the circular shape of the cream application brush and spray beam, roll angle or the robot is free to choice. For brushing operations, there is another type of redundancy due to the circular shape of felt rollers, as illustrated in Fig. 1. Namely, the tool centre point is not restricted to be a fixed point, rather it can be freely chosen at the circumference of the tool. Unfortunately, in general one degree of redundancy is not enough to satisfy simultaneously all secondary tasks - obstacle avoidance, singularity avoidance and preserving the joint angles within their physical limits. More flexibility is offered by the fact that for some tasks it is not necessary to assure strict orientations of the tool. This can be interpreted as two additional degrees of redundancy. In robot trajectory generation, we define primary and secondary task. Primary task is the position of the TCP of the robot. We have multiple secondary tasks, such as a) Maximizing the distance between the robot joins and

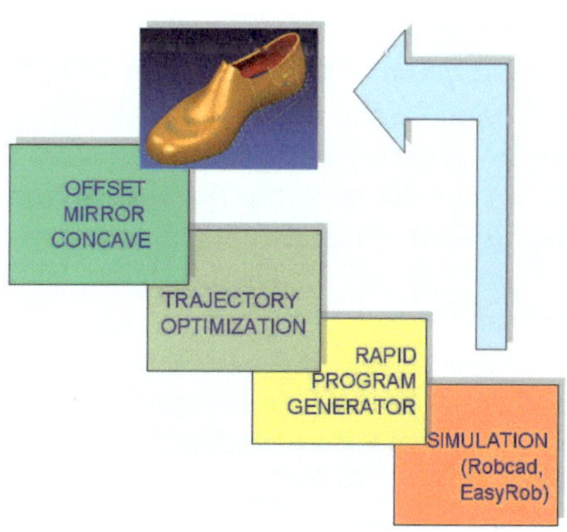

Fig. 9. Batch trajectory generation

the environment objects-obstacles. This task prevents the robot to collide with the obstacles b) Maximizing the distance between the joint position and join limits. This task prevents the robot to come to the join limits. c) Maximizing the distance between the actual and singular pose, which avoids wrist singularity d) Minimizing the difference between the desired and actual tool orientation. Secondary tasks generate robot tool orientation based on the gradient optimization in Jacobian null-space. Since we did not have access to the low level robot control, we implemented virtual mechanism approach as a batch procedure in the trajectory optimization module, as illustrated in the figure 9. Due to the physical limitations of the robot it is possible that the procedure does not converges. In such a case the optimization stops and off-line programming system is used to check and verify the robot configuration and the desired task trajectory. In most cases after the successful accomplishment of the trajectory optimization the verification with off-line programming system is not necessary and the trajectory can be downloaded directly to the robot controller.

7. Humanoid head control

The task of the robot head (Figure 10) is to keep the object in the center of both narrow-angle camera images. The head has to assure proper gaze direction of both eyes (cameras). Therefore, the task has four DOFs, since the gaze direction of each eye is defined by two parameters. A humanoid head typically has more DOFs, e. g. seven on the head of Fig. 10 and the degree of redundancy is three. Gaze direction is a function of a 3-D point in space as well as the position of the eye (see Figure 10). When the head is moved, the position of the eye change and that affects the gaze direction. Thus the task is a function of a point in space as well as a function of the head configuration. The above statement of the problem is not the most common way to describe a task - in general, a task is not a function of a robot configuration. To solve such problems in a systematic way, we have used the virtual mechanism approach (Omrcen & Ude, 2009). Let us explain the virtual mechanism approach on a simple pointing example, where finger points to an object. The task has one DOF and can be defined as an angle of the line from the finger to the object. If the object moves, the angle has to change in order to maintain the correct pointing direction. Similar, when the hand moves, the angle also has to change. The task is therefore a function of the hand and the object position. By adding a virtual prismatic link to the finger, we require that the extension of the finger touches the object. The system now has two DOFs (the angle of the finger and the length of the virtual mechanism). Using that notation, the task can be described as a positioning task; the end of the virtual link has to touch the object. The task now has two DOFs and the description of the task is now only the position of the object and not a function of the hand position. The introduction of the virtual mechanism increases the DOFs of the mechanism by one; however, it also increases the degree of the task. The degree of redundancy remains the same, while the description of the task has now been simplified and systematized: instead of specifying the desired pointing direction, we can consider the problem as a classic inverse kinematics task. In the case of a robot head we define a new virtual mechanism in each eye that points from the robot's eye to a point in space (see Figure 11). Virtual mechanism is a prismatic link, which adds a new degree of freedom to the system. The length of the new link is the distance from the eye to the 3-D point in space. Due to the two new DOFs added to the system, the system now has nine DOFs (one additional per each eye). However, the degree of the task has also increased. The task is not defined as a gaze direction of each eye but as the positioning of the end of the virtual link. Instead of controlling the gaze direction, the task is simplified to a simple

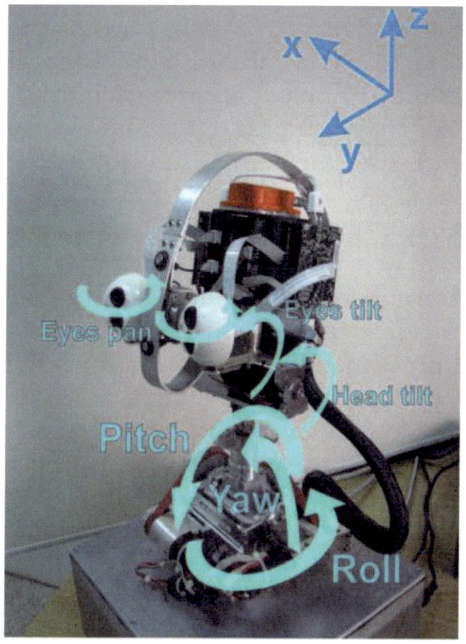

Fig. 10. Humanoid head

position control in space. Degree of redundancy should and does remain the same. Figure 11 schematically shows the kinematics of the head with additional virtual mechanisms.

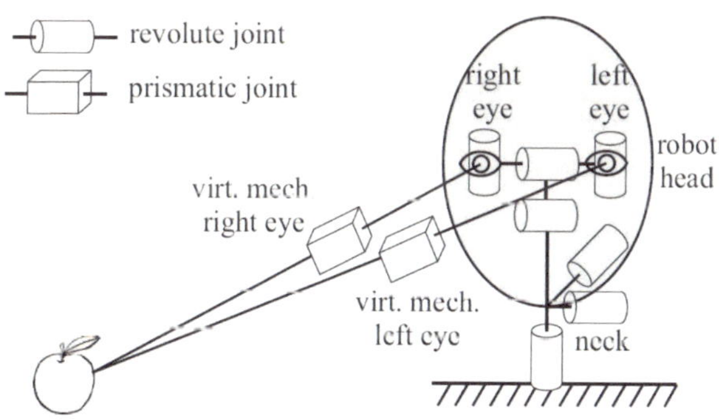

Fig. 11. Schematics of humanoid head with virtual mechanisms

8. Conclusion

In the chapter we deal with the automatic trajectory generation for industrial robots. Automatically generated programs have to consider various limitations of the robot mechanism, such as joint limits, wrist singularity and possible collisions of the robot with the environment. The required flexibility required to solve the above problems is offered by the kinematic redundancy. We proposed a new method of solving the kinematic redundancy which arises from the shape of the work tool. The main benefit of the proposed method is the simplicity and efficiency. It can be used on the existing robot controllers with very moderate changes of the control algorithm. The proposed approach is particularly efficient for the tasks which require automatic trajectory generation, since it helps to generate fault-tolerant trajectories The proposed approach was implemented in various industrial and non-industrial applications. We have outlined three illustrative examples : shoe bottom roughing task, automation of finishing operations in shoe assembly and control of a humanoid head. Two possible modes of implementation were proposed. Implementation of the proposed approach in the control loop allows real time optimization procedure, e.q. obstacle avoidance and to accomplish the given task at the same time. Another, perhaps for the practical implementation even more attractive possibility is to use the proposed approach in the trajectory generation module rather than in the control algorithm. Doing so, we get benefits of the kinematic redundancy due to circular shape of the tool without any modification of the robot controller. This latter approach was successfully implemented in the cell for custom finishing operations in shoe assembly. Unfortunately, in this case we have to deny to the real-time trajectory modifications, which can be only implemented in the control loop.

9. References

Asada, H. & Slotine, J.-J. (1986). *Robot Analysis and Control*, John Wiley & Sons.

Chaumette, F. & Marchand, . (2001). A redundancy-based iterative approach for avoiding joint limits: Application to visual servoing, *IEEE Transactions on Robotics and Automation, 17(5).*

Dulio, S. & Boer, S. (2004). Integrated production plant (ipp): an innovative laboratory for research projects in the footwear field, *Int. Journal of Computer Integrated Manufacturing, 17(7) : 601-611.*

Jatta, F., Zanoni, L., Fassi, I. & Negri, S. (2004). A roughing/cementing robotic cell for custom made shoe manufacture, *Int. J. Computer Intergrated Manufacturing, 17(7) : 645–652.*

Kapoor, C., Pholsiri, C. & Tesar, D. (2004). Manipulator task-based performance optimization., *Proc of DECT'04 ASME Conference, Salt Lake City.*

Khatib, O. (1986). Real-time obstacle avoidance for manipulators and mobile robots, *Int. J. of Robotic Research, 5 : 90 – 98.*

Khatib, O. (1987). A unified approach for motion and force control of robot manipulators:the operational space formulation, *IEEE Trans. on Robotics and Automation, 3(1) : 43 – 53.*

Nemec, B. & Zlajpah, L. (2008). Robotic cell for custom finishing operations, *Int. J. Computer Intergrated Manufacturing, 21(1) : 33–42.*

Nemec, B., Zlajpah, L. & Omrcen, D. (2007). Comparison of null-space and minimal null-space control algorithms, *Robotica, 2007, 25(5):511–520.*

Nenchev, D. N. (1989). Redundancy resolution through local optimization: A review, *J. of Robotic Systems, 6(6) : 769 – 798.*

Oh, Y., Chung, W., Youm, Y. & Suh, I. (1998). Experiments on extended impedance control of redundant manipulator, *Proc. IEEE/RJS Int. Conf. on Intelligent Robots and Systems, : 1320 – 1325, Victoria.*

Omrcen, D. & Ude, A. (2009). Redundancy control of a humanoid head for foveation and 3-d object tracking: A virtual mechanism approach., *submitted to Journal of Advanced Robotics.*

Park, J., Chung, W. & Youm, Y. (1999). Computation of gradient of manipulability for kinematically redundant manipulators including dual manipulators system, *Transactions on Control, Automation and Systems Engineering, 1(1).*

Park, J., Chung, W. & Youm, Y. (2002). Characterization of instability of dynamic control for kinematically redundant manipulators, *Proc. IEEE Conf. Robotics and Automation, : 2400 – 2405, Washington DC.*

Sevier, J., Kapoor, C. & Tesar, D. (2006). Benefitting from underutilized task specific resources for industrial robotic systems, *International Symposium on Robotics and Applications (ISORA) Budapest.*

Robotic Grasping of Unknown Objects

Mario Richtsfeld and Markus Vincze
Institute of Automation and Control
Vienna University of Technology
Gusshausstr. 27-29, Vienna, Austria

1. Introduction

"People have always been fascinated by the exquisite precision and flexibility of the human hand. When hand meets object, we confront the overlapping worlds of sensorimotor and cognitive functions" (Castiello, 2005). In the last few decades, the problem of grasping and manipulation of unknown objects in a fully automatic way has gained increasing importance, due mainly to the wide-spread use of service and rehabilitation robotics (Casals & Merchan, 1999), (Martens & Ruchel, 2001), (Ivlev & Martens, 2005). The grasping task has been studied from a psychological, biological and engineering focus but is still unresolved. There exist different solutions for certain cases; however there is still no general valid solution.

Automatic grasping of unknown objects from a single-view is a difficult problem because the pose and shape of the object are unknown and the possible hand configurations to realize a stable grasp are numerous. We propose a new strategy for modelling and grasping unknown objects based on 3D reconstruction. The system consists of a laser-range scanner, a simulation environment, a 7 DOF robot arm and a multi-fingered robot hand. The object to be grasped is scanned by a laser-range scanner from only one single-view and reconstructed in simulation. During the scanning process the system gets important information about the object structure, shape, place, size, and orientation of the object. We present a method for segmentation of a 2.5D point cloud into parts, assembly of rotationally symmetric parts into objects, and calculation of grasping points and poses. The algorithm checks potential collisions between the gripper, the object to be grasped, all surrounding objects and the table top. The experimental results demonstrate the effectiveness of the proposed method to grasp a wide range of objects.

The outline of the proposed work is as follows: Section 2 introduces our robotic system and its components. Section 3 describes the segmentation of 2.5D point clouds into parts, the assembly of parts into objects and details the merging of clipped rotationally symmetric objects. Section 4 details the calculation of grasping points for rotationally symmetric objects and optimal hand poses for arbitrary objects to grasp and manipulate an object without collision. Section 5 shows the achieved results and Section 6 finally concludes the work.

1.1 Problem Statement and Contribution

The goal of this work is to show a robust way of calculating possible grasps for unknown objects despite of noise, outliers and shadows. From a single-view two shadows appear: one from the camera and another one from the laser which can be caused by specular or reflective surfaces. We calculate collision free hand poses with a 3D model of the used gripper to grasp the objects, as illustrated in Fig. 1. That means that occluded objects can not be analyzed or grasped.

Fig. 1. Detected grasping points and hand poses. The green points display the grasping points for rotationally symmetric objects. The red points show an alternative grasp along the top rim. The illustrated hand poses show a possible grasp for the remaining graspable objects[1].

The problem of automatic 2.5D reconstruction to get practical grasping points and poses consists of several challenges. One of these concerns that an object might be detected as several disconnected parts, due to missing sensor data from shadows or poor surface reflectance. From a single-view the rear side of an object is not visible due to self occlusions, and the front side may be occluded by other objects. The algorithm was developed for arbitrary objects in different poses, on top of each other or side by side with a special focus on rotationally symmetric objects. If objects can not be separated because they are stacked one of each other they are considered as one object. If the algorithm detects rotationally symmetric parts (hypothesizing that the parts belong to the same object) this parts are merged, because this object class can be robustly identified and allows a cylindrical grasp as well as a tip grasp along the top rim (Schulz et al., 2005). For all other objects the algorithm calculates a tip grasp based on the top surface. To evaluate the multi-step solution procedure, we use 18 different objects presented in Fig. 2.

---

[1] All images are best viewed in colour!

Fig. 2. 18 different objects were selected to evaluate the grasp point and grasp pose detection algorithm, from left: 1. Coffee Cup (small), 2. Saucer, 3. Coffee Cup (big), 4. Cube, 5. Geometric Primitive, 6. Spray-on Glue, 7. Salt Shaker (cube), 8. Salt Shaker (cylinder), 9. Dextrose, 10. Melba Toast, 11. Amicelli, 12. Mozart, 13. Latella, 14. Aerosol Can, 15. Fabric Softener, 16. C-3PO, 17. Cat, 18. Penguin.

1.2 Related Work

In the last few decades the problem of grasping novel objects in a fully automatic way has gained increasing importance in machine vision. (Fagg & Arbib, 1998) developed the FARS model, which focuses especially on the action-execution step. Nevertheless, no robotic application has been yet developed following this path. (Aarno et al., 2007) presented an idea that the robot should, like a human infant, learn about objects by interacting with them, forming representations of the objects and their categories.

(Saxena et al., 2008) developed a learning algorithm that predicts the grasp position of an object directly as a function of its image. Their algorithm focuses on the task of identifying grasping points that are trained with labelled synthetic images of a different number of objects.

(Kragic & Bjorkman, 2006) developed a vision-guided grasping system. Their approach was based on integrated monocular and binocular cues from five cameras to provide robust 3D object information. The system was applicable to well-textured, unknown objects. A three fingered hand equipped with tactile sensors was used to grasp the object in an interactive manner. (Bone et al., 2008) presented a combination of online silhouette and structured-light 3D object modelling with online grasp planning and execution with parallel-jaw grippers. Their algorithm analyzes the solid model, generates a robust force closure grasp and outputs the required gripper pose for grasping the object. They consider the complete 3D model of one object, which will be segmented into single parts. After the segmentation step each single part is fitted with a simple geometric model. A learning step is finally needed in order to find the object component that humans choose to grasp it.

(Stansfield, 1991) presented a system for grasping 3D objects with unknown geometry using a Salisbury robotic hand, whereby every object was placed on a motorized and rotated table under a laser scanner to generate a set of 3D points. These were combined to form a 3D model. (Wang & Jiang, 2005) developed a framework of automatic grasping of unknown objects by using a laser-range scanner and a simulation environment. (Boughorbel et al.,

2007) aid industrial bin picking tasks and developed a system that provides accurate 3D models of parts and objects in the bin to realize precise grasping operations, but their superquadrics based object modelling approach can only be used for rotationally symmetric objects. (Richtsfeld & Zillich, 2008) published a method to calculate possible grasping points for unknown objects with the help of the flat top surfaces of the objects based on a laser-range scanner system. However there exist different approaches for grasping quasi planar objects, (Sanz et al., 1999). (Huebner et al., 2008) developed a method to envelop given 3D data points into primitive box shapes by a fit-and-split-algorithm with an efficient minimum volume bounding box. These box shapes give efficient clues for planning grasps on arbitrary objects. Another 3D model based work is presented by (El-Khoury et al., 2007).

(Ekvall & Kragic, 2007) analyzed the problem of automatic grasp generation and planning for robotic hands where shape primitives are used in synergy to provide a basis for a grasp evaluation process when the exact pose of the object is not available. The presented algorithm calculates the approach vector based on the sensory input and in addition tactile information that finally results in a stable grasp. (Miller et al., 2004) developed an interactive grasp simulator "GraspIt!" for different hands, hand configurations and objects. The method evaluates the grasps formed by these hands. At the beginning this work uses shape primitives, by modelling an object as a sphere, cylinder, cone or box (Miller et al., 2003). Their system uses a set of rules to generate possible grasp positions. This grasp planning system "GraspIt!" is used by (Xue et al., 2008). They use the grasp planning system for an initial grasp by combining hand pre-shapes and automatically generated approach directions. Their approach is based on a fixed relative position and orientation between the robotic hand and the object, all the contact points between the fingers and the object are efficiently found. A search process tries to improve the grasp quality by moving the fingers to its neighboured joint positions and uses the corresponding contact points to the joint position to evaluate the grasp quality and the local maximum grasp quality is located. (Borst et al., 2003) show that it is not necessary in every case to generate optimal grasp positions, however they reduce the number of candidate grasps by randomly generating hand configuration dependent on the object surface. Their approach works well if the goal is to find a fairly good grasp as fast as possible and suitable. (Goldfeder et al., 2007) presented a grasp planner which considers the full range of parameters of a real hand and an arbitrary object, including physical and material properties as well as environmental obstacles and forces.

(Recatalá et al., 2008) created a framework for the development of robotic applications on the synthesis and execution of grasps. (Li et al., 2007) presented a data driven approach to realize a grasp synthesis. Their algorithm uses a database of captured human grasps to find out the best grasp by matching hand shape to object shape.

Summarizing to the best knowledge of the authors in contrast to the state of the art reviewed above our algorithm works only with 2.5D point clouds from a single-view. We do not operate on a motorized and rotated table, which is unrealistic for real world use. The segmentation and merging step identifies different objects in the same table scene. The presented algorithm works on arbitrary objects and calculates especially for rotationally symmetric objects grasping points. For all other objects the presented method calculates possible grasping poses based on the top surfaces with a 3D model of the gripper. The algorithm checks potential collision with all surrounding objects. In most cases the shape information recovered from a single-view is too limited (missing rear side of the objects) that we do not attend to calculate force-closure grasps.

2. System Design and Architecture

The system consists of a pan/tilt-mounted red-light laser, a scanning camera and a seven degrees of freedom robot arm from AMTEC robotics[2], which is equipped with a human like prosthesis hand from OttoBock[3], see Fig. 3a.

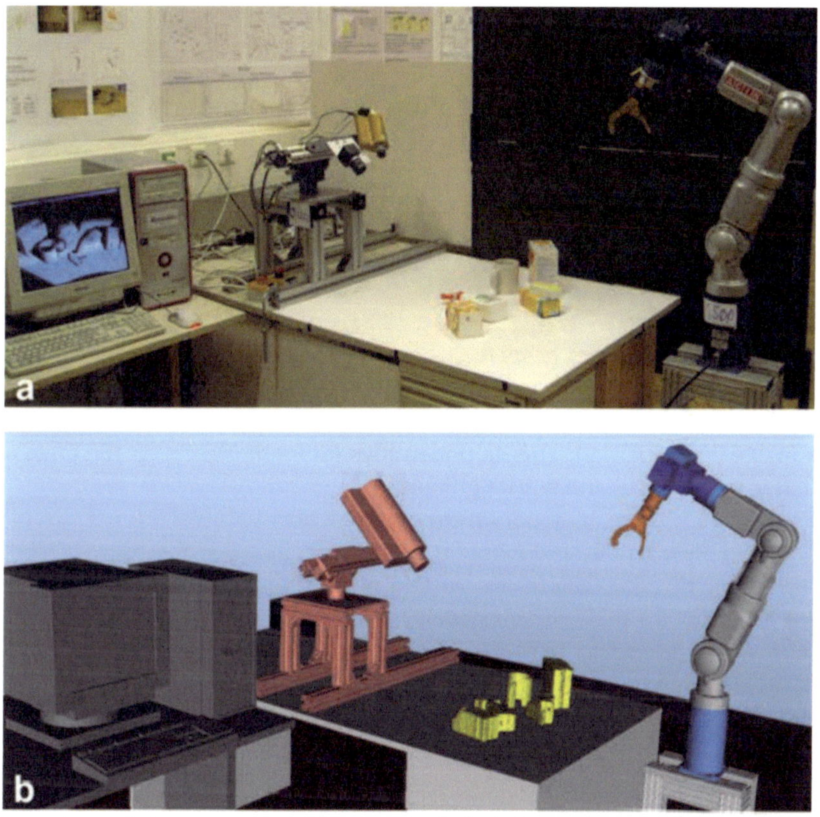

Fig. 3. **a** Overview of the system components and their interrelations. **b** Visualization of the experimental setup by a simulation tool, which is suitable to calculate the trajectory of the robot arm. The closed rear side of the objects on the table by an approximation of 2.5D to 3D is clearly visible.

First, the laser-range system scans the table scene and delivers a 2.5D point cloud. A high resolution sensor is needed in order to detect a reasonable number of points of the objects with sufficient accuracy. We use a red-light LASIRIS laser from StockerYale[4] with 635nm

[2] http://www.amtec-robotics.de/
[3] http://www.ottobock.de/
[4] http://www.stockeryale.com/index.htm

and a MAPP2500 CCD-camera from SICK-IVP[5] mounted on a PowerCube Wrist from AMTEC robotics.

The prosthesis hand has three active fingers: the thumb, the index finger, and the middle finger; the last two fingers are just for cosmetic reasons. The integrated tactile sensors are used to detect the sliding of objects to initialize a readjustment of the pressure of the fingers. It is thought that people will accept this type of gripper rather than an industrial gripper, due to the form and the optical characteristics. The virtual centre between the fingertip of the thumb, the index and the last finger is defined as tool centre point (TCP). The seventh degree of freedom of the robot arm is a rotational axis of the whole hand and is required to enable complex object grasping and manipulation and to allow for some flexibility for avoiding obstacles. There is a defined pose between the AMTEC robot arm and the scanning unit. A commercial path planning tool by AMROSE[6] calculates a collision free path to grasp the object. Before the robot arm delivers the object, the user can check the calculated trajectory in a simulation sequence, see Fig. 3b. Then the robot arm executes the off-line programmed trajectory. The algorithm is implemented in C++ using the Visualization Tool Kit (VTK)[7].

2.1 Algorithm Overview

The grasping algorithm consists of six main steps, see Fig. 4:

- Raw Data Pre-Processing: The raw data points are pre-processed with a smoothing filter to reduce noise to reduce noise.
- Range Image Segmentation: This step identifies different objects on the table or parts of an object based on a 3D DeLaunay triangulation.
- Pairwise Matching: Find high curvature points, which indicate the top rim of an object part, fit a circle to these points, and merge rotationally symmetric objects.
- Approximation of 2.5D Objects to 3D Objects: This step is only important to detect potential collisions by the path planning tool.
  - -Rotationally Symmetric Objects: Add additional points by using the main axis information.
  - -Arbitrary Objects: The non-visible range will be closed with planes, normal to the table plane.
- Grasp Point and Pose Detection:
  - -Grasp Point Detection: Rotationally Symmetric Objects.
  - -Grasp Pose Detection: Arbitrary Objects.
- Collision Detection: Considering all surrounding objects and the table surface as obstacles, to evaluate the calculated hand pose.

---

[5] http://www.sickivp.se/sickivp/de.html

[6] http://www.amrose.dk/

[7] Freely available open source software, http://public.kitware.com/vtk

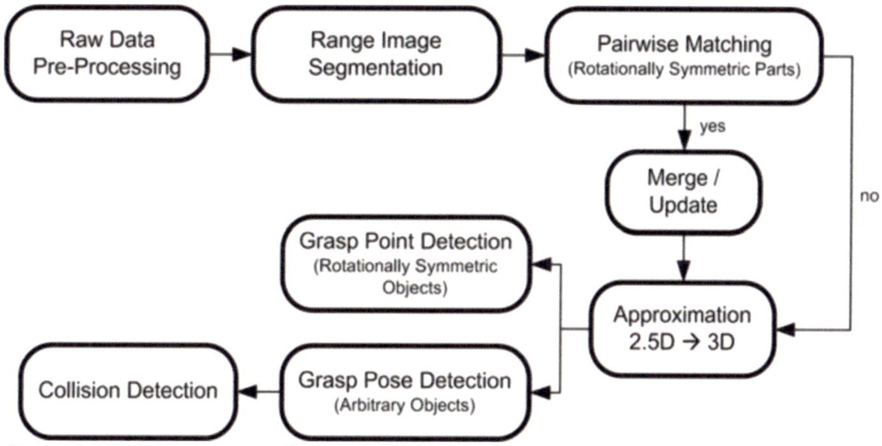

Fig. 4. Overview of the presented grasping algorithm.

3. Range Image Segmentation

The range image segmentation starts by detecting the surface of the table with a RANSAC (Fischler et al. 1981) based plane fit (Stiene et al., 2002). We define an object or part as a set of points with distances between neighbors. For that we build a kd-tree (Bentley, 1975) and calculate the minimum d_{min}, maximum, d_{max}, and average distance d_a between all neighboring points as input information for the mesh generation step. (Arya et al., 1998). The segmentation of the 2.5D point cloud is achieved with the help of a 3D mesh generation based on the triangles calculated by a 3D DeLaunay triangulation (O'Rourke, 1998). Then all segments of the mesh are extracted by a connectivity filter (Belmonte et al., 2004). This step segments the mesh into different components (objects or parts). An additional cut refinement was not arranged. The result may contain an over- or an under segmentation depending on the overlap of the objects as illustrated in Fig. 5.

Fig. 5. Results after the first segmentation step. Object no. 1 is cut into two parts and objects no. 5 and 7 are overlapping. The not perfectly segmented objects are red encircled

After the object segmentation step the algorithm finds the top surfaces of all objects using a RANSAC based plane fit and generates a 2D DeLaunay triangulation, with this 2D surface information the top rim points and top feature edges of every object can be detected, as illustrated in Fig. 6. For the top surface detection the algorithm uses a pre-processing step to find out all vertices[8] of the object with a normal vector in x-direction bigger than in y- or z-direction, n[x] > n[y] and n[x] > n[z], the x-direction is normal to the table plane. The normal vectors of all vertices are calculated with the faces (triangles) of the generated mesh.

Fig. 6. Results after the merging step. The wrongly segmented rotationally symmetric parts of object no. 1 are successfully merged to one object. The blue points represent the top rim of the objects.

3.1 Pairwise Matching

We developed a matching method, which is specifically for rotationally symmetric objects, because this objects can be stable segmented, detected and merged in a point cloud with unknown objects. To detect the top rim circle of rotationally symmetric objects a RANSAC based circle fit (Jiang & Cheng, 2005) with a range tolerance of 2mm is used.

Several tests have shown that this threshold provides good results for our currently used laser-range scanner. For an explicit description, the data points are defined as (p_{xi}, p_{yi}, p_{zi}) and (c_x, c_y, c_z) is the circle's centre with a radius r. The error must be smaller than a defined threshold:

$$\left| \left\| \vec{p} - \vec{c} \right\| - r \right| \leq 2 \tag{1}$$

This operation will be repeated for every point of the top rim. The run with the maximum number n of included points wins.

$$n = \left| \left\{ p \left| \left\| \vec{p} - \vec{c} \right\| - r \right| \leq 2 \right\} \right| \tag{2}$$

[8] In geometry, a vertex is a special kind of point which describes the corners or intersections of geometric shapes and a polygon is a set of faces.

If more than 80% of the rim points of both parts (rotationally symmetric parts) lie on the same circle, the points of both parts are examined more closely with the fit. For that we calculate the distances of all points of both parts to the rotation axis, see Equ. 3, the yellow lines represent the rotation axis, see Fig. 1, object no. 3. If more than 80% of all points of both parts agree, both parts are merged to one object, see Fig. 6, object no. 1.

$$d = \left(\vec{p} - \vec{c}\right) \times \vec{n} \tag{3}$$

3.2 Approximation of 3D Objects

This step is important to detect potential collisions by the path planning tool from AMROSE. In order to avoid wrong paths and collisions with other objects, due to missing model information, because in 2.5D point clouds every object is seen from only one view, but the path planning tool needs full information to calculate a collision free path. During the matching step the algorithm detected potential rotationally symmetric objects and merged clipped parts. With this information, the algorithm rotates only points along the axis by 360° in 5° steps, which fulfil the necessary rotation constraint. This means that only points will be rotated, which have a corresponding point on the opposite side of the rotation axis (Fig. 5, object no. 1) or build a circle with the neighbouring points along the rotation axis, as illustrated in Fig. 1, object no. 3 and Fig. 7, object no. 1. By this relatively simple constraint object parts such as handles or objects close to the rotationally symmetric object will not be rotated. For all other arbitrary objects, every point will be projected to the table plane and with a 2D DeLaunay triangulation the rim points can be detected. These points correspond with the rim points of the visible surfaces. So, the non-visible surfaces can be closed, these surfaces will be filled with points between the corresponding rim points, as illustrated in Fig. 7. Filling the non-visible range with vertical planes may lead to incorrect results, especially when the rear side of the objects is far from vertical, but this step is only to detect potential collisions by the path planning tool.

Fig. 7. Detection of grasping points and hand poses. The green points illustrate the computed grasping points for rotationally symmetric objects. The red points show an

alternative grasp along the top rim, thereby one grasping point is enough for an open object. The illustrated hand poses show a possible grasp for the remaining graspable objects.

4. Grasp Point and Pose Detection

The algorithm for grasp point detection is limited to rotationally symmetric objects and the grasp poses will be calculated for arbitrary objects. After the segmentation step we find out if the object is open or closed, for that we fit a sphere into the top surface. If there is no point of the object in this sphere we consider the object is opened. Then the grasping points of all cylindrical objects can be calculated. For every rotationally symmetric object we calculate two grasping points along the rim in the middle of the object (green coloured points, as illustrated in Fig. 8, object no. 1 and no. 6). If the path planner is not able to detect a possible grasp, the algorithm calculates alternative grasping points along the top rim of the object near the strongest curvature, as illustrated in Fig. 8, object no. 6 as red points. If it is an open object one grasping point is enough to realize a stable grasp near the top rim, as illustrated in Fig. 8 object no. 1. The grasping points should be calculated in such a way that they are next to the robot arm, which is mounted on the opposite side of the laser-range scanner. The algorithm detects the strongest curvature along the top rim with a Gaussian curvature filter (Porteous, 1994).

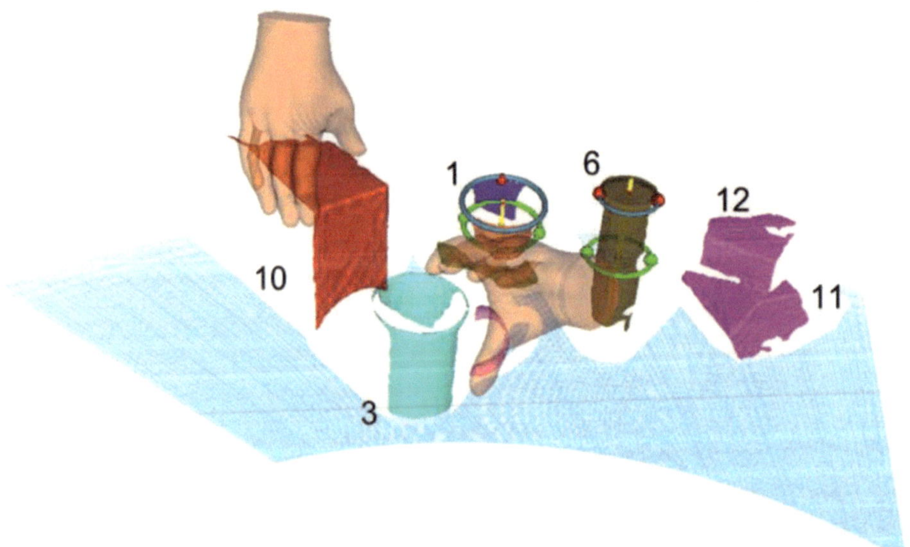

Fig. 8. Detection of grasping points and hand poses. The computed grasping points for rotationally symmetric objects. The red points show an alternative grasp along the top rim, thereby one grasping point is enough for an open object. The illustrated hand poses show a possible grasp for the remaining graspable objects.

To successfully grasp an object it is not always sufficient to locally find the best grasping pose. The algorithm should calculate an optimal grasping pose to realize a good grasp without collision as fast as possible. In general, conventional multidimensional "brut force" search methods are not practical to solve this problem. (Li et al., 2007) show a practical shape matching algorithm, where a reduced number of 38 contact points are considered. Most shape matching algorithms need an optimization step through that the searched optimum can be efficiently computed.

At the beginning the internal centre and the principal axis of the top surface are calculated with a transformation that fits and transforms a sphere inside, see Fig. 9b the blue top surfaces. After the transformation, this sphere has an elliptical form in alignment of the top surface points, hereby also the principal axis is founded. The algorithm transforms the rotation axis of the gripper (defined by the fingertip of the thumb, the index finger, and the last finger, as illustrated in Fig. 9a) along the principal axis of the top surface and the centre (calculated with the fingertips) of the hand c_h will be translated to the centre of the top surface c_{top}, whereby $c_h = c_{top}$ results. The hand will be rotated, so the normal vector of the hand aligns in reverse direction with the normal vector of the top surface. Afterwards the hand is shifted along the normal vectors up to a possible collision with the grasping object.

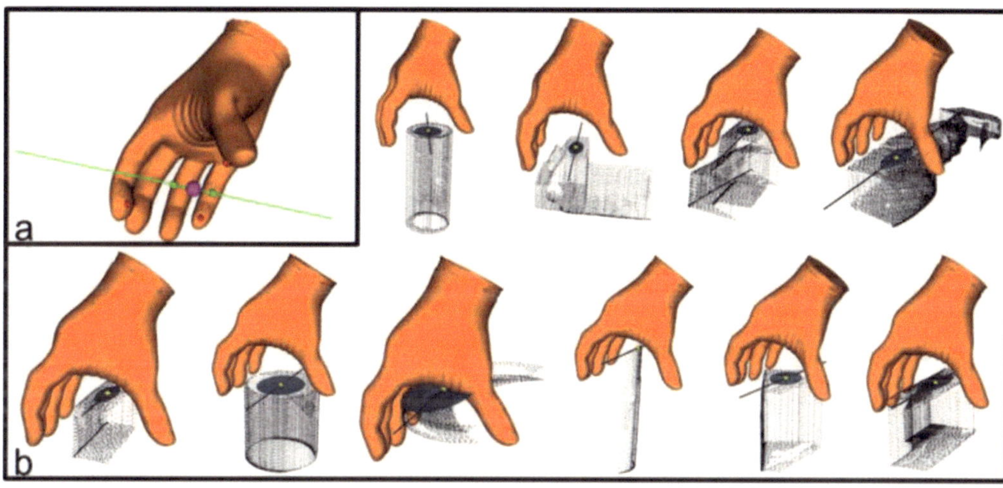

Fig. 9. Detection of grasping poses. **a** The rotation axis of the hand must be aligned with the principal axis of the top surface. **b** First grasping result: The hand was transformed and rotated along the principal axis of the top surface. After this step the algorithm checks potential collisions with all surrounding objects.

4.1 Collision Detection

The calculated grasping pose will be checked for a potential collision with the remaining objects and the table, as illustrated in Fig. 8. The algorithm determines if it is possible to grasp the object with an obb-tree. This method verifies possible points of the objects inside the hand by the calculated pose. If the algorithm detects a potential collision, the calculated pose will not be accepted.

5. Experiments and Results

We evaluated the detected grasping points and poses directly on the objects with an AMTEC robot arm and gripper. The object segmentation, merging, grasp point, pose detection, and collision detection is performed on a PC with 3.2GHz dual core processor and takes an average time of 35sec., depending on the number of the surrounding objects on the table, see Tab.1. The algorithm is implemented in C++ using the Visualization Tool Kit (VTK). Testing of 5 different point clouds for every object in different combination with other objects, the algorithm shows positive results. A remaining problem is that in some cases, interesting parts of shiny objects are not visible for the laser-range scanner, hence our algorithm is neither able to calculate correct grasping points nor the pose of the object. Another problem is that the quality of the point cloud is sometimes not good enough to guarantee a successful grasp, as illustrated in Fig. 10. The success of our grasping point and grasping pose algorithm depends on the ambient light, object surface properties, laser-beam reflectance, and absorption of the objects. For object no. 2 (saucer) the algorithm cannot detect possible grasping points or a possible grasping pose, because of shadows of the laser-range scanner and occlusion with the coffee cup, as illustrated in Fig. 1. In addition this object is nearly impossible to grasp with the used gripper. The algorithm cannot calculate possible grasping poses for object no. 16 (C-3PO), because of inadequate scan data. Finally the used gripper was not able to grasp object no. 15 (fabric softener), because of a slip effect. For all tested objects we achieved an average grasp rate of 71.11%.
In our work, we demonstrate that our grasping point and pose detection algorithm with a 3D model of the used gripper for unknown objects, performs practical grasps, as illustrated Tab. 2.

Calculation Steps	**Time[sec]**
Plane Fit	1.4sec
Mesh Generation	11sec
Mesh Segmentation	0.7sec
Top Surface Detection	0.9sec
Merging Rotationally Symmetric Objects	2.0sec
Approximation of 3D Objects	6sec
Grasp Point Detection	3.5sec
Grasp Pose Detection	6.5sec
Collision Detection	10sec
Sum	**35sec**

Table 1. Duration of Every Calculation Step.

No.	**Objects**	**Rate[%]**
1	Coffee Cup (small)	100%
2	Saucer	0%
3	Coffee Cup (big)	60%
4	Cube	100%
5	Geometric Primitive	100%
6	Spray on Glue	100%
7	Salt Shaker (cube)	100%
8	Salt Shaker (cylinder)	100%
9	Dextrose	80%

10	Melba Toast	100%
11	Amicelli	80%
12	Mozart	60%
13	Latella	100%
14	Aerosol Can	80%
15	Fabric Softener	0%
16	C-3PO	0%
17	Cat	20%
18	Penguin	100%
Overall		**71.11%**

Table 2. Grasping rate of different, unknown objects (each object was tested 5 times).

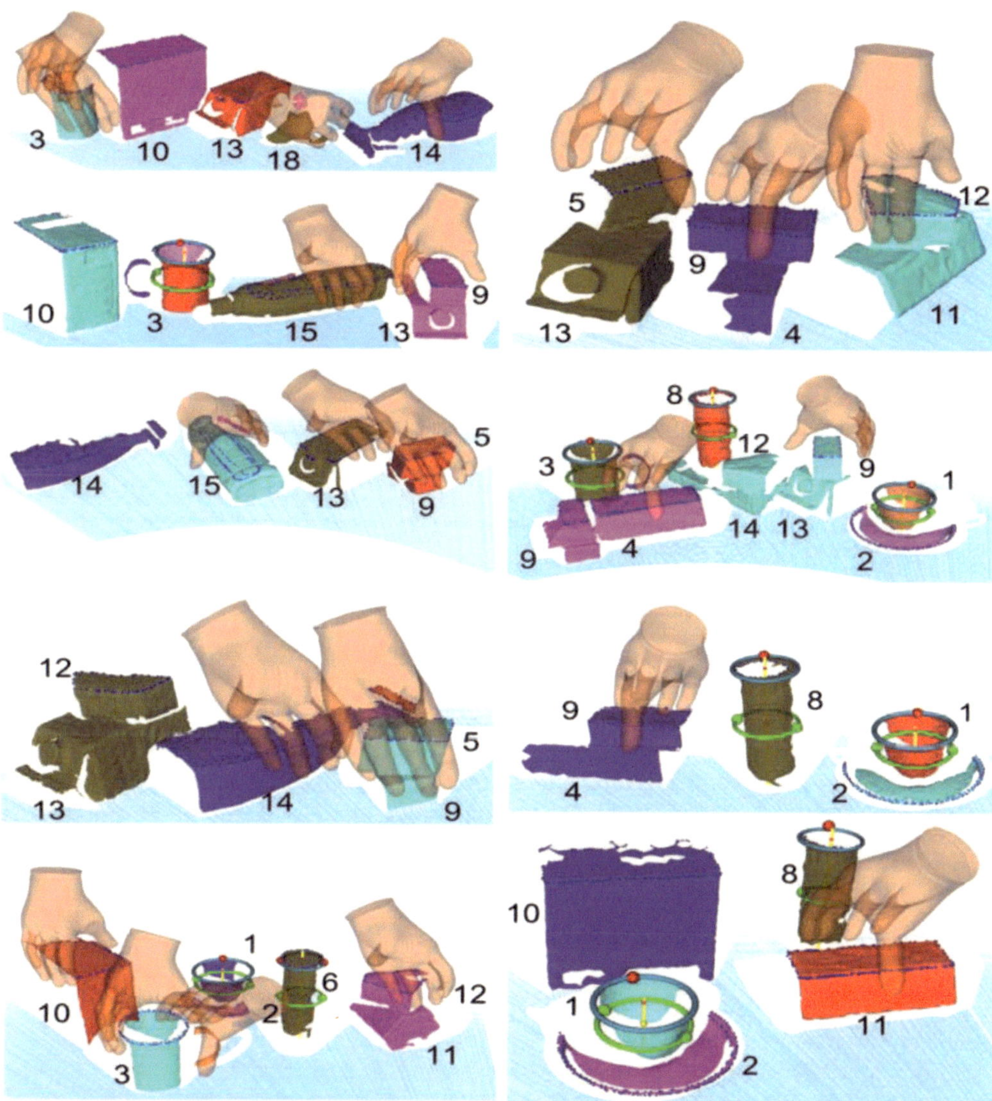

Fig. 10. Examples of detection results.

7. Conclusion and Future Work

We present a method for automatic grasping of unknown objects with a prosthesis hand, by incorporating a laser-range scanner which shows a high reliability. The approach for object grasping is well suited for use in related applications under different conditions and can be applied to a reasonable set of objects.
From a single-view the rear side of an object is not visible due to self occlusions, and the front side may be occluded by other objects. The algorithm was developed for arbitrary objects in different poses, on top of each other or side by side with a special focus on rotationally symmetric objects. If objects can not be separated because they are stacked one of each other they are considered as one object. If the algorithm detects rotationally symmetric parts this parts are merged, because this object class can be robustly identified and allows a cylindrical grasp as well as a tip grasp along the top rim. For all other objects the algorithm calculates a tip grasp based on the top surface, also for on top of each other arranged objects.
In the near future we plan to use a deformable hand model to reduce the opening angle of the hand, so we can model the closing of a gripper in the collision detection step. We will also use geometric hashing (Wolfson, 1997) for the matching step in order to be able to unite faster, several parts of the point cloud.

8. References

Aarno, D., Sommerfeld, J., Kragic, D., Pugeault N., Kalkan, S., Wörgötter, F., Kraft, D., Krüger, N. (2007). Early Reactive Grasping with Second Order 3D Feature Relations. IEEE International Conference on Robotics and Automation, Workshop: From features to actions - Unifying perspectives in computational and robot vision.

Arya, S., Mount, D.M., Netanyahu, N.S., Silverman, R. (1998). An Optimal Algorithm for Approximate Nearest Neighbor Searching in Fixed Dimensions. Journal of the ACM, Vol. 45, No. 6, pp. 801-923.

Bentley, J.L. (1975). Multidimensional Binary Search Trees Used for Associative Searching. Communications of the ACM, Vol. 18, No. 19, pp. 509-517.

Belmonte, Ó., Remolar, I., Ribelles, J., Chover, M., Fernández, M. (2004). Efficiently using connectivity information between triangles in a mesh for real-time rendering. Elsevier Science, Vol. 20, No. 8, pp. 1263-1273.

Besl, P.J., McKay, H.D. (1992). A method for registration of 3-D shapes. IEEE Transactions on Pattern Analysis and Machine Intelligence, Vol. 14, No. 2, pp. 239-256.

Bone, G.M., Lambert, A., Edwards, M. (2008). Automated modelling and robotic grasping of unknown three-dimensional objects. IEEE International Conference on Robotics and Automation, pp. 292-298.

Borst, C., Fischer, M., Hirzinger, G. (2003). Grasping the dice by dicing the grasp. IEEE/RSJ International Conference on Robotics and Systems, pp. 3692-3697.

Boughorbel, F., Zhang, Y. (2007). Laser ranging and video imaging for bin picking. Assembly Automation, Vol. 23, No. 1, pp. 53-59.

Casals, A., Merchan, R. (1999). Capdi: A robotized kitchen for the disabled and elderly people. Proceedings of the 5th European Conference for the Advancement Assistive Technology / AAATE, pp 346–351.

Castiello, U. (2005). The neuroscience of grasping. Nature Reviews Neuroscience, Vol. 6, No. 9, pp. 726-736.
Ekvall, S., Kragic, D. (2007). Learning and Evaluation of the Approach Vector for Automatic Grasp Generation and Planning. IEEE International Conference on Robotics and Automation, pp. 4715-4720.
El-Khoury, S., Sahbani A., Perdereau, V. (2007). Learning the Natural Grasping Component of an Unknown Object. IEEE/RSJ International Conference on Intelligent Robots and Systems, pp. 2957-2962.
Fagg, A.H., Arbib, M.A. (1998). Modeling parietal-premotor interactions in primate control of grasping. Neural Networks, Vol. 11, pp. 1277-1303.
Fischler, M.A., Bolles, R.C. (1981). Random Sample Consensus: A Paradigm for Model Fitting with Applications to Image Analysis and Automated Cartography. Communications of the ACM, Vol. 24, No. 6, pp. 381-395.
Goldfeder, C., Allen, P.K., Lackner, C., Pelossof, R. (2007). Grasp Planning via Decomposition Trees. IEEE International Conference on Robotics and Automation, pp. 4679-4684.
Huebner, K., Ruthotto, S., Kragic, D. (2008). Minimum Volume Bounding Box Decomposition for Shape Approximation in Robot Grasping. IEEE International Conference on Robotics and Automation, pp. 1628-1633.
Ivlev, O., Martens, C. (2005). Rehabilitation robots friend-i and friend-i with the dexterous lightweight manipulator. IOS Press, Vol. 17, pp. 111–123.
Jiang, X., Cheng, D.C. (1999). Fitting of 3D circles and ellipses using a parameter decomposition approach. Proceedings of the Fifth International Conference on 3-D Digital Imaging and Modeling, pp. 103–109.
Kragic, D., Bjorkman, M. (2006). Strategies for object manipulation using foveal and peripheral vision. IEEE International Conference on Computer Vision Systems, pp. 50-55.
Li, Y., Fu, J.L., Pollard, N.S. (2007). Data-Driven Grasp Synthesis Using Shape Matching and Task-Based Pruning. IEEE Transactions on Visualization and Computer Graphics, Vol. 13, No. 4, pp. 732-747.
Martens, C., Ruchel, N. (2001). A friend for assisting handicapped people. IEEE Robotics & Automation Magazine, Vol. 8, pp. 57–65.
Miller, A.T., Allen, P.K. (2004). GraspIt! A Versatile Simulator for Robotic Grasping. IEEE Robotics & Automation Magazine, Vol. 11, No. 4, pp. 110-112.
Miller, A.T., Knoop, S. (2003). Automatic grasp planning using shape primitives. IEEE International Conference on Robotics and Automation, pp. 1824-1829.
O'Rourke, J. (1998). Computational Geometry in C. Univ. Press, Cambridge, 2nd edition.
Porteous, I.R. (1994). Geometric Differentiation. Univ. Press, Cambridge.
Recatalá, G., Chinellato, E., Del Pobil, Á.P., Mezouar, Y., Martinet, P. (2008). Biologically-inspired 3D grasp synthesis based on visual exploration. Autonomous Robots, Vol. 25, No. 1-2, pp. 59-70.
Richtsfeld, M., Zillich, M. (2008). Grasping Unknown Objects Based on $2^1/_2$D Range Data. IEEE International Conference on Automation Science and Engineering, pp. 691-696.
Sanz, P.J., Iñesta, J.M., Del Pobil, Á.P. (1999). Planar Grasping Characterization Based on Curvature-Symmetry Fusion. Applied Intelligence, Vol. 10, No. 1, pp. 25-36.

Saxena, A., Driemeyer, J., Ng, A.Y. (2008). Robotic Grasping of Novel Objects using Vision. International Journal of Robotics Research, Vol. 27, No. 2, pp. 157-173.

Stansfield, S.A. (1991). Robotic grasping of unknown objects: A knowledge-based approach. International Journal of Robotics Research, Vol. 10, No. 4, pp. 314-326.

Schulz, S., Pylatiuk, C., Reischl, M., Martin, J., Mikut, R., Bretthauer, G. (2005). A hydraulically driven multifunctional prosthetic hand. Robotica, Cambridge University Press, Vol. 23, pp. 293-299.

Stiene, S., Lingemann, K., Nüchter, A., Hertzberg, J. (2006). Contour-based Object Detection in Range Images. Third International Symposium on 3D Data Processing, Visualization, and Transmission, pp. 168-175.

Wang, B., Jiang, L. (2005). Grasping unknown objects based on 3d model reconstruction. Proceedings of International Conference on Advanced Intelligent Mechatronics, pp. 461-466.

Wolfson, H.J. (1997). Geometric Hashing: An Overview. IEEE Computational Science and Engineering, Vol. 4, No. 4, pp. 10-21.

Xue, Z., Zoellner, J.M., Dillmann, R. (2008). Automatic Optimal Grasp Planning Based On Found Contact Points. IEEE/ASME International Conference on Advanced Intelligent Mechatronics, pp. 1053-1058.

3

A modeling approach for mode handling of flexible manufacturing systems

Nadia Hamani and Abderahman El Mhamedi
Modélisation et Génie des Systèmes Industriels (MGSI, LISMMA- EA 2336)
IUT de Montreuil-Université de Paris 8
France

1. Introduction

Due to increasing competitiveness, Flexible Manufacturing Systems (FMS) were introduced to overcome the drawbacks of Dedicated Manufacturing Lines (DML) (Koren et al., 1999). Indeed, FMS are able to carry out several parts in small and average series while adapting quickly the production changes demand thanks to their flexibility (Ranky, 1990). Several research works focus on the design of fault tolerant control systems of FMS. However, the design of such systems, in particular the supervision function, is difficult due to increasing flexibility and complexity. Thus, the aim of our research project is to provide a fault tolerant control system dedicated to FMS. This control system ensures on line and real time management of failures. In view of a disturbance, the supervision role is to take necessary decisions to return to normal or accepted operation. The supervision according to our approach is made up of three functions: decision, piloting and mode handling. Mode handling function is the scope of this chapter.
The purpose of this chapter is to present a new modeling approach for mode handling of Flexible Manufacturing Systems (FMS). Based on a review of the modeling methods and the specification formalisms in the existing approaches, we show that the mutual benefit of functional modeling and synchronous languages is very convenient for mode handling problem. We start by introducing the context of our work and the basic concepts of the proposed modeling approach. Then we present the steps of functional modeling and we illustrate them through an example of a flexible manufacturing cell. Functional modeling is completed by generic behavioral specifications representing the states of a subsystem or the whole system. The specification method is modular, hierarchical and supports re-use concept. The established model is generic and well adapted to our control system framework. Mode handling function role within the control system is then studied. This function enables a reactive update of the availability of the resources and functions and the transmission of high level control and reconfiguration orders.
This chapter is organized as follows. After the presentation of the context of our study in Section 1, Section 2 presents the roles of mode handling function within the control system. Section 3 introduces at first the main characteristics and the basic concepts of the proposed modeling method. Then the functional modeling steps are detailed. The behavior

specification of the FMS and its subsystems are then presented. The logical relationships between those subsystems are formalized. The modeling method is illustrated through an example of a flexible manufacturing cell. Finally Section 4 focuses on carrying out mode handling function within the control system framework.

2. Mode handling function

The operator supervising a production system should manage this latter both under nominal and degraded operation. Due to the complexity, he/she cannot apprehend all the constraints in a coherent and optimal way. Thus, mode handling function needs to provide a model of the system which takes into account the operating modes of the whole system and its subsystems. To this aim, it is important to use an adequate modeling method and powerful specification formalisms. The modeling method should be well adapted to the complexity and to the production system characteristics by allowing an easy representation of its operating modes. When setting up the production demand, the decisions are affecting all the subsystems of the production system. Therefore the specification formalism should support the hierarchy and represent several forms of preemption. Moreover, it must be sufficiently formal to allow carrying out formal verification.

Some research works in the literature focus on mode handling of Automated Production Systems (APS). The modeling methods and the specification formalisms used in the existing approaches are compared in (Hamani et al., 2005).

The modeling methods we studied are classified according to many criteria (Hamani et al., 2005). To deal with complexity, the system is decomposed into subsystems. Two approaches are distinguished (Toguyeni et al., 1996a), structural approaches (Biland et al., 1994; Bois et al. et al., 1992; Parayre et al., 1992) and functional ones (Ausfelder et al., 1993; Dangoumau & Craye, 2001; Kermad et al., 1994). The modeling steps, the specification of the modes and the specification formalisms are detailed in (Hamani et al., 2005) for each approach. The role of the established models within the control systems is also studied. In structural approaches, the decomposition is based on structural relationships between the resources of the system. Reconfiguration actions are then taken on the resources, the workstations and the cells. Functional approaches are concerned with the services delivered by the system rather than the resources. The decomposition is based on functional relationships between subsystems. Such relationships enable the implementation of automated reconfiguration procedures.

Our study shows also that the specification formalisms should enable an easy representation of hierarchy, concurrency and preemption. They have to be also with rigorous semantics in order to guarantee some properties using formal proofs.

The main conclusions of our comparative study show that the mutual benefit of a functional modeling approach and a powerful specification model using the graphical synchronous language Safe State Machines (SSM) (André, 2003) is very convenient for APS mode handling. Indeed, functional approaches are well suited for efficient reconfiguration procedures. Besides, the required characteristics of hierarchy, parallelism, and preemption are well represented using synchronous languages instead of common models such as Sequential Functional Charts (SFC) and Petri Nets (PN). The interests of using synchronous formalisms rely also on their strong semantics. They are compiled efficiently and it is possible, using formal techniques, to prove that the behavior of the system respects some properties (Hamani et al., 2007). In addition, synchronous approaches ensure the

preservation of the verified properties between the specification and the implementation phases of the development process (What You Prove Is What You Execute: WYPIWYE) (Berry, 1989).
Mode handling function ensures both an informational role and an operational role within the control system. The information provided by mode handling function about the state of the system and the subsystems, enables updating the models of the control system. The informational, decisional and operational functions of the control system are respectively monitoring (Elkhattabi et al., 1995; Toguyeni et al., 1996b), decision (Berruet et al., 2000), piloting (Tawegoum et al., 1994) and mode handling. The reaction loop in view of a failure is given in the following: the monitoring function detects and localizes the failures in the plant; the decision function provides the new configuration of the system; Mode handling and piloting functions carry out the decisions about the new configuration.
Indeed, concerning its operational role within the control framework, mode handling with piloting implements some actions on low level control and on the plant. Mode and production goals changing are ensured by mode handling function whereas the actions on low level control are carried out both by mode handling and piloting functions. Piloting solves remaining indeterminism resulting from routing flexibility. Berruet (Berruet et al., 2000) introduces several reconfiguration solutions; they are classified according to the difficulty of their implementation by the operational functions (i.e. piloting and mode handling) and the required time.
The interactions of mode handling with all the functions of the control system are necessary to achieve its role. Indeed, mode handling receives as inputs high level control orders and reconfiguration decisions and the failures and provides as outputs the current configuration of the system. This latter characterizes the state of the system and of its subsystems (resources, operations) and the operating parameters. In order to achieve its role mode handling function needs the models representing the behavior of the system. Reconfiguration goals are reached under some performance and time constraints and using some methods.

3. The modeling method

The advantages of functional modeling approaches and synchronous languages for mode handling of APS are shown in the previous section. Our approach is based on a functional modeling and a synchronous reactive approach using SSM (André, 2003).
Besides, this approach is well adapted to FMS because it is based on the mission concept which represents the flexibility of production of such systems. A mission (or a production goal) is a subset of Logical Operating Sequences (LOS) which are carried out simultaneously.
The model represents the subsystems and their operating modes for an existing system and also for a system being designed. The modeling approach is based on generic concepts. For a given system, the predefined functional and behavioral subsystems are instantiated to generate the model.

3.1 Basic concepts

The Logical Operating Sequence (LOS) concept was introduced for the specification of control models in (Cruette et al., 1991). A LOS is a subset of ordered machining functions that the FMS performs on some parts.

A function (f) is the response of a system or a component to a given stimulus when it works normally independently of its environment (Toguyéni et al., 2003). A LOS is noted LOS $f_1...f_n$ or LOS f_i ($i = 1, ..., n$). The functions performed by the FMS are mainly: machining functions, transfer functions and storage functions.

Let us remind the operation concept and the accessibility relations between operations. In an FMS, an operation (Op) is defined as a function carried out by a resource (a component) of the FMS (Berruet et al., 2000; Toguyeni et al., 2003). An operation is noted $Op_{Ri, fi}$ where f_i is the performed function and R_i the resource which implements it. As for functions, several kinds of operations are defined: machining operations, transfer operations and storage operations.

The concept of elementary operation was also introduced (Berruet et al., 2000; Toguyeni et al., 2003). An elementary operation is an operation carried out only once, continuously, i.e. without another alternative during the normal execution of the operation. Several kinds of elementary operations are also defined: elementary machining operations, elementary transfer operations, etc. The concept of elementary operation is important for FMS modeling because it is a final step in the decomposition process.

The accessibility relation represents the link between some resources concerning the parts flow. Thus, the accessibility relation characterizes the existence of a parts flow between two resources (Amar et al., 1992). A resource has a direct accessibility with another resource if a part moves from one resource to another one without going through an intermediate resource. Loading or unloading a machine tool using a robot is a typical example of direct accessibility. Indirect accessibility relation results from transitivity.

The accessibility relation between operations results directly from the existing accessibility between resources. It means a directed flow of parts between two operations (precedence relation).

The Characteristic Area (CA) concept (Amar et al., 1992) characterizes a physical area (machining, transfer or storage areas) mobile or not, able to receive a part. A physical area is being characteristic if it is a machining area or if it is in an external accessibility with a physical area of another production medium. From a functional point of view, a CA is either a machining resource or an access point to a resource. By extension to the CA concept, Toguyéni (Toguyéni et al., 2003) introduced the Main Characteristic Area (MCA) concept. It is a CA which corresponds to a machining area or a storage area used as an input or an output area for the parts.

The MCA enables to introduce the elementary transfer concept proposed in (Toguyéni et al., 2003). An elementary transfer (TrE) is performed by one resource between two Characteristic Areas. It is noted $TrE_{R_i}^{S \to D}$ with S a source CA, D a destination CA, and R_i the transfer resource.

The other concepts we need for the modeling process are introduced in the following subsection.

3.2 The FMS functional model

An FMS produces a set of parts simultaneously. Each part has its own LOS. Usually we desire to change the production goal that is why it is interesting to introduce the mission concept (Hamani et al., 2006).

Definition 1. A mission ($\mathcal{M}$) is a subset of Logical Operating Sequences which are carried out simultaneously.

Each function of a LOS is implemented by one or more operations performed by distinct resources. In order to achieve a mission of the FMS, it is necessary to carry out the corresponding LOS. This needs the implementation of machining functions and also the corresponding machining operations. The corresponding transfer operations should also be performed. Thus, we propose to determine, for each machining area of the FMS, the transfers that ensure successful loading and unloading of a part. The part is loaded from a machining area or from the cell input. Then it is unloaded onto another machining area or onto the cell output. So transfer operations are generic entities which depend only on the plant and do not depend on production goals.
For each MCA corresponding to a machining area, the Access Transfers (TrA) concept is introduced (Hamani et al., 2006). This concept represents the subset of transfer operations that allow loading (unloading) a part from (onto) a machining area.

Definition 2. Access Transfers (TrA) associated with a machining area (or a MCA) are the subset of elementary transfer operations that connect this area with the other areas of the FMS. The TrA corresponding to a machining MCA are noted $TrA_{machining\_MCA}$.
In order to determine TrA, a first step consists in listing symmetrical transfers between MCA representing both sources and destination areas. Then it is necessary to refine these transfers until obtaining elementary transfer operations.
Once the Access Transfers are determined, it is necessary to identify elementary transfers which compose them. If there is a direct accessibility between two MCA then $Tr_{MCA\_source \rightarrow MCA\_destination}$ corresponds to an elementary transfer. If not, it is necessary to refine the transfers between the CAs until obtaining elementary transfers. The possible paths are then established and those which are redundant are linked together with a logical OR.
Once TrA are established for each machining area, they are gathered together with machining operations corresponding to the same area in order to obtain an aggregate operation (Hamani et al., 2006).

Definition 3. An aggregate operation corresponding to a machining area is a subset of elementary machining operations and TrA. An aggregate operation associated with a machining MCA is noted $Op_{MCA\_machining}$.

Aggregate operations are generic entities which represent the possible choices for implementing machining functions by the FMS. So with each function of a LOS is associated its possible achievements. They are aggregate operations for which the machining operation is defined. A machining function is then implemented by an aggregate operation or some aggregate operations performed by distinct resources.

Finally we should determine machining resources and transfer resources which perform the elementary machining and transfer operations of the aggregate operations.
The specification steps of the functional model are described in the following. The modeling method is illustrated using a flexible manufacturing cell (Fig. 1) with two machines M_1 and M_2 and INPUT/OUTPUT buffers.

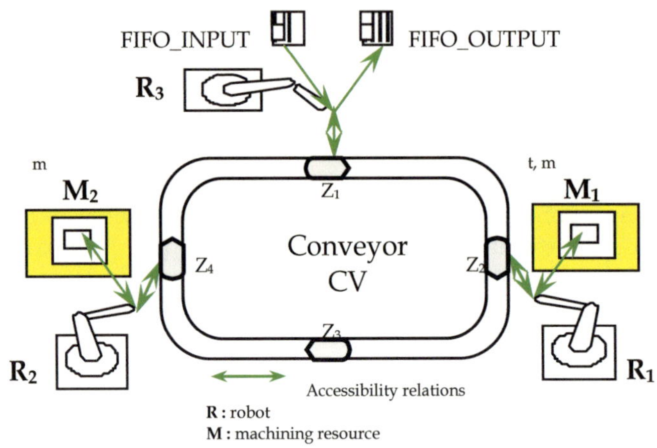

Fig. 1. The machining flexible cell

The machines are loaded with a transport system using three robots R_1, R_2 and R_3 and a conveyer (CV). Moving direction is $Z_1 \rightarrow Z_2 \rightarrow Z_3 \rightarrow Z_4 \rightarrow Z_1$. M_1 is loaded using R_1 and M_2 is loaded using R_2. The parts are loaded onto the conveyor using the robot R_3. The machining functions performed by the system are turning (t) and milling (m). Turning function is carried out by M_1 and milling by M_1 and M_2. The functional requirements provide three missions. The CA of this machining cell are storage areas FIFO_INPUT, FIFO_OUTPUT, conveyor zones Z_1, Z_2, Z_3 and Z_4 and the machining areas M_1 and M_2. The MCA are storage areas FIFO_INPUT, FIFO_OUTPUT and machining areas M_1 and M_2.

1st step: Identification of the entities of the model

- List the missions that the FMS should carry out
- List for each mission its corresponding Logical Operation Sequences
- For each LOS determine the corresponding machining functions
- A machining function is implemented by one or several elementary machining operations. Each one is belonging to an aggregate operation.

- Determine the aggregate operations of the FMS (see the 2nd step)
- For each aggregate operation, identify the resources performing it (see the 3rd step)

According to functional requirements, three missions can be selected: $\mathcal{M}_1$, $\mathcal{M}_2$ and $\mathcal{M}_3$. The corresponding LOSs are the following:
$\mathcal{M}_1$: LOS_1 and LOS_2; $\mathcal{M}_2$: LOS_1, LOS_2 and LOS_{12}; $\mathcal{M}_3$: LOS_1, LOS_2 and LOS_{21}

The machining functions which compose each LOS are the following:
LOS_1: turning; LOS_2: milling; LOS_{12}: turning then milling; LOS_{21}: milling then turning

The turning function is performed by the elementary machining operation $Op_{M1,t}$ belonging to the aggregate operation Op_{M1}. The milling function is performed by the elementary machining operation $Op_{M1,m}$ belonging to the aggregate operation Op_{M1} or by the elementary machining operation $Op_{M2,m}$ belonging to the aggregate operation Op_{M2};

2nd step: Determination of aggregate operations
For each machining area of the FMS:

- Identify elementary machining operations which are performed in this area
- Identify the TrA corresponding to this area
- Gather elementary machining operations together with TrA to obtain aggregate operations

For the machining area M_1:
The elementary machining operations performed by M_1 are $Op_{M1,t}$ and $Op_{M1,m}$;

Access Transfers corresponding to the machining area M_1 are[1]:

$$TrA_{M_1} = AND\,(Tr_{MCA\_source \to M_1},\ Tr_{M_1 \to MCA\_destination}).$$

$$TrA_{M_1} = AND\,[OR\,(Tr_{IN \to M_1},\ Tr_{M_2 \to M_1}),\ OR\,(Tr_{M_1 \to M_2},\ Tr_{M_1 \to OUT})],$$

$$TrA_{M_1} = AND\,\{OR\,[AND\,(TrE_{R_3}^{IN \to Z_1},\ TrE_{CV}^{Z_1 \to Z_2},\ TrE_{R_1}^{Z_2 \to M_1}),\ AND\,(TrE_{R_2}^{M_2 \to Z_4},\ TrE_{CV}^{Z_4 \to Z_1},\ TrE_{CV}^{Z_1 \to Z_2},\ TrE_{R_1}^{Z_2 \to M_1})],\ OR\,[AND\,(TrE_{R_1}^{M_1 \to Z_2},\ TrE_{CV}^{Z_2 \to Z_3},\ TrE_{CV}^{Z_3 \to Z_4},\ TrE_{R_2}^{Z_4 \to M_2}),\ AND\,(TrE_{R_1}^{M_1 \to Z_2},\ TrE_{CV}^{Z_2 \to Z_3},\ TrE_{CV}^{Z_3 \to Z_4},\ TrE_{CV}^{Z_4 \to Z_1},\ TrE_{R_3}^{Z_1 \to OUT})]\};$$

The elementary machining operations $Op_{M1,t}$ et $Op_{M1,m}$, gathered together with TrA_{M1} enable to obtain the aggregate operation $Op_{M1} = AND\,[OR\,(Op_{M1,t}, Op_{M1,m}),\ TrA_{M_1}]$;

3rd step: Determination of the resources which perform elementary operations
For each aggregate operation:

- Associate with each elementary machining operation the resource or the configuration of the resource (in the case of a polyvalent resource) which performs it
- Associate also with each elementary transfer operation the resource(s) performing it, redundant resources are linked with a logical OR

The aggregate operation associated with the machining area M_1 is performed by the following resources:

- The polyvalent machining resource M_1 performs the elementary operations $Op_{M1,t}$ and $Op_{M1,m}$.

For transfer resources:

---

[1] This notation uses the logical AND and OR and uses also three distinct levels: '{' for the first level, '[' for the second level and '(' for the third level.

- R_1 performs the elementary transfer operations $TrE_{R_1}^{Z_2 \to M_1}$ and $TrE_{R_1}^{M_1 \to Z_2}$
- R_2 performs the elementary transfer operations $TrE_{R_2}^{M_2 \to Z_4}$ and $TrE_{R_2}^{Z_4 \to M_2}$
- R_3 performs the elementary transfer operations $TrE_{R_3}^{IN \to Z_1}$ and $TrE_{R_3}^{Z_1 \to OUT}$
- CV performs the elementary transfer operations $TrE_{CV}^{Z_1 \to Z_2}$, $TrE_{CV}^{Z_2 \to Z_3}$, $TrE_{CV}^{Z_3 \to Z_4}$ and $TrE_{CV}^{Z_4 \to Z_1}$

The aggregate operation corresponding to the machining area M_2 is obtained in the same manner.

The functional model is represented with the following entities:

- The missions
- The logical operating sequences
- The machining functions
- The aggregate operations
  - o elementary machining operations
  - o access transfers (subset of transfer operations)
- Transfer resources, machining resources

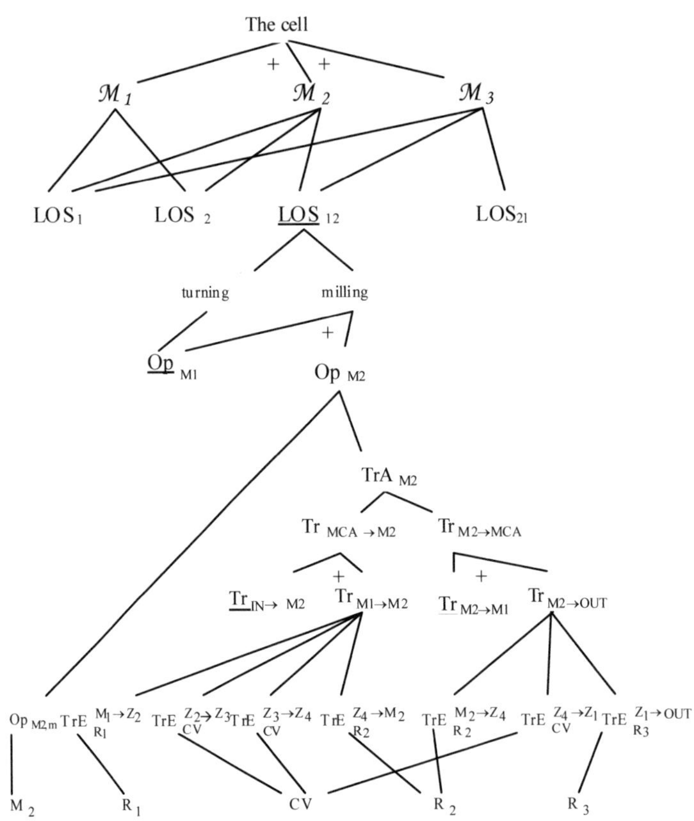

Fig. 2. The functional model of the machining cell

During operation the active functional model is a subset of the potential functional model. The current mission enables to define this subset.
The functional model (AND/OR graph) of the machining cell (Fig. 1) is represented in Fig. 2. The underlined entities are not developed on this figure.

In Fig. 2 the AND nodes do not have any notation, however OR nodes are denoted using +. OR nodes correspond either to an inclusive OR or an exclusive OR according to the constraints given in the functional requirements. For example, an exclusive OR is necessary for safety reason. For instance when we have two machining operations that utilize the same resource, and only one can utilize the resource at any time. The functions turning and milling are examples of entities linked with a logical AND to form LOS_{12}. The transfers $Tr_{IN \rightarrow M2}$ and $Tr_{M1 \rightarrow M2}$ are linked with a logical OR.

3.3 Behavior specification

Functional modeling is completed by behavioral specification. We propose in the following at first generic specifications representing the states of a subsystem (that we call an entity) or the whole system, then the specifications of the relationships between the entities of the model.

1. Specification of the behavior

For an entity or the whole system, states are divided into two board categories. The modes and operating parameters of the system, they are the states required by the operator and in fact he/she could select them. The degradations and the failures of the resources change the availability of the entities of the model, such states are unexpected.

By **required states**, we mean all the modes which are selected before setting up the production. There are one or more families of modes and the criteria of regrouping them into families vary from a system to another, and according to its complexity. The FMS behavior at each instant is then defined by the set of the simultaneously active states in each one of these families of modes. The initial situation corresponds to states initially selected in each family. After selecting the required modes, setting up the production is made through the activation of the selected mission. This involves activating the operations and the resources which take part in the selected mission. The missions introduced in the previous paragraph are now associated with a state transition graph called the graph of mission handling.

Definition 4. The graph of mission handling is the state transition graph whose states correspond to the missions of the FMS and the transitions represent the change of a mission to another.

The **unexpected states** concern the availability of the entities of the system; the monitoring function detects the failures of the resources. The change-of-state of the resources affects the elementary operations and therefore the machining functions, the LOSs, and the selected mission. All the entities of the model are thus characterized by their availability. Several states belonging to distinct families (including the selected mission) define the FMS current mode before setting up the production. We introduce now the current mode concept.

Definition 5. The current mode of FMS is the subset of states that are simultaneously active and belonging to distinct families. The selected mission belongs to this subset.

The entities of the functional model which take part in the selected mission can be activated if they are available. Thus two conditions are necessary to the activation of an entity: belonging to the current mode and being available.

Fig. 3 shows the states of an entity. The activation (deactivation) of an entity is represented by the working mode. The functioning mode represents the availability (unavailability) of an entity.

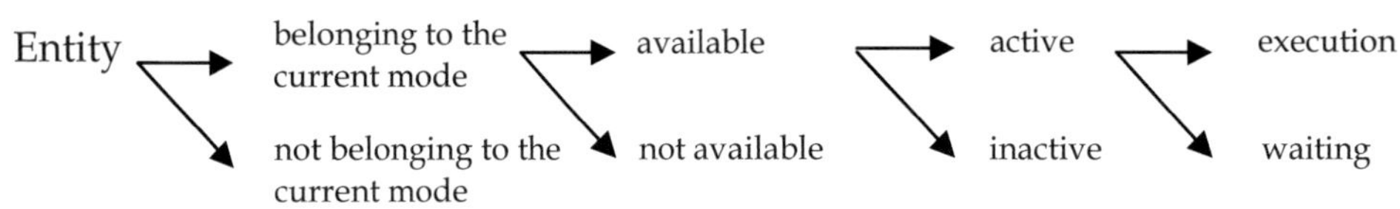

Fig. 3. The states of an entity

In the following the formal specification of the behavior using the synchronous formalism SSM is presented.

Formal specification using Safe State Machine: It is necessary to represent the activation (deactivation) at all the hierarchy layers of the functional model (see subsection 3.2) as well as the availability of the resources and the functions of the FMS.

The model should then handle concurrently the information flows downwards to transmit high level control and reconfiguration orders and upwards to follow up the reports and failures detected by the monitoring function. The reactivity needed for this bidirectional exchange of information, in addition with concurrency and preemption, requires a synchronous approach. Let us remind that synchronous languages were introduced to overcome the problems of modeling reactive systems (Halbawachs, 1993).

The visual synchronous formalism Safe State Machine (SSM or SyncCharts (Synchronous Charts) (André & Péraldi, 1993; André, 1996; André, 2003) is used to specify the behavior of the entities of the functional model. SSM inherits many features from Statecharts (Harel, 1987) but offers several forms of preemption and benefits of more strict semantics fully compatible with that of the synchronous language Esterel (Berry & Gonthier, 1992; Boussinot & De Simone, 1991). SSM supports also, with very rigorous semantics, hierarchy, communication, concurrency and various forms of preemption, which characterize our modeling approach. Instantaneous broadcast of signals is the unique communication medium. The presentation of the syntax of SSM is beyond the scope of this paper (see (André, 1996; André, 2003). In addition, SSM takes advantage of an industrial development environment (Esterel Studio™), which provides necessary tools to the development phases of systems, from specification to implementation. It enables also verifying and validating the specified models (Hamani et al., 2007).

The behavior of each entity of the model, whatever its level, is represented by a SSM including two modes (Fig. 4). At any time we know, if an entity, belonging to the current mode of the FMS, is in normal state e_N, degraded state e_D or out of order state e_OUT according to the functioning mode point of view; and if it is activated e_Active or inactive (in stopping state) e_OFF according to the working mode point of view. We take as initial states, the stopping state e_OFF for the working mode and the normal state e_N for the functioning mode.

There is a relationship between the two represented modes. Indeed, the activation of an entity e using start_e can be made only if the entity is available (i.e. it should not enter the out of order state). This is represented using 'start e and not e_OUT'. As soon as the entity is active, it enters the working state e_ON if the conditions of its activation are appropriate (in Fig. 4, there is no condition, we will see in the next subsection that such conditions depend on the relationship between the entities). The deactivation of the parent entity using stop_e depends also on the deactivation of its child entities (if they exist). The stopping state is effectively reached when the entity ends its activity. Indeed, when the entity reaches the final state (see the state End distinguished by its double outline), the macro-state e_Active is exited through the normal termination (whose tail is a small triangle). Notice that in presence of concurrency which characterizes the FMS model, using final states and normal termination is more suitable for synchronization (André, 2003).

According to the previous specification, the input and the output signals represented in Fig. 4 are the following: for working mode, the input signals start_e and stop_e correspond respectively to the activation and deactivation orders. They lead respectively entering e_ON and e_OFF states. For functioning mode, the input signals e_n, e_d and e_out provoke respectively entering in normal state e_N, degraded state e_D, and out of order state e_OUT. Local signals are also needed for specification. They are signals which are neither controllable nor directly observable from outside.

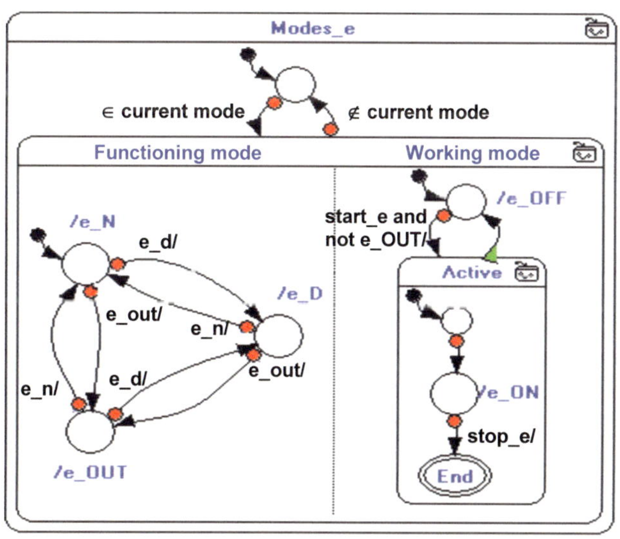

Fig. 4. The SSM representing an entity

2. Specification of the logical relationships

In the relation between a parent entity (level i) and a set of child entities (level i-1), the child entities are linked either with a logical AND or a logical OR. The information exchange between a couple of parent-child entities is given in the following.

- For working mode, the activation or deactivation request (using an input signal or a local signal) is a downward information flow from a parent entity to its child entities. The reports of such requests are propagated upwards.
- For functioning mode, failure events provoke the change-of-state of the resources i.e. the entities at the lower level of the model. For intermediate levels of the model this information flow is propagated upwards from the child entities to their parent entity.

The change-of-state of the entities of the model belonging to successive levels is carried out according to precise rules which depend on the kind of the logical relationship between the entities. If a parent entity has only one child then, for working mode, the activation (deactivation) of the parent entity provokes the activation (deactivation) of its child. The parent entity is effectively in execution if and only if the child entity is in execution. For functioning mode, mode changing of the child entity provokes mode changing of the parent entity.

In the case of a parent entity which has several child entities, state calculations corresponding to the case of two child entities (binary case) using an example of an entity e and its child entities e_1 and e_2 are given in (Hamani et al., 2007). The behavior is obtained using the interactive simulator XES (XEtented Simulator) (Esterel Studio™). For m child entities, the functioning mode is obtained in a recursive way. For working mode, the specifications are obtained in the same way as for the binary case.

3. Modular representation using reference models

The specification of any entity of the model differs according to whether it has one or more child entities or it has no child entity (the leaves of the graph). It depends also on the logical relationship between the child entities. Thus, we have defined reference models, a representation in SSM which is similar to the procedures (or functions) used in programming languages.

The representation of the deep hierarchy of the model as well as the I/O and local signals are taken into account. They are presented in the following paragraphs.

The representation of the model: We represent the model according to its hierarchy (see Fig. 2). The behavior of the entities is obtained by instantiation of the reference models. The behavioral models of the entities belonging to the same level of the hierarchy are orthogonal. The encapsulation mechanism offered by SSM allows linking two successive levels (level i and level i-1). The integration of single behavioral models should respect some communication rules. The main rules are presented in the following.

Communication rules: The relation parent-child defines the existing information exchange and the logical relationships taken into account. The I/O and local signals are declared for each behavioral model of an entity. Nevertheless, it is necessary to respect some rules for a coherent integration of the entire model.

- For functioning mode, only the resources i.e. the leaves of the model have input signals. Indeed, for functioning mode, mode changing is detected only for the resources and then propagated upwards.

- For working mode, only the entities of the top level of the model have input signals which correspond to the orders emitted at supervision level. The entities belonging to the other hierarchical levels of the model receive these requests which are propagated downwards thanks to local signals.
- All input signals used for selecting the configurations of the model (redundancies) should be declared. For example, for the model of Fig. 2, the input signals $C_{M2,m}$, $C_{M1,m}$ allow selecting milling operation on M_2 or on M_1 respectively.

The model (Fig. 2) is specified using 7 generic models and 43 instances of those generic models for nearly 1452 lines of Esterel code generated from the SSM models. This model declares 36 input signals and 212 output signals.

4. Carrying out FMS model handling

The functional model is characterized, on one hand, by functional relationships between the entities belonging to successive levels (vertical interactions); and on the other hand, by direct or indirect accessibility relations between the entities belonging to the same level (horizontal interactions).

Considering the behavior, each entity of this model is characterized by its functioning mode and its working mode. These two modes are constrained: the activation of an entity can be made only if it is available.

The availability of the entities which take part in the realization of the selected mission allows knowing if this mission is realizable or not. The missions we introduced represent the production goals that the system should perform, each function or resource which takes part in this realization must be available.

We explain now the behavior of the mode handler. In the exploitation phase, the behavioral model should handle the information flows downwards to transmit the orders from Human Machine Interface and upwards to propagate the information provided by the monitoring function. As shown in Fig. 5, the downward flow of high level control and reconfiguration orders is propagated through the LOSs corresponding to the selected mission. The flow is then propagated through the operations and the resources which take part in the realization of the selected mission. As for the downward flow, the upward flow transmits the results obtained by the execution of high level control and reconfiguration orders and process the information about the availability of the resources and the operations. Taking into account this information exchange in the control system, the communications of mode handling model with control and monitoring models should be specified.

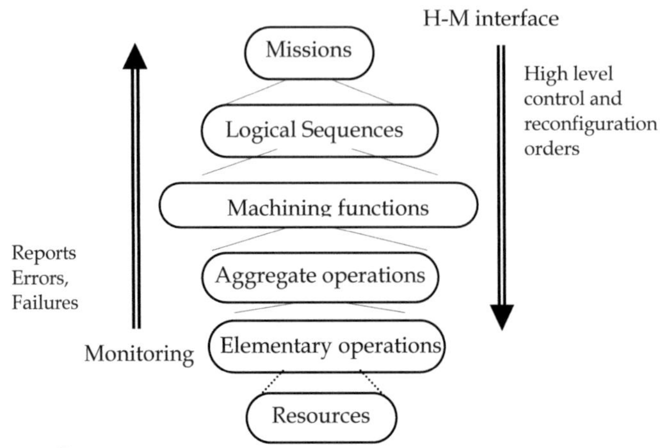

Fig. 5. The information flows

The mode handler is a reactive system (Harel, 1985) which continuously reacts to stimuli (inputs) of the environment by sending back other stimuli (outputs). Thus, mission and working mode changing are transmitted downwards. The change-of-state of functioning mode caused by the failures detected on the plant is transmitted upwards. These failures cause either the degradation (degraded state) or the breakdown (out of order state) of the resources.
Consider the reactions of a SSM associated with an entity of the model. Indeed, each entity e at level i can receive Either a request from level i+1 indicating a mission or a working mode change or a mode changing of one or more child entities of level i–1.

5. Conclusion

We have presented in this paper a new modeling method for FMS mode handling. We introduced the mission concept which corresponds to a subset of LOSs that the system carries out simultaneously. The idea is to decompose and identify the FMS functional entities which take part in the realization of its missions. The functional model is obtained by a modular and hierarchical decomposition leading to the elementary machining and transfer operations. The structural aspect completes this description by associating resources with the elementary operations they perform. The proposed modeling steps are illustrated through an example of a manufacturing cell. This modeling method is characterized by its generic concepts which allow their implementation in the computer aided tool CASPAIM_soft (Ndiaye et al., 2002).
The behavior of the obtained model is specified using the synchronous formalism SSM. Indeed, the strong hierarchy and the concurrency which characterize the functional model as well as the reactivity of information processing, justify using a synchronous approach. The behavior associated with the entities of the FMS model and the logical relationships (AND, XOR and OR) between them are also generic. An instantiation of the predefined models make it possible to obtain the model of a given system. Within the design phase, the reuse of functional entities and predefined behavioral models is one of the main characteristics of our approach. In addition, the behavior specifications using the

synchronous formalism SSM enables to benefit fully from formal analysis tools (Hamani et al., 2007).
Within the exploitation phase, the role of the model is to handle the missions and the configurations. The information flows within the control system framework are provided both by high level control and monitoring function.
Further works aim at first to implement the introduced concepts in the information system CASPAIM_soft. Within the control system, we would like to complete the reaction loop in view of a failure by studying the integration and the communication between all those functions of supervision, monitoring and low level control. Therefore the model should be enhanced in order to allow this information exchange.

6. References

Amar, S.; Craye, E., Gentina, J.C. (1992). Une méthode hiérarchique de spécification et de prototypage des systèmes de production flexibles. *Revue Auto. Prod. Info. Indus*, 26, 5-6, (1992) 483-514.

André, C. & Péraldi, M.A. (1993). *Synchronous approach to industrial process control*, Technical Report RR 93-10, I3S, Sophia-Antipolis, France.

André, C. (1996). Representation and analysis of reactive behaviors: a synchronous approach, *Proceedings of IMACS Multiconference in Computational Engineering in Systems Applications (CESA 96)*, pp. 19-29, Lille, France.

André, C. (2003). *Semantics of Safe State Machines*, Guyancourt, France, http://www.esterel-technologies.com

Ausfelder, C.; Maik, J.P., Kermad, L., Gentina, J.C., Delfieu, D., Moisand, R., Sahraoui, A.E.K. (1993). Integration of operating modes in the control of flexible manufacturing systems combining synchronous and asynchronous approaches, *Proceedings of IEEE International conference on Systems Man and Cybernetics (SMC 93)*, pp. 234-239, Le Touquet, France.

Berruet, P.; Toguyéni, A.K.A., Craye, E. (2000). Towards implementation of recovery procedures for FMS supervision. *Comp. in industry*, 43, (2000) 227-236.

Berry, G. (1989). Real-time programming: general purpose or special purpose languages. *Info. Processing*, 89, (1989) 11-17.

Berry, G.; Gonthier, G. (1992). The Esterel synchronous programming language: design, semantics, implementation. *Science of Comp. Prog.*, 19, 2, (1992) 87-152.

Biland, P.; Deplanché, A.M., Elloy, J.P. (1994). A model for operating modes of computer integrated manufacturing systems, *Proceedings of the European workshop on Integrated Manufacturing Systems Engineering (IMSE 94)*, Grenoble, France.

Bois, S.; Craye, E., Gentina, J.C. (1992). Manager of working modes. *CIM Integ. Des. in Manufa.*, 21, 3, (1992) 201-207.

Boussinot, F. & De Simone, R. (1991). The Esterel language: another look at real time programming, *Proceedings of the IEEE*, pp. 1293-1304.

Cruette, D.; Bourey, J.P., Gentina, J.C. (1991). Hierarchical specification and validation of operating sequences in the context of FMS. *Comp. Integ. Manufa. Sys.*, 4, 3, (1991) 140-156.

Dangoumau, N. & Craye, E. (2001). Modes management for safe production systems, *Proceedings of the 10th IFAC Symposium on Information Control Problems in Manufacturing (INCOM 01)*, Vienna, Austria.

Elkhattabi, S.; Craye, E., Gentina, J.C. (1995). Supervision by the behavior modeling, *Proceedings of the IEEE International Conference on Systems Man and Cybernetics (SMC 95)*, pp. 1416-1422, Canada.
Esterel Studio™ : http://www.esterel-technologies.com
Halbawachs, N. (1993). *Synchronous Programming of Reactive Systems*, Kluwer Academic Publishers, Amsterdam.
Hamani, N.; Dangoumau, N., Craye, E. (2009). Reactive mode handling of flexible manufacturing systems. *Mathematics and Computers in Simulation*, 79, (2009) 1421-1439.
Hamani, N.; Dangoumau, N., Craye, E. (2005). A comparative study of mode handling approaches, *Proceedings of the 35th International Conference on Computers & Industrial Engineering (CIE 05)*, Istanbul, Turkey.
Hamani, N.; Dangoumau, N., Craye, E. (2006). A functional modeling approach for FMS mode handling, *Proceedings of the 12th IFAC Symposium on Information Control Problems in Manufacturing (INCOM 06)*, Saint Etienne, France.
Hamani, N.; Dangoumau, N., Craye, E. (2007). Safe State Machines for reactive mode handling of manufacturing systems, *Proceedings of the International Modeling and Simulation Multiconference (IMSM-IMAACA 07)*, Buenos Aires, Argentina.
Harel, D. (1985). On the development of reactive systems, *Proceedings of NATO Advanced Study Institute on Logics and Models for Verification and Specification of Concurrent Systems*, pp 477-498, NATO ASI series F, Vol. 13, Springer-Verlag.
Harel, D. (1987). StateCharts: a visual approach for complex systems. *Sc. of Comp. Prog.*, 8, 3, (1987) 231-275.
Kermad, L.; Craye, E., Bourey, J.P., Gentina, J.C. (1994). The exploitation modes of flexible manufacturing systems, *Proceedings of IEEE International conference on Systems Man and Cybernetics (SMC 94)*, pp. 1521-1526, San Francisco, USA.
Koren, Y.; Heisel, U., Moriwaki, T., Ulsoy, G., Van Brussel, H. (1999). Reconfigurable Manufacturing Systems. *Annals of the CIRP*, 48, 2, (1999) 527-540.
Ndiaye, D.; Bigand, M., Corbeel, D., Bourey, J.P. (2002). Information system for production engineering: contribution to maintaining consistency of composite data using an object oriented approach. *Int. J. of Comp. Integ. Manufa.*, 15, 3, (2002) 233-241.
Parayre, T.; Sallez, Y., Soenen, R. (1992). Model of the state of operation of automated systems, *Proceedings of the 8th International PROLAMAT conference (PROLAMAT 92)*, pp. 553-575, Tokyo, Japan,
Ranky, P. (1990). *Flexible Manufacturing Systems in CIM*, CIMware Ltd, Guildford, U.K.
Tawegoum, R.; Castelain, E., Gentina, J.C. (1994). Real time piloting of flexible manufacturing systems, *Euro. J. of Opera. Res.*, 78, (1994) 252-261.
Toguyéni, A.K.A.; Elkhattabi, S., Craye, E. (1996a). Functional and/or structural approach for the supervision of flexible manufacturing systems, *Proceedings of IMACS-IEEE multiconference on Computational Engineering in Systems Applications (CESA 96)*, pp. 716-721, Lille, France.
Toguyeni, A.K.A.; Craye, E., Gentina, J.C. (1996b). A framework to design a distributed diagnosis in Flexible Manufacturing Systems, *Proceedings of IEEE International conference on Systems, Man and Cybernetics (SMC 96)*, pp. 2774-2779, Beijing, China.
Toguyéni, A.K.A.; Berruet, P., Craye, E. (2003). Models and algorithms for failure diagnosis and recovery in flexible manufacturing systems, *Int. J. of Flexi. Manufa. Syst.*, 15, 1, (2003) 57-85.

Computed-Torque-Plus-Compensation-Plus-Chattering Controller of Robot Manipulators

Leonardo Acho, Yolanda Vidal and Francesc Pozo
CoDAlab, Departament de Matemàtica Aplicada III,
Escola Universitària d'Enginyeria Tècnica Industrial
de Barcelona, Universitat Politècnica de Catalunya,
Comte d'Urgell, 187, 08036 Barcelona,
Spain

1. Introduction

Robot control is a modern technology that requires of innovation in control theory. The robot system is a complex and nonlinear system involving mechanics, electronics, and computer science. With technological innovation in electronics, more complex controllers can be designed and implemented in robotic systems to conceive a computer controlled robot manipulator. In this sense, a robot system can be viewed as a mechanical arm that operates under computer control in order to have a reprogrammable - and thus multifunctional - manipulator designed to move material, parts, or performing tracking motion for a great variety of tasks. However, there still exists an important challenge: to cope with friction that can degrade the performance of our robot system.

Friction is a natural phenomenon that affects almost all mechanical systems. This phenomenon has been extensively studied for many years, as it is hard to model and, in some situations, hard to predict because of several factors that vary over time (wasting, humidity, and temperature). For these reasons, friction is usually ignored at the controller design stage. Although there are many controllers based on friction models such as (Orlov et al., 2003), (Aguilar et al., 2003), and (Guerra & Acho, 2007), the real implementation of these controllers requires on-line final tuning. In other words, those controllers that were designed by neglecting the friction perturbations have to be robust against them. From the robot control point of view, there have been many controllers based on frictionless robot modeling: PD and P"D" control with gravity compensation, computed-torque plus control, etc. (Kelly et al., 2005). From the engineering point of view, it is of interest to redesign some of these controllers to make them robust against friction perturbations. Friction mitigation is an important topic in the high-precision control of mechanisms (Weiping & Xu, 1994). It is well known that chattering controllers can deal with model uncertainties like friction, (Orlov et al., 2003). Chattering is a fast commuting term that is added to a given controller.

The computed-torque-plus-compensation controller of robot manipulators, that was originally called *computed-torque control with compensation,* has been well documented, *e.g.* (Kelly et al., 2005). According to (Kelly et al., 2005), for the academic robotics community,

the global stability of the closed-loop system with this controller is still an open problem. Here, a chattering term is added to the previous controller to improve the global asymptotic stability. We call it the *computed-torque-plus-compensation-plus-chattering* controller of robot manipulators. Moreover, according to numerical experiments applied to the tracking control of a robot manipulator with two degrees of freedom, this new controller represents an important and robust improvement over the original one, especially when the system is operated under Coulomb friction effects. Lyapunov theory is employed in proving the global uniform asymptotic stability of the closed-loop system.
This work is structured as follows. Section 2 introduces the computed-torque plus compensation controller of robot manipulators. The dynamic notation for an n-degree-of-freedom (n-DOF) robot manipulator is also presented. In Section 3, the chattering version of the computed-torque plus compensation controller is defined. Global uniform asymptotic stability is achieved by invoking Lyapunov theory. Section 4 studies the performance and robustness of the proposed controller and compares it with the performance of the original controller through numerical experiments on a 2-DOF vertical robot manipulator with Coulomb friction. This kind of robot is one that is affected by gravity. Finally, Section 5 states the conclusions.

2. Computed-torque plus compensation control of robot manipulators

Consider the following general equation describing the dynamic of an n-degrees-of-freedom (n-DOF) rigid robot manipulator in joint space:

$$M(q)\ddot{q} + C(q,\dot{q})\dot{q} + G(q) = \tau, \tag{1}$$

where $q \epsilon R^n$ is the vector of generalized coordinates, $\tau \epsilon R^n$ is the vector of external torques, $M(q) \epsilon R^{nxn}$ is the positive-definite inertia matrix, $C(q,\dot{q})\dot{q} \epsilon R^n$ is the vector of Coriolis and centrifugal torques, and $G(q) \epsilon R^n$ is the vector of gravitational torques. The equation for the computed-torque control plus compensation is given by (Kelly et al., 2005):

$$\tau = M(q)\left[\ddot{q}_d + K_v\dot{\tilde{q}} + K_p\tilde{q}\right] + C(q,\dot{q})\dot{q} + G(q) - C(q,\dot{q})v, \tag{2}$$

where K_p and K_v are symmetric positive-definite design matrices, $\tilde{q}(t) = q_d(t) - q(t)$ denotes the position error vector, and thus $\dot{\tilde{q}}(t) = \dot{q}_d(t) - \dot{q}(t)$ is the velocity error vector. $q_d(t)$ is the given reference trajectory vector which is assumed to be smooth and bounded in its first and second time derivatives. Finally, $v \epsilon R^n$ is obtained by filtering $\tilde{q}$ and $\dot{\tilde{q}}$ (Kelly et al., 2005):

$$v = -\frac{bp}{p+\lambda}\dot{\tilde{q}} - \frac{b}{p+\lambda}\left[K_v\dot{\tilde{q}} + K_p\tilde{q}\right], \tag{3}$$

where $p = \frac{d}{dt}$ is the differential operator, and λ and b are scalar positive constants given by the designer. For simplicity, we can set $b = 1$ as in (Kelly et al., 2005). The above equation can be expressed as follows:

$$\dot{v} + \lambda v = -[\ddot{\tilde{q}} + K_v\dot{\tilde{q}} + K_p\tilde{q}]. \tag{4}$$

The controller in equations (2) and (4) applied to the robot system in equation (1) satisfies the next motion control objective (Kelly et al., 2005), that is,

$$\lim_{t\to\infty} \tilde{q}(t) = 0. \tag{5}$$

3. Computed-torque-plus-compensation-plus-chattering control of robot manipulators

We now introduce the chattering version of the computed-torque-plus-compensation controller:

$$\tau = M(q)\left[\ddot{q}_d + K_v\dot{\tilde{q}} + K_p\tilde{q}\right] + C(q,\dot{q})\dot{q} + G(q) - C(q,\dot{q})v - K_\alpha sgn(v), \tag{6}$$

where v is obtained from equation (4), $K_\alpha = diag\{k_a\}$, with $k_a > 0$; $sgn(v)^T = [\ sgn(v_1)\ \ sgn(v_2)\ \ \cdots\ \ sgn(v_n)\ ]$, and $v^T = [v_1\ \ \cdots\ \ v_n]$. The function $sgn(\cdot)$ is the signum function, which is 1 if its argument is positive, -1 if it is negative, and 0 if it is zero. The closed-loop system in equations (1), (4) and (6) yields

$$M(q)\left[\ddot{\tilde{q}} + K_v\dot{\tilde{q}} + K_p\tilde{q}\right] - C(q,\dot{q})v - K_\alpha sgn(v) = 0, \tag{7}$$

which, after invoking equation (4), produces

$$M(q)[\dot{v} + \lambda v] + C(q,\dot{q})v + K_\alpha sgn(v) = 0. \tag{8}$$

Consider now the following nonnegative Lyapunov function, which is also used in (Kelly et al., 2005),

$$V(t,v) = \frac{1}{2}v^T M(q)v = \frac{1}{2}v^T M(q_d - \tilde{q})v\ . \tag{9}$$

Its time derivative is

$$\dot{V}(t,v) = \frac{1}{2}v^T\dot{M}(q)v + v^T M(q)\dot{v}\ . \tag{10}$$

Solving equation (8) for $M(q)\dot{v}$ and substituting it in equation (10), we arrive at

$$\dot{V}(t,v) = v^T\left[\frac{1}{2}\dot{M}(q) - C(q,\dot{q})\right]v - \lambda v^T M(q)v - v^T K_\alpha sgn(v),$$

where the term $v^T[\frac{1}{2}\dot{M}(q) - C(q,\dot{q})]v$ can be canceled thanks to the fact that $\frac{1}{2}\dot{M}(q) - C(q,\dot{q})$ is a skew-symmetric matrix. Thus,

$$\dot{V}(t,v) = -\lambda v^T M(q)v - v^T K_\alpha sgn(v).$$

On one hand, there exists a real positive number α such that

$$\|M(q)\|_2 \le \alpha\|I\|_2,$$

and using the Cauchy-Schwartz inequality, it follows that

$$V(t,v) = \frac{1}{2}v^T M(q)v \le \frac{\alpha}{2}\|v\|_2^2.$$

Using that $\|v\|_2 \le \|v\|_1$, we obtain

$$V(t,v) \le \frac{\alpha}{2}\|v\|_2^2 \le \frac{\alpha}{2}\|v\|_1^2,$$

or equivalently,

$$k\sqrt{V(t,v)} \le \|v\|_1, \quad \text{where } k = \sqrt{\frac{2}{\alpha}}. \tag{11}$$

On the other hand,

$$\dot{V}(t,v) = -\lambda v^T M(q)v - v^T K_\alpha sgn(v) \le -k_a v^T sgn(v) \le -k_a\|v\|_1, \tag{12}$$

and substituting (11) in equation (12) we arrive at,

$$\dot{V}(t,v) \le -k_a\|v\|_1 \le -k_a k\sqrt{V(t,v)}; \quad \Rightarrow \quad \dot{V}(t,v) + k_a k\sqrt{V(t,v)} \le 0,$$

thus, there exists a settling time, t_s, such that $\lim_{t\to t_s} v(t) = 0$ and $v(t) = 0$ for all $t \ge t_s$. For details, see Theorem 4.2 on finite-time stability in (Bhat & Bernstein, 2000). From equation (4) and using that $v(t) = 0$ (and $\dot{v}(t) = 0$) for all $t \ge t_s$, we have $\ddot{\tilde{q}} + K_v\dot{\tilde{q}} + K_p\tilde{q} = 0$, which is a linear time-invariant and asymptotically stable system. In summary, we have obtained the following main result stated in Theorem 1.

Theorem 1.- *The controller in equations (6) and (3) (or (4)) global-uniformly-asymptotically stabilizes the robot system described in equation (1) at the equilibrium point* $(\tilde{q}, \dot{\tilde{q}}, \upsilon) \equiv 0$.

Remark 1.- *Although the closed-loop system contains discontinuity terms in the right-hand side, its solution is continuous and locally Lipschitz everywhere except at the origin. Hence, every set of initial conditions in* $R^n \backslash \{0\}$ *has a unique solution in forward time on a sufficiently small time interval. The chattering appears at the origin. This justifies the use of Lyapunov theory for this special case of non-smooth dynamical systems.*

4. Numerical experiments

The performance of the controller specified in Theorem 1 is compared with that of the computed-torque plus compensation controller in equations (2) and (4). Consider a 2-DOF robot manipulator moving in a vertical plane (see Figure 1). The characterization of this manipulator is taken from (Berghuis & Nijmeijer, 1993),

$$M(q) = \begin{bmatrix} 8.77 + 1.02\cos(q_2) & 0.76 + 0.51\cos(q_2) \\ 0.76 + 0.51\cos(q_2) & 0.62 \end{bmatrix},$$

$$C(q, \dot{q}) = \begin{bmatrix} -0.51\sin(q_2)\dot{q}_2 & -0.51\sin(q_2)(\dot{q}_1 + \dot{q}_2) \\ 0.51\sin(q_2)\dot{q}_1 & 0 \end{bmatrix},$$

$$G(q) = g\begin{bmatrix} 7.6\sin(q_1) + 0.63\sin(q_1 + q_2) \\ 0.63\sin(q_1 + q_2) \end{bmatrix},$$

where g is the gravity acceleration. Moreover, let us assume that the robot system is subject to a Coulomb friction perturbation, that is, the robot with added friction is given by (Orlov et al., 2003)

$$M(q)\ddot{q} + C(q, \dot{q})\dot{q} + G(q) + F(\dot{q}) = \tau,$$

where $F(\dot{q}) = K_f sgn(\dot{q})$ is the friction force vector (which can be seen as the un-modeled dynamics). We use $K_f = diag\{9.9\}$, and, to complete the numerical experimental platform, we set $K_p = diag\{100\}$, $K_v = diag\{50\}$, and $\lambda = 10$, for the original controller, and $K_\alpha = diag\{10\}$ for the proposed controller. We set the reference trajectory vector, $q_d(t) = [q_{d1}(t) \quad q_{d2}(t)]^T = [\pi + 0.5\sin(2\pi t) \quad 0.5\pi + 0.5\sin(2\pi t)]^T$. The simulation results are shown in Figures 2 and 3. From these two figures, it is clear that the proposed chattering controller represents an important performance improvement.

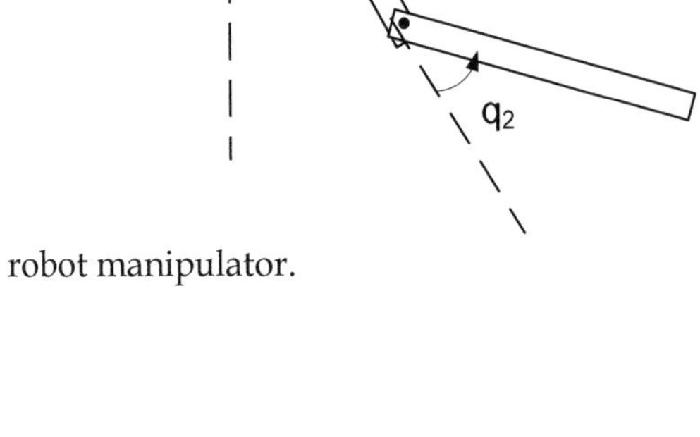

Fig. 1. 2-DOF vertical robot manipulator.

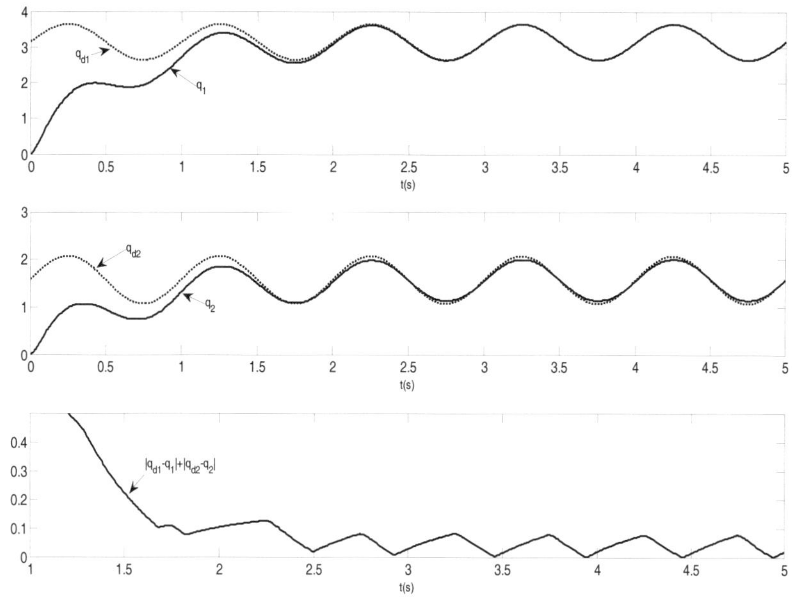

Fig. 2. Simulation results on the computed-torque plus compensation controller.

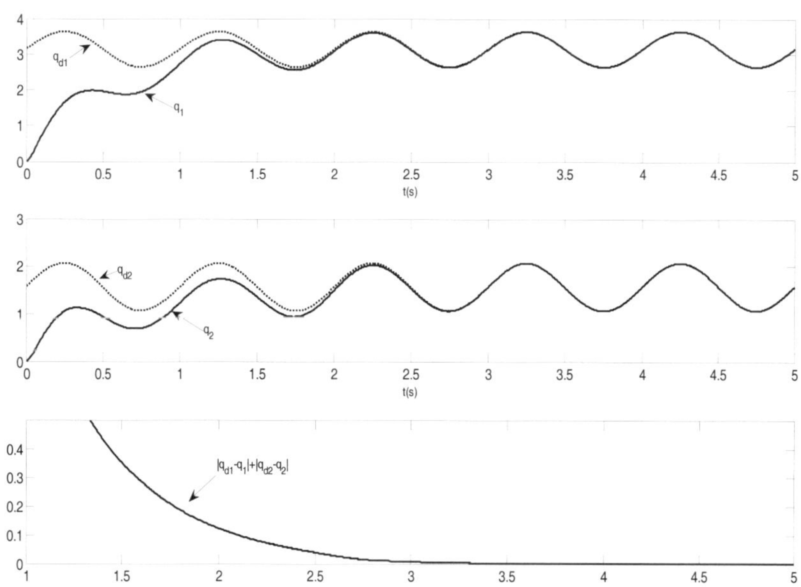

Fig. 3. Simulation results on the computed-torque-plus-compensation-plus chattering controller.

Figure 2 shows the system trajectories and their comparison with respect to the desired ones. The graph of $|q_{d1} - q_1| + |q_{d2} - q_2|$ versus time captures the 1-norm error position. Here, an oscillating error is obtained because of friction. In some applications, this tracking error can be unacceptable. For instance, repeatability (the measure of how close a manipulator can return to a previously taught point) is perturbed, as well as accuracy (the measure of how close the manipulator can approach a given point within its workspace). However, using our controller (Figure 3), the oscillatory error behavior is precluded, thus improving the repeatability and accuracy performance. Moreover, the tracking error shown in Figure 3 can be inside of the controller resolution (the smallest increment that the controller can sense). When this happens, our controller rejects completely the effects of friction on the robot system. Figures 4 and 5 show the control signals for both cases. We can appreciate that both control signals are alike. Only small chattering appears in our case. This chattering has small amplitude ant it is not persistent, like the chattering that appears, for instance, in (Orlov et al., 2003).

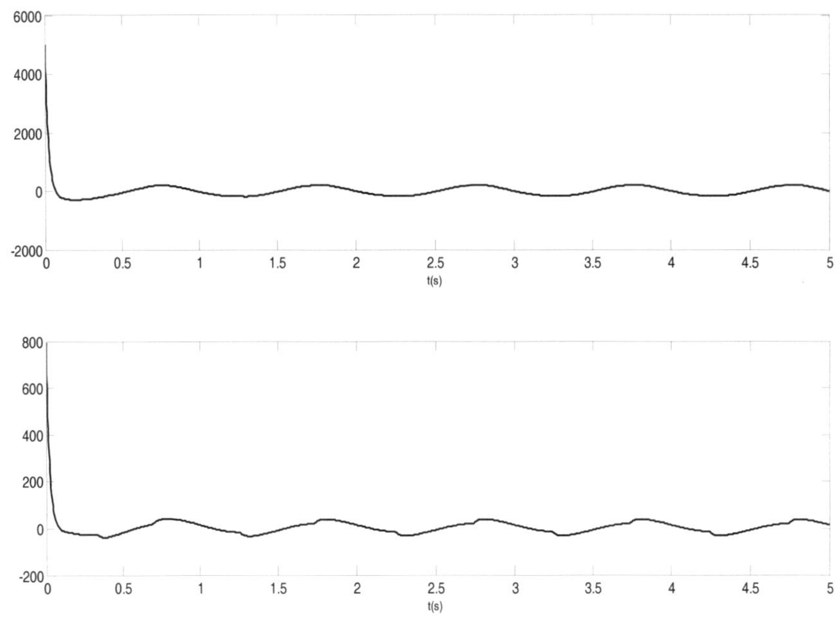

Fig. 4. Simulation results on the computed-torque plus compensation controller: the applied torque (N-m) to the first link (top) and the applied torque (N-m) to the second link (bottom).

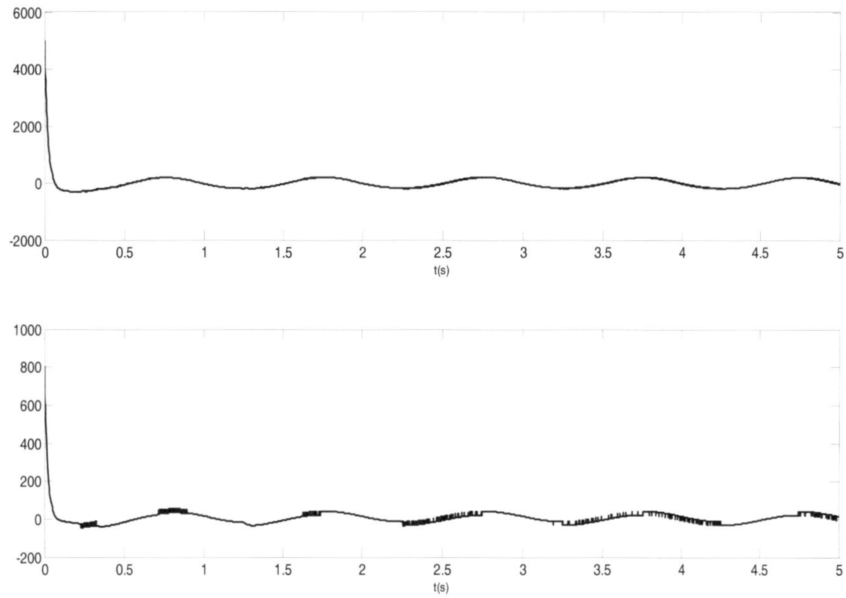

Fig. 5. Simulation results on the computed-torque-plus-compensation-plus chattering controller: the applied torque (N-m) to the first link (top) and the applied torque (N-m) to the second link (bottom).

Let us test the controllers performance by means of a more general case of perturbation. Consider that the robot system is subject to external perturbation; that is, consider the system:

$$M(q)\ddot{q} + C(q,\dot{q})\dot{q} + G(q) + F(\dot{q}) = \tau + d(t),$$

where $d(t) \epsilon R^2$ is a bounded external perturbation. This perturbation can be introduced into the robot system, for instance, when working on a ship since wave motion induces vertical force perturbation. Let us set $d^T(t) = [\sin(t) \quad \sin(2t)]$. Simulation results are shown in Figures 6, 7, 8 and 9. When the proposed controller is used, the tracking error between the system trajectory and the reference trajectory is clearly improved for the second joint. Thus, when the external perturbation is present, our controller outperforms the original one.

5. Conclusion

A modified version of the computed-torque plus compensation controller was designed by adding a chattering term. Because of this chattering term, the new robot controller outperforms the original one, especially when the robot is subject to Coulomb friction perturbations. Moreover, this new controller facilitates the proof of global stability of the closed-loop system, and also improves the repeatability and accuracy of the robot control system. From the control design point of view, our chattering controller has the following *sliding mode control interpretation*. It is well known that sliding motion occurs when the trajec-

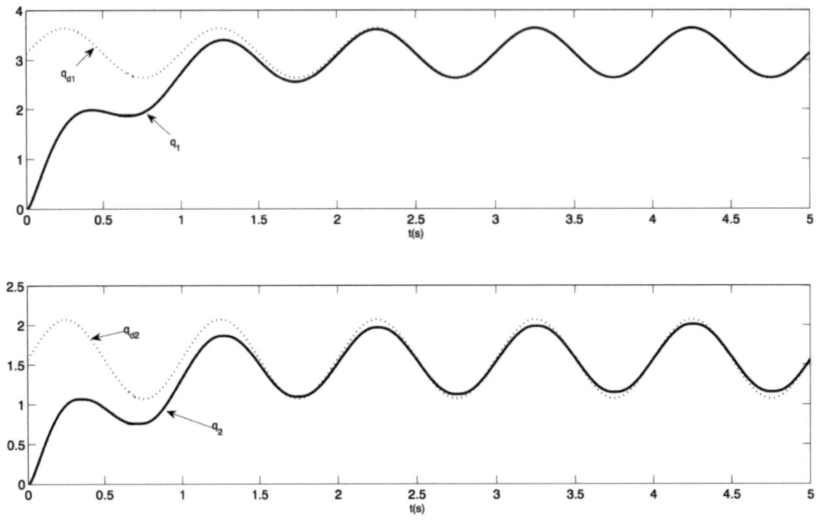

Fig. 6. Simulation results on the computed-torque plus compensation controller.

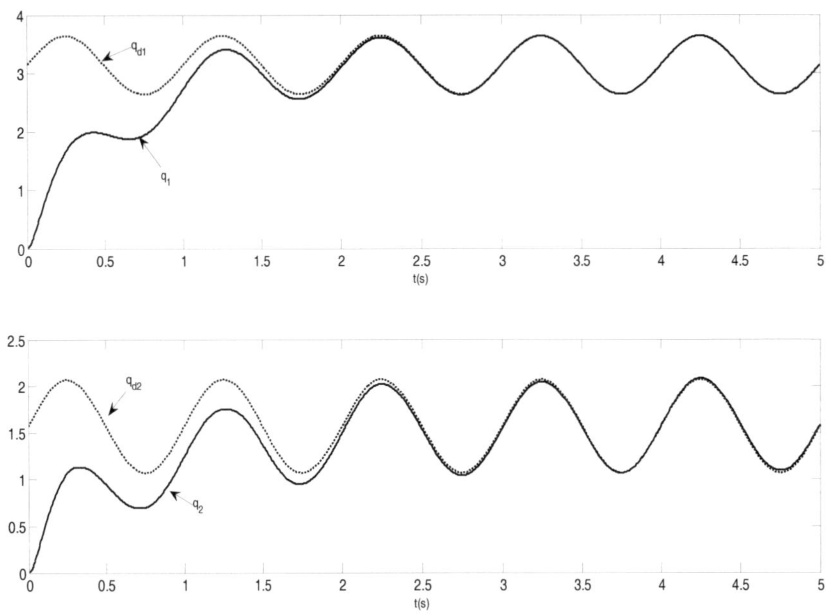

Fig. 7. Simulation results on the computed-torque-plus-compensation-plus chattering controller.

tory of the system is driven (in finite time) towards a sliding surface, where the system has a reduced order behavior, and forced to remain on it where some stability property is satisfied. See, for instance, (Edwards & Spurgeon, 1998), (Perruquetti & Barbot, 2002), and (Spong & Vidyasagar, 1989). Our chattering controller drives the system trajectory, in finite time, to the condition where the non-linear robot system has a linear-time-invariant asymptotically stable behavior given by $\ddot{\tilde{q}} + K_v\dot{\tilde{q}} + K_p\tilde{q} = 0$. This is an important contribution of our chattering control, which is impossible to be fulfilled with the original computed-torque plus compensation controller. So, our controller is in fact a sliding mode controller but designed in an implicit form.

6. References

Aguilar, L.; Orlov, Y. & Acho, L. (2003). Nonlinear H-infinity control of non-smooth time-varying systems with application to friction mechanical manipulators. *Automatica*, Vol. 39, 1531-1542.

Berghuis, H. & Nijmeijer, H. (1993). Global regulation of robots using only position measurements. *Systems and Control Letters*, Vol. 21, 289-293.

Bhat, S. & Bernstein, S. (2000). Finite-time stability of continuous autonomous systems. *SIAM Journal of Control Optimization*, Vol. 38, No. 3, 751-766.

Edwards, C. & Spurgeon, K. (1998). *Sliding Mode Control: Theory and applications*,

Guerra, R. & Acho, L. (2007). Adaptive control for mechanism with friction. *Asian Journal of Control*, Vol. 9, No. 4, 422-425.
ISBN 978-0824706715, USA.

Kelly, R.; Santibáñez, V. & Loría, A. (2005). *Control of Robot Manipulators in Joint Space*, Springer-Verlag, ISBN 1852339942, 9781852339944, USA.

Orlov, Y.; Alvarez, J.; Acho, L. & Aguilar, T. (2003). Global position regulation of friction manipulators via switched chattering control. *International Journal of Control*,

Perruquetti, W. & Barbot, J. P. (2002). *Sliding Mode Control in Engineering*, CRC Press,

Spong, W. S & Vidyasagar, M. (1989). *Robot Dynamics and Control*, John Wiley and Sons, ISBN 0-471-50352-5, Republic of Singapore.
Taylor & Francis Ltd, ISBN 0-7484-0601-8, UK.
Vol. 76, No. 14, 1446-1452.

Weiping, L. & Xu, C. (1994). Adaptive high-precision control of positioning tables. Theory and experiments. *IEEE Transactions on Control Systems Technology*, Vol. 2, No. 3, 265-270.

5

Geometric and Threshold Calibration Aspects of a Multiple Line-Scan Vision System for Planar Objects Inspection

Andrei Hossu and Daniela Hossu
University Politehnica of Bucharest, Faculty of Control and Computers
Romania

1. Introduction

The vision system proposed as support of this chapter is dedicated for inspection and localization of flat glass parts, in a robot-based automation of the unloading and packing stages in the flat glass industry. This vision system belongs to the class of the artificial Vision Systems dedicated for analyzing objects located on a moving scene (conveyor).
The Industrial Vision System described in the paper is designed for silhouette inspection of planar objects (it is a pure 2D Vision System, the volumetric characteristics of the analyzed objects being not relevant for the application).
Analyzing the functional system requirements can be identified a sum of characteristics that have to be achieved by the system, from which the most critical ones and also most relevant for this paper, are:

The response time - especially because this Vision System belongs to the class dedicated analyzing objects located on a moving scene (other parts are coming under camera) and also because it is part of an automation process were all the following application partners' components are piped along the conveyor.

The accuracy of the analysis results. The main purpose of including this Vision System into the automation system is to analyze and to provide decisional results on the inspection of the glass plates. The aspects to be analyzed are the accuracy of the edges and corners (resulted from the cutting process and/or from the previous handling process of plates).

The paper is focused on the *geometric calibration* and *threshold calibration* aspects of a multiple line-scan camera vision system (in particular a dual line-scan camera system).
In our specific vision system application the size of the image that has to be processed is very large. This is caused by the size of the inspected parts: lengthwise the conveyor up to 6500 mm and the width of the area of interests is 4000 mm in conjunction with the accuracy requirements (which is leading to a resolution of the acquired image of about 0.5 mm/pixel).
The major research and development efforts were to define, implement and test, for both geometric calibration and threshold calibration processes, methods with minimal negative

impact on the critical requirements of the vision system (the response time and the system accuracy).
The methods defined for the both calibration processes have to maintain an acceptable system Set-up time and also to provide the ability of moving the most of the vision system computational effort from on-line to off-line processing stage.

2. The Automation System Description

In Figure 1 is presented the architecture of this automation system. This architecture is often utilized in industrial applications (in palletizing of moving objects systems).

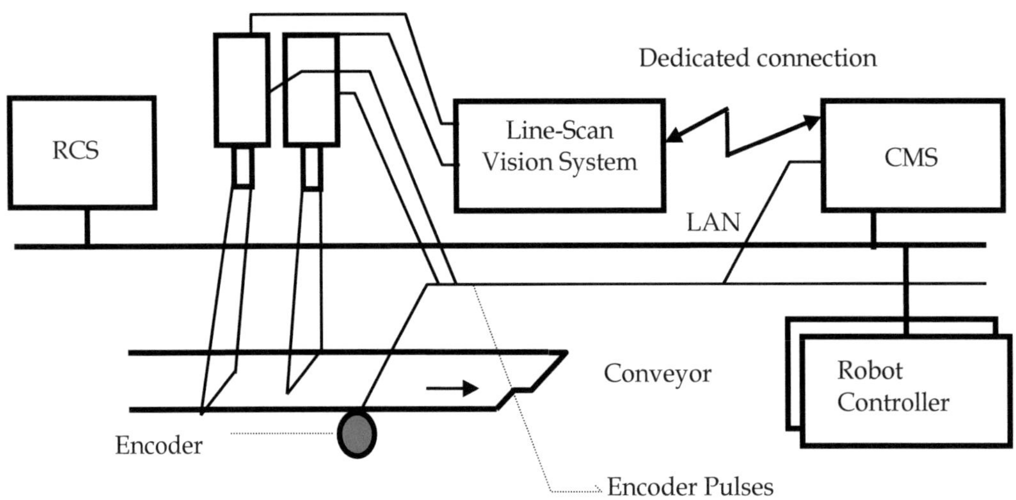

Fig. 1. The robot-based automation system for inspecting and handling moving glass plates from a conveyor

2.1 The Structural Aspects of the Automation System

The structural aspects of the automation system architecture are (Hossu & Hossu, 2008-c):

Active Elements: Control Management System (CMS), Routing Control System (RCS), Vision System and Robotic Cells.

Passive Elements: Conveyor, glass plates.

Infrastructure: Communicational Links: Vision System - CMS, CMS - Robots Controllers, CMS - RCS.

General assumptions: The plates are connected to the conveyor (the same speed and direction).

2.2 The Functional Aspects of the Automation System

The Routing Control System has to provide for CMS the Routing Data - a description of the possible destinations (one or more of the robotic cells) of each plate in the moment the plate is passing the Decision Point of the Vision System. The role of the Vision System is to inspect the cutting accuracy and the shape parameters of every plate. The vision system is analyzing the information provided by a Line Scan Acquisition System (a dual line scan camera system) in conjunction with the information provided by an encoder connected to the transport conveyor. The Vision Data, containing the data resulted from the inspection process, together with the data describing the location of the plate, are transmitted to CMS in the moment the vision system processing time ended. The moment (time-based) is called Vision Decision Point. Both sets of data (Routing Data and Vision Data) are merged by CMS. CMS will take the decision to send the pick plate command to a certain robotic cell only if Vision Data describe the plate having cutting accuracy and shape parameters inside the accepted tolerances for a certain packing destination and also if the plate is routed to that certain destination.

3. The Description of the Vision System

This system belongs to the class of the artificial Vision Systems dedicated for analyzing objects located on a moving scene (conveyor).
The Vision System main task is to inspect glass plates transported by a conveyor.
From the structural system architecture and its working environment we could identify a set of its general intrinsic characteristics, from which the most relevant in this point, are:

The system is using line-scan camera / cameras for the image acquisition and an encoder for estimating the motion of the object by measuring the motion of the transport support (the speed of the conveyor).

The image is obtained by reflection of the light from a linear light source (fluorescent) on the surface of the analyzed objects.

The plates have the same speed and direction as the conveyor, and the orientation of the conveyor is known relative to the acquisition line and constant in time.

4. Geometric Calibration Aspects of a Multiple Line-Scan Vision System for Planar Objects Inspection

This class of the Artificial Vision Systems dedicated for analyzing objects located on moving scenes (conveyor) presents some specific characteristics relative to the Artificial Vision Systems dedicated for static scenes. These characteristics are identified also on the image geometric calibration process (Borangiu, et al., 1995), (Haralick & Shapiro, 1992).
In Figure 2 is presented the model of the image obtained from a dual line-scan camera Vision System.
For this class of the Artificial Vision Systems we could identify as relevant for the geometric calibration process the following characteristics (Hossu, 1999):

The obtained image has significant geometric distortions on (and only on) the image sensors direction. The geometric distortions are along the acquisition line, but not from one line to the other.

There is an overlapped image area between the two cameras. The end of the acquisition line of the 1st camera is overlapping the beginning of the acquisition line of the 2nd camera. This overlapping area is significant in size and is a constant parameter estimated during the artificial vision system installation process.

There is a lengthwise conveyor distance between the acquisition lines of the two cameras. This distance is also a constant parameter and its value is also estimated during the system installation process.

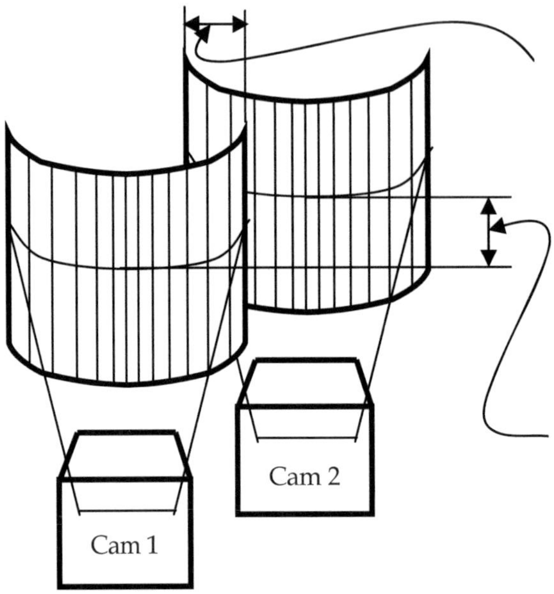

Fig. 2. The geometric distortions of the image acquired with a dual line-scan camera Vision System.

4.1 The Pattern based Calibration Tool

For the calibration process we adopted the method of using a Pattern based Calibration Tool.

This Pattern based Calibration Tool represent a set of blobs with a priori known dimensions and locations for the real world (millimeters and not image pixels) (Croicu, et al., 1998).

The outcome of using this type of calibration technique was to obtain the following:

Estimation with the highest accuracy of the scene model parameters on the direction of the distortions.

Estimation of the size of the overlapped image area for both cameras.

The parallelism of the two acquisition lines is obtained during the installation process, using the support of the Calibration Tool.

Achieving a high accuracy of mounting the cameras in such a way to obtain the perpendicularity of the acquisition lines on the moving direction of the scene (of the conveyor).

Achieving a high accuracy on the distance lengthwise the conveyor of the acquisition lines (the acquisition lines of Camera 1 relative to the acquisition lines of Camera 2). The shape and the dimensions of the pattern adopted for the Calibration Tool force this characteristic.

4.2 The Calibration Tool Description

In Figure 3 is presented the pattern adopted for the Calibration Tool used for the dual line-scan camera Vision System (the dimensions are presented in millimeters) (Croicu, et al., 1998)., (Hossu, et al., 1998).

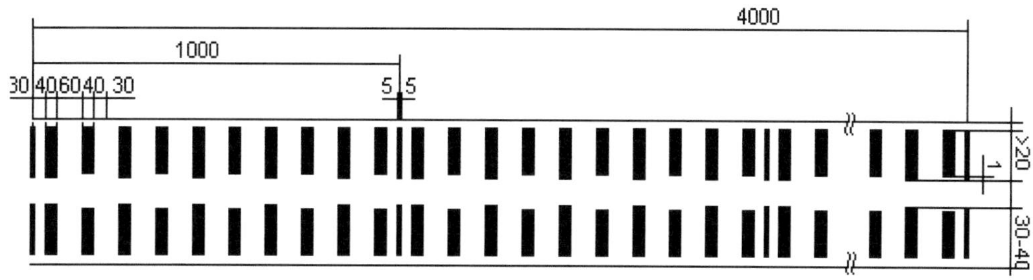

Fig. 3. The pattern of the Calibration Tool used for the dual line-scan camera Vision System.

The characteristics of the adopted Pattern are:

The pattern contains dark blobs (marks) placed on a bright background (with a high level of light intensity for the image).

The pattern is symmetrical on the vertical direction (lengthwise the conveyor). The two cameras have the acquisition lines parallel one each other but located on different position on the conveyor (due to the lighting system adopted - built from two fluorescent tubes used for obtaining the image from the reflection on the object surface). 1st Camera will have the acquisition line located on the top edge of the lower section of the pattern, and the 2nd Camera will locate its acquisition line on the bottom edge of the upper section of the pattern.

The pattern is partially homogenous on the horizontal axis (the direction crosswise the conveyor, the direction of the distortions)

The pattern contains a characteristic of a small difference (1 mm.) between the even and the odd marks. This will force the mounting process of the cameras to be very accurate in obtaining the parallelism of the acquisition lines of the cameras and also the perpendicularity on the conveyor direction.

4.3 Experimental Results of the Calibration Process

In Figure 4 are presented the results obtained from the Calibration process performed on the 1st Camera. (Hossu & Hossu 2008-c)

The Excel Cell used as support for representing the results of the Calibration on the 1st Camera contains the following:

The 1st column (called Mark) contains the number of the corresponding Mark existing in the pattern.

The 2nd column (called Cam1) contains the values of the coordinates of the marks on the Calibration Tool. These values are obtained from the "real world", from direct measuring of the Pattern applied on the Calibration Tool (represented in millimeters).

The 3rd column (called Pixel) represents the coordinates of the existing Marks on the image. These coordinates are represented in pixel number.

For both cameras we chose a Polynomial Trend of order 3 for the approximation of the conversion function of the image coordinates values (pixels) into the conveyor coordinates values (millimeters). Using this type of trend the calibration method will provide results with a maximum approximation error inside the accuracy requirements of the particular Vision System.

a1	b1	c1	d1
1.448567E-09	-9.128585E-06	0.5088785	338.946

Mark	Cam1	Pixel	Estimated
1	429.1	175	427.7278
2	529.2	380	531.0810
3	629.3	576	629.3081
4	729.5	777	729.5128
5	829.6	979	829.7479
6	929.8	1181	929.5853
7	1033.1	1390	1032.5400
8	1133.3	1595	1133.2616
9	1233.6	1799	1233.3085
10	1333.9	2004	1333.7361
11	1434.7	2210	1434.6181
12	1535.1	2415	1535.0504
13	1635.4	2621	1636.0883
14	1735.8	2825	1736.3342
15	1836.2	3028	1836.3487
16	1936.6	3231	1936.6953
17	2042.2	3443	2041.9242
18	2142.6	3644	2142.1760
19	2243.1	3844	2242.4668
20	2343.7	4046	2344.3760

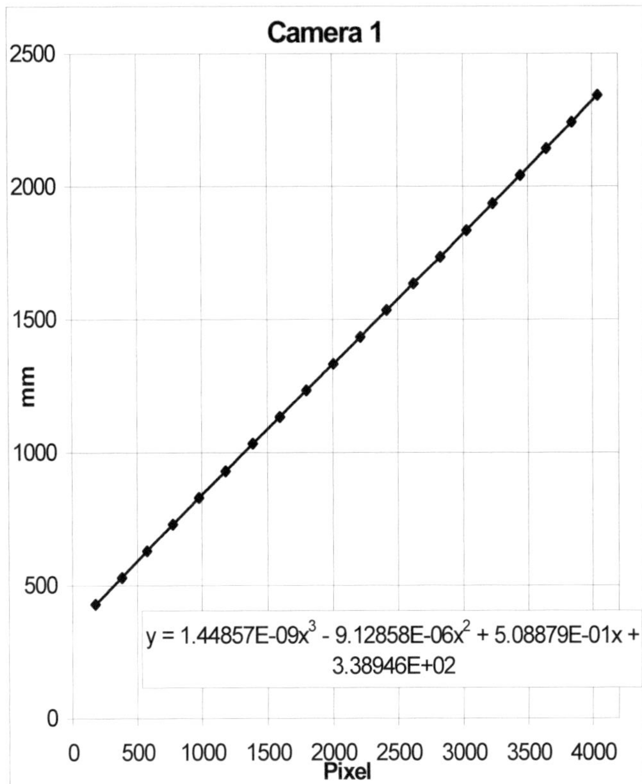

Fig. 4. Experimental results of the calibration process on the 1st camera.

On the top and right side of the Excel Cell (the first two rows and the last four columns) are stored the parameters of the 3rd Order Polynomial estimated as trend.

The last column (called Estimated) contains the estimated of the marks coordinates values (on the conveyor), obtained from applying the Polynomial trend of order 3.

In the bottom right side of the Excel Cell is presented the graphical chart of the trend of the coordinates values of the marks on the conveyor related to the pixel coordinates.

In the figure the following notations are used:

x coordinate value in image - (pixels),

y coordinate value on the objects scene (on conveyor) - (mm).

$$y = a_1 x^3 + b_1 x^2 + c_1 x + d_1 \text{ for the 1st Camera} \quad (1)$$

$$y = a_2 x^3 + b_2 x^2 + c_2 x + d_2 \text{ for the 2nd Camera} \quad (2)$$

In Figure 5 is presented the graphical chart of the evolution of the distortions estimated on the acquisition lines direction. The two cameras are covering around 4 meters wide view. The figure represents the behavior of the two acquisition cameras: in the area from 400 mm to 2250 mm it is represented the behavior of the 1st camera and in the area from 2250 mm to 4400 mm it is represented the behavior of the 2nd camera.

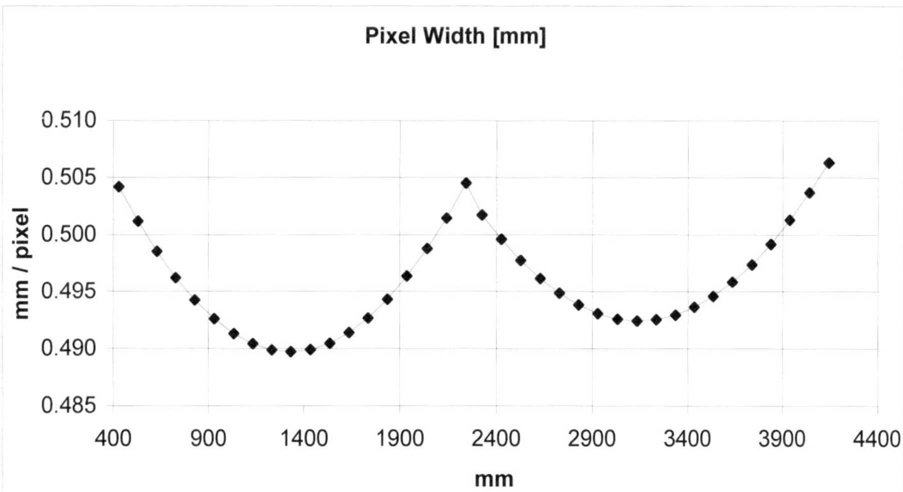

Fig. 5. The distortions estimated on the acquisition lines direction

We can notice the distortions are affecting the pixel width of the image form 0.49 mm/pixel to 0.507 mm/pixel. Ignoring this variation of the pixel width along the image acquisition line, would lead to accumulation of very high errors in some areas of the image, due to the fact the amount of pixels contained in an acquisition line is high (8096).

4.4 Using the Geometric Calibration Results

The most important achievement from estimating the scene parameters is the estimation of the image geometric distortions.

The end of the distortions estimation process leads to obtaining a table for conversion the image coordinates (pixels) in scene coordinates (millimeters) for each of all the 8096 pixels of the acquisition lines. This lookup table will contain 8096 values (of floating-point type) representing the real values of the scene (conveyor) coordinates of each pixel.

The CPU effort of the image processing algorithm will be minimal on converting the image coordinates into conveyor coordinates, using the pixel coordinate as the index of the offline-built lookup table.

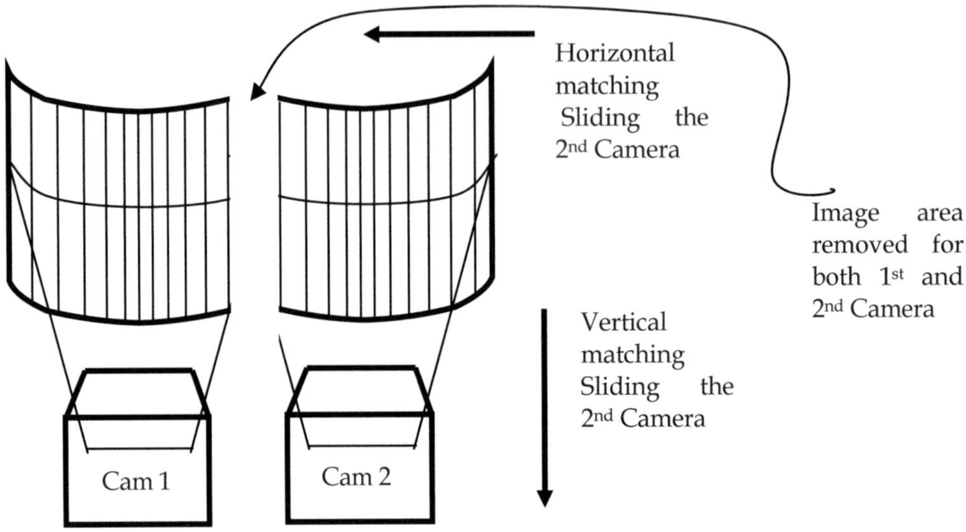

Fig. 6. Using the calibration results for a dual line-scan camera system.

In order to minimize the processing time of the image, the polynomial estimation is not used on-line. The polynomial estimation is used for building the pixel to millimeters lookup table, in the offline stage of the presented method. In fact using a polynomial trend of 3rd order or other type of trend is not relevant for the response time performances of the vision system but only for obtaining the required approximation error. This stage, being offline and using a relative small amount of data, another more sophisticated approximation method could be used.
In Figure 6 are presented the ways the last steps of the calibration process are performed.

5. Threshold Calibration Aspects of a Multiple Line-Scan Vision System for Planar Objects Inspection

For this specific vision system we could identify as relevant for the threshold calibration process the following structural characteristics:

- The lighting environment (provided by the light source) is stable in time (from one acquisition line to the other). This is obtained using a high frequency controller for the control of the fluorescent lamp of the light source. This assumption of stability in time of the lighting environment is true for a short and medium segment of the lifetime of the light source.
- The image is acquired on variable acquisition frequency. The consequence of this is that the intensity level of the same image element (pixel) under the same lighting conditions is dependent on the conveyor speed.

Taking into account the critical requirements of achieving the real-time characteristic (minimal response time), our choice was for a Binary Image Processing System. In order to obtain the binary image from the gray level image, we adopted a threshold method.
Threshold methods are defined as starting from the analyze of the values of a function *T* of the type (Gonzales & Wintz, 1981), (Borangiu, 2004), (Hossu, 1999):

$$T = T\,[x, y, p(x, y), f(x, y)] \tag{3}$$

Where:

$f(x, y)$ - represents the intensity value of the image element located on the co-ordinates (x, y),

$p(x, y)$ - represents the *local properties* of the specific point (like the average intensity of a region centered in the co-ordinates (x, y)),

T - is the *image threshold*

The goal is to obtain from an original gray level image, a binary image $g(x, y)$ defined by:

$$g(x,y) = \begin{cases} 1 & \text{for } f(x,y) > T \\ 0 & \text{for } f(x,y) \le T \end{cases} \tag{4}$$

For T a function only of $f(x, y)$, the obtained threshold is called *global threshold.*

In the case of T a function of both $f(x, y)$ and $p(x, y)$, the obtained threshold is named *local threshold.*

In the case of T a function of all $f(x, y)$, $p(x, y)$, x and y, the threshold is a *dynamic threshold.*

5.1 Image Segmentation using Global Threshold

Gray level histogram represents the probability density function of the intensity values of the image (Gonzales & Wintz, 1981), (Borangiu, 2004), (Hossu, 1999).

In order to simplify the explanations, we suppose the image histogram of the gray levels is composed from two values combined with additive Gaussian noise:

The first segment of the image histogram corresponds to the background points - the intensity levels are closer to the lower limit of the range (the background is dark).

The second segment of the image histogram corresponds to the object points - the intensity levels are closer to the upper limit of the intensity range (the objects are bright).

The problem is to estimate a value of the threshold T for which the image elements with an intensity value lower than T will contain background points and the pixels with the intensity value greater than T will contain object points, with a minimum error. For a real image, the partitioning between the two brightness levels is not so simple and also not so accurate. The partitioning is fully accurate only if the two modes of the bimodal histogram are not overlapped. The classification is defined as the process of the distribution of the pixels in classes. The goal of the process of finding a segmentation threshold is the minimization of the error of classification. The optimum segmentation threshold is located in the intersection position of the two normal distributions.

The estimation of the error of classification is obtained from the area of the overlapped segments:

$$E = \frac{A + B}{\textit{image size}} \tag{5}$$

Suppose the image contains two intensity level values affected with additive Gaussian noise. The mixture probability density function is:

$$p(x) = P_1 p_1(x) + P_2 p_2(x) \tag{6}$$

Where:

x - the random value representing the intensity level,

$p_1(x)$, $p_2(x)$ – are the probability density functions,

P_1, P_2- are the a priori probabilities of the two intensity levels ($P_1 + P_2 = 1$).

For the normal distribution case on the two brightness levels:

$$p(x) = \frac{P_1}{\sqrt{2\pi}\sigma_1} \exp\frac{(x-\mu_1)^2}{2\sigma_1^2} + \frac{P_2}{\sqrt{2\pi}\sigma_2} \exp\frac{(x-\mu_2)^2}{2\sigma_2^2} \quad (7)$$

Where:

μ_1, μ_2 - are the mean values of the two brightness levels (the two modes),

σ_1, σ_2 - are the standard deviations of the two statistical populations.

Suppose the background is darker than the object. In this case $\mu_1 < \mu_2$ and defining a threshold T, so that all pixels with intensity level below T are considered belonging to the background and all pixels with level above T are considered object points. The probabilities of misclassification an object point (classifying an object point as a background point) and similarly of misclassification a background point (classifying a background point as an object point) are:

$$E_1(T) = \int_{-\infty}^{T} p_2(x)dx \quad (8)$$

$$E_2(T) = \int_{T}^{+\infty} p_1(x)dx$$

The probability of error is given by:

$$E(T) = P_1 E_2(T) + P_2 E_1(T) \quad (9)$$

To find the threshold value for which the error is minimum, $E(T)$ is differentiate with respect to T:

$$P_1 p_1(t) = P_2 p_2(t) \quad (10)$$

Applying the result to the Gaussian density we obtain:

$$AT^2 + BT + C = 0 \quad (11)$$

Where:

$$A = \sigma_1^2 - \sigma_2^2 \quad (12)$$

$$B = 2(\mu_1\sigma_2^2 - \mu_2\sigma_1^2)$$

$$C = \sigma_1^2\mu_2^2 - \sigma_2^2\mu_1^2 + \sigma_1^2\sigma_2^2 \ln\frac{\sigma_1 P_1}{\sigma_2 P_2}$$

If the standard deviations are equal, a single threshold is sufficient:

$$T = \frac{\mu_1 + \mu_2}{2} + \frac{\sigma^2}{\mu_1 - \mu_2} ln\frac{P_2}{P} \quad (13)$$

If the probabilities are equal $P_1 = P_2$ the threshold value is equal with the average of the means.

A way of checking the validity of the assumption of bimodal histogram is to estimate the mean-square error between the mixture density, $p(x)$ and the experimental histogram $h(x_i)$.

$$M = \frac{1}{N}\sum_{i=1}^{N}[p(x_i) - h(x_i)]^2 \tag{14}$$

Where: N - number of possible levels of the image (usually $N = 256$).
The binary image is obtained changing the color attribute of each pixel according to its intensity level relative to the segmentation threshold. Characteristics of the global threshold methods (Borangiu, et al., 1995), (Hossu & Andone, 2005) for these classes of vision applications:

- The assumption that both classes have the same standard deviation is not correct in our vision system environment. The background pixels (dark) have much smaller standard deviation than the object pixels.
- The assumption the classes (two levels) have the same a priori probabilities in many applications is also not acceptable for this vision system application.

In the case of the artificial vision systems dedicated to object recognition for industrial applications there is a large amount of a priori information about the image that has to be processed. Better results of estimation of the distribution of the image elements of the scene (background image, without the objects) can be obtained. Usually, in robotic applications, the illumination environment is known and controlled and also the object classes with a probability of apparition in the image are known. In many robotic application an estimation of the ratio between the area of the objects to be analyzed and the total area of the image scene, can be made with good results (a better estimation than the assumption of $P_1 = P_2 = 0.5$ could be obtained).

5.2 Image Segmentation using Dynamic Threshold

There are some classes of scenes of artificial vision systems where using the global threshold methods is not acceptable:

- The case of the applications where the lighting system does not supply a uniform intensity all over the analyzed surface.
- Segments of the image (or sometimes image elements) do not have the same behavior in the same lighting conditions.

For these types of images, in order to obtain a binary image, the most often used are dynamic threshold methods. The methods are based on the local analyze of the image. The algorithm of the estimation of the dynamic threshold consist of:

- The original image is divided in regions of a prescribed size.
- For each region it is estimated the histogram
- For each histogram it is estimated the error induced from the assumption of bimodal histogram (a histogram built from two normal distributions)
- If the value of the error is less than an acceptable value, the global threshold for the region is estimated.
- If the value of the error is too big (the histogram is too far from a bimodal histogram) the threshold value is estimated from the interpolation of the neighbors' region threshold values (for which the assumption of a bimodal histogram is considered acceptable).

In the final stage, a second interpolation process is applied: for each image element is assigned a threshold value $T(x, y)$ from the interpolation of the values of the neighbor image elements.

This method is called dynamic threshold method because the value of the resulted threshold for each image element is dependent of the position of the element in the image - $T(x, y)$.

Characteristics of the dynamic threshold methods:

Lack of processing time consumption - each element of the image is used at least two times (the method requires multiple-pass of the image) in different steps of the algorithm (and the number of the elements is very large).

Estimation of the acceptable error value (or the validation of the bimodal histogram assumption) is a complex process.

To choose the size of the image regions we have to take into account:

Large size of the region makes the method to loose the dynamic threshold characteristics and to fail into a global threshold method

Small size of the region makes to loose the statistical characteristic of the population of the image elements contained by the analyzed region (and the accuracy of the results is poor).

The last comment on the method is the fact that this method does not solve the problem of the non-uniformity of the illumination system or of the acquisition sensor.

5.3 Image Segmentation using Statistical Estimated Local Threshold

The set of types of image intensity level distortions identified for this class of vision systems (Hossu, et al., 1998), (Hossu. & Hossu, 2008-b) contains:

Illumination non-uniformity (obtaining a uniform intensity of the light on the whole area of the scene where the image is analyzed - usually 2 m - it is practical impossible).

Sensor and optics non-linearity - for linear cameras with a large number of pixels per row and also on optics with wide ranges, can be identified areas of nonlinear behavior of the image sensor and/or lenses.

Sensor cells non-uniformity - in cameras with CCD sensor, the cells present a different response on sensitivity at light intensity related to their neighbors.

In Figure 7 are presented the image intensity level distortions.

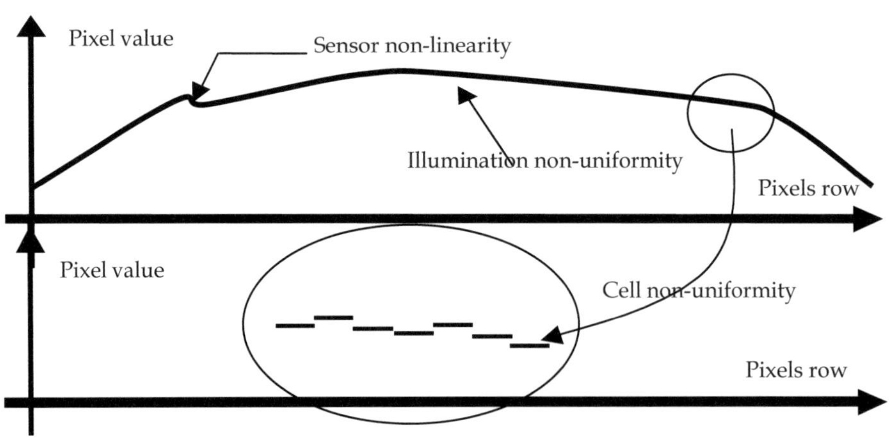

Fig. 7. Image intensity level distortions for CCD linear camera acquisition.

The main problem of the methods presented before represents the assumption that the image is a statistical population obtained from the addition of two ore more distributions (in the general accepted case normal distribution).

This assumption on the distribution of the intensity levels has the starting point the assumption that the insertion point of the noise is located on the transmission level of the information. In other words, the assumptions is that:

The acquired image is an ideal image (with only two gray levels: the gray level of the scene pixels and the gray level of the pixels corresponding to the object)

Then a global noise is applied, transforming the two levels in two normal distributions.

The assumption is false and using it we are analyzing a histogram, which is far away of two normal distributions, and from here the results are distorted. In reality the noise on the intensity level has its insertion point on the acquisition level and not on image transmission level. Intensity source has the meaning of intensity signal on the acquisition element and not only the lighting system. This implies the fact that the noise on the intensity source represents the whole chain of: lighting sources noise, reflective characteristics of the object surface and reflective characteristics of the scene surface, the sensitivity characteristics of the sensor and the vision system environment.

The vision system environment (the industrial environment the system it is dedicate for) is causing the most significant temporal intensity level distortions. This noise on intensity level is caused mainly by mechanical components of the system environment - causing vibrations of the mechanical support of the lighting source and/or vibrations of the object to be inspected (which are varying the reflection angle of the light).

For the image segmentation we adopted a local threshold estimated from temporal built statistical populations of every element in the image.

In the general case (an *array image*) an image represents a data set of:

$$\{f(x, y) \mid x \in [0,N], y \in [0,M]\} \quad (15)$$

Where:

N represents the number of image elements per row (number of image columns),

M represents the number of image elements per column (number of image rows).

In the *linear image* case, this data set becomes:

$$\{f(x) \mid x \in [0,N]\} \quad (16)$$

Where:

N represents the number of image elements per row (number of image columns).

Moving the insertion point of the noise we obtain: In the general case (an *array image*) an image represents a data set of:

$$\{f_i(x, y) \mid x \in [0,N], y \in [0,M], i \in [0,L]\} \quad (17)$$

Where:

L represents the number of the image frames (the size of the statistic population analyzed),

N represents the number of image elements per row (number of image columns),

M represents the number of image elements per column (number of image rows).

In the *linear image* case, this data set becomes:

$$\{f_i(x) \mid x \in [0,N], i \in [0,L]\} \quad (18)$$

Where:

L represents the number of the image frames (the size of the statistic population analyzed),

N represents the number of image elements per row (number of image columns).

In this way several temporal built statistical populations (from intensity levels of the same image element on a set of image frames acquired on different moments) replace the spatial built statistical population (made from image elements of the same image). The method of temporal histogram has the result the fact that each element of this set of histograms represents a bimodal histogram with two not overlapped modes (in case of a correct acquisition environment). It can be also introduced an estimation of the quality of the acquisition and image segmentation process using the estimation of the misclassification error analyzing the parameters of the two normal distributions. The method offers also the capacity of identification of the areas where some modifications should be done (on the lighting system) in order to improve the quality of the acquisition and image segmentation process. The lack of the proposed method is the memory consumption (it has to be built N x M different histograms in array acquisition, or N - in linear acquisition case). This problem is not so restrictive because at the end only the threshold values have to be stored and not the whole histograms. Another restriction is the fact that the method requires a large number of image frames acquired for construction of the statistical populations (in application set-up time). In the case of the systems dedicated to industrial applications usually this does not represent a real problem. This type of applications does not require a system response in condition of a small number of image frames a priori acquired. The vision systems dedicated to industrial applications can take the advantage on the fact that the image environment does not change a lot in time. In this way it can be initially reserved a certain time for acquiring a large enough number of image frames in order to be able to identify the permanent characteristics of the environment. All the intensity level distortions present permanent characteristics. This method has to be used in the case of the artificial vision systems dedicated to applications where the errors on image segmentation are not acceptable. In the applications dedicated exclusively to shape recognition the errors are accepted in a predefined range.

5.4 Local Threshold Values and the Acquisition Frequency

In moving scene applications, in order to maintain a constant resolution of the vision system along the direction of the scene movement, it is necessary the ratio between the acquisition frequency (the image lines rate - in the case of a line scan camera) and the scene speed to be constant. The acquisition frequency determines the exposure time of the CCD sensor cells. It can be notice an important influence of the speed (of the conveyor) on the intensity level of the same image element in the same lighting environment.

In Figure 8 are presented the experimental results obtained analyzing the influence on the intensity levels (for both: bright object and dark background) of the speed of the conveyor (acquisition frequency). The results were obtained on a statistical population from an image element on each measured speed. The second column represents the measured speed of the scene (conveyor) - V [m/min]. The 3rd to 8th columns represent image intensity levels estimated from the analyzed statistical population (temporal histogram). The values from the Threshold column are the image threshold values obtained from a global optimum

temporal threshold method applied on the histogram built for each analyzed level of the speed.
In Figure 9 are presented graphical the explanations on the meanings of the data involved in the analysis of the influence of the speed (acquisition frequency) on the intensity levels. The artificial vision system benefits from these results using a relation between the value of the image threshold and the speed V of the scene.

$$T = T(x, V) \tag{19}$$

Because of the response time restrictions imposed to the artificial vision system, instead of using an explicit expression of the estimated function $T(x, V)$, a search method in an a priori filled table (at set-up time) is more appropriate.
The size of the table is 256 (the number of the possible values of the image thresholds), containing floating-point values of the speed of the conveyor (acquisition frequency) for which the value of the threshold has to be changed.

	V [m/min]	Object Min+	Object Max	Object Min-	Scene Min+	Scene Max	Scene Min-	Threshold
1	20.7	221.532847	203.810219	168.364964	66.459854	44.306569	22.153285	110.766423
2	25.9	174.338624	160.391534	132.497354	52.301587	34.867725	17.433862	87.169312
3	31.1	147.510373	135.709544	112.107884	44.253112	29.502075	14.751037	73.755187
4	36.2	130.479452	120.041096	99.164384	39.143836	26.095890	13.047945	65.239726
5	41.3	118.513120	109.032070	90.069971	35.553936	23.702624	11.851312	59.256560
6	46.4	109.644670	100.873096	83.329949	32.893401	21.928934	10.964467	54.822335
7	51.4	102.927928	94.693694	78.225225	30.878378	20.585586	10.292793	51.463964
8	56.5	97.474747	89.676768	74.080808	29.242424	19.494949	9.747475	48.737374
9	61.6	93.040293	85.597070	70.710623	27.912088	18.608059	9.304029	46.520147
10	66.7	89.363484	82.214405	67.916248	26.809045	17.872697	8.936348	44.681742
11	72.5	85.877863	79.007634	65.267176	25.763359	17.175573	8.587786	42.938931

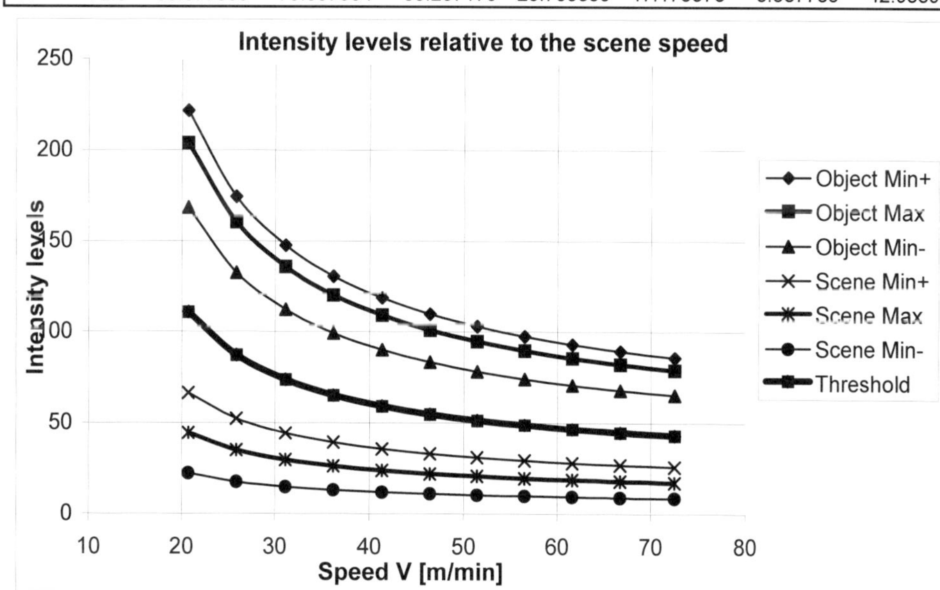

Fig. 8. The influence on the intensity levels of the acquisition frequency.

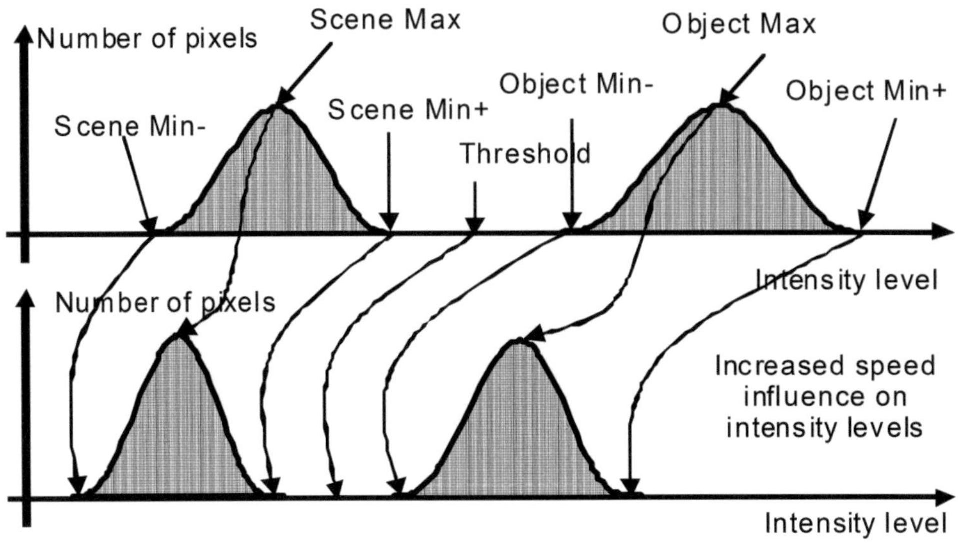

Fig. 9. The influence of the speed on the intensity levels

6. Conclusions

The proposed *geometric calibration* method is specific for multiple line-scan camera artificial vision applications for inspecting planar parts on moving scene. The system is a pure 2D Vision System. However the height of the object is varying in time (from one set of parts to another). Due to the fact the distance between the cameras and the objects is changing, the measuring results are affected. The proposed geometric calibration method allows the Vision System to self-adjust the calibration parameters for a known change in height of the objects, without affecting both the system accuracy and the response-time performances.

The presented geometric calibration method is moving all the computational effort from the vision system real-time to the off-line stage (setup stage of the system). The only process that remains quasi-on-line is the self-adjusting of calibration parameters on objects height change. This is done on the change production time (this is a time when the vision system is idle). The presented method doesn't cover the calibration aspects on the direction lengthwise the conveyor. This is not a trivial aspect for the real industrial parts-transportation systems (the objects are not following exactly the conveyor speed - causing a lower accuracy of the vision system on the lengthwise conveyor direction).

In the same way, the presented *threshold calibration* method is specific for line-scan camera artificial vision applications for inspecting planar parts on moving scene. For this class of vision systems, dedicated for inspection and measurement industrial applications the error on object segmentation process is not acceptable. In this case, classic methods of global and dynamic threshold are not applicable. Starting from these methods, for the gray level image segmentation we adopted a statistical estimated local threshold. In the industrial environment the vision system is dedicated for, slow but significant modification of the lighting environment during the lifetime of the light source is a normal process of aging. In

the future research an efficient method has to be developed to provide the vision system the ability of self-adjusting to these environment changes.
In its implementation, the vision system subject of this paper has the acquisition frequency dependent on the speed of the moving scene (the speed of the conveyor). The paper proposes a processing time efficient method to estimate the modification on the threshold values for the case of an *error free* vision system in the case of variation of the acquisition frequency.

7. Acknowledgment

Parts of this paper are reprinted from 1. Hossu, A., Hossu, D. ,2008, A new approach of gray images binarization for artificial vision systems with threshold methods , *ICINCO 2008,* Vol. RA-2, pp. 11-16; and 2. Hossu, A., Hossu, D. ,2008, Calibration aspects of multiple line-scan vision system application for planar objects inspection, *ICINCO 2008,* Vol. RA-2, pp. 36-40.

8. References

Borangiu, Th. (2004) *Intellingent Image Processing in Robotics and Manufacturing,* The Publishing House of the Romanian Academy, ISBN: 973-27-1103-5, Bucharest.

Borangiu, Th, Hossu A., Croicu, A. (1995) - *RobotVisionpro Machine Vision Software for Industrial Training and Applications.* Version 2.2, Cat. #100062, Amsterdam, Tel Aviiv, New Jersey, ESHED ROBOTEC, Tel - Aviv.

Croicu, A., Hossu, A., Dothan, E., Ellenbogen, D., Livne, Y. (1998)- ISCAN-Virtual Class based Architecture for Float Glass Lines, *IsoCE'98,* Sinaia

Haralick, R., Shapiro, L. (1992) *Computer and Robot Vision,* Addison-Wesley Publishing Company, ISBN: 0-201-10877-1 (v. 1), ISBN 0-201-56943-4 (v. 2).

Hossu, A. (1999), *Robot Adaptation to the Environment by Artificial Vision,* Ph. D. Thesis, University Politehnica of Bucharest.

Hossu A., Andone D. (2005) *Artificial Vision Systems for Robotic Applications – case studies;* Ed. Printech, ISBN 973-718-230-8, Bucharest

Hossu, A., Croicu, A., Dothan, E., Ellenbogen, D., Livne, Y.(1998) - *ISCAN Cold-Side Glass Inspection System for Continuous Float Lines,* User Manual, Rosh-Haayn.

Hossu, A., Hossu, D. (2008-a) A new approach of gray images binarization for artificial vision systems with threshold methods , *ICINCO 2008, 5th International Conference on Informatics in Control, Automation and Robotics, Proceedings of the fifth International Conference on Informatics in Control, Automation and Robotics,* Vol. RA-2, pp. 11-16, Funchal, Madeira - Portugal, May 11-15, 2008

Hossu, A., Hossu, D. (2008-b) Calibration aspects of multiple line-scan vision system application for planar objects inspection, *ICINCO 2008, 5th International Conference on Informatics in Control, Automation and Robotics, Proceedings of the fifth International Conference on Informatics in Control, Automation and Robotics,* Vol. RA-2, pp. 36-40, Funchal, Madeira - Portugal, May 11-15, 2008.

Hossu, D., Hossu, A. (2008-c) Temporal match of multiple source data in an Ethernet based industrial environment, *ICINCO 2008, 5th International Conference on Informatics in Control, Automation and Robotics, Proceedings of the fifth International Conference on*

Informatics in Control, Automation and Robotics, Vol. RA-1, pp. 140-144, Funchal, Madeira - Portugal, May 11-15, 2008.

Gonzales, R. C., Wintz, P. (1981) *Digital Image Processing,* Addison-Wesley Publishing Company, ISBN: 0-201-02596-5, ISBN: 0-201-02597-3 pbk

6

Robot-Based Inline 2D/3D Quality Monitoring Using Picture-Giving and Laser Triangulation Sensors

Chen-Ko Sung, Robin Gruna, Minzi Zhuge and Kai-Uwe Vieth
Fraunhofer Institute IITB, Fraunhoferstrasse 1, 76131 Karlsruhe Germany

1. Introduction

Robots with multiple intelligent sensors will be increasingly used in the future for demanding production and assembly tasks. A substantial reason for the hesitant use of picture-giving sensors for the monitoring of inspection is the insufficient flexibility of the used monitoring concepts in relation to changing setting of tasks. At present either one (user-specific) sensor or more sensors with static arrangement for a certain task of inspection are used (Malamas et al., 2003; Ranky, 2003). This rigid approach is unsuitable for the inspection of variant products.

An especially attractive area of application is the inline 2D/3D quality monitoring of complex, large-area production parts such as the aircraft fuselage or parts of the bodies of man-made objects. For example hundred different mounting parts on an aircraft fuselage (about 4 m x 10 m in size) must be inspected, whether proper parts have been attached correctly. A special transport system like a monorail conveyor will probably be needed for the transportation of large objects. Such transport systems do not allow a precise positioning. The test object is free-hanging over the ground. At present no existing inspection system can do automatically the total quality monitoring of such production parts well.

Another very important task for the quality monitoring in factories is the inspection of work piece surfaces and structural defects. Scratches, dents, pores, deformations, recesses, increases, slivers, blisters and other kinds of defects on or below the surface should be detected and classified. Different kind of hardware (sensors, light sources, grip systems, robots and so on) and software (models, algorithms, data bases and so on) at present must be used, in order to mastering this task. That is surely a wrong way that we use numerous robots and inspection station in different places for the quality monitoring of such products.

For these reasons, we develop and examine a new concept for 2D/3D metric and logical quality monitoring with a more accurate, flexible, economical and efficient inspection centre. At the same time the measuring methods with the new modern technology are also improved. With the new concept the testing periods and investment in plants can be substantially decreased.

2. System overview

A four-step concept has been developed and realised at IITB for the flexible inline 2D/3D quality monitoring with the following characteristics (Fig. 1):

- Multiple short-range and wide-range sensors;
- Cost reduction of investment at plants without reduction of the product quality;
- Large flexibility regarding frequently changing test tasks;
- Low operating cost by minimisation of the test periods.

The robot-based system uses an array of test-specific short-range and wide-range sensors which make the inspection process more flexible and problem-specific. To test this innovative inline quality monitoring concept and to adapt it to customised tasks, a development and demonstration platform (DDP) was created (Fig. 2). It consists of an industrial robot with various sensor ports - a so-called "sensor magazine" - with various task-specific, interchangeable sensors (Fig. 3) and a flexible transport system.
All sensors are placed on a sensor magazine and are ready to use immediately after docking on the robot arm. The calibration of all sensors, robot calibration and the hand-eye calibration have to be done before the test task starts. The central projection for the camera calibration has been used.

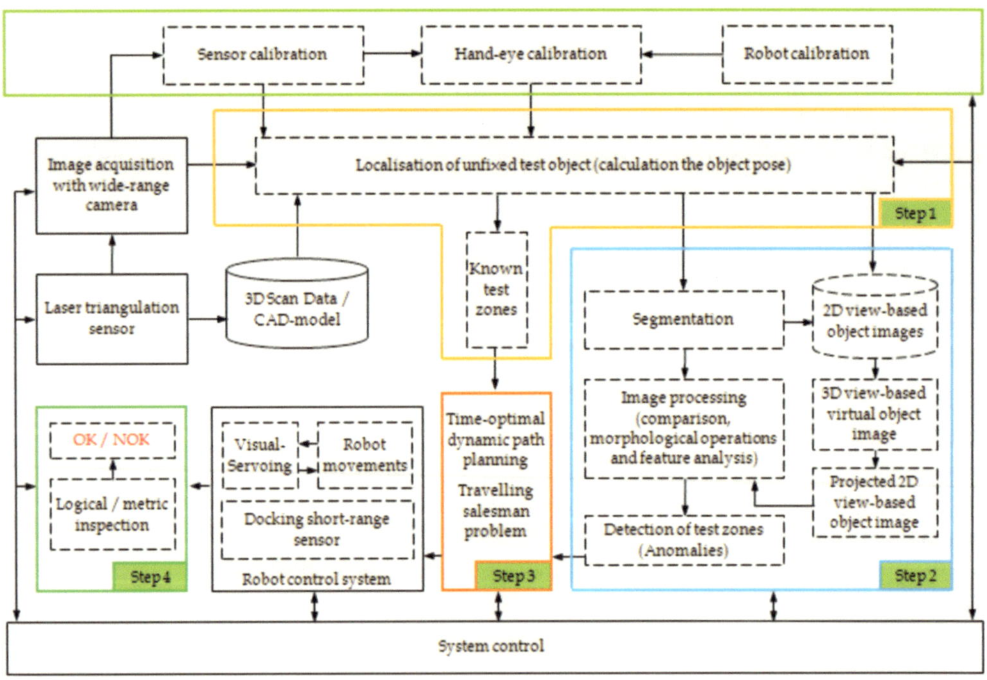

Fig. 1. System overview: A four-step concept for the flexible inline quality monitoring.

The four steps for a flexible inline quality monitoring which are described in the following sections are:

- Localisation of unfixed industrial test objects;
- Automatic detection of test zones;

- Time-optimal dynamical path planning;
- Vision-based inspection.

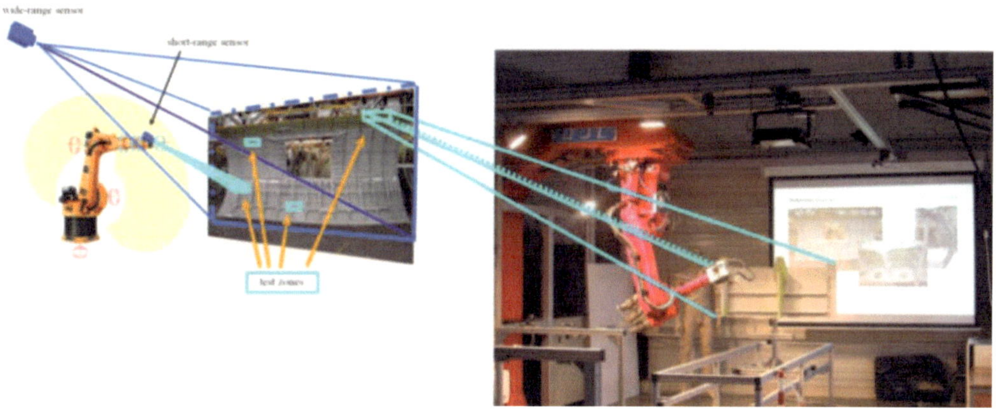

Fig. 2. Quality monitoring of aircraft fuselages with wide- and short-range inspection sensors. Left: Inspection station and test environment. The movement of production pieces is carried out by monorail conveyors which do not allow precise positioning. Right: Development and demonstration platform (DDP).

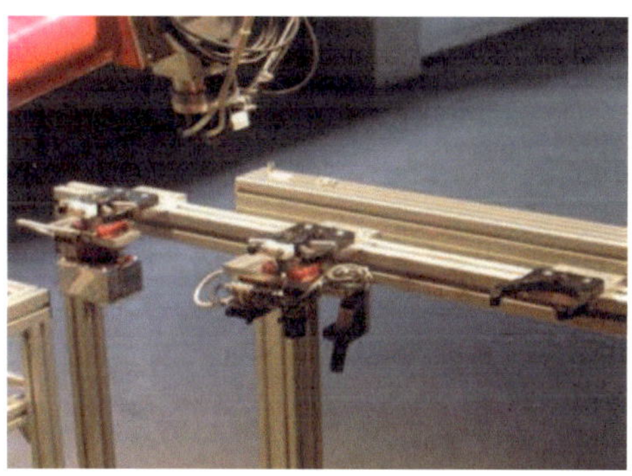

Fig. 3. Sensor magazine.

2.1 Localisation of unfixed industrial test objects

As the first step of the presented quality monitoring chain, the exact position of a production piece is determined with a wide-range picture-giving sensor (Fig. 2), which is - depending on the object size - mounted in an adequate object distance, i.e. not necessarily fixed on an inspection robot's end-effector.

A marker-less localisation calculates the exact object position in the scene. This procedure is based only on a 3D CAD-model of the test object or at least a CAD-model which represents a composition of some of its relevant main parts. The CAD-model contours are projected into the current sensor images and they are matched with sub-pixel accuracy with corresponding lines extracted from the image (Müller, 2001).

Fig. 4 shows a localisation example. The CAD-model projection is displayed in yellow and the object coordinate system in pink colour. The red pixels close to the yellow projection denote corresponding image line pixels which could automatically be extracted from the image plane. The calculated object pose (consisting of three parameters for the position in 3D scene space as well as three parameters for the orientation, see the red text in the upper part of the figure) can easily be transformed into the global scene coordinate system (displayed in green colour).

Known test zones for detail inspection as well as associated sensor positions and orientations or required sensor trajectories (cf. section 2.2 and 2.3) can be defined with respect to the object coordinate system in an inspection preceding step. All the object based coordinates will be transformed online into the global scene coordinate system or the robot coordinate system with respect to the localisation result, i.e. with respect to the position and orientation of the test object in the scene. The red, T-shaped overlay in Fig. 4 shows an example for an optimal 3D motion trajectory (see the horizontal red line which is parallel to the object surface) together with the desired sensor's line of sight with respect to the object surface (the red line which points from a position in the middle of the trajectory towards the test object).

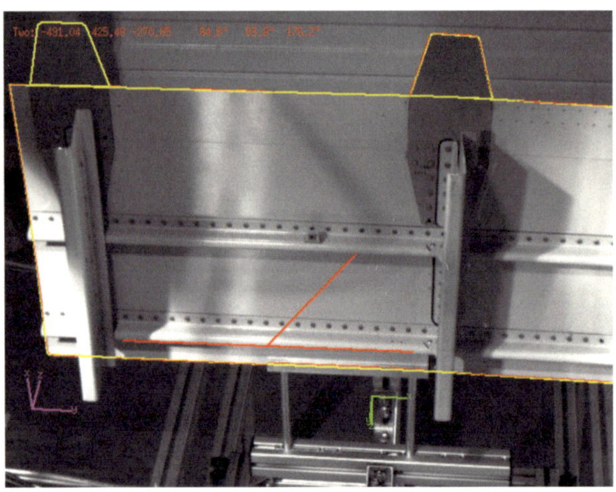

Fig. 4. Localisation of an object to be inspected and computation of an initial optimal inspection trajectory.

2.2 Automatic detection of test zones

Two approaches can be applied to find automatically anomalies on a test object. One is model-based comparison between the CAD-model projection and the extracted image features (edges, corners, surfaces) to detect geometric differences (Veltkamp & Hagedoorn, 2001). Another one resembles probabilistic alignment (Pope & Lowe, 2000) to recognize unfamiliar zones between view-based object image and test image.

In this second step, we used purely image-based methods and some ideas of the probabilistic alignment to achieve a robust inline detection of anomalies under the assumption that the object view changes smoothly. The same wide-range camera for object localisation was used for this step.

Using the result of object localisation to segment an object from an image, a database with 2D object images can be built up in a separate learning step. We postulated that the views were limited either of the front side or the back side of the test object with small changes of viewing angles and furthermore postulated that we had constant lighting conditions in the environment.

We used the calibration matrix and the 2D object images to create a 3D view-based virtual object model at the 3D location where an actual test object was detected. The next process was to project the view-based virtual object model into the image plane. The interesting test zones (anomalies, Fig. 5) where detailed inspections were needed (see section 2.3 and 2.4) were detected within the segmented image area by the following steps:

- Comparison between the projected view-based object image and the actual test image;
- Morphological operations;
- Feature analysis.

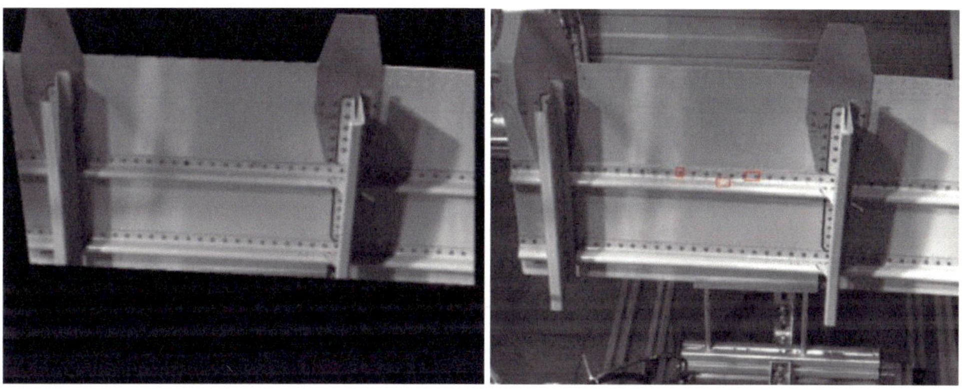

Fig. 5. Left: One of the segmented object images in the learning step. Only the segmented area in an image is relevant for the detection of anomalies. Right: The automatic detected test zones are marked with red rectangles (overlays).

2.3 Time-optimal dynamical path planning

In the third step, an optimised inspection path plan is generated just in time, which is then carried out using various inspection-specific short-range sensors (e.g. cameras, feeler, etc.). All the interesting test zones or the regions of interest (ROIs) have been found in the second step, but the path plan is not perfect yet. A time-optimal path has to be found from the supervising system.

The problem is closely related to the well known travelling salesman problem (TSP), which goes back to the early 1930s (Lawler et al., 1985; Applegate et al., 2006). The TSP is a problem in discrete or combinatorial optimisation. It is a prominent illustration of a class of problems in computational complexity theory which are classified as NP-hard (Wikipedia, 2009).

The total number of possible paths is calculated by: $M = (n-1)!/2$. The definition of the TS-problem is based on the following assumptions:

- Modelled as a graph with nodes and edges;
- Graph is complete, this means that from each point there is a connection to any other point;

- The graph can be symmetric or asymmetric;
- The graph is metric, that means it complies the triangle inequality $C_{ij} \leq C_{ik} + C_{kj}$ (e.g. Euclidian metric, maximum metric).

Looking at the algorithms for solving TS-problems, there exist two different approaches: Exact algorithms which guarantee a global optimal solution and heuristics, where the solution found is only locally optimal.

The most accepted exact algorithms which guarantee a global optimum are Branch-and-Cut Method, Brute-Force and Dynamic Programming. The major disadvantage of the exact algorithms mentioned above is the time consuming process finding the optimal solution. The most common heuristic algorithms used for the TSP are:

- Constructive heuristics: The Nearest-Neighbour-Heuristic chooses the neighbour with the shortest distance from the actual point.

The Nearest-Insertion-Heuristic inserts in a starting path additional points;

- Iterative improvement: Post-Optimisation-methods try to modify the actual sequence in order to shorten the overall distance (e.g. k-opt heuristic).

A heuristic algorithm with the following boundary conditions was used:

- The starting point has the lowest x-coordinate;
- The Nearest- Neighbour-Constructive heuristics look for the nearest neighbour starting with the first node and so on;
- The iterative improvement permutes single nodes or complete sub graphs randomly;
- Terminate, if there was no improvement after n tries.

The optimised path planning discussed above was tested at the DDP with a realistic scenario. Given a work piece of 1 by 0.5 square meter, the outputs of the second step (see section 2.2) are 15 detected ROIs, which belong to the same error class. This would lead to a total number of about 43.6 billion possible different paths.

Starting with a 1st guess as outlined with an associated path length set to 100 %, after 15 main iteration loops the path lengths drops down to nearly 50 % of the first one, and no better achievement could be found (Fig. 6). The calculation time for the iterated optimal path was less than 1 s. on a commercial PC, Intel Pentium 4 with 3 GHz, and took place while the robot moved to the starting position of the inspection path.

2.4 Vision-based inspection

In the fourth step, the robot uses those sensors which are necessary for a given inspection path plan and guides them along an optimal motion trajectory into the previously-identified ROIs for detailed inspection. In these ROIs a qualitative comparison of the observed actual topography with the modelled target topography is made using image-processing methods. In addition, quantitative scanning and measurement of selected production parameters can be carried out.

For the navigation and position control of the robotic movement with regard to the imprecisely- guided production object as well as for the comparison of the observed actual topography with the target topography, reference models are required.

These models, using suitable wide-range and short-range sensors, were scanned in a separate learning step prior to the generation of the automated inspection path plan. Two sensors have been used for our work: A laser triangulation sensor is used (Wikipedia, 2009) for the metric test task (Fig. 7) and a short-range inspection camera with a circular lighting is used for the logical test task. For a fuselage, for example, it can be determined if construction elements are missing and/or if certain bore diameters are true to size.

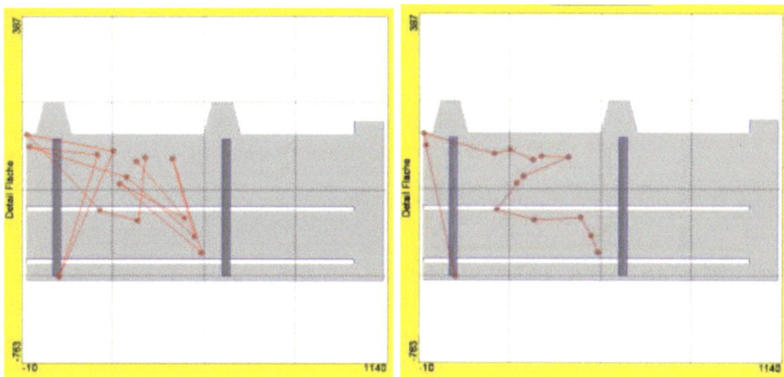

Fig. 6. Left: initial path; Right: final path.

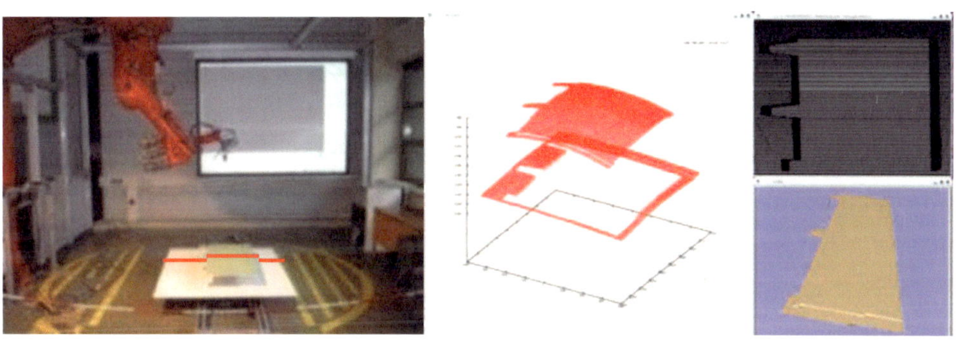

Fig. 7. A laser line scanning technique captures the structure of a 3D object (left part) and translates it into a graphic model (right part).

By using the proposed, robot-based concepts of multiple sensor quality monitoring, the customary use of expensive 3D CAD-models of the test objects for high-precision CNC controlled machine tools or coordinate inspection machines becomes, in most instances, unnecessary. The quality of the results of the metric test task is therefore strongly dependent on the quality of the calibration of the laser triangulation sensor which will be discussed next in chapter 3.

An intelligent, sensor-based distance-control concept (Visual-Servoing-Principle) accurately controls the robot's movements with regard to the work piece and prevents possible collisions with unexpected obstacles.

3. 3D Inspection with laser triangulation sensors

The test objects like aircraft fuselages consist of a large ensemble of extended components, i.e., they are 3D objects. For inline 3D quality monitoring of so-called metric objects, the sensor magazine contains a laser triangulation sensor. The sensor presented here is currently equipped with two line laser projectors but is not necessarily reduced to two light sources. The usage of two or more sources yields a predominant shadow-free quality monitoring for most inspection tasks. Thus, the inspection path of the sensor can be reduced for metric objects principally by a factor of two or more compared to the usage of one line laser. Before going into details, a short overview of 3D measurement techniques is given as the sensor magazine could also contain other 3D sensors, of course. Depending on the requirements of the inspection task the corresponding optical technique has to be chosen.

3.1 From 2D towards 3D in-line inspection

As described in the previous Section, 2D computer vision helps to roughly localize the position of the object to be inspected. Then the detailed quality inspection process starts which can be performed and actually should be performed with 2D image processing where possible. For many inspection tasks, traditional machine vision based systems are not capable of detecting defects because of the limited information provided by 2D images. For this reason, optical 3D measurement techniques have been gaining an increased importance in industrial applications because of their ability to capture shape data from objects.
Geometry or shape acquisition can be accomplished by several techniques, e.g., shape from shading (Rindfleisch, 1966), phase-shift (Sadlo et al., 2005), Moiré-approach, which dates back to Lord Rayleigh (1874), Stereo-/Multi-View-Vision (Breuckmann, 1993), tactile coordinate metrology, time-of-flight (Blanc et al., 2004), light sectioning (Shirai & Suwa, 1971), confocal microscopy (Sarder & Nehorai, 2006), interferometric shape measurement (Maack et al., 1995).
A widely adopted approach is laser line scanning or laser line triangulation. Because of its potentiality low cost and the ability to optimize it for high precision and processing speed, laser triangulation has been frequently implemented in commercial systems which are then known as laser line scanners or laser triangulation sensors (LTSs). A current overview about triangulation based, optical measurement technologies is given in (Berndt, 2008).
The operating principle of laser line triangulation is of actively illuminating the object to be measured with a laser light plane, which is generated by spreading out a single laser beam using a cylindrical lens. By intersecting the laser light plane with the object, a luminous laser line is projected onto the surface of the object, which is then observed by the camera of the scanning device. The angle formed by the optical axis of the camera and the light plane is called angle of triangulation. Due to the triangulation angle, the shape of the projected laser line as seen by the camera is distorted and is determined by the surface geometry of the object. Therefore, the shape of the captured laser line represents a profile of the object and can be used to calculate 3D surface data. Each bright pixel in the image plane is the image of a 3D surface point, which is illuminated by the laser line. Hence, the 3D coordinate of the illuminated surface point can be calculated by intersecting the corresponding projection rays of the image pixels with the laser light plane.
In order to capture a complete surface, the object has to be moved in a controlled manner through the light plane, e.g. by a conveyer belt or a translational robot movement, while multiple laser line profiles are captured by the sensor. In doing so, the surface points of the object as seen from the camera can be mapped into 3D point data.

3.2 Shadow-free laser triangulation with multiple laser lines

There are, however, certain disadvantages shared by all laser scanners and which have to be taken into account when designing a visual inspection system. All laser line scanning systems assume that the surface of an inspection object is opaque and diffusely reflects at least some light in the direction of the camera. Therefore, laser scanning systems are error-prone when used to scan shiny or translucent objects. Additionally, the object colour can influence the quality of the acquired 3D point data, since the contrast of the projected laser line must be high enough to be detectable on the surface of the object. For this reason, the standard red HeNe laser (633 nm) might not always be the best choice, and laser line projectors with other wavelengths have to be considered for different inspection tasks. Furthermore, the choice of the lens for laser line generation is crucial when the position of the laser line should be detected with sub-pixel accuracy. Especially when the contrast of the captured laser line is low, e.g., due to bright ambient light, using a Powell lens for laser line generation can improve measurement accuracy, compared to the accuracy obtained with a standard cylindrical lens (Merwitz, 2008).

An even more serious problem associated with all triangulation systems is missing 3D point data due to shadowed or occluded regions. In order to measure 3D coordinates by triangulation, each surface point must be illuminable by the laser line and observable by the camera. Occlusions occur if a surface point is illuminated by the laser line but is not visible in the image. Shadowing effects occur if a surface point is visible in the image but is not illuminated by the laser line. Therefore, both effects depend on the camera and laser setup geometry, and the transport direction of the object. By choosing an appropriate camera-laser geometry, the amount of shadowing effects and occlusion can be reduced, i.e., by choosing a smaller angle of triangulation. However, with a smaller angle of triangulation also measurement accuracy decreases and in most cases, shadowing effects and occlusion cannot be eliminated completely without changing the setup of camera and laser.

To overcome this trade-off and to be able to capture the whole surface of an object without the need of changing the position of camera or laser, various methods can be applied. One solution is to position multiple laser triangulation sensors in order to acquire multiple surface scans from different viewpoints. By aligning the individual scans into a common world coordinate system, occlusion effects can be eliminated. Obviously, the main disadvantage of this solution is additional hardware costs arising from costly triangulation sensors. In the case of robot-based inspection, missing 3D data can also be reduced by defining redundant path-plans, which allows capturing multiple surface scans from different points of view of a single region to be inspected. This approach would make path-planning more complex and would lead to a longer inspection time.

In order to avoid the aforementioned disadvantages, a 3D measurement system with single triangulation sensor but multiple laser line projectors is presented, which keeps inspection time short and additional costs low. Due to new advances in CMOS technology, separate regions on a single triangulation sensor can be defined. Each one is capable of imaging a single projected laser line. Furthermore, processing and extracting image coordinates of the imaged laser profiles is done directly on the sensing chip, and thus extremely high scanning frame rates can be achieved. The scan data returned from such a smart triangulation sensor is organised as two-dimensional array; each row containing the sensor coordinates of a captured laser line profile. Thus, the acquired scan data has still to be transformed into 3D

world coordinates, using the calibrated camera position and laser light plane orientation in a common world coordinate system (see Section 3.4).
In the presented system, such a smart triangulation sensor is used in combination with two laser line generators, where each laser line illuminates a separate part of the sensor's field of view. Therefore, by scanning the surface of an object, scans from the same point of view but different light plane projection directions are acquired. By merging the individual scans of each laser, shadowing effects can be minimised and the 3D shape of a measurement object can be captured with a minimised amount of missing 3D data. This step is performed for both, the creation of a reference model and the subsequent inline inspection of the production parts.

3.3 3D inspection workflow

Fig. 8 gives an overview of the steps required for a 3D inspection task. In the following, the individual steps of the data acquisition and processing workflow are described in more detail.

3.4 Sensor calibration and 3D point data acquisition

As mentioned before, the scan data returned by the triangulation sensor are related to sensor coordinates, describing the position of individual laser line profiles which were captured during the scanning process. In order to get calibrated measurements in real world coordinates, the laser triangulation sensor has to be calibrated. This yields a transformation from sensor coordinates (x, y) into world coordinates (X, Y, Z) which compensates for nonlinear distortions introduced by the lens and perspective distortions caused by the triangulation angle between laser plane and the optical axis of the sensor. Therefore, the calibration procedure can be divided into the following steps:

- Camera calibration;
- Laser light plane calibration;
- Movement calibration of the object relative to the measurement setup.

For camera calibration, extrinsic and intrinsic parameters have to be determined. Extrinsic parameters define the relationship between the 3D camera coordinate system and the 3D world coordinate system (WCS). For example, the z-axis of the camera coordinate system coincides with the optical centre of the camera, i.e., with the optical axis of the lens.
The intrinsic parameters are not dependent on the orientation of the camera expressed in world coordinates. Moreover, the intrinsic parameters define the transformation of the 3D camera coordinate system (metric) and the 2D image coordinate system (ICS) (pixel). Thus, they describe the internal geometry like focal length f, optical centre of the lens c, radial distortion k, and tangential distortion p. The effect of parameters k and p are visualised in Fig. 9.

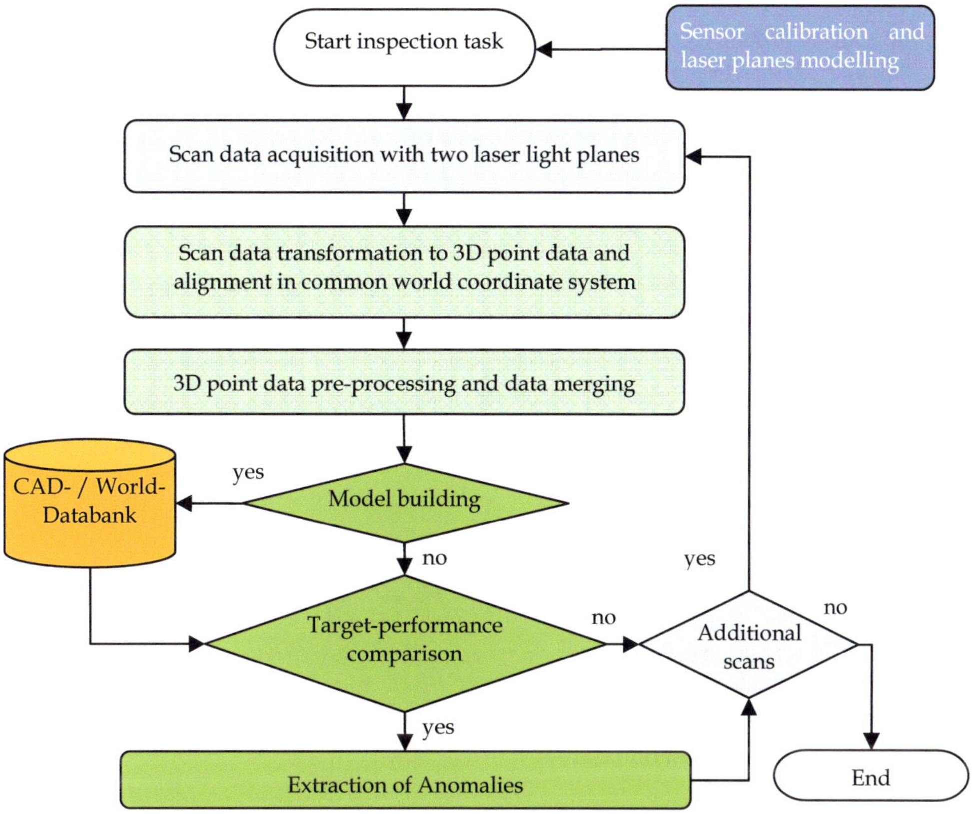

Fig. 8. The 3D inspection workflow depicts the major elements described in this chapter.

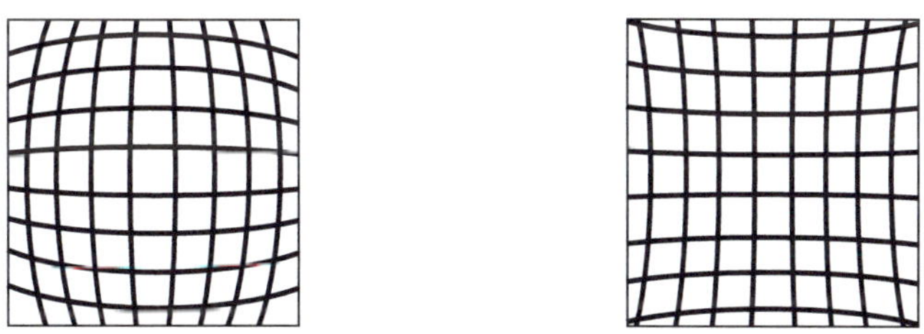

Fig. 9. The left pattern depicts the effect of radial distortion, whereas the right pattern shows the effect of tangential distortion.

We perform the calibration according to Zhang (2000) which is based on the pinhole camera. For this method an image of a planar chess board is taken with at least two different positions. The developed algorithm computes the projective transformation of the 2D image coordinates of the extracted corner points of the chess board by using the n different images and its 3D coordinates. Therewith, the extrinsic and intrinsic parameters of the camera are gained with a linear least-square method. Afterwards a non-linear optimisation is applied based on maximum-likelihood criteria using the Levenberg-Marquardt algorithm. By doing

so, the error of back projection is minimised. The distortion coefficients are determined according to Brown (1971) and are optimised as mentioned above.
Zhuge (2008) describes that a minimum of 2 images are required depicting a 3x3 (4 corners) chess board. Improving numerical stability a chess board with more squares and more pictures is to be recommended. We used a 7x9 chess board (48 corners, Fig. 10) and checked the change of the parameters as a function of the number of images used as input for computing the intrinsic and extrinsic parameters. The extrinsic parameters are expressed in terms of rotation and translation in order to put into coincidence point of origins of image coordinates and world coordinates.

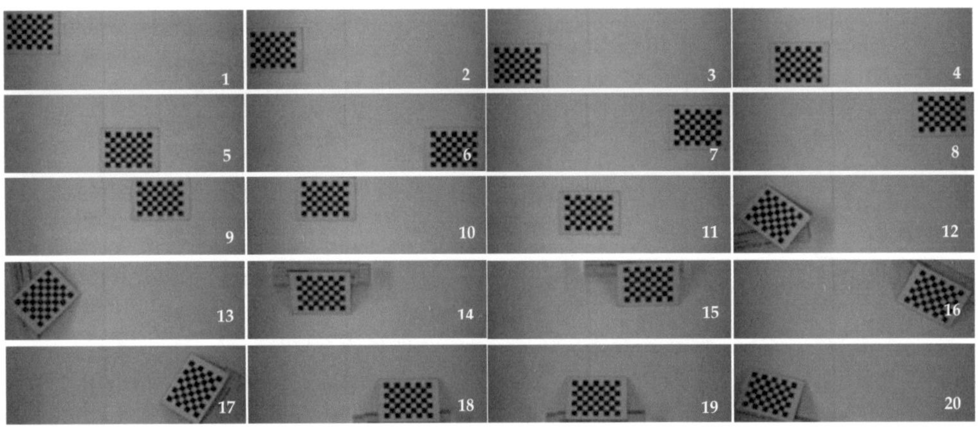

Fig. 10. Images for the camera calibration.

Table 1 shows the exemplary results using 7, 12, and 20 images as input for calculating the intrinsic and extrinsic parameters. The results, for example with arbitrarily selected 7 pictures, are totally wrong. If the positions and the views of the chess board are well distributed in selected pictures, the results become then better and better. The best choice for the current investigation is marked in Table 1 with red. It is not necessary to use 20 pictures for the camera calibration. For low measuring accuracy or for a smaller sensor chip the number of pictures can be reduced to 12 or less.
In order to test the quality of the estimated camera parameters, the 3D world coordinate corners of the chess board are projected onto the image with the computed camera parameters. The smaller the deviation is between the 2D coordinates of the back projected corners and the 2D coordinates that correspond to the 3D world coordinates the better is the computation of the parameters. To determine the extrinsic parameters it is recommended to use images where the chess board is centred. Finally the x- and y-axis of the ICS are brought into line with the X- and Y-axis of the WCS if the camera parameters are perfectly calculated. In Fig. 11 the first row and the first column of the green dots depict the X- and Y-axis of the WCS. If the camera parameters would have been computed perfectly, the WCS axes and the ICS axes would coincide.
After lens correction, image coordinates can be mapped to world coordinates using the orientation of the light plane in the world coordinate system, and the relative movement of the object between two acquired scans. Since the robot-based inspection system allows for an accurate tracking of the triangulation sensor in any scanning direction, no calibration of

the linear positioning by the robot is needed. To determine the orientation of the light plane, several methods have been proposed (Teutsch, 2007), which essentially use at least three data points from the projected light plane to approximate a best-fit plane into the obtained points.

No. of pictures	Intrinsic camera parameters			
	c_x, c_y [Pixel]	f_x, f_y [Pixel]	k_x, k_y [a.u.]	p_x, p_y [a.u.]
7(03-09)	622, 160	576, 575	-0.014, -0.001	0.003, 0.001
12(¹)	725, 229	2940, 2930	0.041, -3.034	0.005, 0.003
20(01-20)	718, 240	2933, 2924	0.041, -2.974	0.006, 0.003

No. of pictures	Extrinsic camera parameters					
	θ_x [rad]	θ_y [rad]	θ_z [rad]	t_x [mm]	t_y [mm]	t_z [mm]
7(03-09)	0.01	0.01	0	-134.3	-35.0	120.3
12(¹)	0.04	0.02	0	-154.6	-49.2	617.0
20(01-20)	0.05	0.02	0	-153.0	-51.4	615.4

Table 1. Compare the exemplary results of the estimated camera parameters with different positions and views of the chess board used for the camera calibration. The best one is marked with red. *f*: focal length; *c* : optical centre of the lens; *k*: radial distortion; *p*: tangential distortion; t_x, t_y, t_z: shift to the X-, Y- or Z-direction; θ_x, θ_y, θ_z: rotation to the X-, Y- or Z-axis.
[1]: the selected pictures ware 1, 3, 6, 8, 12, 13, 14, 15, 16, 17, 18 and 19.

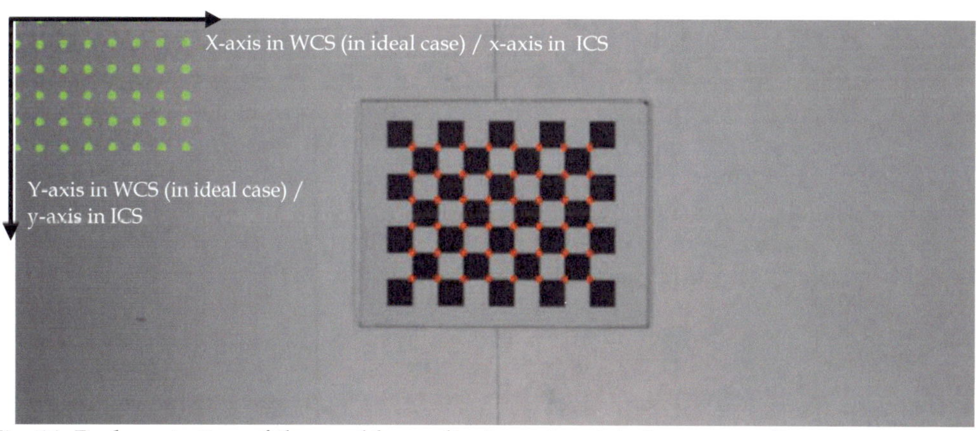

Fig. 11. Back projection of the world coordinate system into image coordinate system.

3.5 Laser calibration, pre-processing, and data merging

In this work, a laser calibration method is used that maps scan data acquired with different lasers to a common world coordinate system in two steps. In doing so, the orientations of the light planes of the lasers are not determined explicitly but are given implicitly by two 3D projective transformations, one for each laser. First, the scan data of each laser is augmented by a third coordinate, which comprises the translational movement of the object between two single laser line scans. This yields two perspective-distorted 3D point data sets for each laser. In a second step, perspective transformations are applied to the data sets, which correct for the perspective-distortions and align the data in a common world coordinate system (see lower Fig. 12).

These perspective transformations are modelled by systems of linear equations which can be solved by established 3D point correspondences. To identify corresponding 3D point pairs in the common world coordinate system and in the perspective-distorted 3D point data of each laser, a pyramid-shaped calibration target is used. Since planes are mapped to planes by perspective distortions, the faces of the pyramid can be easily identified in the distorted 3D point data. To this end, surface normals are computed for each 3D point in a local neighbourhood. By comparing the direction of the surface normals, the 3D points can be segmented into four point sets, each belonging to a pyramid face. For each point set, a least-square best-fit plane is computed. By intersecting the pyramid planes and the measurement plane, five intersection points are obtained which serve as feature points for establishing point correspondences.

From the obtained point correspondences, two perspective transformations are computed by solving the system of linear equations for each laser. Since the point correspondences are established from the same feature points in the world coordinate system, both transformations compensate for the perspective distortion and align the 3D point data sets into the world coordinate system (see Fig. 12).

Before the perspective transformations have been computed, the acquired scans of each laser should be mapped in real world 3D point data in a common world coordinate system. However, the 3D point data of both lasers is maintained in separate data structures. In order to take advantage of the complementary missing data regions in the 3D point data sets, it is desirable to merge the aligned point data into a unified data structure. Due to the sequential nature of the laser line scanning process, captured 3D point data for each laser are sorted by their acquisition direction. Since most point data processing techniques, e.g., methods for surface reconstruction, take advantage of sorted data, this property should be maintained while the data is merged.

The problem of merging the two 3D point data sets therefore reduces to merging two sorted sequences into a single sorted sequence (FUSION, 2008). By stepwise comparing and choosing 3D points from the data sets, it is ensured, that the merged list is still sorted. Furthermore, 3D points whose coordinates do not agree with their acquisition direction are identified and dropped. These 3D points are assumed to stem from interreflections of the projected laser lines and the surface of the object, and therefore might lead to wrong measurements.

In most cases, dropping 3D data points does not significantly decrease data quality for subsequent processing steps since the acquired raw 3D point data are very dense. However, there exist surface regions where data acquisition is very difficult due to interreflections and larger regions of missing data can occur. In such regions, bilinear interpolation is used to estimate missing surface data. Fig. 13 shows how missing data gaps are closed.

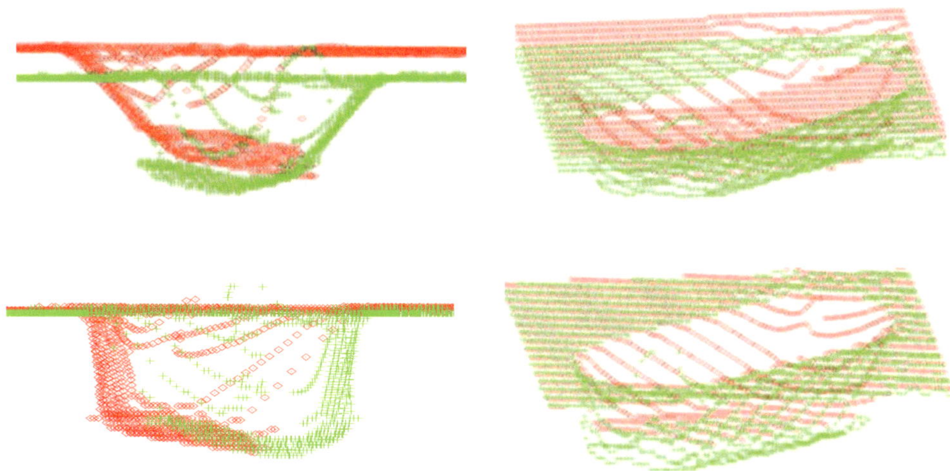

Fig. 12. 3D point data of a scanned mould. Upper: Unaligned 3D point data acquired with two laser lines, projected from the left (green) and right (red) side of the mould. Lower: By calculating and applying perspective transformations to the perspective-distorted 3D point data of each laser, the data becomes aligned in a common world coordinate system.

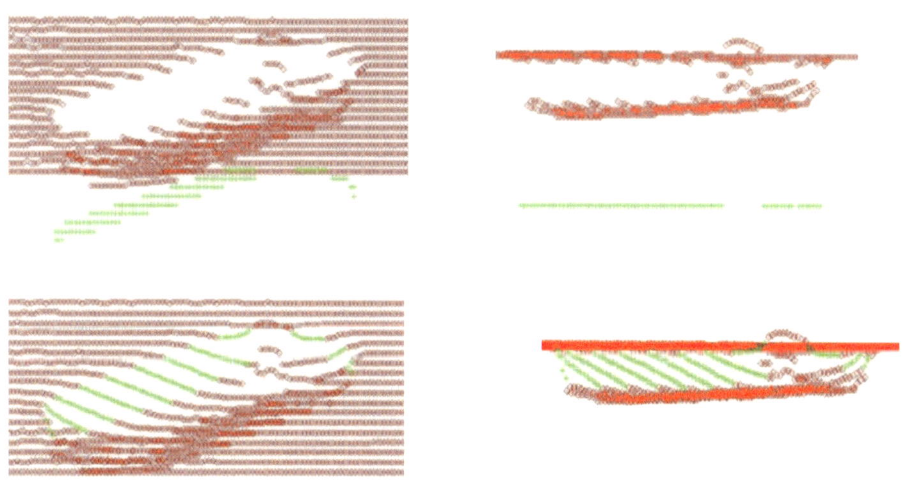

Fig. 13. Upper: Green dots are the missing scan data. Lower: The missing scan data gaps are closed.

3.6 Model building

The final goal of the data acquisition of the 3D laser triangulation sensor is to compare either CAD data or a reference model based on scanned data with acquired data of the test object. Therefore, this step is depicted at the end of the inspection pipeline (Fig. 8). Once a model is created in the sense that it meets the requirements for being comparable with the test object it can be saved in a data base to be picked up from there if needed. The usefulness, i.e., meeting of requirements of the model depends on the task of the quality check. A task could be to determine the volume of a complete object or to determine the volume only at a certain location of an object. Another task could be to scan the surface of an object with a position-dependent precision. Hence, the model includes locally varying tolerances. Here we follow the creation of a model, a so-called reference model, yielded by scanning a good test object which is later compared with a test object having an artificial error.

The transformation matrix of the calibration of the reference measurement is used for the comparison of future test scans. Additionally, a registration of the reference measurement and the test scans are performed and needed for comparison. The registration corresponds to the alignment of the reference model and the test data. The data of the two lasers, as supplied by the camera, could have identical indices associated with different 3D coordinates. Thus, the point cloud is re-indexed in the WCS and assigned to a new mesh. An interpolation is performed for all points of each square of the mesh in order to compute the value of the centre point. In accordance to the resolution of the camera, the size of the mesh is adjusted. On the one hand, it has to be taken into account that small meshes contain less or no data points which results in an increasing number of data gaps. On the other hand, if the meshes are too big, the determination of the volume becomes less precise.

Fig. 14. The pyramid was scanned with and without the red artificial error.

We used for the data acquisition a sample rate of 0.2 mm along all coordinate axes and each mesh square had a size of 0.4x0.4 mm. The test object was a pyramid (Fig. 14) which is 150 mm wide at the basis and a maximal height of 20 mm. The sensor has got 1500x500 pixels. During the data acquisition signal noise was observed which was not negligible. Due to observations, it can be stated that the noise caused an uncertainty of less than ± 0.1 mm. Hence, a threshold of ± 0.1 mm is applied for a cluster process.

In the cluster process the model data and the test data are compared with the aim to suppress data which result from measurement uncertainty, and hence to get the volume difference of the model and the test object. The cluster in Fig. 15 is bounded by bold lines and does not exist in the beginning, but is generated after the comparison. If the criteria described below are fulfilled a cluster is the outcome. For this comparison the neighbours of a data point are taken into account and two criteria determine the clustering:

- If two of eight neighbours (red dots in Fig. 15) of the data point considered (green dot in Fig. 15) and the data point itself have a height difference that is larger than the absolute value of 0.1 mm, the data point is assigned to a cluster and not regarded as measurement uncertainty. Fig. 15 shows on the left the model data (blue dot) used for comparison. The cluster is depicted by bold black borders.
- If the height difference of one neighbour and the considered data point is larger than the absolute value of 0.1 mm and the neighbour belongs to the cluster area, the data point is assigned to the cluster to be built. This situation is shown in Fig. 15 on the right.

Fig. 16 shows the difference plot of the model data and the test data, i.e., the difference of the pyramid with and without red coding dot label. Red depicts positive deviations whereas green shows negative deviations.

Fig. 17 shows the result of the difference plot after clustering of the model data and the test data. The red coding dot label stands out clearly due to clustering.

The computation of the volume yields a good result for the red coding dot label on one side of the pyramid. Fig. 17 shows the result in detail in graphical form from a side view and from a top view.

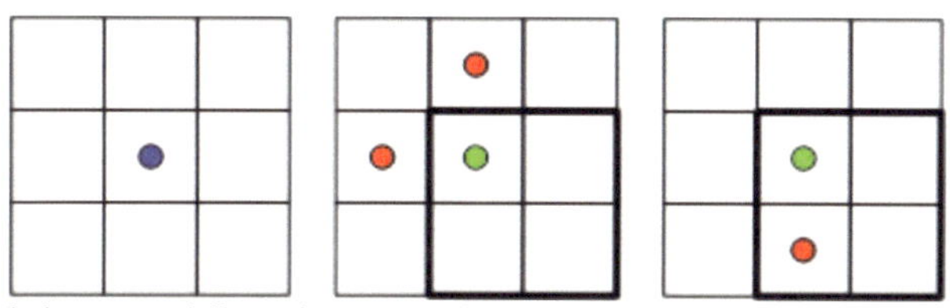

Fig. 15. Left: Model data. Middle: Cluster with at least two neighbours. Right: Cluster with a neighbour in the cluster.

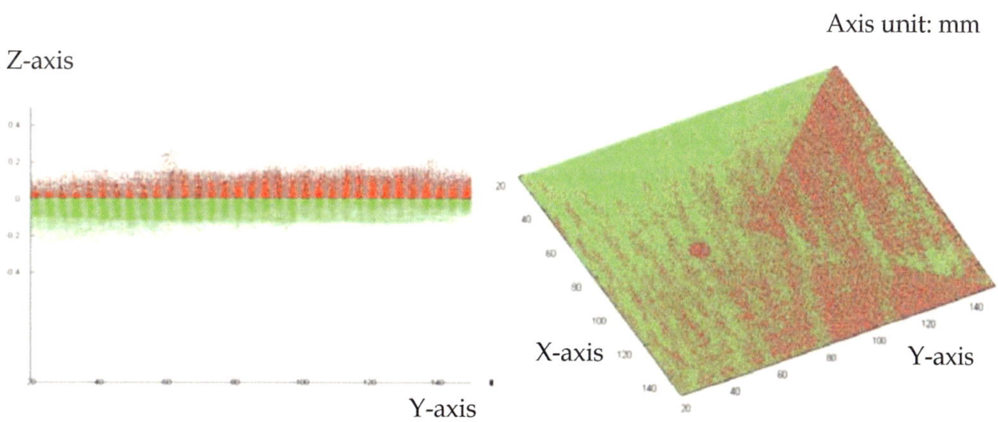

a) View from aside b) Top view

Fig. 16. Difference plot of model and test object.

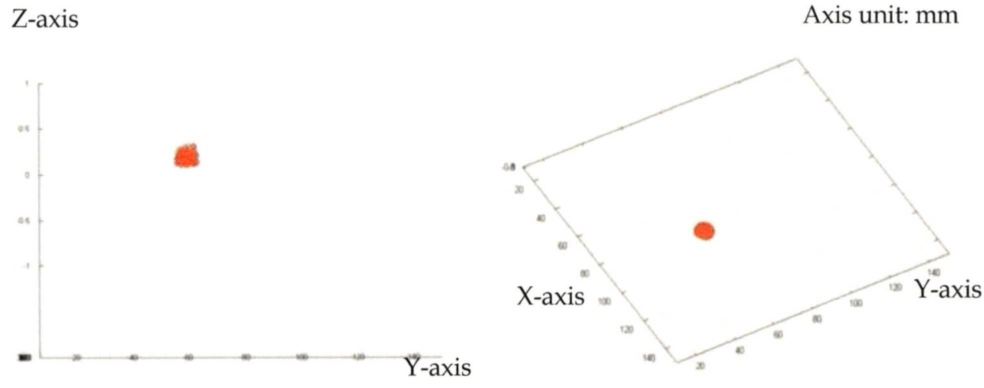

a) View from aside   b) Top view

Fig. 17. Difference plot of model and test object after clustering.

4. Conclusions

This paper has presented a robot-based inspection centre that uses wide-range sensors and an array of test-specific short-range which make the logical quality monitoring and the metric inspection process more accurate and flexible. Path optimisation and use of a sensor magazine supplement the robot-based inspection centre such that its flexibility yields problem-specific application coupled with economical and efficient improvements; these improvements are methodologically and operationally inherent. Specifically, using multiple projected laser lines the inspection time can be strongly reduced, additional costs kept low, and the quality of the inspection results is even still higher.

As Fraunhofer has got a strong focus on applied research, one of our main goals for the creation of the development and demonstration platform is to investigate, to optimize and to demonstrate to our potential cooperation partners how the inspection centre can be applied to reduce effort and to increase accuracy and flexibility. It can be used in the robot-based coordination of short and wide-range monitoring, for the introduction of learning capable evaluation processes, as a tool for visualizing results and for user interaction, as well as for the flexible networking and integration of various wide and short-range sensors. Depending on the test task the presented four steps can be used completely or partly in a meaningful combination. Its modular and hence flexible structure is a modern basis for future requirements and/or add-ons. This structure is especially needed, because nowadays fabrication is characterised by continuous, flexible product change.

Further inspection sensors, which are based on other measurement principles, will be developed on the sensor magazine and made available for the inspection on and below the surfaces. Due to other research work at our institute the sensor magazine especially could be equipped with devices used for deflectometric methods (Werling et al., 2007) as it is a promising inspection technique for shiny and mirror-type surfaces.

5. References

Applegate, D. L.; Bixby, R. E.; Chvátal, V. & Cook, W. J. (2006). *The Traveling Salesman Problem: A Computational Study*. Princeton University Press. ISBN 978-0-691-12993-8.

Berndt, D. (2008). *Optische 3-D-Messung in der industriellen Anwendung,* PhD thesis, Universität Magdeburg.

Blanc, N.; Oggier, T.; Weingarten, G. G.; Codourey, A. & Seitz, P. (2004). Miniaturized smart cameras for 3D-imaging in real-time, *Proc. IEEE Sensors '04*, Vol. 1, pp. 471-474.

Breuckmann, B. (1993). *Bildverarbeitung und optische Messtechnik in der industriellen Praxis,* Franzis Verlag, München.

Brown, D.C. (1971). Close-range Camera Calibration. *Photogrammetric Engineering,* Vol. 37, No. 8, pp. 855-866.

FUSION 2008, (2008). *11. International Conference on Information Fusion*, Cologne, Germany, [www] http://www.fusion2008.org.

Malamas, E.N.; Petrakis, E.G.M.; Zervakis, M.; Petit, L. & Legat, J-D. (2003). A Survey on Industrial Vision Systems, Applications and Tools, *Image and Vision Computing,* Elsevier Science, Amsterdam, Vol. 21, No. 2, pp. 171-188.

Maack, T.; Notni, G. & Schreiber, W. (1995). Three coordinate measurement of an object surface with a combined two-wavelength and two-source phase shifting speckle interferometer, *Opt. Commun.*, Vol. 115, pp. 576-584.

Ranky, P.G. (2003). Novel Automated Inspection Methods, *Tools and Technologies*, Assembly Automation, Vol. 23, No. 3, Emerald Publishing Ltd., England, pp. 252-256.

Lawler, E. L.; Lenstra, J. K.; Rinnooy Kan, A. H. G. & Shmoys, D. B. (1985). *The Travelling Salesman Problem. A Guided Tour of Combinatorial Optimization.* Wiley, Chichester 1985. ISBN 0-471-90413-9.

Müller, Th. (2001). *Modellbasierte Lokalisation und Verfolgung für sichtsystemgestützte Regelungen*, PhD thesis, Universität Karlsruhe (TH).

Pope, A. R. & Lowe, D. G. (2000). Probabilistic Models of Appearance for 3D Object Recognition. *International Journal of Computer Vision*, Vol. 40, No. 2, pp. 149–167.

Rayleigh, L. (1874). On the manufacture and theory of diffraction gratings, *Philosophical Magazine,* Vol. 47, pp. 81-93, 193-204.

Rindfleisch, T. (1966). Photometric method for lunar topography, *Photogrammetric Engineering*, Vol. 32, No. 2, pp. 262-277.

Sadlo, F.; Weyrich, T.; Peikert, R. & Gross, M. (2005). A practical structured light acquisition system for point-based geometry and texture, *Proc. Point Based Graphics, Eurographics '05*, pp. 89-145.

Sarder, P. & Nehorai, A. (2006). Deconvolution methods for 3-d fluorescence microscopy images, *IEEE Signal Processing Magazine*, Vol. 23, pp. 32-45.

Shirai, Y. & Suwa, M. (1971). Recognition of polyhedrons with a range finder, *Proc. of the 2nd Int. Joint Conf. on Artificial Intelligence*, London, UK, pp. 80-87.

Teutsch, C. (2007). *Model-based Analysis and Evaluation of Point Sets from Optical 3D Laser Scanners,* PhD thesis, Magdeburger Schriften zur Visualisierung, Shaker Verlag.

Veltkamp, R.C. & Hagedoorn, M. (2001). State-of-the-art in shape matching. In: *Principles of Visual Information Retrieval*, M. Lew (Ed.), pp. 87-119, Springer, ISBN 1-85233-381-2.

Vogelgesang, J. (2008). *Fusion der Tiefeninformation mehrerer Lasertriangulationssensoren, Studienarbeit,* Universität Karlsruhe (TH) und Fraunhofer Institut Informations- und Datenverarbeitung.

Werling, S., Balzer, J. & Beyerer, J. (2007). A new approach for specular surface reconstruction using deflectometric methods, *Informatik 2007. Informatik trifft Logistik,* R. Koschke (Ed.), Beiträge der 37. Jahrestagung der Gesellschaft für Informatik e.V. (GI), 24.-27. Sep. 2007, Bremen, GI-Edition - Lecture Notes in Informatics (LNI) - Proceedings 109, Vol. 1, pp. 44-48, ISBN 978-3-88579-203-1.

Wikipedia, 3d scanner (2009). [www] http://en.wikipedia.org/wiki/3d_scanner.

Wikipedia, NP-hard (2009). [www] http://en.wikipedia.org/wiki/NP-hard.

Zhang, Z. (2000). A Flexible new Technique for Camera Calibration, *IEEE Transactions On Pattern Analysis and Machine Intelligence,* Vol. 22, No. 11, pp. 1330-1334.

Zhuge, M. (2008). *Validierung des Füllvolumens im Blister mittels Laserlichtschnitt,* Studienarbeit, Universität Karlsruhe (TH) und Fraunhofer Institut Informations- und Datenverarbeitung.

7

Prospective polymer composite materials for applications in flexible tactile sensors

M.Knite and J.Zavickis
Riga Technical University
Latvia

1. Introduction

The purpose of this paper is to give a review of the polymer composite materials specially developed for application in strain and pressure sensors that can be used for elaboration of flexible tactile sensing systems. Our recent achievements in design, processing and investigation of physical properties of elastomer and nanostructured carbon composites as prospective materials for mechanical (pressure, strain) indicators are also presented.

The material used most often for design of pressure sensors is the piezoelectric ceramics made separated from the material or structure being monitored. There is a demand for new flexible large-area sensors that can be embedded, for example, into the flexible skin material of robotic fingers and used for sensing multiple locations. High elasticity polymer matrix based materials are still of interest. All polymer-composites materials elaborated for mechanical sensing can be generally classified as electrically active (electronic response) and optically active (photonic response) materials. An example of optically active polymeric composite transducer for tactile sensation is based on optical fiber with Bragg grating embedded in polydimethilsiloxane (Heo et al., 2006). The basic principle of this transducer lies in the monitoring of the wavelength shift of the returned Bragg-signal as a function of the strain or force. Comprehensive picture of current status of micro- and nano-structured flexible optical fiber sensors with particular reference to surface plasma resonance fiber sensors and photonic crystal fiber sensors is given by (Lee et al., 2009).

In present paper we are going to concentrate to electrically active polymer composites. One of well-known methods to obtain an electrically active polymer composite is adding of micro-size and nano-size conductive particles or structures to a polymer matrix. Conductive polymer-composite for strain sensing can be obtained when particles of good conductors (carbon black, graphite powder, carbon fibres or undersized particles of metals) are implanted in to an insulating polymer matrix. A continuous insulator-conductor transition is observed in two-component systems at gradual increase of the number of randomly dispersed conductor particles in an insulator matrix. Most often such transitions called percolation transitions are described by the model of statistical percolation (Staufer & Aharony, 1992) (Roldughin & Vysotskii, 2000). The volume concentration of conductor particles V_C at which the transition proceeds is called percolation threshold or critical point. According to the statistical model, conductor particles, in the vicinity of V_C, assemble in clusters the correlation radius ξ (average distance between two opposite particles of a cluster) diverges as

$$\xi \sim |V - V_C|^{-\nu} \tag{1}$$

upon approaching V_C (ν - critical indices) (Staufer & Aharony, 1992).
In the vicinity of percolation threshold, electric conductivity of the composite changes as:

$$\sigma \sim |V - V_C|^{t}, \tag{2}$$

here t - critical indices (Roldughin & Vysotskii, 2000). If such conductive composite have been mechanically stressed, the ξ and as follows σ would change. This is the reason of tensoresistance (the resistance changes versus tensile strain) and piezoresistance (the resistance changes versus compressive strain) effects. The changes of electric resistance with strain and pressure can be explained on microscopic level as a result of change (destruction or formation of conductive micro-channels) of the percolation structure of conductive particles network.
New interesting properties are expected in case the composite contains dispersed nano-size conducting particles (Yang & Sheng 2000; Wolf, 2004). Polymer – electro-conductive nanostructure composites (PENC) offer attractive alternatives for developing new generation of flexible large-size sensors because of their superior mechanical and electrical properties.
The main task of this paper is to focus directly to the PENC developed for application in strain and pressure sensors. From the thermodynamic point of view the principle of strain sensing is based on the shift of percolation threshold, for example, under tensile strain as shown in Figure 1 (experimental points for construction of curves are taken from our previous works (Knite et al., 2004 a) (Knite et al., 2002)). One can see that due to the shift of percolation threshold under 30% strain the electric resistance changes more than 10^4 times for composite with 10 mass parts of nano-structured carbon black. The best strain sensitivity can be expected in the percolation region (9-11 mass parts of filler) for relaxed PENC. Thus, one may expect the maximum sensitivity of PENC materials to thermodynamic forces near the percolation threshold of electric conductivity.

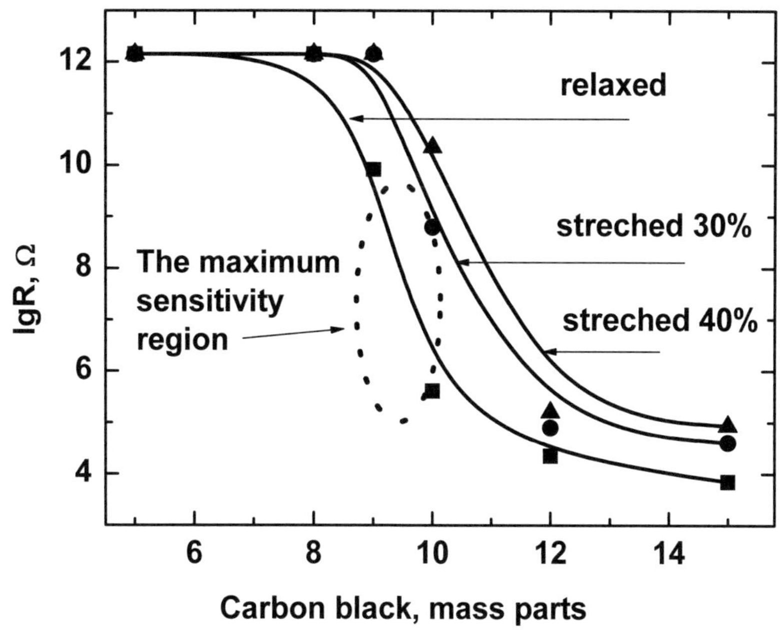

Fig. 1. Shift of the percolation threshold under tensile strain in polyisoprene/high structure carbon nanoparticles composite. The experimental points are taken from the papers (Knite et al., 2002) (Knite et al., 2004 a)

2. Prior work

To date there are several approaches to design PENC for application in strain and pressure sensors. One can subdivide all known such PENC in to composites with insulator matrix and different conducting filler: 1) metal nanoparticles (Ishigure et al., 1999) (Zhang et al., 2000); 2) conductive polymer nanoparticles (Xie & Ma, 2000) (Flandin et al., 2001); 3) carbon microcoils (Yang et al., 2006); 4) graphite nanosheets (Chen et al., 2007); 5) carbon black nanoparticles (Aneli et al., 1999) (Dohta et al., 2000) (Das et al., 2002) (Job et al., 2003) (Flandin et al., 2000) (Knite et al., 2002) (Knite et al., 2003) (Knite et al., 2004 a) (Zavickis et al., 2008); 6) carbon nanotubes (Farajian et al., 2003) (Dharap et al., 2004) (Knite et al., 2007b).

2.1 Composites with insulating polymer matrix and metal filler

Several series of polymer-conductor micro- and nano-composites were prepared from epoxy, silicone rubber, polyethylene, and polypropylene as matrix, and metal, graphite and conducting ceramics as filler materials and examined for applications such as pressure sensor by (Ishigure et al., 1999). Particle size was within 0.1 μm and 600 μm. Particles has different shapes: spherical, plate-like, angular, dendritic, somewhat dendritic and aggregated spherical. Different preparation procedures were used (Ishigure et al., 1999):

1) epoxy composites were prepared by stirring with Cu or Sb-doped SnO_2 filler in acetone and subsequent curing;
2) silicone rubber composites were prepared by hand-mixing with Cu, Ni, Sb-doped SnO_2 or $La_{0.5}Sr_{0.5}CoO_3$ filler, and with a small amount of hardener and subsequent curing;
3) thermoplastic polymer composites were mixed with Sb-doped SnO_2 or graphite filler by laboplastomill at 140 °C for polyethylene, and at 180 °C for polypropylene. The mixture was molded into disk under 20 MPa at 140 °C for polyethylene, and under 100 MPa for polypropylene. Some polymer-conductor composites showed a piezoresistive effect near the critical threshold of the percolation curve. This effect was large in composites prepared from polymers such a silicone rubber having a low Young's modulus. The largest piezoresistive effect was found in silicone rubber composites with 15 vol % fine (100 nm) Sb-doped SnO_2 particles: resistance diminishes about 10^3 times with 5 MPa pressure (Ishigure et al., 1999).

The piezoresistance and its time dependence of conductor-filled polymer composites were investigated also by (Zhang et al., 2000). The conducting fillers were Sn-Pb alloy powder (mean particle diameter 100 nm), Cu powder (200 nm) and Al powder (700 nm). Polyethylene (PE), polystyrene (PS) and epoxy were used as polymer matrixes. Epoxies were modified by adding dibutylphthalate. The addition of this additive was to decrease the compressive modulus of epoxy. For the samples of Sn-Pb/PE, Cu/PE, Al/PE, Sn-Pb/PS, Cu/Ps, and Al/Ps, the conducting filler and polymer powder mixtures were mixed in a satellite ball mill (QM-1SP) at 200 rpm for 11 h, then the powder mixtures were compression moulded in a matched metal die at 170 °C for 20 min to form the 4 mm-thick sheets. For the epoxy matrix samples, the epoxies (added with the hardener and additive) were mixed with the conducting fillers for 10 min by hand, than cured at 90 °C for 3h (Zhang et al., 2000). Mechanical properties, such as the compressive modulus and the creep behaviour were measured. For the convenience of the piezoresistance measurement and to get excellent conductive contact, brass electrodes of 1 cm^2 were adhered at each side by silver paste. The piezoresistance measurements were performed under uniaxial presses. It was found that the relative resistance decreases with an increase of applied stress. Authors (Zhang et al., 2000) suppose that this process is affected by the applied stress, filler particle diameter, filler volume fraction, matrix compressive modulus, and potential barrier height between the adjacent particles. The deformation of polymer matrix under a fixed stress increases with the time, which is the basic reason of time dependence of piezoresistance. The results of (Zhang et al., 2000) show that the increases of applied stress, filler particle diameter, and polymer matrix creep enhances the time dependence of piezoresistance, while the increase of filler volume fraction weakens it.

Authors from the Durham University (Bloor et al., 2005) showed that the nickel particle/silicone matrix composites QTC™ produced by Peratech Ltd has unusual properties, namely a very high resistance even when the quantity of nickel powder incorporated in the composite is above the percolation threshold because the adhering polymer separates the metal particles. The resistance of such composite is found to be extremely sensitive to any mechanical deformation. Thus unlike other PENC the resistance falls when the material is stretched, or bent or twisted. Modest applied forces produce resistance changes of many orders of magnitude. The exponential dependence of sample resistance on deformation indicates that the principal conduction mechanism is determined by carriers tunneling between the filler particles (Bloor et al., 2005).

Conductive nanocomposites prepared using styrene butadiene rubber as the polymer matrix and nanosized powder of cooper-nickel (Cu-Ni) alloy as the filler (Mohanraj et al., 2007) show weak piezoresistance effect (resistance decreases only about one order under pressure 30 kPa). The positive temperature coefficient of resistance (PTC) on the contrary is very high. The DC resistivity changes of 8 orders of magnitude when temperature rises from 300K to 400K. The appreciable PCT effect is explained by the predominant breakdown of the conducting network structure due to higher thermal expansion of rubber matrix compared to the filler (Mohanraj et al., 2007). It is not understandable for us why thermal expansion of matrix affects the percolation structure of conductive particles much more effective than mechanical compression.

Consequently, for composites with insulating polymer matrix and metal filler the piezoresistance effect always is considered as negative ($(\Delta R)/R_0<0$). Unique is the negative tensoresistance effect (when the resistance falls if the material is stretched) for nickel particle/silicone matrix composite (Bloor et al., 2005).

2.2 Composites with insulating polymer matrix and conductive polymer nanoparticles

A polyaniline (PAn)/styrene-butadiene-styrene (SBS) triblock copolymer conductive elastomeric composites were recently investigated by (Xie & Ma, 2000). The composites were prepared by *in situ* emulsion polymerization of aniline in the presence of SBS using dodecylbenzene sulfonic acid (DBSA) as an emulsifier and a dopant. The product was melting processed (MP), solution processed (SP), or secondary doped with *m*-cresol (SSP). The results of conductivity measurements showed that for the MP and SP samples conductivity increases with extension, whereas for the SSP sample when PAn content is lower than percolation threshold, conductivity diminishes with increasing extension, but when the PAn content exceeds the percolation threshold value, conductivity followed an empirical equation with a maximum value. All differences are related to their different morphological structures (Xie & Ma, 2000).

New strain sensing materials have recently been processed (Flandin et al., 2001) by mixing an insulating latex of a stryrene-butil acrylate copolymer with a colloidal suspension of intrinsic conducting polymer (polypyrolle) particles. This work has dealt with the AC electrical properties measurements under large strain of composites. A good correlation between electrical conductivity and mechanical stress was measured. The macroscopic conductivity and stress evolution were very smooth, due to the nanoscopic size of the fillers. Real part of the conductivity at low frequency versus deformation is explainable in terms of damage in the percolating network. Imaginary part of the conductivity was found to increase at high frequency for volume fractions close to percolation threshold. Authors (Flandin et al., 2001) related this to the creation of new capacitors with very high values (tiny distance) when the network is broken between two fillers. Different numerical tools have been developed by (Flandin et al., 2001) that may be applied to many composites.

Investigations (Barra et al., 2008) of electrical conductivity and electromechanical properties of a polystyrene- block- poly(ethylene-ran-butylene)- block- polystyrene copolymer (SEBS)/ polyaniline doped with dodecylbenzenesulfonic acid (PAn.DBSA) blends show the insulator - conductor transition not as sharp as those found in carbon black conducting systems and the percolation threshold was lower than 20 wt.% of PAn.DBSA. The effect of conductivity as a function of compression force then was evaluated. For the filler content below the percolation threshold, the conducting PAn domains are separated by a polymer matrix insulation layer and the electrical conductivity is low, therefore no significant changes in relative electrical conductivity under compressive stress were observed. Above the percolation threshold, the conducting polymer particles are much closer, which contributes to an increase in the electrical conductivity of the blends. Under compressive stress, the contact between the conducting particles increases and, consequently, the relative electrical conductivity of the polymer blend increases significantly. The best compressive stress sensitivity (Barra et al., 2008) have been reached for blends with 30 wt.% of PAn.DBSA, there is a 15-fold increase in the relative conductivity with compressive stress up to 200 MPa, reaching a constant value above this pressure. Thus, the SEBS/PAn.DBSA blend samples (70/30) are prospective active element for future flexible tactile systems (Barra et al., 2008).

2.3 Composites with insulating polymer matrix and conductive spring shaped carbon microcoils

Composites filled with spring shaped carbon microcoils (CMCs) seems to be promising materials for tactile sensing (Yang et al., 2006) (Yoshimura et al., 2007). It's reported (Yang et al., 2006) that until now three kinds of CMCs have been used in tactile sensors: (1) conventional double-helix CMCs: the separation between coil wires are quite small; (2) super-elastic double-helix CMCs: the separation between coil wires are quite large; (3) single-helix CMCs. The electrical properties of these three kinds of CMCs change with the extension or contraction and this property is the foundation of the CMCs tactile sensors (Yang et al., 2006). It was shown (Yang et al., 2006) that polysilicone/single-helix CMCs composites have a higher sensitivity than composites filled with double-helix CMCs. The CMCs with average fiber diameter of 0.7 μm and coil diameter of 5 μm were dispersed into silicone rubber (Shin-Etsu Chemical, KE-1842) matrix by (Yoshimura et al., 2007). The CMCs used in this study were prepared by the chemical vapor deposition process using acetylene as a carbon source at 700-800 ^{0}C (Yoshimura et al., 2007). The mechanism of electrical resistivity change in CMCs/silicone-rubber composites was investigated and discussed in comparison with that in the composites involving vapor grown carbon nanofibers (VGCFs) or carbon blacks (CBs). The carbon content was given as volume fraction, using the density of CMCs (1.9 g/cm^3), VGCFs (2.0) and CBs (1.8) (Yoshimura et al., 2007). The measurements of electrical resistance of test examples were carried out under compressive as well as tensile stress applied using tensile tester (Instron, 4505) at constant speed of 10 mm/min. The values obtained were converted into resistivity under the condition of Poisson's ratio = 0.3. The addition of relatively small amount of CMCs effectively made the material semiconductive. The electrical resistivity of the CMCs/silicone-rubber composites increased with the compressive or tensile strain, and a change in the resistivity was found to be much higher that of the other composites involving VGCFs and CBs. The authors (Yoshimura et al., 2007) have not given any specific data of CBs - only SIGMA-ALDRICH, 05-1530-5. It is believed these CBs to be a low structure microsized CBs because in case of HSCB (Knite et al., 2004) approximately the same strain sensitivity has been achieved as for CMCs/silicone-rubber composites (Yoshimura et al., 2007). The higher sensitivity of CMCs/rubber composites according to the authors (Yoshimura et al., 2007) was attributable to the synergetic effects of the changes in resistivity of the conductive paths and CMCs themselves. We believe that better sensing properties of CMCs/rubber composites (Yoshimura et al., 2007) relate with much higher structure of CMCs compared to CB used.

2.4 Composites with insulating polymer matrix and graphite nanosheets

Novel conductive nanocomposites with low percolation threshold were prepared by adding conductive graphite nanosheets (GNs) to high density polyethylene (Lu et al., 2005). GNs with average sheet diameter 5-20 μm and 30-80 nm in thickness were prepared by ultrasonic powdering technique (Lu et al., 2005). The HDPE and GNs were first wet-mixed to achieve a uniform dispersion and after mixed on a two-roll mill. The compound was further compressed by hot press. The piezoresistivity measurements were performed under uniaxial pressures both at low- and high-pressure range. It reveals that the piezoresistitvity strongly depends on the GNs concentration. Authors (Lu et al., 2005) stated that such piezoresistive material could be used as pressure sensor as the electrical component.
A novel pressure sensing nanocomposite with remarkable and reversible piezoresistivity is successfully fabricated by dispersing homogenously conductive GNs in a silicone rubber (SR) matrix (Chen et al., 2007). Authors have used as filler the same GNs with average sheet diameter 5-20 μm and 30-80 nm in thickness and have got super-sensitive positive piezoresistivity under very low pressure, in finger -pressure range. Authors (Chen et al., 2007) described the piezoresistive behaviour by the tunnelling models as well as used the models for change of the number conducting paths that are similar to model developed in (Knite et al. Sensors, 2004) for polyisoprene/high structure carbon black composites. In the same way as (Knite et al., 2004 a) stated for nanostructured carbon black the authors (Chen et al., 2007) made conclusions that stronger adhesion of the GNs to the polymer matrix compared to the cohesion between the nanosheets themselves lead to excellent piezoresistive properties of composite.

2.5 Composites with insulating polymer matrix and carbon black

Electric conductivity measurements of carbon black-filled rubbers under low and high stretching deformations as well as all-round and uniaxial compression at both low and high pressure were carried out by (Aneli et al., 1999). Relaxation processes were also investigated. High-molecular elastomer polymethylvinilmethylsiloxan and low-molecular elastomer polydimethylsiloxan were composed with P357E and P803 carbon black using the additive vulcanization technique. The carbon particle size and specific surface area of carbon black are not given. Electrodes were embedded into sample during vulcanisation of rubbers. It was found that a significant increase of the interaction between macromolecules of the elastomer and filler particles induces a sharp growth of rubber tensoresistivity. The maximum changes of electric resistivity did not exceed 10^2 times under stretching strain till 160 %. The maximum changes of electric resistivity did not exceed 5 times under pressure up to 20 MPa. Addition to that, the changes of resistivity was not reversible - significant hysteresis effects were observed.
Scientists from the Okayama University (Dohta et al., 2000) demonstrate an application of a flexible strain sensor to a pneumatic rubber hand. The flexible strain sensor is composed of the silicone rubber and electro-conductive paints. The electro-conductive paint (FUJICURA Co, Ltd.: DOTITE RA-3) is painted thinly on the rubber of 20 mm in length and 3 mm in width. In order to increase the durability, the electro-conductive paints are coated with thin silicone rubber. The operating principle of the flexible strain sensor is as follows: when the sensor is expanded longitudinally, the density of carbon in the electro-conductive paint increases in the direction of the thickness and lowers in the direction of the length. Consequently the electric resistance increases by 20 kΩ under 100 % strain. Two kinds of

flexible strain sensors are installed in the pneumatic rubber hand. One is a displacement sensor that can detect a bending displacement of soft gripper to confirm the grasping of an object and to recognize a size of this object. The other is a tactile sensor that can detect contact force and contact position to recognize the stiffness and shape of the object. The variation of electrical resistivity of carbon black and short carbon fiber (SCF) filled rubber composites against the degree of constant strain rate have been studied by (Das et al., 2002). Ethylene-propylene-diene rubber (EPDM) and ethylene vinyl acetate copolymer (EVA) was taken as matrix, and conductive carbon black Vulcan XC-72 and Carbon India Ltd. Carbon fibers, grade RK 12 as filler. Chosen matrix and filler were mixed in Brabender plasticorder (PLE-330) under identical conditions of mixing time (5 min), temperature (90 °C) and rotor speed (60 rev.min^{-1}). The mixes were sheeted in a laboratory size two-roll mixing mill. The optimum cure times at 170 °C for these composites were determined using Monsanto rheometer R-100S. The mixes were than cured at 170 °C in an electrically heated press under identitical pressure of 5 Mpa. These cured sheets were kept at room temperature for 24h maturation before testing. The Zwick universal testing machine was coupled with a volume resistivity measurement set up for continuous reading the variation of resistivity of the sample during extension and retraction cycles. Composites (Das et al., 2002) exhibit appreciable irreversibility in the variation of the electrical resistivity against strain during extension-retraction cycles.

An ethylene-octene (EO) elastomer was the matrix in studies of (Flandin et al., 2000). The fillers were a carbon fiber (CF) with diameter 10 μm and aspect ratio 20, a low structure carbon black (LSCB) with particle diameter 300 nm and surface area 8 m^2/g, and two high structure blacks (HSCB) Printex 30 (particle diameter 27 nm and surface area 80 m^2/g) and Condutex 975 Ultra (particle diameter 21 nm and surface area 242 m^2/g) (Flandin et al., 2000). Blending was carried out in Haake Rheomix 600 mixing head. The polymer was mixed without filler for 1 min until melted and the filler was added slowly over a period 10 min. Specimens were deformed in an Instron 1123 testing machine at a strain rate of 0.2 %/min. Electrical properties were measured with a four-terminal technique using a Keythley model 220 as current source and a Keythley model 619 for voltage and current measurements. Composites with CF and LSCB did not maintain good mechanical properties, generally exhibited an increase in resistivity with strain, and exhibited irreversible changes in both mechanical and electrical properties after extension to even low strains. It was found that the composite with HSCB exhibited an unusual decrease in resistivity at low strains and reversibility in the resistivity upon cyclic deformation. For an explanation of the unusual properties of EO with HSCB authors (Flandin et al., 2000) proposed model that incorporates the multifunctional mobile physical crosslinks of the EO matrix and dynamic filler-matrix bonds. For HSCB, rotation and translation, or possibly shape changes of the asymmetric aggregates, can reduce resistivity. Authors (Flandin et al., 2000) imagine that the very small particles of HSCB are less constrained during elongation if they are incorporated into a network of mobile junctions than if the network has fixed junctions. Therefore, deformations can produce new conducting pathways by particle reorientation.

By (Job et al., 2003) a natural rubber (NR) and carbon black (CB) were compounded, aiming the development of composites with good mechanical properties, processability and electrical conductivity for use as pressure sensors. The carbon particle size and specific surface area of carbon black was between 60 to 200 nm and 16-24 m^2/g respectively (LSCB).

Composites were prepared by casting. First, a solution of 25 ml of ammonium hydroxide and CB was prepared under stirring for 15 minutes. After 10 ml of the stabilized latex was mixed to the carbon black solution under stirring for 10 minutes. Finally, composites were prepared by casting the obtained solution onto glass substrate and then heated at 65 °C, for 6 hours, for evaporation of volatile components. A linear dependence of conductivity on the applied pressure is observed for CB contents up to 20 wt % in the pressure range of 0 to 1.6 Mpa, tending to saturation for higher values of pressure. For contents of CB less than 10 % the results are scattered and no clear dependence of the conductivity on the pressure was found.

A polyisoprene/very high structure carbon nanoparticles (surface area 950 m^2/g) composites (PHSCNC) were prepared and investigated by (Knite et al., 2002) (Knite et al., 2004 a). Plates of 12 mm diameter were cut for electrical resistivity vs pressure measurements of 1 mm thick 20 cm x 20 cm sheets vulcanized at high pressure. To study dependence of electric resistance of the PHSCNC on stretching force and stretch deformation 15 cm × 1.5 cm samples were cut. Copper foil electrodes were glued on both sides at the ends and each pair of electrodes was shortcut by copper wiring. On a relaxed sample the distance between electrodes l_0 did not exceed 50 mm. Sandpaper was glued on the electrodes to fasten the samples in the stretching machine. A modified conductive tapping mode AFM Nanoscope III (Dimensions 3000, Digital Instruments) was used to investigate the electrical conductivity on local nano-size spots on the surface of the composites. The AFM cantilever tips were of standard silicon nitride. The AFM tip and cantilever were coated with a 5 nm thick Cr adhesive layer and a 15 nm thick Au layer to get conductive tip. A typical tip radius measured with a transmission electron microscope was 30 nm. A bias of 10 - 1000 mV was applied between the tip and the sample. With a contact mode conductive atomic force microscope authors (Knite et al., 2002) (Knite et al., 2004 a) succeeded in obtaining a topographic picture of the sample surface, as well as a nanoscale map of cross-sections of the electro-conductive channels and an insulating matrix of the same local spot. The current distribution inside the conductive pattern was not uniform. The size of sub-channels was around 30 nm.

Of all the composites examined (Knite et al., 2002) (Knite et al., 2004 a) the best results were obtained for the samples with 10 mass fractions of carbon nanoparticles, which apparently belonged to the region of percolation phase transition. The electric resistance of these samples changed more than 4 orders upon a 40% stretch and more than 3 orders upon a 0.30 Mpa pressure. After the samples were released, the resistance practically returned to its previous value. The reversibility and the giant changes of electric resistance under both tensile and compressive strain were explained due to comparatively higher mobility of high structure nanoparticles compared to low structure particles as well as stronger adhesion of carbon nanoparticles to the polymer matrix compared to cohesion between nanoparticles themselves (Knite et al., 2004 a). The growth of electric resistance with uniaxial stretching and pressure can be explained as a result of destruction of the structure of the carbon electro-conductive nano size channel network. At low stretching deformation the experimental data has good coincidence with model of tunneling conductance (Knite et al., 2004 a). The AC conductivity measurements (Knite et al., 2004 b) also verify the tunneling model of conductance. For description of the experimental results at high deformation both - the destroying of conductive network and decrease of conducting path have to be taken into account. In case of uniaxial pressure destruction of electrically conductive channels is caused by deformation of the polyisoprene nanocomposite perpendicular to the direction of pressure. The stated fact that the resistance grows rapidly and reversibly with both tensile and compressive strain can be used also for elaboration of chemical sensors (Knite et al., 2007a) based on swelling of the composite matrix in ambience of different chemical substances.

The effect of a plasticizer on the change of the electrical resistance under deformation at stretch were studied in polyisoprene composites containing dispersed nano-size conducting carbon particles at concentrations in the vicinity of the percolation threshold by (Knite at al, 2003) (Knite at al, 2005) (Knite et al., 2006). The addition of plasticizer to the initial materials increases the strain sensitivity of the composite and broadens the carbon concentration interval of the percolation threshold. The observed improvements of the tenso-resistive response are explained by decrease of cohesion forces between carbon nanoparticles and higher mobility of the carbon nano-particles in the elastomer matrix in the presence of the plasticizer.

2.6 Composites with insulating polymer matrix and conductive carbon nanotubes

Carbon nanotubes (CNT) change their electrical properties when subjected to strain. Calculated (Farajian et al., 2003) theoretically the I-V characteristics, showed that the current of metallic tube decreases with increased bending, while that of semiconducting tube increases. Possible application to nanoelectromechanical sensors and switches is discussed.

An attempt to use the strain sensing capability of single-walled carbon nanotubes (SWCNT) on the nanoscale to develop a strain sensor on the macroscale was made by (Dharap et al., 2004). The carbon nanotube film is produced by mixing SWCTNs with 0.25 mg/ml N,N-dimethylformamide (DMF). The film (10 μm thick) was composed of mechanically entangled randomly oriented nanotube bundles, to which it has isotropic electric properties. The carbon nanotube film was attached to a brass plate with PVC film between them. A four point probe was used to measure voltage changes in the carbon nanotube film. The brass specimen was subjected to tension as well as compression cycles. It can be concluded that there is nearly linear relationship between the measured change in voltage and the strains in the carbon nanotube films (Dharap et al., 2004).

Recently the interest rise about elaboration of polymer/multi-walled carbon nanotube (MWCNT) composites for strain sensing (Knite et al., 2007b) (Bokobza, 2007) (Pham et al., 2008) (Hu et al., 2008) (Li et al.,, 2008) (Knite et al., 2008) (Knite et al., 2009).

The purpose of (Knite et al., 2007b) paper was the design, elaboration and investigation of the polyisoprene/multiwall carbon nanotube (MWCNT) composites for application in strain sensors as well as to compare they with the polyisoprene/high structure carbon black composites elaborated and prepared by the same technology. Polyisoprene composites containing dispersed nano-size particles, in this case two types, multiwall carbon nanotubes (MWCNT) Aldrich 636835 as well as highly structured carbon black (HSCB) Printex XE2 were prepared by "solution method" as follows. The matrix composition was treated with chloroform providing: 1) an increase of the nano-particles mobility and 2) better dispersion of the nano-particles within the matrix. Prepared matrix composition was allowed to stand for swelling ~ 24 h. Nano-size carbon black is carefully grinded with a small amount of solvent in a china pestle before addition to the polyisoprene matrix. Polyisoprene matrix solution and nano-size carbon structure concentrated product was mixed in the mixer with small glass beads at room temperature for 15 min. Product is poured out into little aluminium foil box and stand ~ 24 h, dried at 40 °C for more than 12 h and vulcanized under high pressure at 160°C for 20 min (Knite et al., 2007b). The size of MWCNT: OD = 60-100 nm, ID = 5-10 nm, length = 0.5-500 µm, BET surface area: 40-300 m^2/g. The average particle size of HSCB was 30 nm and its dibutyl phthalate (DBP) absorption is 380 ml/100g. Its surface area was 950 m^2/g.) (Knite et al., 2007b). It has been shown that the use of high structure carbon nanoparticle as filler of a polyisoprene matrix provided about 6 orders of magnitude a reversible change in the electric resistivity at large (40%) stretch. The use of multiwalled carbon nanotube as filler of a polyisoprene matrix provided about 4 orders of magnitude a non-reversible change in the electric resistivity at large (40%) stretch. The maximum change at stretch for both type of composites were observed near the percolation threshold. The reversible change in electric resistivity of high structure carbon nanoparticle - polyisoprene composite even at large stretch can be explained by the high porosity of carbon agglomerates providing better adhesion to polymeric globules and mobility of nanoparticles, which is not observed in the case of a poorly structured carbon black. The non-reversible change in electric resistivity of multiwalled carbon nanotube - polyisoprene composite at large stretch we explain with entangled structure of the nanotube. Authors (Knite et al., 2007b) stated that polyisoprene/MWCNT composite can be used for small tensile strain sensing but polyisoprene/HSCB composite are preferable for large tensile strain sensing. The compressive strain sensitivity of above mentioned polyisoprene/MWCNT composites were proved in papers (Knite et al., 2008) (Knite et al., 2009).

In the review of (Bokobza, 2007) the mechanical and electrical properties of composite based on styrene-butadiene rubber (SBR) filled with MWCNT which have lengths and diameters in the range of 0.1-5 µm and 10-50 nm, respectively. Electrical measurements carried out under uniaxial stretching deformation display an increase in resistivity as a result of a breakdown of conductive path and an orientation of the filler particles (Bokobza, 2007).

The possibility of using MWCNT films purchased from Nano-Lab Inc. (Newton, MA, USA) as strain sensors have been studied by (Li et al., 2008). Uniaxial load-unload tensile test was carried out to investigate the static sensing property. Results indicated that the change in resistance of the MWCNT film was proportional to the applied strain. It was indicated that MWCNT film is potentially useful for structural health monitoring and vibration control applications (Li et al., 2008).

Researchers from the United States (Pham et al., 2008) report about the development of conductive, MWCNT-filled, polymer composite films that can be used as strain sensors with tailored sensitivity. The films were made from polymethylmethacrylate (PMMA) matrices and MWCNT obtained from Aldrich by two different methods - dry blended film fabrication or solution based fabrication. The electric surface resistivity of the films was measured in situ as the specimen was loaded in tension. The increasing resistivity with increasing tensile strain is explained due to the reduction in conductive network density and increase in inter-nanotube distances induced by applied strain. The highest sensitivity achieved in this study was almost an order of magnitude greater than conventional resistance strain gauges (Pham et al., 2008).

2.7 Summary on prior work

From the available papers regarding polymer/MWCNT composites for strain sensing one can conclude that attention is devoted only to investigations of tensile strain sensing properties. Almost in all papers the report was about the best sensitivity of MWCNT composites in comparison with conventional resistance strain gauges. Thus more attention should be paid to elaboration of polymer/MWCNT composites for compressive strain sensing.

Of all the composites examined, elastomer/(carbon nanostructure) composites shows the best electromechanical properties as flexible large area materials for strain and stress sensing. To reveal the strain sensing mechanisms further investigations of these composites are required. We present in next paragraphs an attempt to use the HSCB as well as MWCNT to devise an all flexible composite for macro-scale pressure indicators (relative pressure difference sensors) or robotic tactile elements.

3. Design principles of the structure of polymer/carbon nanostructure composites for pressure strain sensing

Based on the review of other authors, we have developed four simple principles, which should be obeyed to obtain maximum sensitivity of multifunctional elastomer-carbon nanocomposites:

1) Polyisoprene (natural rubber) of the best elastic properties has to be chosen as the matrix material;

2) High-structured carbon nano-particles (HSNP) providing a fine branching structure and a large surface area (better adhesion to polymer chains compared to LSNP) or MWCNT should be taken as a filler. Because of a higher mobility of HSNP compared with LSNP the electro-conductive network in the elastomer matrix in this case is easily destroyed by very small tensile or compressive strain. We suppose this feature makes the elastomer–HSNP composite an option for more sensitive tactile elements in robots.

3) The highest sensitivity is expected in the percolation region of a relaxed polyisoprene composite. The smallest mechanical strain or swelling of the composite matrix remarkably and reversibly increases resistance of such a composite. The sharper is the percolation transition of insulator/conductive particle composite the higher should be the compressive stress sensitivity of sensing element.

4) The investigation of development of percolative structure during curing process could be very suitable for finding out the optimal vulcanization time of the PHSCNC with the best compressive strain sensing properties.

4. The investigation of development of percolative structure in PHSCNC during curing process

To investigate a development of carbon nanoparticle cluster percolative structure during vulcanization process the test samples with different levels of vulcanization were prepared and the character of their piezoresistivity was established and compared. Measurements of mehano-electrical properties as well as SEM investigations were carried out.

First of all PHSCNC samples with 9 and 10 mass parts of filler have been prepared. The mixing was done using cold rolls. To obtain good electrical connection with samples, clean sandpapered brass foil mould inserts were used on both sides of the samples. The previous research approved them to be the most suitable for this need because brass forms permanent electro-conductive bonding with the PHSCNC during vulcanization. To provide optimal processing parameters, first the optimal complete curing time of the composite was ensured using MonsantoRheometer100 rubber rheometer and appeared to be 40 minutes for current rubber composition. Disk shape PHSCNC samples 18mm in diameter (Figure 2) with 9 and 10 mass parts of filler were made using different curing times in range from 1 to 40 minutes. 40 minutes corresponds to complete vulcanization of PHSCNC and 1 minute was the smallest possible time to obtain the desired shape of the sample. During "pre-research" the original method was developed to measure samples initial electrical resistivity "in-situ" in the curing mould. The results claimed that electrical resistivity of PHSCNC dramatically drops exactly during the vulcanization (Figure 3). This fact made us to assume, that the development of percolative electrocondutive structure of filler nanoparticles is happening during the vulcanisation although.

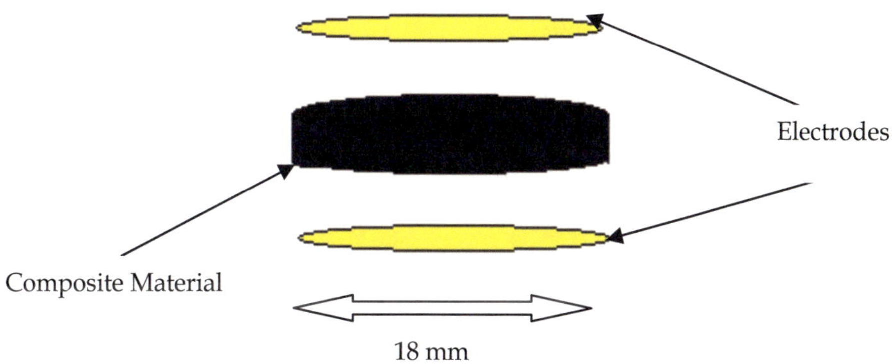

Fig. 2. Schematic structure of the PHSCNC sample with embedded brass foil electrodes

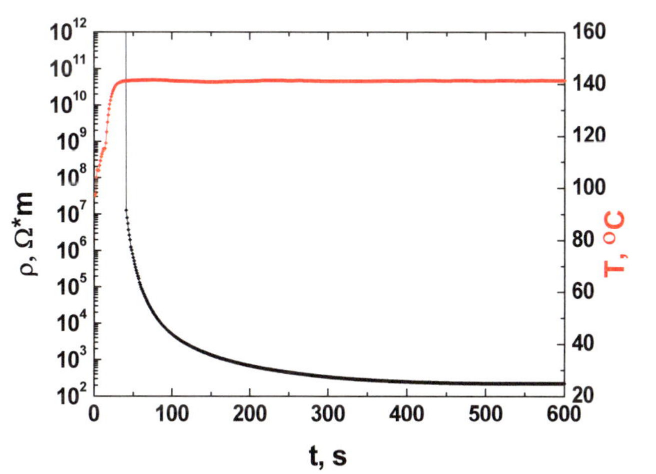

Fig. 3. The change of specific electrical resistivity (black) and temperature (red) as a function of time for PHSCNC sample with 9 mass parts of carbon.

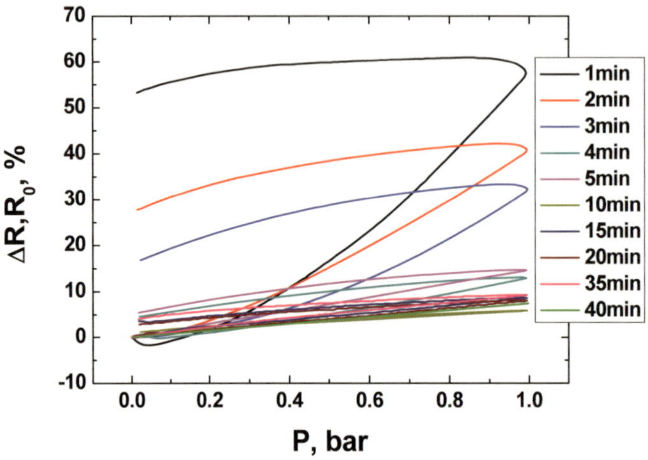

Fig. 4. The piezoresistance of PHSCNC samples with 10 mass parts of carbon black which are made using different curing times from 1 to 40 minutes.

The piezoresistive properties of PHSCNC samples were determined using Zwick/Roell Z2.5 universal material testing machine, equipped with HBM 1kN load cell and HBM Spider8 data acquisition module. This allowed the measurements of mechanical and electrical properties to be taken simultaneously. This testing was done using variable external operational pressure from 0 to 1 bar, with speed of $1x10^{-2}$ bar·s^{-1}. The piezoresistive properties of samples were determined and evaluated as shown in Figure 4.

To ensure our previous assumption, SEM investigation was made on incompletely vulcanized samples, fractured in liquid nitrogen. Technically, the smallest possible vulcanization time here was 3 minutes from 40 which corresponds to 7,5% of complete vulcanization time. The SEM picture of this sample is shown if Figure 5. It was compared with SEM image of PHSCNC sample cured for 15 minutes, which corresponds to 35,5% of complete vulcanization time shown in Figure 6. Comparing these pictures it can be seen, that sample with less vulcanization time has more uniform structure of conductive filler particles (opaque dots all over the image). On other hand in sample with more vulcanization time the conductive filler particles has formed entangled or forked structure. With reference to (Balberg, 2002), exactly the entangled structure of carbon agglomerates is responsible for unique conductive properties of percolative concentrations in polymer matrices.

The results indicate that the balance between the maximum piezoresistivity and more complete relaxation of initial electrical resistivity of sample is critical. If one of them is greater, the other starts to lack useful dimensions and vice versa. The optimum vulcanization time was found out to be at least the 12% of the time necessary for complete vulcanization.

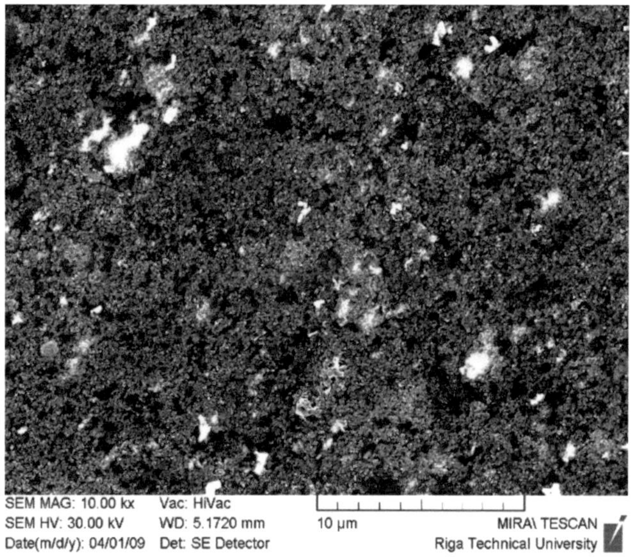

Fig. 5. The SEM image of liquid nitrogen fractured surface of PHSCNC sample with 10 mass parts of carbon black, cured for 7,5% of time necessary for complete vulcanization. No structurization of carbon black aggregates.

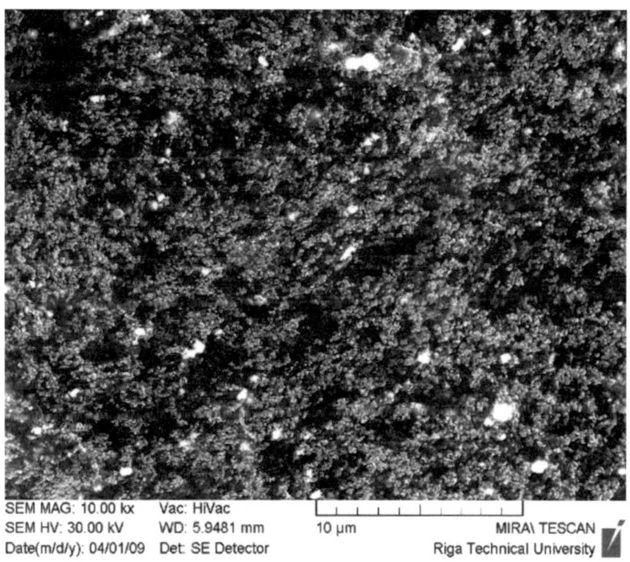

Fig. 6. The SEM image of liquid nitrogen fractured surface of PHSCNC sample with 10 mass parts of carbon black, cured for 35,5% of time necessary for complete vulcanization. The structurization of carbon aggregates (opaque dots) are clearly visible.

5. All-elasto-plastic polyisoprene/nanostructured carbon pressure sensing element with glued conductive rubber electrodes

To obtain completely flexible tactile sensing elements of large area (relative to rigid piezoelectric sensors) a layer of the active PENC composite is fixed between two conductive rubber electrodes by means of specially elaborated conductive rubber glue.

5.1 Preparation of samples and organisation of experiment

The PHSCNC was made by rolling high-structured PRINTEX XE2 (DEGUSSA AG) nano-size carbon black and necessary additional ingredients - sulphur and zinc oxide - into a Thick Pale Crepe No9 Extra polyisoprene (MARDEC, Inc.) matrix and vulcanizing under 30 bar pressure at 150 °C for 15 min. The mean particle size of PRINTEX XE2 is 30 nm, DBP absorption - 380 ml/100 g, and the BET surface area - 950 m^2/g.
The polyisoprene - carbon nanotube (PCNT) composites containing dispersed multi-wall carbon nanotubes (MWCNT) were prepared as follows. The size of MWCNT: OD = 60-100 nm, ID = 5-10 nm, length = 0.5-500 μm, BET surface area: 40-300 m^2/g. To increase the nano-particles mobility and to obtain a better dispersion of the nano-particles the matrix was treated with chloroform. The prepared matrix was allowed to swell for ~ 24 h. The MWCNT granules were carefully grinded with a small amount of solvent in a china pestle before adding to the polyisoprene matrix. Solution of the polyisoprene matrix and the concentrated product of nano-size carbon black were mixed with small glass beads in a blender at room temperature for 15 min. The product was poured into a small aluminum foil box and let to stand for ~ 24 h, dried at 40 °C and vulcanized under high pressure at 160°C for 20 min (Knite et al., 2008).
Discs of 16 mm in diameter and 6 mm thick were cut from the vulcanized PHSCNC sheet. Conductive polyisoprene - HSCB (30 mass parts) composite electrodes were prepared and fastened to the disc with special conductive adhesive (BISON Kit + 10 mass parts of HSCB) as shown in Figure 7.

Fig. 7. Picture of completely flexible strain sensing element made of PHSCNC with conductive rubber electrodes.

Aluminum electrodes were sputtered on opposite sides of the sensing element (20 × 11.5 × 2.4 mm) made of the PCNT composite as shown in Figure 8. Electrical resistance of samples was measured vs mechanical compressive strain and pressure on a modified Zwick/Roell Z2.5 universal testing machine, HQ stabilized power supply, and a KEITHLEY Model 6487 Picoammeter/Voltage Source all synchronized with HBM Spider 8 data acquisition logger. Resistance R of the composites was examined with regard to compressive force F and the absolute mechanical deformation Δl in the direction of the force. Uniaxial pressure and relative strain were calculated respectively.

Fig. 8. Picture of a strain sensing element made of PCNT composite with sputtered Al electrodes.

5.2 Experimental results and discussion

The percolation thresholds of PHSCNC and PCNT composites were estimated at first. Of all the composites examined, the best results were obtained with samples containing 14.5 mass parts of MWCNT and 10 mass parts HSCB, apparently belonging to the region slightly above the percolation threshold. Dependence of electrical resistance on uniaxial pressure first was examined on a PHSCNC disc without the flexible electrodes. Two brass sheets 0.3 mm thick and 16 mm in diameter were inserted between the disc and electrodes of the testing machine.

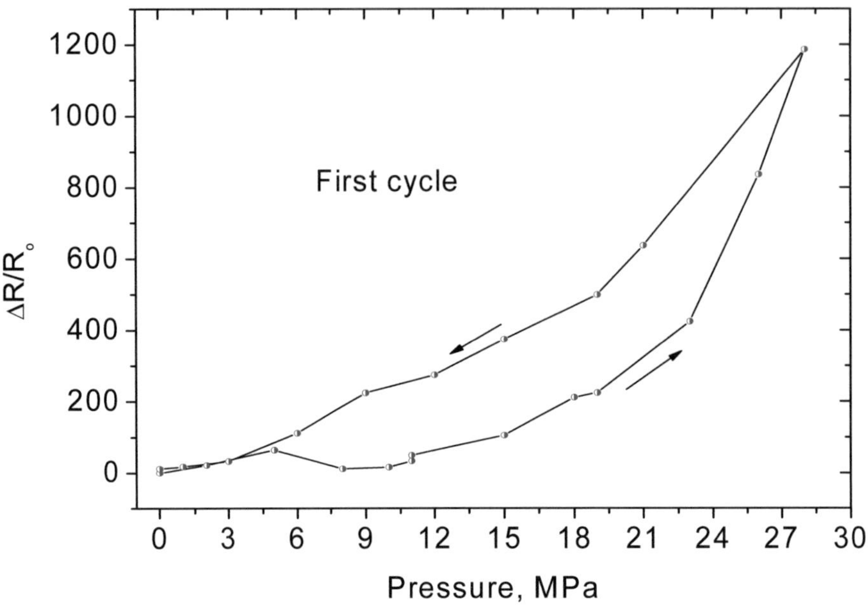

Fig. 9. Electrical resistance (in relative units) of an element (without flexible electrodes) of PHSCNC containing 10 mass parts of HSCB as function of pressure. T = 293 K.

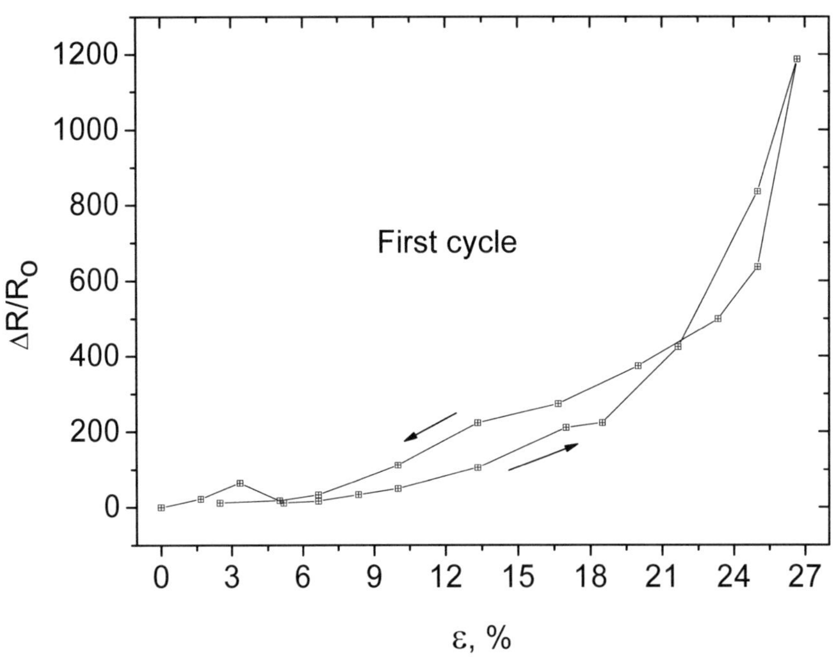

Fig. 10. Electrical resistance (in relative units) of an element (without flexible electrodes) of PHSCNC containing 10 mass parts of HSCB as function of compressive strain ε. T = 293 K.

The piezoresistance effect in PHSCNC is reversible and positive ($(\Delta R)/R_0>0$) (Figure 9 and Figure 10).
As a next the measurements of the piezoresistance effect observed in an element of PHSCNC with flexible electrodes attached is illustrated in Figure 11 and Figure 12 showing that the piezoresistance effect decreases approximately 10 times but remains positive.
The positive effect can be explained by transverse slippage of nano-particles caused by external pressure leading to destruction of the conductive channels.

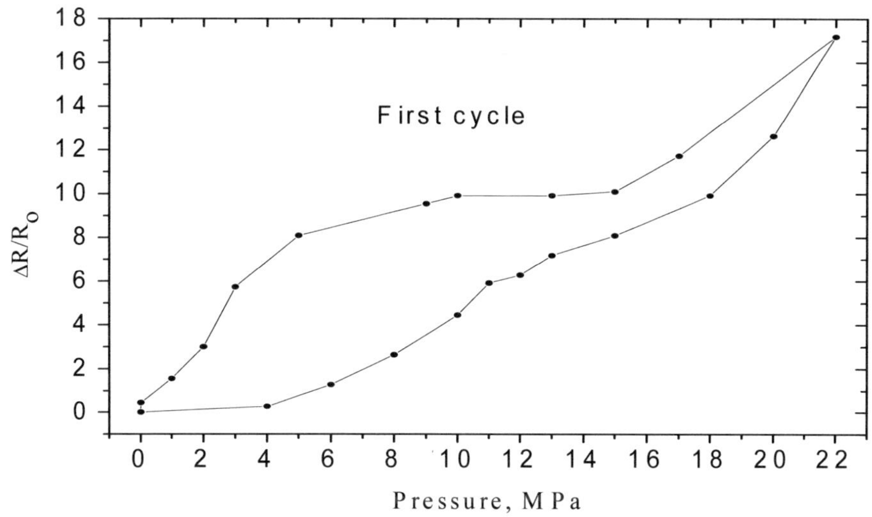

Fig. 11. Electrical resistance (in relative units) of an element (with flexible electrodes) of PHSCNC containing 10 mass parts of HSCB as function of pressure. T = 293 K.

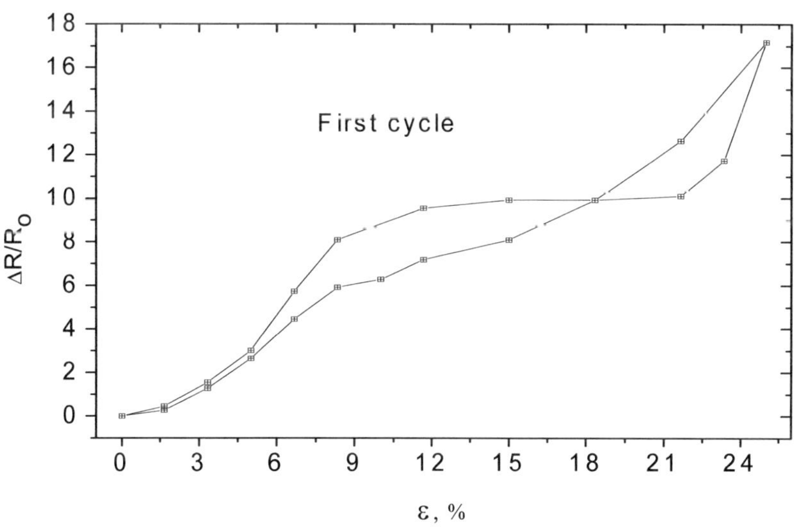

Fig. 12. Electrical resistance (in relative units) of an element (with flexible electrodes) of PHSCNC containing 10 mass parts of HSCB as function of compressive strain ε. T = 293 K.

As seen from Figures 13, 14 and 15, the electrical resistance of the sensing element of PCNT composite decreases monotonously with small uniaxial pressure and compressive strain. In this case the piezoresistance effect is considered as negative ($(\Delta R)/R_0<0$). For larger values of uniaxial pressure and compressive strain the piesoresistive effect becomes positive but compared with a PHSCNC sensing element with flexible electrodes the piezoresistance effect of the PCNT composite sensing element – the absolute value of $(\Delta R)/R_0$ (Figures 9 and 10 and Figures 11 and 12) is more than 10 times smaller. Thus, the PHSCNC is more sensitive to mechanical action than the PCNT composite. The latter exhibits a more monotonous dependence of electrical resistance under small compressive strain.
Moreover, only insignificant changes of disposition of the curve were observed during 20 cycles (Figure 15). We explain the negative piezoresistance effect by formation of new conductive channels of MWCNT under external pressure.

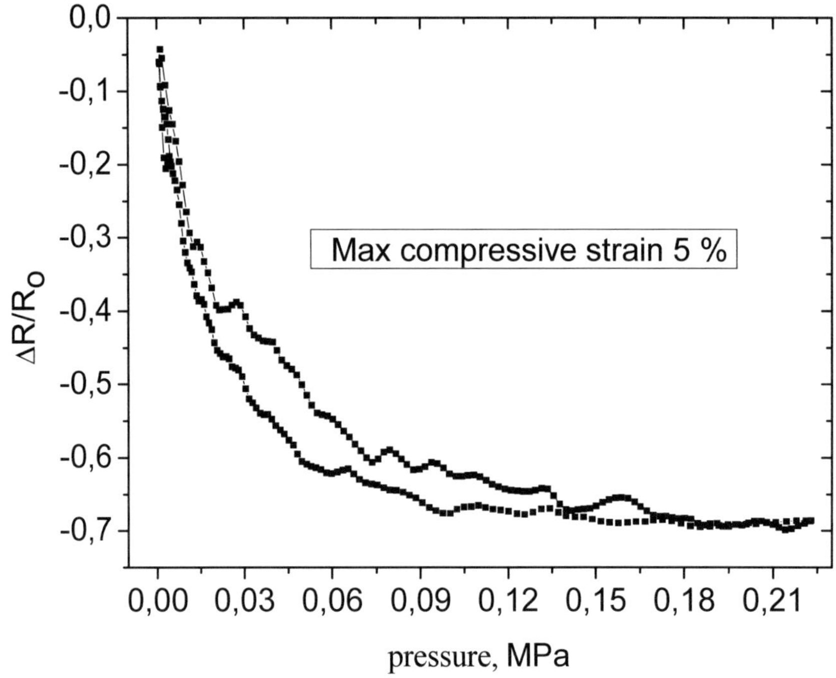

Fig. 13. Electrical resistance (in relative units) of an element (with Al electrodes) of PCNT composite containing 14.5 mass parts of MWCNT as function of pressure. T = 293 K.

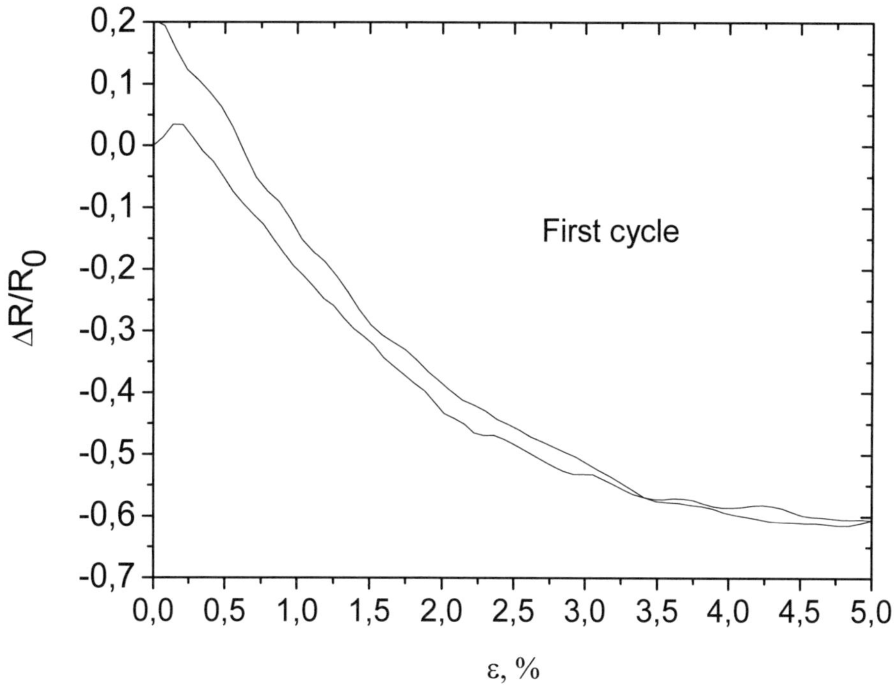

Fig. 14. Electrical resistance (in relative units) of an element (with Al electrodes) of PCNT composite containing 14.5 mass parts of MWCNT as function of compressive strain ε. T = 293 K.

Consequently, the PHSCNC could be a prospective material for pressure-sensitive indication while the PCNT composite can be considered as a prospective material for pressure sensors.

5.3 Conclusions on all-elasto-plastic polyisoprene/nanostructured carbon pressure sensing

Completely flexible sensing elements of polyisoprene - high-structured carbon black and polyisoprene - multi-wall carbon nanotube composites have been designed, prepared and examined. The first composite having a permanent drift of its mean electrical parameters is found to be a prospective material for indication of pressure change. The other composite has shown good pressure sensor properties being capable to withstand many small but completely stable and reversible piezoresistive cycles.

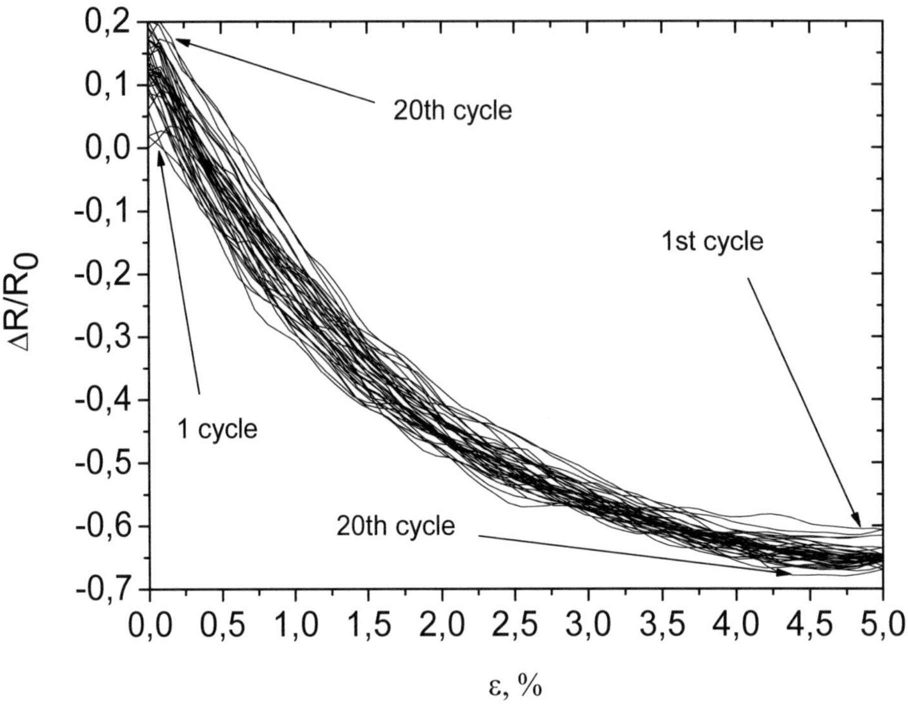

Fig. 15. Electrical resistance (in relative units) of an element (with Al electrodes) of PCNT composite containing 14.5 mass parts of MWCNT as function of compressive strain ε. 20 loading cycles. T=293 K.

6. All-elasto-plastic polyisoprene/nanostructured carbon pressure sensing element with vulcanized conductive rubber electrodes

In this paragraph our recent success in the design, processing and studies of properties of vulcanized foliated composite sensor element is reported.

6.1 Preparation of samples and organisation of experiment

The polyisoprene - nano-structured carbon black composite was made by rolling high-structure PRINTEX XE2 (DEGUSSA AG) nano-size carbon black (CB) and necessary additional ingredients (sulphur and zinc oxide) into a Thick Pale Crepe No9 Extra polyisoprene (MARDEC, Inc.) matrix and vulcanizing under 3 MPa pressure at 155 °C for 20 min. The mean particle size of PRINTEX XE2 is 30 nm, DBP absorption - 380 ml/100 g, and the BET surface area - 950 m^2/g.

The sensor element was made as follows. Two blends of polyisoprene accordingly with 30 and 10 phr (parts per hundred rubber) carbon black have been mixed. Initially 30 phr of PRINTEX have been used for obtaining PENC composite electrodes, but the tests of mechanical and electrical properties showed, that electrodes made from PENC composites with 20 phr of PRINTEX were as much conductive as 30 phr carbon black/polyisoprene electrodes but had better elasticity as well as superior adhesion to active element. Three semi-finished rounded sheets made from mentioned above two PENC composite blends have been formed and fitted onto special steel die. Those are two sheets for conductive electrodes (30 phr CB) and one sensitive sheet (10 phr CB) for pressure-sensing part. Each of these three sheets were separately pre-shaped under 3 MPa pressure and 110°C temperature to obtain disk shape. This operation lasted for 10 minutes. After that the components were cooled and cleaned with ethanol. Further, all three parts were joined together in one sensor element and were placed into the steel die and vulcanized under pressure of 3 MPa and 155° C temperature for 20 minutes vulcanization (previous attempts (Knite et al., 2008) to create sensor element with conductive glue were shown to be relatively ineffective due to later sample dezintegration). To study mechano-electrical properties small brass foil electrodes were inserted into die before vulcanization. Finally, disc shape sensor 50 mm in diameter and 3 mm thick was obtained. From this preparation we cut out useful sensor elements for testing (Figure 16). The Brass foil electrode extensions shown in this picture are necessary only to make soldered wire connection for resistivity measurements.

Fig. 16. The accomplished all-elasto-plastic sensor element with brass foil electrode extensions.

A modified Zwick/Roell Z2.5 universal testing machine, HQ stabilized power supply and a KEITHLEY Model 6487 Picoammeter/Voltage Source was used for testing mechanical and electrical properties of sensor elements. All devices were synchronized with the HBM Spider 8 data acquisition logger. Resistance R versus compressive force *F* was examined. Uniaxial pressure was calculated respectively.

6.2 Experimental results and discussion

Before testing the accomplished sensor element, we measured the electrical properties of separate vulcanized electrode layers. We also separately tested the mechano-electrical properties of vulcanized active element layer to see whether it has expected sensing capabilities. The active element of the sensor (nano-structured carbon black composite with 10 phr) belongs to the region of the percolation threshold (specific electrical resistance $\rho = 12$ $\Omega \cdot m$). The specific resistance for flexible electrodes is in the order of 0.1 $\Omega \cdot m$, which is noticeably above the percolation threshold.

Let's look closer at the conductive properties of sensors. Measurement results for electrical resistance versus pressure for small pressure range are given in Figure 17.

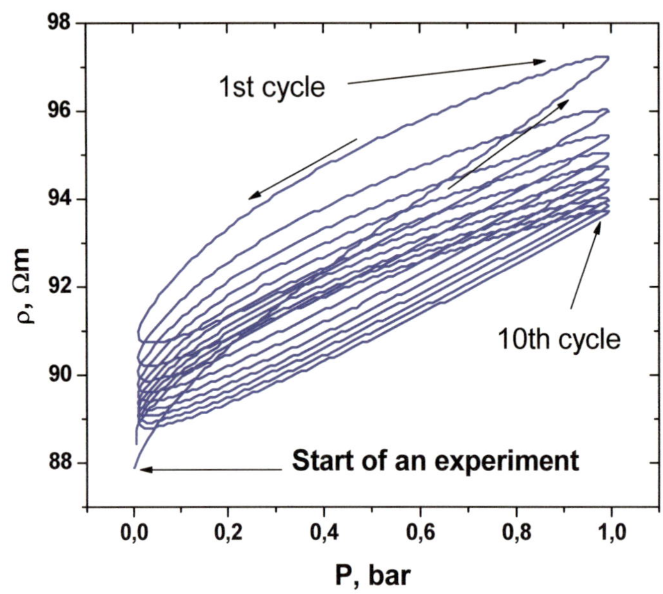

Fig. 17. Electrical resistance of the all-elasto-plastic sensor element as function of cyclic pressure (pressure range 0 to 1 bar, T = 294 ^{0}K)

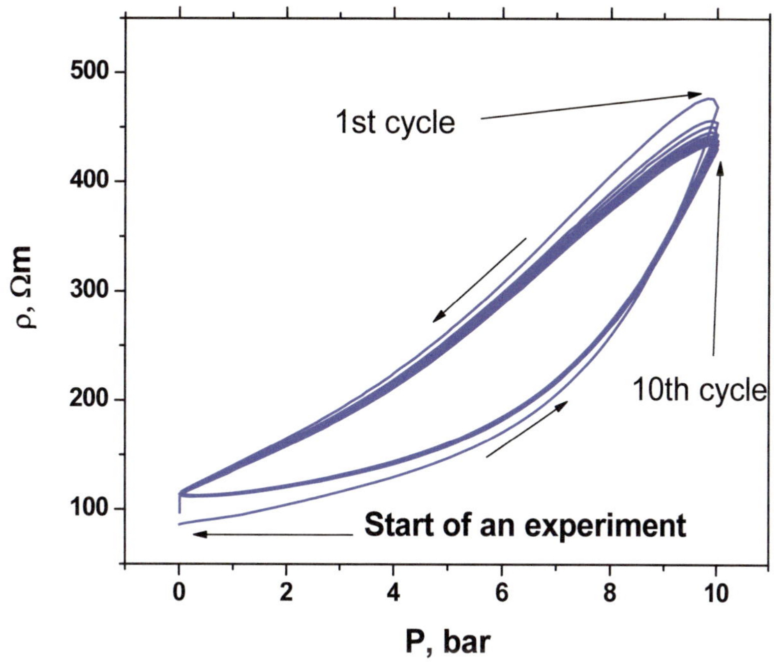

Fig. 18. Electrical resistance of the all-elasto-plastic sensor element as function of cyclic pressure (pressure range 0 to 10 bar, T = 294 ºK)

Measurement results for relatively large pressure range are shown in Figure 18. The observed positive piezoresistance effect can be explained by transverse slip of nano-particles caused by external pressure leading to disarrangement of the conductive channels.

Because of higher mobility of HSNP compared to LSNP the electro-conductive network in the elastomer matrix is easily disarranged by very small tensile, compressive or shear strain. We suppose this feature makes the elastomer–HSNP composite an option for flexible sensitive tactile elements for robots and automatics.

The scanning electron microscopy (SEM) was used to check the quality of joined regions of three PENC sheets of the AEP sensor element. SEM micrographs of fracture surface of the sensor element are shown in Figure 19. To prepare the sample for SEM investigations the sensor element was fractured in liquid nitrogen. Good joint quality of all three PENC sheets can be clearly visible in SEM images with different scales. Pale regions correspond to electrically more conductive PENC composite with 30 phr CB and dark regions cover the PENC composite with 10 phr CB. The pale particles, which are visible in the bottom picture, are carbon nano-particles.

A functional model of low-pressure-sensitive indicator was made. The block diagram of pressure indication circuit is shown on Figure 20. The sensor is connected to power supply (PS) via resistor (R) and to the input of amplifier (Amp). Transistor-based two-stage amplifier includes integrating elements. These elements are necessary to avoid noise from

induced currents and to flatten the wavefronts. The first stage amplifies the signal in linear mode. The second stage works in saturation mode. The output of the amplifier is connected to the comparator (Comp), which forms sharp wavefronts.
These signals are passed to the differential circuit and they form a sharp pulse, which is passed further to the one-shot multivibrator (OSM).
The duration of the pulse of the OSM is adjustable. The OSM is necessary to form the determined length of pulse which is independent from AEP sensor element deformation time. The output of OSM is connected to performing device PD (indicator/counter or actuator). Current setup allowed us to use AEP sensor element as a external pressure sensitive switch to temporary turn on any external electrical equipment (ambient illumination, for example), connected through our device to conventional 220V AC power source.

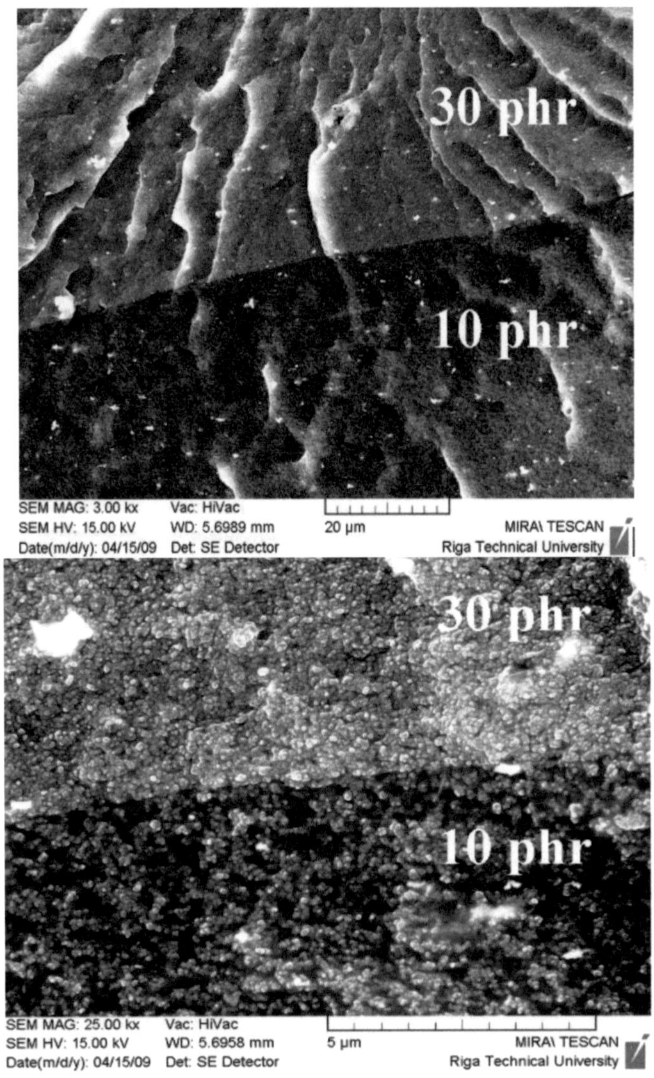

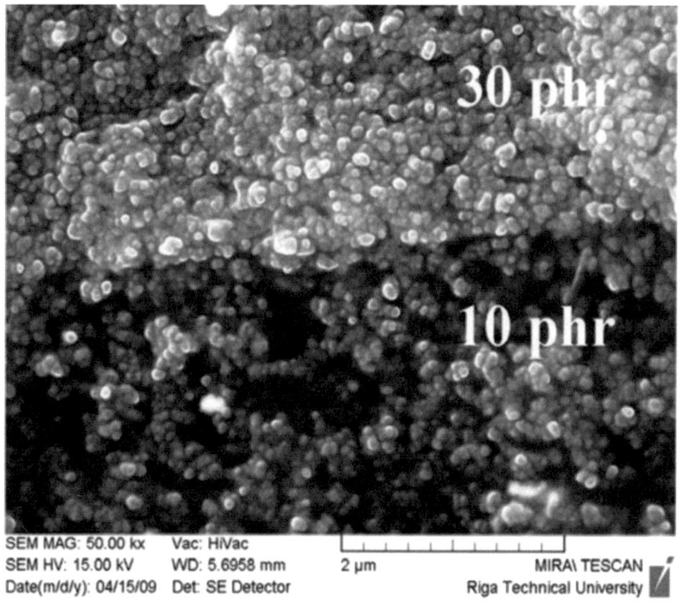

Fig. 19. SEM micrographs of sensor element. Sensor element was frozen in liquid nitrogen and then broken in two. One of the broken sides is shown in different scales: 20 μm, 5 μm and 2 μm. Boundary between two PENC composite layers with 10 and 30 phr (parts per hundred rubber) carbon black are shown.

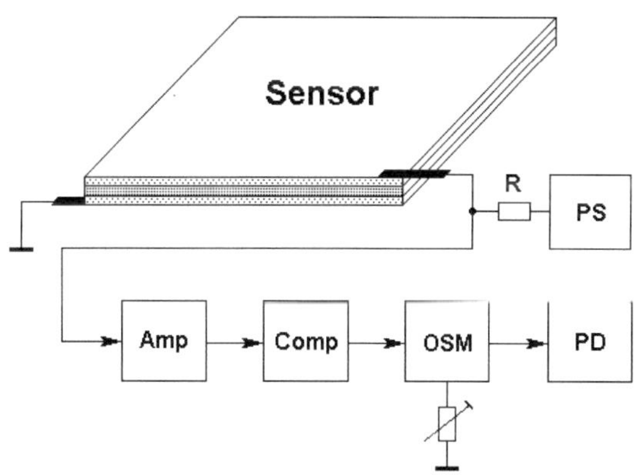

Fig. 20. Block diagram of pressure-sensitive indication circuit with completely elasto-plastic sensing element

6.3 Conclusions on all-elasto-plastic polyisoprene/nanostructured carbon pressure sensing element with vulcanized conductive rubber electrodes

Completely flexible polyisoprene - high-structured carbon black all-elasto-plastic sensing element has been designed, prepared and examined.

The sensor element was composed of two electrically conductive composite layers (electrodes) and piezoresistive PENC layer (active element) between them. A method for curing three-layer hybrid composite for pressure sensing application was developed. The

joining in-between conductive flexible electrodes and sensitive sensor material was remarkably improved. The piezoresistive behaviour of the polyisoprene/high structured carbon black has been explained by the tunnelling model.
Hybrid three-layer polyisoprene/high-structure carbon black composite has shown good pressure sensing properties. Functioning model of low-pressure-sensitive indication circuit which can turn on suitable actuator has been made.

7. Acknowledgements

The authors acknowledge R.Orlovs for electronic support, V.Teteris, J.Barloti and V.Tupureina for technical help as well as fruitful discussions and the master students G.Mallefan, S.Zike and G.Podins for assistance in carrying out some of the experiments. This work has been supported by National Program "Material Science"as well as partly by the European Social Fund within the National Programme „Support for the carrying out doctoral study programm's and post-doctoral researches" project „Support for the development of doctoral studies at Riga Technical University".

8. References

Aneli, J.N., Zaikov, G.E., Khananashvili, I.M., 1999. Effects of mechanical deformations on the structurization and electric conductivity of electric conducting polymer composites. *Journal of Applied Polymer Science*, 74: 601-621.

Balberg, I., 2002. A comprehensive picture of the electrical phenomena in carbon black-polymer composites. *Carbon*, 40: 139-143.

Barra, G.M.O., Matins, R.R., Kafer, K.A., Paniago, R., Vasques, C.T., Pires, A.T.N., 2008. Thermoplastic elastomer/polyaniline blends: evaluation of mechanical and electromechanical properties, *Polymer Testing*, 27: 886-892.

Bloor, D., Donnelly, K., Hands, P.J., Laughlin, P., Lussey, D., 2005. A metal-polymer composite with unusual properties, *Journal of Physics D: Appl.Phys.*, 38: 2851-2860.

Bokobza, L., 2007. Multiwall carbon nanotube elastomeric composites: a review, Polymer, 48: 4907-4920.

Chen, L., Chen, G., Lu, L., 2007. Piezoresistive behaviour study on finger-sensing silicone rubber/graphite nanosheet nanocomposites, *Advanced Functional Materials*, 17: 898-904.

Das, N.C., Chaki, T.K., Khastgir, D., 2002. Effect of axial stretching on electrical resistivity of short carbon fibre and carbon black filled conductive rubber composites. *Polymer International*, 51: 156-163.

Dharap, P., Li, Z., Nagarjaiah, S., Barrera, E.V., 2004. Nanotube film based on single-wall carbon nanotubes for strain sensing, *Nanotechnology*, 15: 379-382.

Dohta, S., Ban, Y., Matsushita, H., 2000. Application of a flexible strain sensor to a pneumatic rubber hand. *Proc. of 6th Triennal International Symposium on Fluid Control, Measurement and Visualization*, Canada, Sherbrooke, 87.

Farajian, A.A., Yakobson, B.I., Mizuseki, H., Kawazoe, Y., 2003. Electronic transport through bent carbon nanotubes: nanoelectromechanical sensors and switches. *Physical Review B*, 67 205423-1 – 205423-6.

Flandin, L., Brechet, Y., Cavaille, J.Y., 2001. Electrically conductive polymer nanocomposites as deformation sensors. *Composites science and technology*, 61: 895-901

Heo, J.S., Chung, J.H., Lee, J.J., 2006. Tactile sensor arrays using fiber Bragg grating sensors, *Sens.Actuator A*, 126: 312-327.

Hu, N., Yoshifumi, K. Yan, C., Masuda, Z., Fukunaga, H., 2008. Tunneling effect in a polymer/carbon nanotube composite strain sensor, *Acta Materialia*, 56: 2929-2936.

Ishigure, Y., Ijima, S., Ito, H., Ota, T., Unuma, H., Takahashi, M., Hikichi, Y., Suzuki, H., 1999. Electrical and elastic properties of conductor-polymer composites. *Journal of Materials Science*, 34: 2979-2985.

Job, A.E., Oliveira, F.A., Alves, N., Giacometti, J.A., Mattoso, L.H.C., 2003. Conductive composites of natural ruber and carbon black for pressure sensors. *Syntetic metals*, 135-136: 99-100

Knite, M., Ozols, K, Zavickis, J., Tupureina,V., Klemenoks,I., Orlovs,R., 2009. Elastomer - Carbon Nanotube Composites as Prospective Multifunctional Sensing Materials. *Journal of Nanoscience and Nanotechnology*, 9: 3587-3592.

Knite, M., Podins, G., Zike, S., Zavickis, J., Tupureina, V., 2008. Elastomer - carbon nanostructure composites as prospective materials for flexible robotic tactile sensors. In *Proc. of 5th International Conference on Informatics in Control, Automation and Robotic*, 1: 234-238.

Knite, M., Klemenok, I., Shakale, G., Teteris, V., Zicans, J., 2007. Polyisoprene-carbon nano-composites for application in multifunctional sensors, *Journal of Alloys and Compounds*, 434-435: 850-853, a.

Knite, M., Tupureina, V., Fuith, A., Zavickis, J., Teteris, V., 2007. Polyisoprene - multi-wall carbon nanotube composites for sensing strain, *Materials Science & Engineering C*, 27: 1125-1128, b.

Knite, M., Hill, A., Pas, S.,J., Teteris, V., Zavickis, J., 2006. Effects of plasticizer and strain on the percolation threshold in polyisoprene-carbon nanocomposites: positron annihilation lifetime spectroscopy and electric resistance measurements, *Materials Science & Engineering C*, 26: 771-775

Knite, M., Tupureina, V., Dzene, A., Teteris,V., Ķiploka, S., Zavickis, J., 2005. Influence of plasticizer on the improovement of strain sensing effect in polymer-carbon nano-composites, *Chemical Technology*, 36: 5-10.

Knite, M., Teteris, V., Kiploka, A., Kaupuzs , J., 2004. Polyisoprene-carbon black nanocomposites as strain and pressure sensor materials, *Sens.Actuator A*, 110: 142-149, a.

Knite, M., Teteris, V., Aulika, I., Kabelka, H., Fuith, A., 2004. Alternating-current properties of elastomer-carbon nanocomposites, *Advanced Engineering Materials*, 6: 746-749, b.

Knite, M., Teteris, V., Polyakov, B., Erts, D., 2002. Electric and elastic properties of conductive polymeric nanocomposites on macro- and nanoscales. *Materials Science & Engineering C*, 19: 5-19.

Lee, B., Roh, S., Park, J. 2009. Current status of micro- and nanostructured fiber sensors, *Optical Fiber Technology*, 15: 209-221.

Li, X., Levy, C., Elaadil, L., 2008. Multiwalled carbon nanotube film for strain sensing, *Nanotechnology*, IOP Publishing, 19: 7 pp.

Lu, J., Chen, X., Lu, W., Chen, G., 2006. The piezoresistive behaviours of polyethilene/foliated graphite nanocomposites, *European Polymer Journal*, 42: 1015-1021.

Mohanraj, G.T., Dey, P.K., Chaki, T.K., Chakraborty, A., Khastgir, D., 2007. Effect of temperature, pressure, and composition on DC resistivity and AC conductivity of conductive styrene-butadiene rubber-particulate metal alloy nanocomposites, *Polymer Composites*, DOI 10.1002/pe: 696-704.

Pham, G., Park, Y.B., Liang, Z., Zhang, C. Wang, B., 2008. Processing and modeling of conductive thermoplastic/carbon nanotube films for strain sensing, *Composites: part B*, 39: 209-216.

Roldughin, V.I, Vysotskii, V.V., 2000. Percolation properties of metal-filled polymer films, structure and mechanisms of conductivity, *Progress in Organic Coatings*, 39: 81-100.

Staufer, D. and Aharony, A., 1992. *Introduction in to percolation theory*, Washington: Taylor & Francis, 198.

Wolf, E.L., 2004. *Nanophysics and Nanotechnology*. Weinheim: WileY-VCH Verlag GmbH & KgaA, 174.

Xie, H.,Q., Ma, Y.,M., 2000. Change of conductivity of polyaniline/(styrene-butadiene-styrene) triblock copolymer composites during mechanical deformation. *Journal of Applied polymer Science*, 77: 2156-2164.

Yang, S., Chen, X., Motojima, S., 2006. Tactile sensing properties of protein-like single helix carbon microcoils, Letters to the Editor/ Carbon, 44: 3348-3378.

Yang, S., Sheng, P., 2000. *Physics and Chemistry of Nanostructured Materials*. London and New York: Taylor and Francis, 241.

Yoshimura, K., Nakano, K., Miyake, T., Hishikawa, Y., Kuzuya, C., Katsuno, T., Motojima, S. 2007. Effect of compressive and tensile strains on the electrical resistivity of carbon microcoil/silicone-rubber composites, *Carbon*, 45: 1997-2003.

Zavickis, J., Knite, M., Teteris, V., Zike, S., Tupureina, V., 2008. Polyisoprene-nanostructured carbon black composite for pressure sensors - processing and mechano-electrical properties, *Proc. of the International Scientific Conference "Material science and manufacturing technology" MITECH 2008*, Czech Republic, Prague, 235-240.

Zhang, X.W., Pan, Y., Zheng, Q., Yi X. S., 2000. Time dependence of piezoresistance for the conductor filled polymer composites. *Journal of Polymer Science B.*, 38: 2739-2749.

8

Simultaneous Localization and Mapping (SLAM) of a Mobile Robot Based on Fusion of Odometry and Visual Data Using Extended Kalman Filter

André M. Santana† and Adelardo A. D. Medeiros‡
† Federal University of Piauí – UFPI
Department of Informatics and Statistics – DIE
Teresina - Piauí – Brasil
‡ Federal University of Rio Grande do Norte – UFRN
Department of Computer Engineering and Automation - DCA
Natal – Rio Grande do Norte - Brasil

1. Introduction

The term SLAM is used as an abbreviation for Simultaneous Localization and Mapping, and was originally developed by Leonard & Durrant-Whyte (1991) based on previous work by Smith et al. (1987). In the SLAM problem, a mobile robot uses its sensors to explore the environment, gains knowledge about it, interprets the scenario, builds an appropriate map and then calculates its location relative to it. The maps can be illustrated in several ways, such as occupation grids and characteristic maps. We are interested in the second illustration. A detailed theoretical description about the topic can be found in Durrant & Bailey (2006).

In addition to perception reliability, for the general acceptance of applications, the technologies used must provide a solution at a reasonable cost, that is, the components must be inexpensive. A solution is to use optical sensors in the robots to solve environment perception problems. Due to the wide use of personal digital cameras, cameras on computers and cell phones, the price of image sensors has decreased significantly, making them an attractive option. Furthermore, the cameras can be used to solve a series of key problems in robotics and in other automatized operations, as they provide a large variety of environmental information, use little energy, and are easily integrated into the robot hardware.

The main challenges are to take advantage of this powerful and inexpensive sensor to create reliable and efficient algorithms that can extract the necessary information for the solution of problems in robotics. When cameras are used in robots as the main sensors for solving SLAM problems, the literature uses the term visual SLAM to denote this process, the objective of this study.

The major challenges in visual SLAM are: a) how to detect characteristics in images; b) how to recognize if a detected characteristic is or is not the same as one previously detected; c) how to decide if a newly detected characteristic will or will not be adopted as a new landmark; d) how to calculate the 3D position of the landmarks from 2D images; and e) how to estimate the uncertainty associated with the calculated values. In general, all of these aspects must be resolved. However, in special situations, it is possible to develop specific strategies to overcome all of these problems. This is the aim of this work.

The system that will be presented shows a visual SLAM technique equipped for flat and closed environments with floor lines. This is not a very limiting pre-requisite, as many environments such as universities, shopping malls, museums, hospitals, homes and airports, for example, have lines as floor components.

The algorithm used in visual SLAM is based on the Extended Kalman Filter (EKF), to allow the robot to navigate in an indoor environment using odometry and pre-existing floor lines as landmarks. The lines are identified using the Hough transform. The prediction phase of EKF is done using the geometric model of the robot. The update phase uses the parameters of the lines detected by the Hough transform directly in Kalman's equations without any intermediate calculation stage. Using existing lines as landmarks reduces the total complexity of the problem as follows: a) lines can be easily detected in images; b) floor lines are generally equally well spaced, reducing the possibility of confusion; c) since the number of lines in the images are not very large, each new line detected can be defined as a new landmark; d) a flat floor is a 2D surface and thus there is a constant and easy-to-calculate conversion matrix between the image plane and the floor plane, with uncertainties about 3D depth information; and e) after processing the number of pixels in the image that belong to the line is a good reliability measure of the landmark detected.

2. Related Work

Recent extensions to the overall SLAM problem have focused on the possibility of using cameras instead of sonar or laser. Examples are the works of Davison et al. (2002) , Jung (2004) and Herath et al. (2007), using stereo vision, as well as the studies conducted by Davidson et al. (2004) and Kwok et al. (2005), using a single camera.

Mansinghka (2004) presented a visual SLAM for dynamic environments using the SIFT transform and optical flow. Estrada et al. (2005) proposed a hierarchical mapping method that enables the obtaining of accurate metric maps of large environments in real time. The lower level of the map is composed of a set of local maps that are statistically independent. The upper part of the map is an adjacency graph whose arches are labeled with the relationship between the location of the local maps and a relative estimate of these local maps is kept at this level in a stochastic relation.

A solution using geometric information of the environment is proposed by Chen (2006). He reports that redundancy in SLAM may reinforce the reliability and accuracy of the characteristics observed and for this reason the geometric primitives common in indoor environments, lines and squares, for example, are incorporated into an Extended Kalman Filter (EKF) to raise the knowledge level of the characteristic observed.

Frintrop et al. (2006) introduced a new method to detect landmarks that consists of a biologically inspired attention system to detect contrasting regions in the image. This approach enables the regions to be easily redetected, thereby providing more ease of communication. Dailey & Parnichkun (2006) used stereo vision for a visual SLAM based on a particle filter. Choi et al. (2006) used an approach based on sonar information with stereo vision in an extended Kalman filter. In this work, object recognition is accomplished with Harris corners, using SIFT and RANSAC to eliminate false-positives.

Automatic recognition and a record of objects as visual landmarks are proposed by Lee & Song (2007). SIFT transform and contour algorithms are used to distinguish objects in the background of the image. When objects are detected and considered adequate for robot navigation, they are stored for later use to correct position.

Clemente et al. (2007) demonstrated for the first time that SLAM with a single camera providing input data can reach a large scale outdoors while functioning in real time. Jing Wu & Zhang (2007) conducted a study on a camera model for visual SLAM. The focus of this work is on how to model optical sensor uncertainty and how to build probabilistic components of the camera model. The deterministic component of the camera calibration process, an intrinsic parameter, is used to re-project the error. The errors are then found according to bivariate Gaussian distribution and the measure of covariance can be calculated when the characteristics are at different distances from the camera.

More recent studies in visual SLAM address several specific points. Esteban (2008) attacked the problem of illumination for omnidirectional vision. Steder et al. (2008) studied visual SLAM for aerial vehicles and Angeli et al. (2008) investigated the closed-loop problem. The latter used color information to resolve the closed-cycle problem with a voting system.

Lemaire & Lacroix (2007) proposed the use of 3D lines as landmarks. They report the following advantages of using 3D lines: first, these primitives are very numerous in indoor environments; second, in contrast to sparse point maps, which are only useful for location purposes, a relevant segmentation map provides information on the structure of the environment. Also using lines, in this case, vertical, Fu et al. (2007) carried out a study on the fusion of laser and camera information in an extended Kalman filter for SLAM. In this work, the lines are extracted from the image using Canny. Ahn et al. (2007) built a map with characteristics of 3D points and lines for indoor environments. Considering the last three studies, our approach differed by using 2D straight landmarks existing on the floor of the environment. In addition, the extraction of the characteristics is based on the Hough transform.

3. Proposed System

The system proposed in this study shows an adequate visual SLAM technique for flat and closed environments with pre-existing floor lines and is an evolution of the robot location study conducted by Santana et al. (2008).

The algorithm used in visual SLAM is based on the extended Kalman filter (EKF) to allow the robot to navigate in indoor environments using odometry and pre-existing floor lines as landmarks. The lines are identified using the Hough transform. The prediction phase of the EKF is done using the geometric model of the robot. The update phase uses the parameters of the lines detected by the Hough transform directly in the Kalman equations without any intermediate calculation stage. Figure 1 shows the scheme of the proposed system.

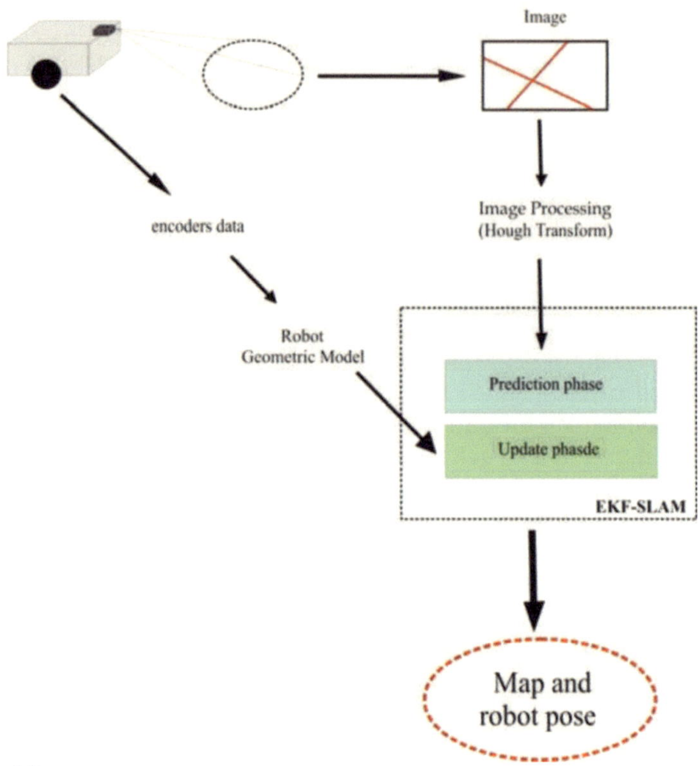

Fig. 1. Proposed System.

3.1 Theoretical Background

Extended Kalman Filter

In this work, the Extended Kalman Filter (EKF) deals with a system modeled by System (1), whose variables are described in Table (1). ε_t and δ_t are supposed to be zero-mean Gaussian white noises.

$$\begin{cases} s_t = p(s_{t-1}, u_{t-1}, \varepsilon_{t-1}) \\ z_t = h(s_t) + \delta_t \end{cases} \tag{1}$$

At each sampling time, the EKF calculates the best estimate of the state vector in two phases: a) the prediction phase uses System (2) to predict the current state based on the previous state and on the applied input signals; b) the update phase uses System (3) to correct the predicted state by verifying its compatibility with the actual sensor measurements.

$$\begin{cases} \bar{\mu}_t = p(\mu_{t-1}, u_{t-1}, 0) \\ \bar{\Sigma}_t = G_t \Sigma_{t-1} G_t^T + V_t M_t V_t^T \end{cases} \tag{2}$$

$$\begin{cases} K_t = \bar{\Sigma}_t H_t^T (H_t \bar{\Sigma}_t H_t^T + Q_t)^{-1} \\ \mu_t = \bar{\mu}_t + K_t(z_t - h(\bar{\mu}_t)) \\ \Sigma_t = (I - K_t H_t)\bar{\Sigma}_t \end{cases} \quad (3)$$

where:

$$G_t = \frac{\partial p(s,u,\varepsilon)}{\partial s}\bigg|_{s=\mu_{t-1}, u=u_{t-1}, \varepsilon=0} \quad (4)$$

$$V_t = \frac{\partial p(s,u,\varepsilon)}{\partial \varepsilon}\bigg|_{s=\mu_{t\_1}, u=u_{t-1}, \varepsilon=0} \quad (5)$$

$$H_t = \frac{\partial h(s)}{\partial s}\bigg|_{s=\mu_{t-1}} \quad (6)$$

s_t	state vector (order n) at instant t
$p(.)$	non-linear model of the system
u_{t-1}	input signals (order l), instant $t - 1$
ε_{t-1}	process noise (order q), instant $t - 1$
z_t	vector of measurements (order m) retourned by the sensors
$h(.)$	non-linear model of the sensors
δ_t	measurement noise
μ_t μ_t	mean (order n) of the state vector $\mathbf{s}_t$, before and after the update phase
Σ_t Σ_t	covariance (n x n) of the state vector $\mathbf{s}_t$
$\mathbf{G}_t$	Jacobian matrix (n x n) that linearizes the system model $p(.)$
$\mathbf{V}_t$	Jacobian matrix (n x q) that linearizes the process noise ε_t
$\mathbf{M}_t$	covariance (q x q) of the process noise ε_t
$\mathbf{K}_t$	gain of the Kalman filter (n x m)
$\mathbf{H}_t$	Jacobian matrix (m x n) that linearizes the model of the sensors $h(.)$
$\mathbf{Q}_t$	covariance matrix (m x m) of the measurement noise δ_t

Table 1. Symbols in Equations (1), (2) and (3).

EKF-SLAM

In SLAM, besides estimating the robot pose, we also estimate the coordinates of all landmarks encountered along the way. This makes necessary to include the landmark coordinates into the state vector. If ${}^{i}c$ is the vector of coordinates of the *i-th* landmark and there are k landmarks, then the state vector is:

$$s_t = \left[x_t y_t \theta_t \, {}^{1}c_t^T \ldots {}^{k}c_t^T \right]^T \quad (7)$$

When the number of marks (k) is a priori known, the dimension of the state vector is static; otherwise, it grows up when a new mark is found.

3.2 Modeling

Prediction phase: Process Model

Consider a robot with diferential drive in which $\Delta\theta_R$ and $\Delta\theta_L$ are the right and left angular displacement of the respective wheels, according to Figure 2.

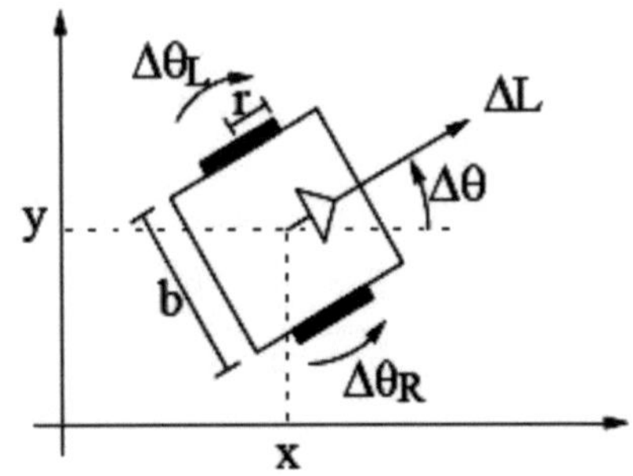

Fig. 2. Variables of the kinematic model.

Assuming that the speeds can be considered constant during one sampling period, we can determine the kinematic geometric model of the robot's movement (System 3.8):

$$\begin{cases} x_t = x_{t-1} + \dfrac{\Delta L}{\Delta\theta}\left[sin(\theta_{t-1} + \Delta\theta) - sin(\theta_{t-1})\right] \\ y_t = y_{t-1} - \dfrac{\Delta L}{\Delta\theta}\left[cos(\theta_{t-1} + \Delta\theta) - cos(\theta_{t-1})\right] \\ \theta_t = \theta_{t-1} + \Delta\theta \end{cases} \quad (8)$$

in which:

$$\begin{cases} \Delta L = (\Delta\theta_R r_R + \Delta\theta_L r_L)/2 \\ \Delta\theta = (\Delta\theta_R r_R + \Delta\theta_L r_L)/b \end{cases} \quad (9)$$

where ΔL and $\Delta\theta$ are the linear and angular displacement of the robot; b represents the distance between wheels and r_R and r_L are the radii of the right and the left wheels, respectively. When $\Delta\theta \to 0$, the model becomes the one in Equation (10), obtained from the limit of System (8).

$$\begin{cases} x_t = x_{t-1} + \Delta L\cos(\theta_{t-1}) \\ y_t = y_{t-1} + \Delta L\sin(\theta_{t-1}) \\ \theta_t = \theta_{t-1} \end{cases} \quad (10)$$

Adopting the approach advocated by Thrun et al. (2005), we consider odometric information as input signals to be incorporated to the robot's model, rather than as sensorial measurements. The differences between the actual angular displacements of the wheels ($\Delta\tilde{\theta}_R$ and $\Delta\tilde{\theta}_L$) and those ones measured by the encoders $\Delta\tilde{\theta}_R$ and $\Delta\tilde{\theta}_L$ are modeled by a zero mean Gaussian white noise, accordingly to System (11).

$$\begin{cases} \Delta\theta_R = \Delta\tilde{\theta}_R + \varepsilon_R \\ \Delta\theta_L = \Delta\tilde{\theta}_L + \varepsilon_L \end{cases} \quad (11)$$

The measured $\Delta\tilde{L}$ and $\Delta\tilde{\theta}$ are defined by replacing ($\Delta\theta_R$ and $\Delta\theta_L$) by ($\Delta\tilde{\theta}_R$ and $\Delta\tilde{\theta}_L$) in Equations (9). If the state vector s is given by Equation (7), the system model $p(\cdot)$ can be obtained from Systems (8) or (10) and from the fact that the landmarks coordinates ${}^ic$ are static:

$$p(\cdot) - \begin{cases} x_t = \cdots \\ y_t = \cdots \\ \theta_t = \cdots \\ {}^1c_t = {}^1c_{t-1} \\ \vdots \\ {}^kc_t = {}^kc_{t-1} \end{cases} \quad n = 3 \mid k \cdot \text{order}({}^ic) \tag{12}$$

$$u_t = \left[\Delta\tilde{\theta}_R \Delta\tilde{\theta}_L\right]^T \; l = 2 \tag{13}$$

$$\varepsilon_t = \left[\varepsilon_R \varepsilon_L\right]^T \; q = 2 \tag{14}$$

The **G** and **V** matrices are obtained by deriving the $p(\cdot)$ model using Equation (4) and Equation (5) respectively:

$$G = \begin{pmatrix} 1 & 0 & g_{13} & 0 & \cdots & 0 \\ 0 & 1 & g_{23} & 0 & \cdots & 0 \\ 0 & 0 & 1 & 0 & \cdots & 0 \\ 0 & 0 & 0 & 1 & \cdots & 0 \\ \vdots & \vdots & \vdots & \vdots & & \vdots \\ 0 & 0 & 0 & 0 & \cdots & 1 \end{pmatrix} \tag{15}$$

$$V = \begin{pmatrix} v_{11} & v_{12} \\ v_{21} & v_{22} \\ r_R / b & -r_L / b \\ 0 & 0 \\ \vdots & \vdots \\ 0 & 0 \end{pmatrix} \tag{16}$$

where,

$$g_{13} = \frac{\Delta\tilde{L}}{\Delta\tilde{\theta}}\left[\cos(\theta_{t-1} + \Delta\tilde{\theta}) - \cos(\theta_{t-1})\right] \quad g_{23} = \frac{\Delta\tilde{L}}{\Delta\tilde{\theta}}\left[\sin(\theta_{t-1} + \Delta\tilde{\theta}) - \sin(\theta_{t-1})\right]$$

and considering e $r_R = r_L = r$:

$$v_{11} = V_1\cos(\beta) + V_2[\sin(\beta) - \sin\theta_{t-1})] \quad v_{12} = -V_1\cos(\beta) + V_3[\sin(\beta) - \sin\theta_{t-1})]$$
$$v_{21} = V_1\sin(\beta) - V_2[\cos(\beta) - \cos\theta_{t-1})] \quad v_{22} = -V_1\sin(\beta) - V_3[\cos(\beta) - \cos\theta_{t-1})]$$
$$\beta = \theta_{t-1} + \frac{r(\Delta\tilde{\theta}_R - \Delta\tilde{\theta}_L)}{b} \quad V_1 = \frac{r(\Delta\tilde{\theta}_R + \Delta\tilde{\theta}_L)}{2(\Delta\tilde{\theta}_R - \Delta\tilde{\theta}_L)}$$
$$V_2 = \frac{-b\Delta\tilde{\theta}_L}{(\Delta\tilde{\theta}_R - \Delta\tilde{\theta}_L)^2} \quad V_3 = \frac{b\Delta\tilde{\theta}_R}{(\Delta\tilde{\theta}_R - \Delta\tilde{\theta}_L)^2}$$

It is well known that odometry introduces accumulative error. Therefore, the standard deviation of the noises ε_R and ε_L is assumed to be proportional to the module of the measured angular displacement. These considerations lead to a definition of the matrix **M** given by Equation (17).

$$M = \begin{pmatrix} (M_R|\Delta\tilde{\theta}_R|)^2 & 0 \\ 0 & (M_L|\Delta\tilde{\theta}_L|)^2 \end{pmatrix} \tag{17}$$

Update phase: Sensor Model

The landmarks adopted in this work are lines formed by the grooves of the floor in the environment where the robot navigates. The system is based on a robot with differential drive and a fixed camera, as in Figure 3.

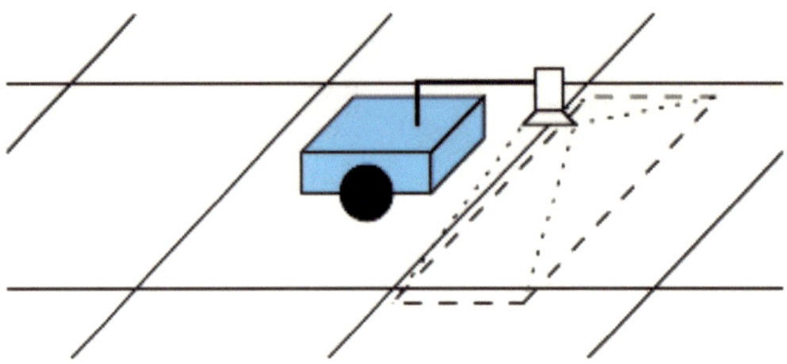

Fig. 3. Robotic system.

The landmarks are detected by processing images using the Hough transform. The detected lines are described by the parameters ρ and α in Equation (18).

$$\rho = x\cos(\alpha) + y\sin(\alpha) \tag{18}$$

Figure 4 shows the geometric representation of these parameters: ρ is the module and α is the angle of the shortest vector connecting the origin of the system of coordinates to the line.

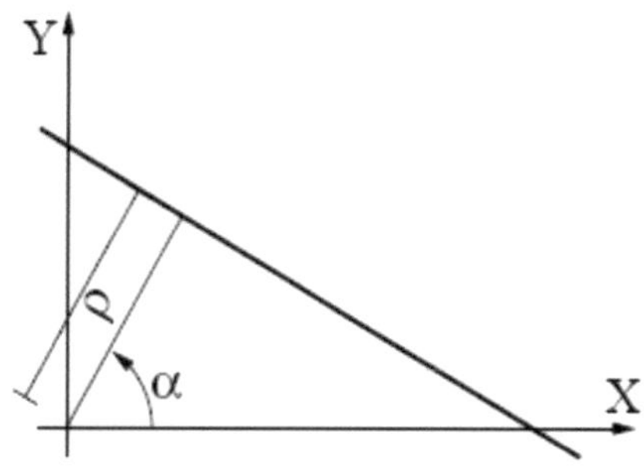

Fig. 4. Line parameters ρ and α.

We define a fixed coordinate system (F) and a mobile one (M), attached to the robot, both illustrated in Figure 5. The origin of the mobile system has coordinates (x_M^F and y_M^F) in the fixed system. θ_M^F represents the rotation of the mobile system with respect to the fixed one.

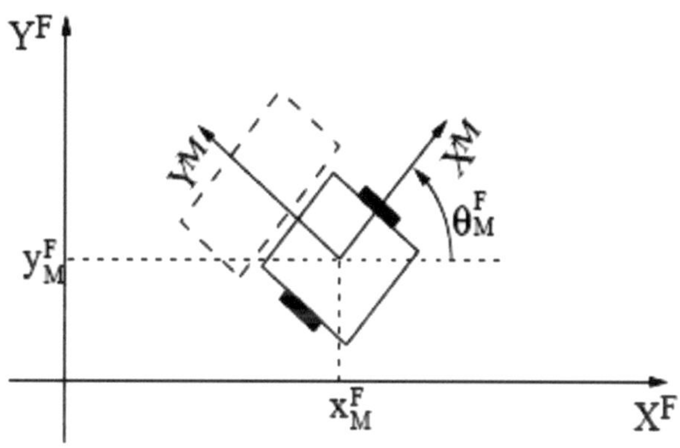

Fig. 5. Mobile and fixed coordinate systems.

One should note that there is a straight relation among these variables (x_M^F, y_M^F, θ_M^F) and the robot's pose (x_t, y_t, θ_t), which is given by Equations (19).

$$x_t = x_F^M \quad y_t = y_F^M \quad \theta_t = \theta_F^M + \pi / 2 \tag{19}$$

Each line on the floor is described by two static parameters ($\tilde{\rho}, \tilde{\alpha}$). The map to be produced by the SLAM process is composed of a set of these pairs of parameters. So, the ${}^{i}c$ vector of coordinates of the *i-th* landmark that appears in Equations (7) and (12) is given by Equation (20):

$$^{i}c = \begin{bmatrix} ^{i}\rho^{F} \\ ^{i}\alpha^{F} \end{bmatrix} \tag{20}$$

At each step the robot captures an image and identifies the parameters ($\tilde{\rho},\tilde{\alpha}$) of the detected lines. These image parameters are then converted to the corresponding parameters ($\tilde{\rho}^M,\tilde{\alpha}^M$) in the mobile coordinate system *M* attached to the robot, using the camera parameters. The vector of measurements $\mathbf{z}_t$ to be used in the update phase of the EKF algorithm (Equation 3) is defined by Equation (21):

$$z_t = \begin{bmatrix} \tilde{\rho}^M \\ \tilde{\alpha}^M \end{bmatrix} \tag{21}$$

To use information directly obtained by image processing ($\tilde{\rho}^M,\tilde{\alpha}^M$) in the update phase of the EKF-SLAM, one must deduct the sensor model *h(.)*, that is, the expected value of these parameters in function of the state variables.
We use the relation between coordinates in the *M* and *F* systems (Sistem 22) and Equation (18) in both coordinate systems (Equations 23 and 24):

$$\begin{cases} x^F = \cos(\theta_F^M)x^M - \sin(\theta_F^M)y^M + x_M^F \\ y^F = \sin(\theta_F^M)x^M + \cos(\theta_F^M)y^M + x_M^F \end{cases} \tag{22}$$

$${}^i\rho^F = x^F\cos({}^i\alpha^F) + y^F\sin({}^i\alpha^F) \tag{23}$$

$$\rho^M = x^M\cos(\alpha^M) + y^M\sin(\alpha^M) \tag{24}$$

By replacing Equations (22) in Equation (23), doing the necessary equivalences with System (24) and replacing some variables using Equations (19), we obtain the Systems (25) and (26), which represent two possible sensor models *h(.)* to be used in the filter. To decide about which model to use, we calculate both values of α^M and use the model which generate the value closer to the measured value $\tilde{\alpha}^M$.

$$\begin{cases} \rho^M = {}^i\rho^F - x_t\cos({}^i\alpha^F) - y_t\sin({}^i\alpha^F) \\ \alpha^M = {}^i\alpha^F - \theta_t + \pi/2 \end{cases} \tag{25}$$

$$\begin{cases} \rho^M = -{}^i\rho^F + x_t\cos({}^i\alpha^F) + y_t\sin({}^i\alpha^F) \\ \alpha^M = {}^i\alpha^F - \theta_t - \pi/2 \end{cases} \tag{26}$$

The sensor model is incorporated into the EKF through the **H** matrix (Equation 6), given by Equation (27):

$$H = \begin{pmatrix} -\cos^i\alpha^F & -\sin^i\alpha^F & 0 & \cdots & 1 & 0 & \cdots \\ 0 & 0 & -1 & \cdots & 0 & 1 & \cdots \end{pmatrix} \tag{27}$$

The final columns of the H matrix are almost all null, except for the columns corresponding to the landmark in the vector state that matchs the detected line on the image.

3.3 Matching

A crucial aspect of the SLAM algorithm is to establish a match between the detected line on the image and one of the landmarks represented in the state vector. To choose the correct landmark, we first calculate the predicted values of (ρ^F,α^F) using the measured values of ($\tilde{\rho}^M,\tilde{\alpha}^M$) and the model in Equations (25), if $\tilde{\alpha}^M \geq 0$, or in Equations (26), if $\tilde{\alpha}^M < 0$. Then, these predicted values are compared to each one of the values $({}^{i}\rho^{F}, {}^{i}\alpha^{F})$ in the state vector. If the difference between the predicted values and the best $({}^{i}\rho^{F}, {}^{i}\alpha^{F})$is small enough, a match was found. If not, we consider that a new mark was detected and the size of the state vector is increased.

3.4 Image processing

Detection of lines

Due to the choice of floor lines as landmarks, the technique adopted to identify them was the Hough transform (Gonzalez & Woodes, 2007). The purpose of this technique is to find imperfect instances of objects within a certain class of shapes by a voting procedure. This voting procedure is carried out in a parameter space, from which object candidates are obtained as local maxima in a accumulator grid that is constructed by the algorithm for computing the Hough transform. In our case, the shapes are lines described by Equation (18) and the parameter space has coordinates (ρ , α). The images are captured in grayscale and converted to black and white using a threshold level, determined off-line. Figure 6 shows a typical image of the floor and the lines detected by the Hough transform.

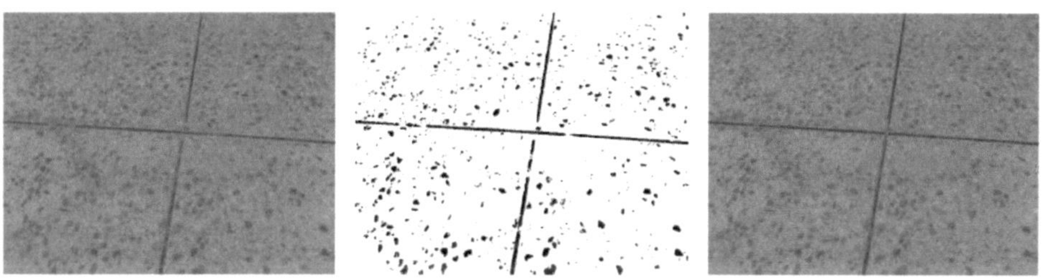

Fig. 6. Detection of lines.

From images to the world

We assume that the floor is flat and that the camera is fixed. So, there is a constant relation (a homography **A**) between points in the floor plane *(x , y)* and points in the image plane *(u , v)*:

$$s \cdot \begin{pmatrix} u \\ v \\ 1 \end{pmatrix} = A \cdot \begin{pmatrix} x \\ y \\ 1 \end{pmatrix} \tag{28}$$

The scale factor s is determined for each point in such a way that the value of the third element of the vector is always 1. The homography can be calculated off-line by using a pattern containing 4 or more remarkable points with known coordinates (see Figure 7). After detecting the remarkable point in the image, we have several correspondences between point coordinates in the mobile coordinate system M and in the image. Replacing these points in Equation (28), we obtain a linear system with which we can determine the 8 elements of the homograpy matrix **A**.

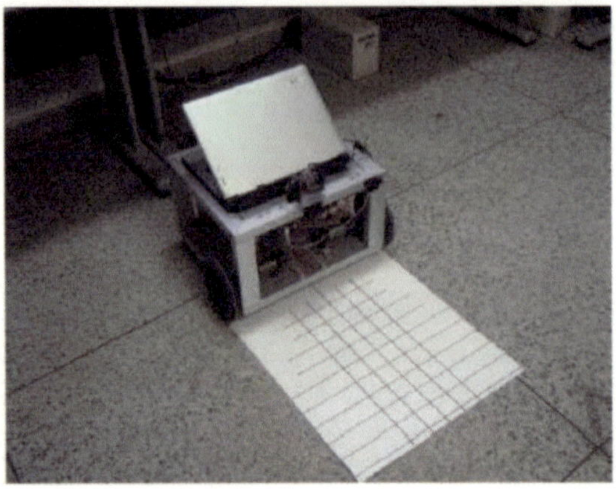

Fig. 7. Calibration pattern.

Once calculated the homography, for each detected line we do the following:

1. using the values of (ρ, α) in the image obtained by the Hough transform, calculate two point belonging to the image line;
2. convert the coordinates of these two points to the mobile coordinate system *M*;
3. determine $(\tilde{\rho}^M, \tilde{\alpha}^M)$ of the line that passes through these two points.

4. Results

The experiments were carried out using the Karel robot, a reconfigurable mobile platform built in our laboratory (see Figure 7). The robot has two wheels that are driven by DC motors with differential steering. Each motor has an optical encoder and a dedicated card based on a PIC microcontroller that controls local velocity. The cards communicate with the computer through a CAN bus, receiving the desired wheel velocities and encoder data.

The computer that controls the robot is a notebook with a USB-to-CAN bridge and a color webcam connected to its USB bus. The camera captures 160 x 120 images (Figure 6) and each image is processed at 300 ms.

The experiment was conducted in an environment with pre-existing floor lines and proceeded as follows: the robot executes a forward movement, turns around and tries to return to the departure point.

Figure 8 shows the position of the lines, the approximate trajectory of the robot, the comparative size of the camera's field of view and the "name" of each straight line. The initial position of the robot is approximately (3, 3.1, 110°), but it is not used in the SLAM algorithm: the robot assumes that its initial position is (0, 0, 0°).

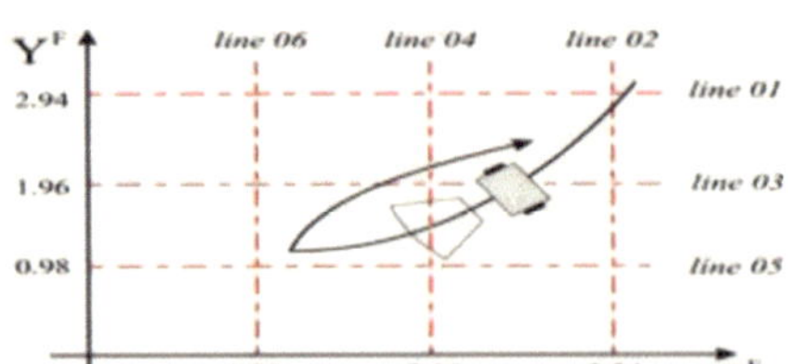

Fig. 8. Description of the experiment performed.

The name of the lines is related to the temporal order in which the robot identified them. Figure 9 shows the moment of identification of each straight line. Initially, the robot identifies two lines (Figure 9-a); advancing it identifies the third (Figure 9-b) and a little further on identifies the fourth (Figure 9-c). The fifth line is only seen when the robot has already negotiated around half of the trajectory (Figure 9-d) and the sixth is only identified when the robot begins to rotate on its own axis (Figure 9-e)..

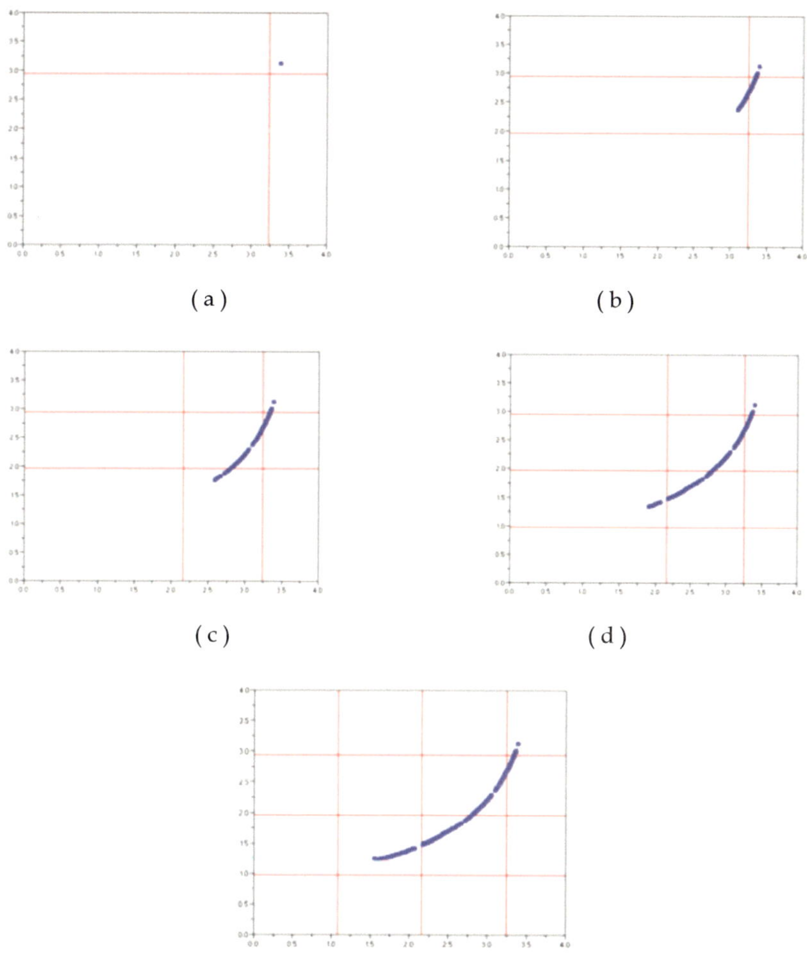

Fig. 9. Temporal process of identifying the floor lines.

Considering the total movement of the robot, we found that it identifies lines 56% of the time (one line 65% of the time and two lines 35% of the time), as shown in Figure 10. This figure shows that at a particular instant the robot identified three lines. This incorrect information originated in an image where the shadow of an obstacle was identified as a line.

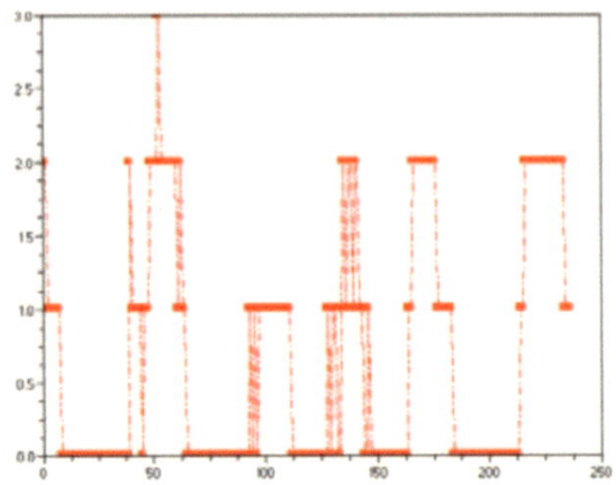

Fig. 10. Temporal process of identifying floor lines.

The process of matching all the experience was successful in 93% of the cases, as illustrated in Table 2. Figure 11 shows the straight lines identified during the robot's trajectory.

Lines	*Identifications*	*Correct*	*Incorrect*	*Hit rate*
Line 01	2	2	0	100%
Line 02	29	28	1	96,50%
Line 03	44	41	3	93,20%
Line 04	34	32	2	94,10%
Line 05	23	22	1	95,60%
Line 06	13	11	2	84,65%

Table 2. Matching.

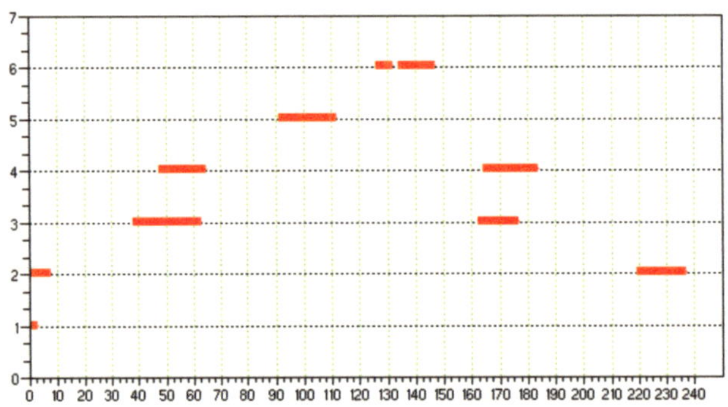

Fig. 11. Straight lines identified during the robot's trajectory.

Based on line 3, the line most seen during the experiment, Figure 12 shows the behavior of the p and α variance for this line. It can be seen that the behavior is in accordance with the theory and is always minimized when it and other lines are observed.

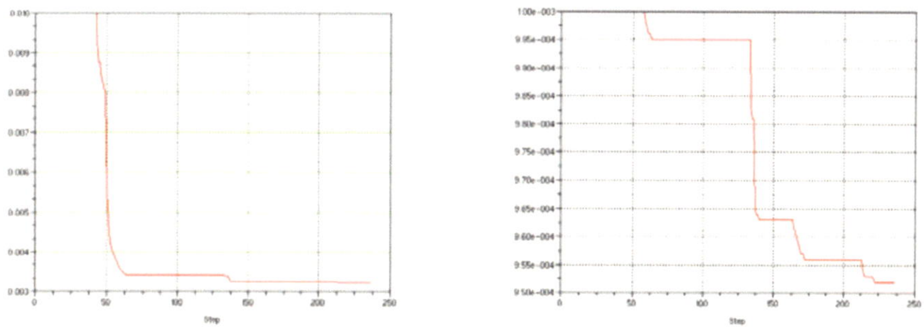

(a) (b)

Fig. 12. Variance behavior: (a) p of line 3 and (b) a of line 3.

The coordinates x, y and θ and their respective variances are shown in Figure 13..

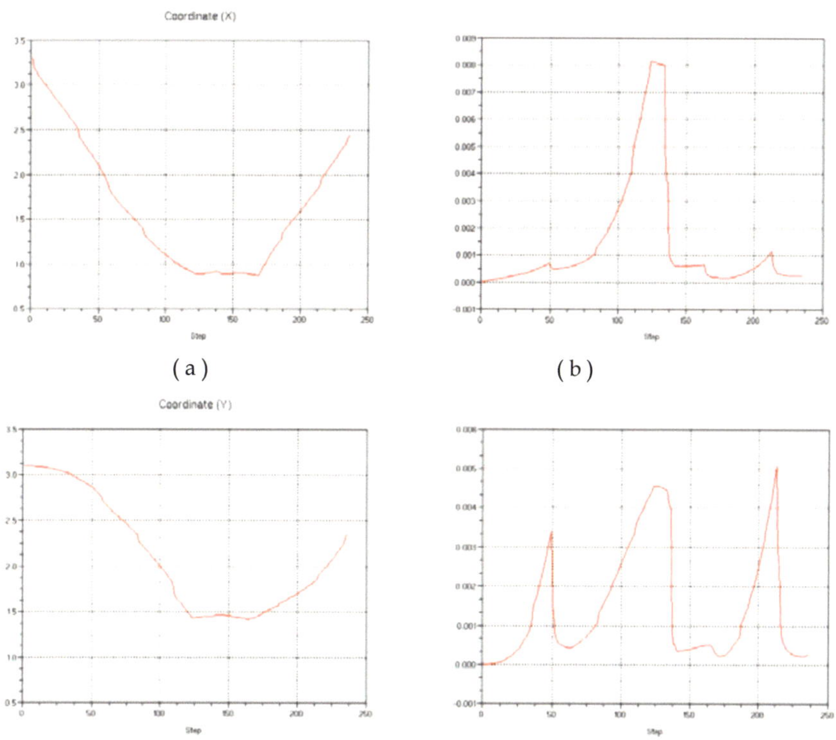

(a) (b)

(c) (d)

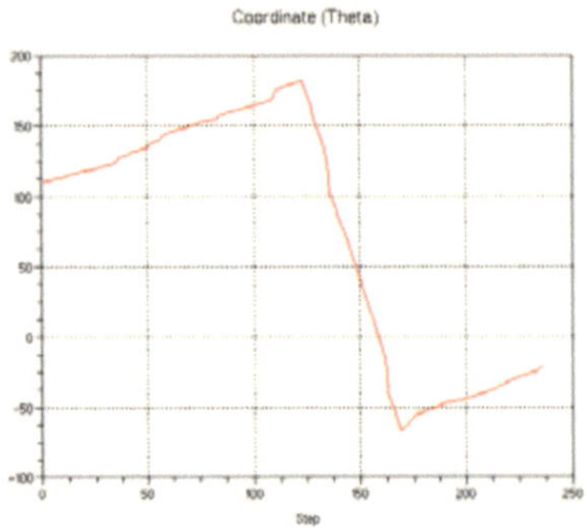

(e)

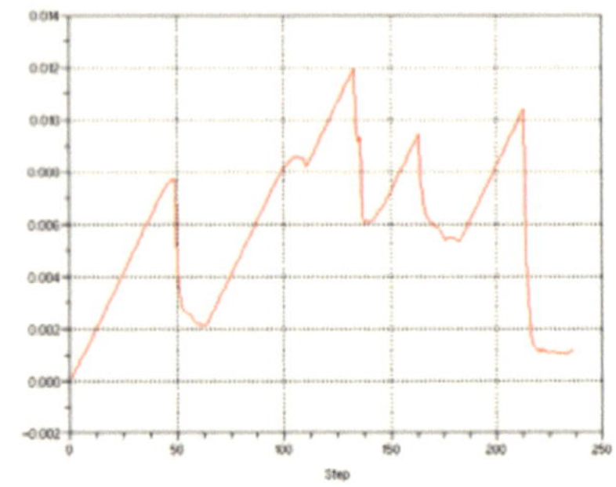

(f)

Fig. 13. (a) coordinate X, (b) variance in X, (c) coordinate Y, (d) variance in Y, (e) coordinate θ, (f) variance in θ.

Table 2 shows the values of the landmarks detected after around 250 iterations. The values are expressed in the system of coordinates, assuming that the initial position of the robot is (0, 0, 0°).

Actual		*Calculated*		*Corrected*	
ρ	α	ρ	α	ρ	α
1.08	0	1.88	108	1.13	1.7
2.16	0	0.82	108	2.19	1.3
3.24	0	0.11	110	3.26	0
0.98	90	2.57	23	0.97	87
1.96	90	1.24	22	1.95	87
2.94	90	0.20	22	2.93	88

Table 2. Feature map.

5. Conclusions and Perspectives

This chapter presents a new perspective for solving SLAM problems using optical sensors, given that they are increasingly cheaper and provide a wealth of information on the environment.

The main contribution of this work is an optical sensor modeling that enables the use of parameters obtained by the image processing algorithm directly in the Kalman filter equations without intermediate phases to calculate position or distance.

The values presented in Table 2 demonstrate that the proposed approach can obtain good results even with a relatively small number of samples.

As future works, we intend to improve the real-time properties of the image processing algorithm, by adopting some of the less timeconsuming variants of the Hough transform. Another required improvement is to deal with line segments with finite length, incorparating this characteristics to the step of matching lines.

6. References

Ahn, S.; Chung, W. K. & Oh, S. (2007), Construction of Hybrid Visual Map for Indoor SLAM, *IEEE/RSJ International Conference on Intelligent Robots and Systems* .

Angeli, A.; Filliat, D.; Doncieux, S. & Meyer, J. (2008), Fast and Incremental Method for Loop-Closure Detection Using Bags of Visual Words, *IEEE Transaction on Robotics,* Vol. 24, No. 5, pp. 1027-1037.

Chen, Z. (2006), A Visual SLAM Solution Based on High Level Geometry Knowledge and Kalman Filtering, *IEEE CCECE/CCGEI,* pp. 1283-1286 .

Choi, Y. & Oh, S. (2006), Grid-based Visual SLAM in Complex Environment, *IEEE/RSJ International Conference on Intelligent Robots and Systems* .

Clemente, L.; Davison, A.; Reid, I.; Neira, J. & Tardos, J. (2007), Mapping Large Loops with a Single Hand-Held Camera', *Proceedings of Robotics: Science and Systems* .

Dailey, M. N. & Parnichkun, M. (2006), Simultaneous Localization and Mapping with Stereo Vision', *International Conference on Control, Automation, Robotics and Vision*.

Davison, A. J. & Murray, D. W. (2002), Simultaneous Localization and Map-Building Using Active Vision, *IEEE Transactions on Pattern Analysis and Machine Intelligence,* pp. 865-880.

Davison, A. J; Cid, Y. G. & Kita, N. (2004), Real-time 3d SLAM with Wide-Angle Vision, *I5th IFAC/EURON Symposium on Intelligent Autonomous Vehicles.*

Durrant-Whyte, H. & Bailey, T. (2006). Simultaneous Localization and Mapping: Part I, *IEEE Robotics & Automation Magazine,* pp. 99-108.

Esteban, I. (2008), Visual Trajectory Based SLAM, *Master Thesis,* Universiteite van Amsterdan.

Estrada, C.; Neira, J. & Tards, J. D. (2005), Hierarchical SLAM: Realtime Accurate Mapping of Large Environments, *IEEE Transactionson Robotics,* pp. 588-596.

Frintrop, S.; Jensfelt, P. & Christensen, H. I. (2006), Attentional Landmark Selection for Visual SLAM, *IEEE/RSJ Int. Conference on Intelligent Robots and Systems* .

Fu, S.; Liu, C.; Gao, L. & Gai, Y. (2007), SLAM for Mobile Robots Using Laser Range Finder and Monocular Vision, *IEEE Int. Conference on Robotics and Automation* .

Gonzalez, R. C. & Woodes, R. E. (2007). *Digital Image Processing,* Prentice Hall, ISBN: 13168728X, USA.

Herath, D. C.; Kodagoda, K. R. S. & Dissanayake, G. (2007), Stereo Vision Based SLAM Issues and Solutions, *IEEE International Conference on Robotics and Automation* .

Jing Wu, J. & Zhang, H. (2007), 'Camera Sensor Model for Visual SLAM, *Conference on Computer and Robot Vision* .

Jung, I. K. (2004), Simultaneous Localization and Mapping in 3D Environments with Stereovision, *PhD Thesis,* Institut National Polytechnique - Toulouse - France.

Kwok, N. M.; Dissanayake, G. & Ha, Q. P. (2005), Bearing-only SLAM Using a Sport Based Gaussian Sum Filter, *IEEE International Conference on Robotics and Automation* .

Lemaire, T. & Lacroix, S. (2007), Monocular SLAM as a Graph of Coalesced Observations, *IEEE Int. Conference on Robotics and Automation,* pp.2791-2796 .

Lee, Y. & Song, J. (2007), Autonomous selection, Registration, and Recognition of Objects for Visual SLAM in Indoor Environments', *International Conference on Control, Automation and Systems* .

Leonard, J. J. & Durrant-Whyte, H. F. (1991), Mobile Robot Localization by Tracking Geometric Beacons, *IEEE Transactions on Robotics and Automation,* Vol. 1, No. 3, pp. 376-382 .

Mansinghka, V. K. (2004), Towards Visual SLAM in Dynamic Environments, *Springer* .

Steder, B.; Grisetti, G.; Stachniss, C. & Burgard, W. (2008), Visual SLAM for Flying Vehicles, *Academic Publish* .

Santana, A. M.; de Souza, A. A. S.; Britto, R. S.; Alsina, P. J.; Medeiros, A. A. D. Localization of a Mobile Robot Based on Odometry and Natural Landmarks Using Extended Kalman Filter. *International Conference on Informatics in Control, Automation and Robotics*. Funchal, Madeira, Portugal, 08/2008.

Smith, R. & Cheesman, P. (1987), On the Representation of Spatial Uncertainty, *The International Journal of Robotics Research* , Vol. 5, No. 4, pp. 56-68 .

Thrun, S.; Burgard, W. & Fox, D. (2005). *Probabilistic Robotics,* MIT Press, ISBN: 9780262201629, USA.

9

Distributed Estimation of Unknown Beacon Positions in a Localization Network

Mikko Elomaa and Aarne Halme
Helsinki University of Technology
Finland

1. Introduction

The localization of a mobile robot is of utmost importance for efficient task execution. The field has been well studied and many different methods have been proposed. These methods can be roughly divided into two categories: methods relying only on robot's own sensors and methods using external infrastructure such as beacons, landmarks, etc.

The most common approach to the localization of a mobile robot is the method called Simultaneous Localization and Mapping (SLAM). The idea of SLAM is to build a map by observing static features in the surrounding environment. In order to position the features correctly with respect to each other, the location of the sensors has to be estimated when making the observations. Same sensors are used for both mapping of the features and localization of the robot, thus the uncertainties in both mapping and pose estimate are correlated. The complexity of the SLAM problem has been shown to be $O(N^2)$ with N being the number of landmarks in the map (Nerurkar and Roumeliotis, 2007). Several different variations of the SLAM method exist to better address different problems. A solution based on extended Kalman filter was introduced already two decades ago (Smith et al., 1990). Newman and Ho address the loop-closing problem by combining observations of visual features with laser scanner data (Newman and Ho, 2005). In (Williams et al., 2002) submaps are used in order to keep computational burden from increasing over time. Fast SLAM algorithms have been developed to cope with a large amount of landmarks (Montemerlo, 2002) or real-time constraints (Montemerlo et al., 2003).

A robot relying only on its own sensors, such as wheel encoders, gyroscopes, laser scanner, etc., can work in environments without any external infrastructure. However, when moving in an unknown environment, the uncertainty in the robot's position estimate increases over time because the robot does not have any absolute reference for its position.

The benefit of using external infrastructure is that a robot can always estimate its position within bounded error. The robot measures its position relative to a known location, e.g. a landmark, and the resulting position estimate does not depend on the history of the robot's trajectory. The error in the robot's position estimate only depends on the measurement error and the error in the position data of the landmark. Thus the error in robot's position estimate is bounded by the limits of measurement error and landmark position error.

A well-known example of a localization system with external infrastructure is the Global Positioning System (GPS). Whenever a GPS device is turned on, it can calculate its pose in the global coordinate frame provided that at least three satellites (beacons) are visible. However, in an indoor environment the metallic structures of the building usually prohibit the use of GPS for localization. Similar kind of localization systems can be built by placing beacons inside the building. A commonly used approach is to measure the distance between robot and beacons with ultrasound sensors (Priyantha et al., 2000).
The major drawbacks of external infrastructure are deployment/maintenance costs and the need for initial localization of each beacon before it can be used. Our key interest is in developing a distributed calibration method for a localization network. The localization network is formed by sparsely deploying inexpensive beacons serving as landmarks.

2. Cooperative approach

The simplest approach to cooperative localization is to have a pair of robots moving together and improving their individual pose estimates using relative measurements between the robots. If one robot can measure its distance and relative bearing angle to a target robot, it can estimate the position of the target robot relative to its own pose. If the robot can also measure the relative heading angle of the target robot, it can estimate the relative pose of the target robot. An example of this kind of approach is a robot with a laser scanner tracking another robot with a three plane target mounted on it (Rekleitis et al., 2003). This concept can be extended to a group of robots, in which at least one robot has a suitable sensor for tracking the other robots (Kurazume et al., 1994; Howard et al., 2003; Spletzer and Taylor, 2003). Another approach based on relative measurements in a group mimics formation flying. Robots drive in a formation while each robot detects changes in the poses of the other robots. The shape of the formation affects considerably the accumulated uncertainty in the robot poses (Hidaka et al., 2005; Andersson and Nygards, 2008).
The strict constraint of robots staying visible for the sensors of the other robots may be too restricting in some scenarios. Then the robots can explore independently but can share localization information only when they meet. This method is effective if at least one of the robots sharing data has been able to keep the position uncertainty fairly low. (Howard et al., 2004; Chang et al., 2007; Zhou and Roumeliotis, 2006). Some approaches use only the landmarks on the overlapping parts of the local maps constructed by the different robots in order to align the maps (Thrun and Liu, 2003; Martinez-Cantin et al., 2007). With this kind of algorithm each robot has to have a suitable localization system in order to build a sufficiently accurate local map.
Our approach uses a group of simple and inexpensive robots for distributed localization of landmarks that form a localization network. Each robot only needs wheel encoders and one sensor that can identify a landmark and measure a bearing angle to the landmark. With these sensors, the robots should be able to estimate the positions of the landmarks and simultaneously localize themselves within bounded error. The landmarks can be deployed sparsely because only one landmark needs to be visible for our algorithm to work. When no landmarks are visible, the robot relies on its odometry sensors. For cooperation between the robots, exact identification of each landmark is required. This can be done with artificial landmarks having unique identification numbers as explained in the case study at the end of this chapter.

As the landmark positions are unknown at the beginning, the robots have to work together in order to correctly localize the landmarks. At this stage, the localization problem is closely related to the SLAM problem. When the robots are configuring the system and localizing the landmarks, the main source of error is the accumulated odometry error which, in a group of robots, is assumed to have a mean close to zero. Thus, when the location estimates of several independent robots on a common object are combined, the location estimate of the object converges towards the correct position. The fusing of information in cooperative localization can be based on Kalman filter.

3. Theory

In this subchapter the robot's sensors for landmark detection are assumed to be located at the origin of the robot's coordinate frame. The positive y-axis of the robot coordinate frame points forward marking the direction of the heading angle, θ. The heading angle of the robot in global coordinate frame is expressed in compass format i.e. the heading angle increases when the robot turns clockwise. The zero heading angle is fixed towards the direction of the positive y-axis. The bearing angle measurements (λ) are expressed as an angle relative to the robots heading angle. The bearing angle increases towards clockwise direction.

3.1 Location estimate

If the robot is equipped with sensors to measure both distance and heading to a landmark, the estimate of the landmark location can be calculated after each measurement. The distance is measured directly and the absolute bearing angle can be calculated with equation 3. The relative displacement in global coordinate frame can then be solved with equation 4. However, if the robot can only measure the bearing angle to the landmark, an estimate can be calculated for the landmark location provided that the robot has basic odometry sensors. In bearing-only localization, the robot is required to make at least two bearing angle measurements to the landmark from different locations in order to calculate an estimate of the landmark's location.

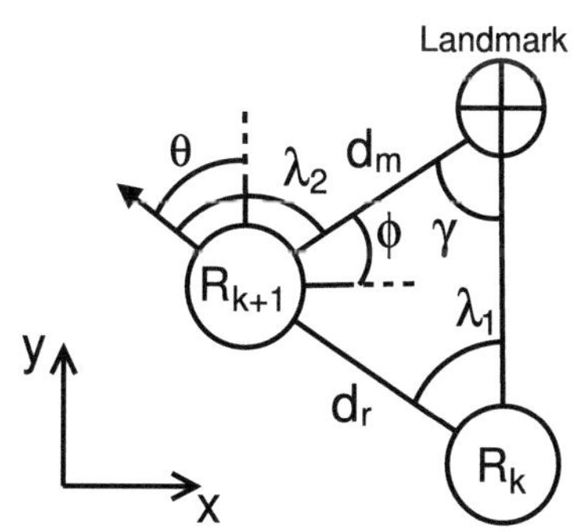

Fig. 1. Landmark localization

Figure 1 shows a robot making two bearing angle measurements at locations R_k and R_{k+1}. The distance d_r between the two locations is measured with wheel encoders. The two bearing angle measurements λ_1 and λ_2 are measured with a suitable sensor. The distance d_m can be solved as follows:

$$\gamma = \lambda_2 - \lambda_1 \tag{1}$$

$$d_m = \frac{d_r \cdot \sin(\lambda_1)}{\sin(\gamma)} \tag{2}$$

The estimated heading angle of the robot, θ, can be based on wheel encoders if no additional sensor data is available. The robot's heading angle together with a bearing angle measurement yield an azimuth angle ϕ.

$$\phi = \frac{\pi}{2} - \theta - \lambda_2 \tag{3}$$

The distance estimate d_m and the azimuth angle estimate ϕ define the displacement of the landmark in polar coordinates relative to the origin of the robot's local coordinate frame. Hence, the displacement expressed in the global Cartesian coordinates can be calculated as follows:

$$\begin{aligned} dx_m &= d_m \cdot \cos(\phi) \\ dy_m &= d_m \cdot \sin(\phi) \end{aligned} \tag{4}$$

The calculated displacement estimates in x- and y-coordinates have uncertainties, which depend on the relative distance d_m, azimuth angle ϕ and the angle difference γ between the two bearing angles. Equation 5 shows a general method for calculating the uncertainties in the landmark coordinates. C_1 and C_2 are constant parameters depending on the accuracy of the measurement system.

$$\begin{aligned} \sigma &= \phi + \frac{\lambda_1 - \lambda_2}{2} \\ r_x &= C_1 \cdot d_m \cdot |\sin(\phi)| + C_2 \left(\left(\frac{\cos(\sigma)}{\sin^2(\gamma)} \right)^2 + \sin^2(\sigma) \right) \\ r_y &= C_1 \cdot d_m \cdot |\cos(\phi)| + C_2 \left(\left(\frac{\cos(\sigma + \pi/2)}{\sin^2(\gamma)} \right)^2 + \sin^2(\sigma + \pi/2) \right) \end{aligned} \tag{5}$$

Each of the uncertainties for x- and y-coordinates consists of two parts. The first part deals with the effect of the error in the azimuth angle. The longer the distance d_m, the bigger the position error caused by the azimuth angle error will be. Also, the weight of the uncertainty in the coordinate axes, x and y, depend on the azimuth angle. The second part deals with the actual measurement uncertainty. Each bearing angle measurement has some angle uncertainty, forming a sector in which the landmark probably lies. Two bearing angle measurements form two sectors which intersect to create a diamond shaped region where the landmark probably lies. For simplicity, this region can be presented as an ellipse with its origin at the intersection of the lines representing the bearing angle measurements and with its major axis at angle σ with respect to x-axis (Figure 2). The uncertainties for x- and y-coordinates are depicted as the distances from the origin to the points where the coordinate axes intersect with the ellipse. The mutual angle γ between the two bearing angle measurements affects the ratio between the minor and major axis. For perpendicular bearing angle measurements, the ratio is one and the uncertainty is represented by a circle.

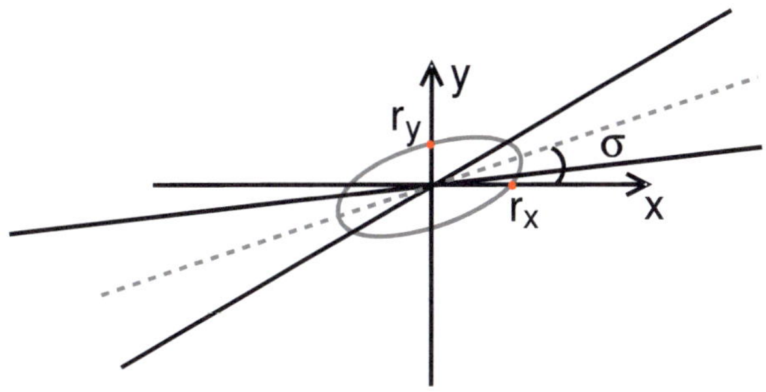

Fig. 2. Displacement measurement uncertainty

With each new angle measurement, the robot can calculate displacement estimates for the landmark using all the previous angle measurements. Thus, after two measurements the robot has one estimate for the landmark location while n measurements will give 1+2+3+...+(n-1) estimates. Figure 3 shows displacement estimates for five bearing angle measurements.

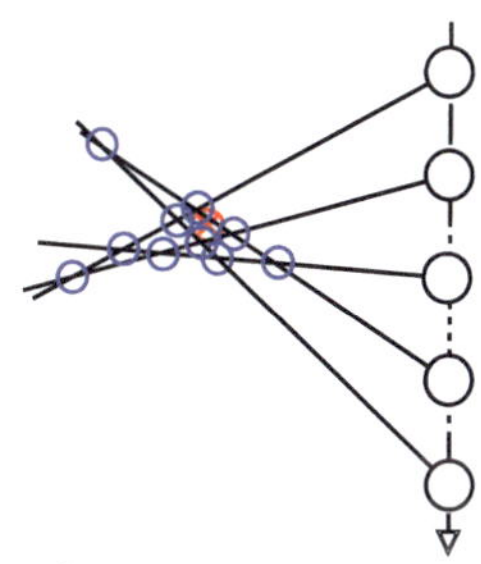

Fig. 3. Displacement estimates of five bearing angle measurements

The estimates based on almost parallel bearing angle measurements have considerable error, but their uncertainties are also large so the effect of such measurements on the final landmark position estimate is small.

3.2 Kalman filter

In (Maybeck, 1979), Maybeck defines Kalman filter as an optimal recursive data processing algorithm. Kalman filter is an application of the Bayesian probabilistic estimation method. In Bayesian scenario, we are estimating, from our prior information and data, the unknown constant value that the parameter had when the data were taken. Kalman filter is a recursive estimator, so all the prior information is contained in the previous estimate. The standard Kalman filter can be used with linear models. Non-linear models can be linearized using the extended Kalman filter. The Kalman filter maintains the first two moments of the state distribution (Welch and Bishop, 1995)

$$E[x_k] = \hat{x}_k \tag{6}$$

$$E\left[(x_k - \hat{x}_k)(x_k - \hat{x}_k)^T\right] = P_k \tag{7}$$

When using a discrete Kalman filter the motion model is presented with a discrete time state-space representation, where $\boldsymbol{\omega}_k$ and $\boldsymbol{\upsilon}_k$ represent the process and measurement noises respectively.

$$\mathbf{x}_{k+1} = \mathbf{A}_k \mathbf{x}_k + \mathbf{B}_k \mathbf{u}_{k+1} + \boldsymbol{\omega}_k \tag{8}$$

$$\mathbf{z}_k = \mathbf{H}_k \mathbf{x}_k + \boldsymbol{\upsilon}_k \tag{9}$$

In the case where matrices **A**, **B** and **H** are constants the general form for Kalman filter equations for the estimate are shown in Table 1. $\mathbf{Q}_k$ represents the process noise covariance and $\mathbf{R}_k$ represents the measurement noise covariance.

a priori state	$\hat{\mathbf{x}}_k^- = \mathbf{A}\hat{\mathbf{x}}_{k-1} + \mathbf{B}\mathbf{u}_k$
a priori estimate error covariance	$\mathbf{P}_k^- = \mathbf{A}\mathbf{P}_{k-1}\mathbf{A}^T + \mathbf{Q}_k$
a posterior state	$\hat{\mathbf{x}}_k = \hat{\mathbf{x}}_k^- + \mathbf{K}_k\left(\mathbf{z}_k - \mathbf{H}\hat{\mathbf{x}}_k^-\right)$
a posterior estimate error covariance	$\mathbf{P}_k = \left(\mathbf{I} - \mathbf{K}_k\mathbf{H}\right)\mathbf{P}_k^-$
Kalman gain	$\mathbf{K}_k = \mathbf{P}_k^-\mathbf{H}^T\left(\mathbf{H}\cdot\mathbf{P}_k^-\mathbf{H}^T + \mathbf{R}_k\right)^{-1}$

Table 1. Discrete Kalman filter equations

3.3 Extended Kalman filter

In bearing only localization, the only measurement is the bearing angle to the target. The robot displacement between measurements can be considered as external control, so it is not calculated as a function of the state vector **x**. When calculating the position of the robot or the landmark based on bearing angle measurements, non-linear equations are needed. The Extended Kalman Filter (EKF) can use the non-linear measurement equation, *h*, but the measurement matrix H has to be the Jacobian matrix composed of partial derivatives of *h* with respect to the state vector **x**. In addition the a priori estimate is calculated with a non-linear function *f*, which involves the heading angle of the robot. The a priori covariance matrix is calculated using the matrix **A**, which is a Jacobian matrix of partial derivatives of *f* with respect to the state vector **x**. The equations for Extended Kalman filter are shown in Equations 10-14

$$\hat{\mathbf{x}}_k^- = f\left(\hat{\mathbf{x}}_{k-1}, \mathbf{u}_k, 0\right) \tag{10}$$

$$\mathbf{P}_k^- = \mathbf{A}\mathbf{P}_{k-1}\mathbf{A}^T + \mathbf{Q}_k \tag{11}$$

$$\hat{\mathbf{x}}_k = \hat{\mathbf{x}}_k^- + \mathbf{K}_k\left(\mathbf{z}_k - h\left(\hat{\mathbf{x}}_k^-, 0\right)\right) \tag{12}$$

$$\mathbf{P}_k = \left(\mathbf{I} - \mathbf{K}_k\mathbf{H}\right)\mathbf{P}_k^- \tag{13}$$

$$\mathbf{K}_k = \mathbf{P}_k^-\mathbf{H}^T\left(\mathbf{H}\cdot\mathbf{P}_k^-\mathbf{H}^T + \mathbf{R}_k\right)^{-1} \tag{14}$$

3.4 Localization

In robot positioning, the state-space model can estimate the robot's position based on the motion measured by wheel encoders. Due to wheel slippage and measurement errors, the model gives increasingly bad position data i.e. position error increases over time. With Kalman filter, a correction to this position data can be applied if external positioning data such as triangulation or trilateration to beacons is available. The Kalman filter combines the data from the encoders (the motion model) and external measurements to beacons or landmarks. The process noise covariance matrix **Q** and the measurement noise covariance matrix **R** are used to tune the Kalman filter for best possible performance in the operation environment. The process noise increases the uncertainty of the robot's position estimate on each step the Kalman filter is updated. The amount of the process noise depends on the quality of the sensors, such as wheel encoders, used for measuring the movement of the robot directly. The measurement noise depends on the quality of the sensors used for obtaining the external positioning data.

The same concept works for bearing only measurements to a single landmark. We can also estimate the location of the landmark relative to the location of the robot. Instead of calculating directly a position estimate for the landmark as described in section 3.1, we calculate an estimate for the bearing angle to the landmark given the current estimates of the robot and landmark positions. Then we compare the estimated bearing with the latest bearing angle measurement and correct both robot and landmark position estimates. In this scenario the state vector **x** has five elements: the x- and y-coordinates of the robot (x_r, y_r), the heading angle of the robot θ and the x- and y-coordinates of the latest visible landmark (x_t, y_t). Only one landmark at a time can be visible for the robot. The zero angle of robot's heading is towards the global positive y-axis and it increases in the clockwise direction. The robot is assumed to be stationary while measuring bearing angle to the landmark. The robot movement between measurements is obtained from the odometry sensors. The movement can be represented by a control vector **u**, and no actual motion model is needed. That simplifies the equations of the a priori state. The landmarks are static, thus they do not have a motion model and their control vector is zero.

$$\mathbf{x} = \begin{bmatrix} x_r \\ y_r \\ \theta \\ x_t \\ y_t \end{bmatrix} \qquad \mathbf{P} = \begin{bmatrix} \mathrm{var}_{rx} & \mathrm{cov}_{rxy} & \mathrm{cov}_{rx\theta} & \mathrm{cov}_{rxtx} & \mathrm{cov}_{rxty} \\ \mathrm{cov}_{rxy} & \mathrm{var}_{ry} & \mathrm{cov}_{ry\theta} & \mathrm{cov}_{rytx} & \mathrm{cov}_{ryty} \\ \mathrm{cov}_{rx\theta} & \mathrm{cov}_{ry\theta} & \mathrm{var}_{\theta} & \mathrm{cov}_{\theta tx} & \mathrm{cov}_{\theta ty} \\ \mathrm{cov}_{rxtx} & \mathrm{cov}_{rytx} & \mathrm{cov}_{\theta tx} & \mathrm{var}_{tx} & \mathrm{cov}_{txy} \\ \mathrm{cov}_{rxty} & \mathrm{cov}_{ryty} & \mathrm{cov}_{\theta ty} & \mathrm{cov}_{txy} & \mathrm{var}_{ty} \end{bmatrix}$$

Prediction equation $x_k^- = f(\hat{x}_{k-1}, u_k, 0)$ and Jacobian matrix **A** are

$$\hat{\mathbf{x}}_k^- = \hat{\mathbf{x}}_{k-1} + \begin{bmatrix} d_r \cdot \sin(\theta_{k-1}) \\ d_r \cdot \cos(\theta_{k-1}) \\ 0 \\ 0 \\ 0 \end{bmatrix} \qquad \mathbf{A} = \begin{bmatrix} 1 & 0 & d_r \cdot \cos(\theta_{k-1}) & 0 & 0 \\ 0 & 1 & -d_r \cdot \sin(\theta_{k-1}) & 0 & 0 \\ 0 & 0 & 1 & 0 & 0 \\ 0 & 0 & 0 & 1 & 0 \\ 0 & 0 & 0 & 0 & 1 \end{bmatrix} \tag{15}$$

The measurement equation is

$$h(\hat{\mathbf{x}}_k^-, 0) = atan2(x_t - x_r, y_t - y_r) - \theta_r \tag{16}$$

The Jacobian matrix **H** is

$$\mathbf{H} = \begin{bmatrix} \frac{-\Delta y}{\Delta y^2 + \Delta x^2} & \frac{\Delta x}{\Delta y^2 + \Delta x^2} & -1 & \frac{\Delta y}{\Delta y^2 + \Delta x^2} & \frac{-\Delta x}{\Delta y^2 + \Delta x^2} \end{bmatrix} \tag{17}$$

where $\Delta x = x_t - x_r$ and $\Delta y = y_t - y_r$.

Combining the measurements of a robot with the Kalman filter results in better estimate of the relative position of the landmark since the effect of measurement errors decreases. However, position estimate of the landmark in the global coordinate frame is based on the estimate of the robot's own pose. The error in robot's pose is thus directly inherited by the position estimate of the landmark. To cope with this, we need measurements from multiple independent sources, e.g. from different robots. A single robot can always turn back when its pose uncertainty grows too big. However, even if it would return back to the starting point and start localization again its pose estimate would be subjected to the same structure related error as in previous runs. Thus, calibration runs made by one robot are not completely independent.

3.5 Distributed landmark localization

When a single robot estimates the position of a landmark the estimate is biased by the errors in the robot's own position and heading angle estimate. Filtering the different measurements of a single robot helps to decrease the effect of measurement errors but not the effect of the error in the robot's own pose estimate. However, if different robots are estimating the position of a landmark, the position estimate is based on several independent groups of measurements. When multiple robots pass a landmark, they each have different error and uncertainty in their own position estimate. This is due to the mechanical differences between the robots and due to the differences in the actions they have taken so far.

The robots have to use the same global coordinate frame in order to cooperate. If the robots start at a common location, they would share a common view of the global coordinate frame right from the start. Otherwise, the robots need to use a hierarchical method to determine which global coordinate frame they should adapt to when detecting landmarks that have already been located by robots using different global coordinate frame. The previous estimate of the position of the landmark and the related uncertainty have to be available for a robot in calculating new estimates. This can be done by storing the information to the memory of the landmark or by broadcasting all new estimates to other robots. The landmark position estimate contains all the information provided by the robots that have already passed the landmark. Thus, each robot passing the landmark benefits from the position information of all the previous robots.

4. Simulation

A simulation model for the multi-robot localization system was built in order to get an idea of the effects of different measurement errors. The simulations were run on Matlab. The system presented here simulates a scenario where a single landmark is localized by a group of robots passing the landmark one by one. All the robots estimate their own pose in a common (global) coordinate frame. Each robot has random error in its own position and heading angle estimate. Each bearing angle measurement also contains a random error. After each bearing angle measurement, the Kalman filter calculates new estimates for the location of the landmark and the pose of the robot.

4.1 Simulation parameters

The simulation model permits setting of several parameters. The robot pose error depends on remaining error after previous localization i.e. when robot passed the previous landmark and the accumulated error while robot has been moving. Both are set separately for each robot as random numbers, but the upper bounds of the errors can be set with parameters.

When a robot arrives near a landmark, the uncertainty in the robot's pose estimate (var_{rx}, var_{ry}, var_{θ}) is initialized with a calculated pose uncertainty. The pose uncertainty and the accumulated error depend on a random number representing the distance the robot has travelled since previous localization. The position and heading angle errors inherited from previous localization have small, fixed uncertainties.

E_{prev}	8 cm	Upper limit for random error in robot's position after previous localization
E_{accu}	24 cm	Upper limit for driving distance related random error in robot's position
θ_{prev}	±2°	Upper limit for random error in robot's heading after previous localization
θ_{accu}	±4°	Upper limit for driving distance related random error in robot's heading
var_{xy_init}	1000 cm^2	Upper limit for robot's initial position uncertainty
$var_{\theta\ init}$	0.14 rad^2	Upper limit for robot's initial heading uncertainty

Table 2. Robot odometry parameters

$D_{landmark}$	80 cm	Landmark detection distance. Within this distance the robot can always detect the landmark
D_{step}	10 cm	Bearing angle measurement interval i.e. distance between two angle measurements
λ_{limit}	±5°	Upper limit for random error in bearing angle measurement
w_{xy}	0.5 cm	Robot position related model noise for Kalman filter
w_{θ}	0.04°	Robot heading angle related model noise for Kalman filter
r	5.7°	Measurement noise for Kalman filter

Table 3. Localization parameters

The landmark detection is only possible in certain range. This range can be defined with a parameter. Other parameters are the distance the robot moves between bearing angle measurements, the upper limit of the error in the bearing angle measurement, model noise for the robot and the measurement noise. In these simulations, we consider a scenario where the landmark is a passive Radio Frequency Identification (RFID) tag and the bearing angle is measured by a turning antenna mounted on the robot. These constraints give us estimates of the realistic detection range and bearing angle measurement error.

While a robot is within the detection distance from the landmark, it measures bearing angles to the landmark and calculates new estimates for the position of the landmark as well as for its own pose. Simulations were run in order to discover how different error sources affect the accuracy of the landmark and robot position estimates. Each simulation consisted of 1000 independent runs, each having 20 robots passing a landmark. As a result, we get the average errors in the landmark and robot position estimate. The different plots in Figure 4 show the error in the position estimate of the landmark after ten robots have passed (landmark10) and after 20 robots have passed (landmark20). They also show the error in the position estimate of the 10th and 20th robot after they have passed the landmark (robot10 and robot20). The fifth line shows the heading angle error of the tenth robot (heading10). The scale for the heading angle error is on the right side of the plot.

4.2 Simulation results

For each simulation, one error parameter was chosen as a variable while the rest had the values defined in Table 2. Figure 4a shows the effect of the error in the bearing angle measurement. The maximum error on the x-axis represents the upper limit of the bearing angle error in degrees to either direction i.e. maximum error of 5 degrees means that the measured bearing angle is within ±5° of the correct angle. As can be seen, the error in the measured bearing angle has very little effect on the landmark or robot localization. The simulated error in the bearing angle measurement has zero mean. Thus, when the number of measurements increases, the mean of the measurement errors approaches zero. While passing the landmark each robot makes approximately 14 bearing angle measurements to the landmark, when the detection distance of the landmark is set to 80 cm. The position estimate of the landmark after ten robots have passed is based on 140 measurements but each robot's own pose correction is based on approximately 14 measurements. This explains why the robot pose estimate is more sensitive to the measurement error than the landmark position estimate.

The accumulated heading angle error has a significant effect on the robot's pose estimate as can be seen in Figure 4b. When localizing with only one landmark, the error in the heading angle of the robot impairs considerably the position estimate of the robot. The x-axis shows the upper limit for the accumulated heading angle error. For example, with upper limit of 10 (±10° to be exact), the expected accumulated heading angle error is ±5°. In addition, the robot is expected to have initial heading angle error of ±1° after previous localization. With these values, the 10th robot had an average heading angle error of 3° after passing the landmark. This shows that only a small correction to the robot's heading angle error can be made while passing a single landmark. However, the position estimate of the landmark depends on the estimates of several different robots with independent heading angle errors. Thus, the position estimate of the landmark is not so sensitive to the errors in the heading angles of individual robots.

The accumulated position error of the robot depends on the distance the robot has been driving after the position update. The accumulated error is the product of the upper limit and two random numbers between one and zero. The first random number represents the distance the robot has been driving after the last position update. The second random number represents an unknown factor causing error in the robot's odometry system. The two random numbers together cause the average accumulated error to be one fourth of the maximum error. The error is calculated separately for both coordinate axes. For maximum error of 20 cm, the expected error is ±5 cm in both position coordinates.

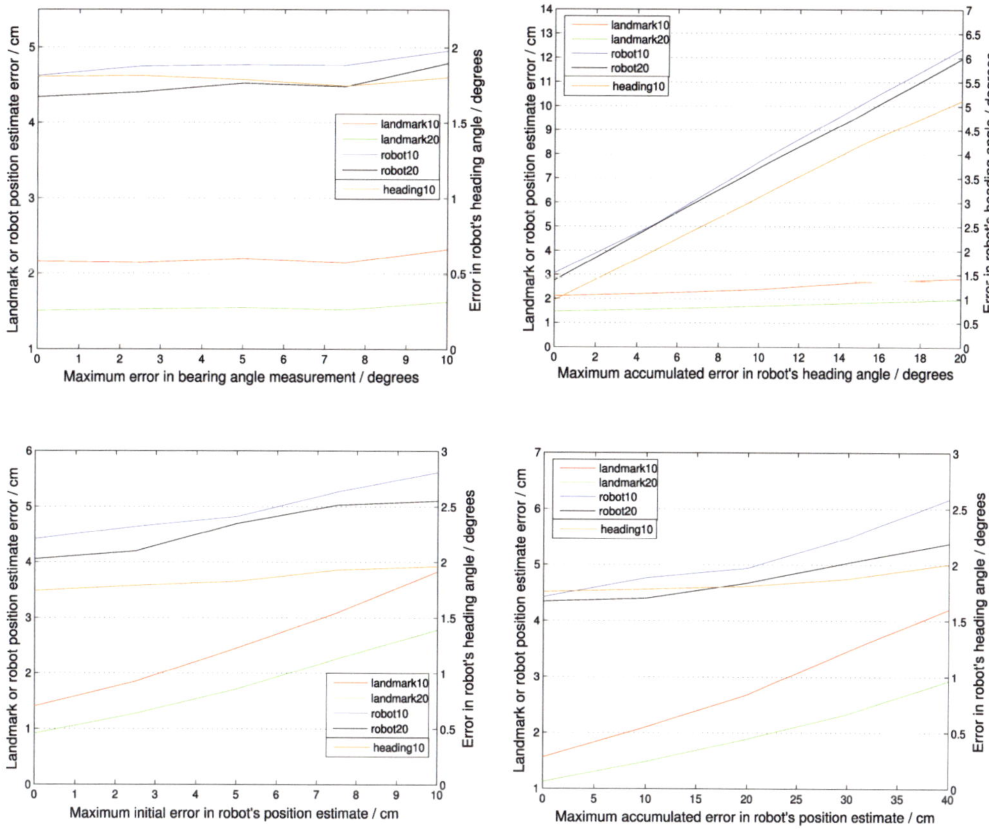

Fig. 4. (a) Effect of bearing angle measurement error. (b) Effect of accumulated heading angle error. (c) Effect of robot's initial position error. (d) Effect of robot's accumulated position error

Figures 4c and 4d show that the position error has a more significant effect on the error in the landmark position estimate than the aforementioned angle errors. The error in the robot's position estimate after passing the tag also increases with increasing initial position error, but the heading angle error seems to be a more significant factor. The correction of the robot's heading angle is slightly slower with bigger initial position error.

Figure 5 shows the convergence of the landmark position as the number of contributing robots increase. The different plots in Figure 5a show the effect of the magnitude of the accumulated error in robot's heading angle. In Figure 5b the effect of accumulated error in robot's position estimate is presented. As before the indicated accumulated error (e.g. 0, 5, 10, 15 or 20 degrees in Figure 5a) is the upper limit of the error, i.e. with a limit of ±20° the expected accumulated heading angle error is ±10°.

The error in landmark position estimate is significant after the initial estimate of the first robot passing it, because the error in the robot's position estimate is inherited by the landmark. However, when more robots contribute to the estimate, the average error decreases rapidly and after five robots, it is already at quite reasonable level. As the number of robots contributing to the estimate and thus, the number of measurements, increases, the uncertainty of the landmark position estimate decreases. When the uncertainty of the landmark position decreases each new measurement has smaller and smaller effect on the landmark position estimate and the estimate becomes more stable.

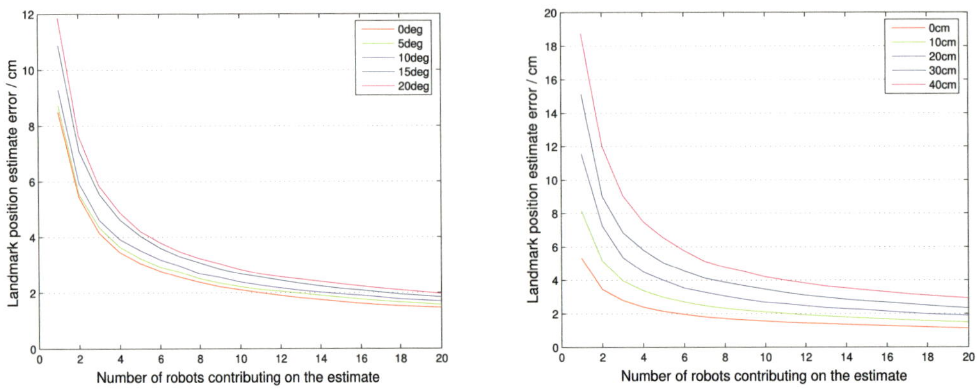

Fig. 5. Effect of accumulated (a) heading angle, (b) position error on landmark position estimate

5. Case Example

This case example is not a real multi-robot experiment, but gives an idea how the concept would work in practice. Only two robots are used for the measurements, each one making several independent runs. The case example uses RFID technology for the localization system. Passive RFID tags serve as landmarks and the robot is equipped with an RFID interrogator (reader). A turning antenna is connected to the RFID interrogator. The bearing angle to the landmark is estimated by turning the interrogator antenna and recording the direction in which the RFID tag responded.

5.1 Experimental setup

For this experiment we are using an UHF RFID system which operates at the 867 MHz frequency. The system is EPC Class 1 GEN 2 compatible. One RFID tag is used as a landmark. It is attached to a wall along a two meter wide corridor. The dimensions of the tag sticker are 97 x 15 mm. The detection distance of the tag is usually around one meter, but it can be up to two meters in an optimal environment.
The RFID interrogator is mounted on a robot. The dimensions of the interrogator are 66 x 33 x 6 mm and it has a Compact Flash connector for interfacing to a USB or a serial port. A three element Yagi-Uda antenna is connected to the interrogator. The antenna has one main lobe with a beam width of 68 degrees on the horizontal plane and 115 degrees on the vertical plane. The main lobe gain is 8 dBi and the front-to-back ratio is -14 dB. The dimensions of the antenna are 149 x 138 x 7 mm. In this prototype the antenna is mounted on a hobby servo which can turn the antenna 360 degrees.
The robots are equipped with wheel encoders. The encoders are connected to a microcontroller, which constantly integrates the relative pose of the robot. The two robots

are not alike. One is a relatively small with wheel base of 17 cm. It moves slowly, but has an accurate odometry system. The other robot has a 30 cm wheel base. It is fast, but the odometry is less accurate.

A reference point (0, 0) is marked on the floor of the corridor. The landmark position is measured to be (2.26, 0.74). For each run a robot is manually placed at the reference point with its heading angle approximately parallel to the global x-axis. After the manual placement there is a small initial error on the robot's pose estimate.

The robot drives forward on small steps (10-30 cm) and stops to detect the landmark. The antenna is turned a full circle on steps of 0.1 rad. The first and the last angle at which the landmark is detected are recorded and the bearing to the landmark is estimated as half way between these two angles. On each successful bearing angle estimate the robot's current pose and the estimated bearing angle to the landmark are recorded.

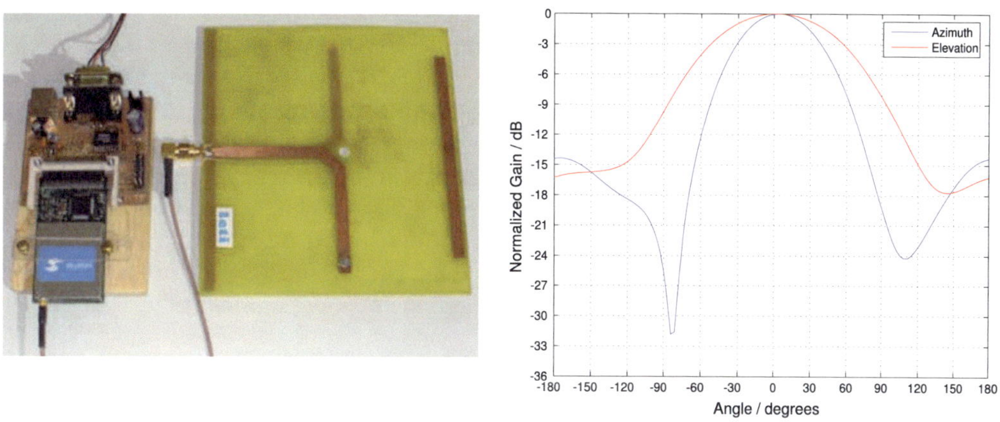

Fig. 6. (a) RFID interrogator and the antenna, (b) antenna gain pattern

5.2 Sensor accuracy

The measurement accuracy of the RFID system is determined by making 20 bearing angle measurements to a stationary tag from a single location. This procedure was repeated for three different angles; tag on the left side of the robot, the robot facing the tag and the tag on the right side of the robot. In all three cases the standard deviation was approximately 0.09 rad. In all the measurements the tag antenna was perpendicular to the line between the tag and the robot, so the polarisation was optimal when the reader antenna was pointing towards the tag. When a robot moves with respect to the tag the polarisation is not always optimal. This is expected to deteriorate the measurement accuracy as the energy transfer from the reader antenna to the passive tag gets more uncertain.

The robot pose estimate is based on wheel encoders only. The big robot has encoder resolution of 500 pulses per rotation of the wheel. The wheel diameter is 150 mm, so the odometry resolution is 0.942 mm/pulse. The encoder is directly connected to the wheel axle. There is no free play between the encoder and the wheel, but the number of pulses per rotation is relatively small. The small robot has encoder resolution of 2688 pulses per rotation of the wheel. The wheel diameter is 159 mm, so the odometry resolution is 0.186 mm/pulse. The encoder is connected to the motor shaft. This gives more pulses per one rotation of the wheel, but backlash between the wheel and the motor shaft may cause inaccurate readings.

5.3 Results

Both robots made ten runs storing the measurement data into files. Matlab was then used to calculate position estimate for the landmark. The algorithms used were the same as in the simulation runs. The initial uncertainty in the robot position estimate was fixed to 0.05 m^2. The model noise was 0.005 m and measurement noise 0.3 rad.

When the first robot detects the landmark for the first time there is no estimate of the position of the landmark and the Kalman filter cannot be used. The initial estimate is simply based on the first bearing angle and the expected detection range of one meter. This is a very rough estimate and has an uncertainty of 1 m^2. With the rest of the measurements the extended Kalman filter is used to correct the estimate.

The order in which the measurements are used affects the position estimate of the landmark. The 20 measurement runs were used in random order to estimate the position of the landmark. This procedure was repeated 1000 times. Figure 7 shows the average error in the landmark position estimate as a function of the number of robots contributing on the estimate. After 10 robots the average error of the position estimate of the landmark was 5.1 cm and the maximum error was 18.4 cm. After 20 robots the average error was less than 2 cm and the maximum error was 5.6 cm. The average distance between two bearing angle measurements was longer than in the simulation runs where it was fixed to 10 cm. In the simulation runs each robot made approximately 14 bearing angle measurements while passing the landmark. In this laboratory experiment the average number of measurements was 7.7 per robot.

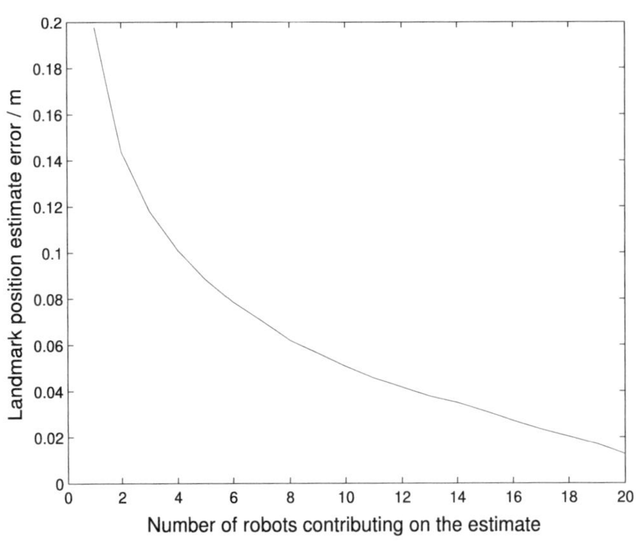

Fig. 7. Convergence of the landmark position estimate

Using the recorded poses of the robot and the known position of the landmark, estimates for correct bearing angle measurements were calculated and compared with the recorded bearing angle measurements. The calculated root mean squared error of the bearing angle measurements was 0.27 rad. This is based on the assumption that the pose estimates for a robot are accurate. The pose of a robot is based on odometry as no another reference was available.

The bigger root mean squared error in the bearing angle measurements and the smaller number of measurements per robot explains why the landmark position estimate converges somewhat slower in the case example than in the simulation. In the case example the total number of measurements after 18 robots is about the same as the number of measurements in a simulation run after 10 robots. Then the position error of the landmark position estimate is approximately 2 cm in both cases. Altogether the case example behaves as can be expected based on the simulations.

6. Conclusion

In this chapter, a multi-robot approach to the calibration of a localization system has been introduced. The bearing-only measurements are shown to be sufficient for single landmark localization. The simulations indicate that the algorithm is fairly robust. Localization of the landmarks can be achieved even in the presence of considerable errors in bearing angle measurements or in the position estimate of an individual robot. A significant error in the heading angle estimate of a robot was noticed to cause problems with the simultaneous localization of the robot, but the landmark localization was still accurate.

The case example shows that the concept can be used with real robots. The RFID technology provides a possibility to use passive landmarks which are maintenance free, easy to deploy and inexpensive. Nonetheless they provide unique identification of the landmark and a possibility to store some information for other entities passing by.

The results achieved in the simulation study as well as in the case example encourage us to further test the concept. Next we plan to use three robots to localize multiple consecutive landmarks in order to determine how fast the localization network starts to support the robots in practice. A reference system will be used to record the error in the robots' own position estimates. This will tell us how well the robots can correct their own position while simultaneously localizing the landmarks.

7. References

Lars A. A. Andersson and Jonas Nygards. On utilizing geometric formations for minimizing uncertainty in 3 robot teams. Technical report, Robotics & Autonomous Mechanical Systems, Linköping University, 2008.

H. J. Chang, C. S. G. Lee, Y. C. Hu, and Yung-Hsiang Lu. Multi-robot slam with topological /metric maps, *International Conference on Intelligent Robots and Systems*, 2007.

Y. S. Hidaka, A. I. Mourikis, and S. I. Roumeliotis. Optimal formations for cooperative localization of mobile robots, *International Conference on Robotics and Automation*, 2005.

A. Howard, M. J. Mataric, and G. S. Sukhatme. Putting the 'i' in 'team': an egocentric approach to cooperative localization, *International Conference on Robotics and Automation*, 2003.

Andrew Howard, Lynne E. Parker, and Gaurav S. Sukhatme. The sdr experience: Experiments with a large-scale heterogenous mobile robot team. In *9th International Symposium on Experimental Robotics*, 2004.

R. Kurazume, S. Nagata, and S. Hirose. Cooperative positioning with Multiple robots, *International Conference on Robotics and Automation*, 1994.

R. Martinez-Cantin, J. A. Castellanos, and N. de Freitas. Multi-robot marginal-slam. *IJCAI Workshop on Multi-Robotic Systems for Societal Applications*, 2007.

P. S. Maybeck. *Stochastic Models, Estimation and Control*, pages pp. 1–16. Academic Press, New York, 1979.

MichaelMontemerlo. Fastslam: A factored solution to the simultaneous localization and mapping problem. In *Proceedings of the AAAI National Conference on Artificial Intelligence*, pages 593–598. AAAI, 2002.

Michael Montemerlo, Sebastian Thrun, Daphne Koller, and Ben Wegbreit. Fastslam 2.0: An improved particle filtering algorithm for simultaneous localization and mapping that provably converges. In *Proceedings of the Int. Conf. on Artificial Intelligence (IJCAI*, pages 1151–1156, 2003.

E. D. Nerurkar and S. I. Roumeliotis. Power-slam: A linear-complexity, consistent algorithm for slam. In *International Conference on Intelligent Robots and Systems, 2007. IROS 2007*, pages 636–643, 2007.

P. Newman and Kin Ho. Slam-loop closing with visually salient features, *International Conference on Robotics and Automation*, 2005.

Nissanka B. Priyantha, Anit Chakraborty, and Hari Balakrishnan. The cricket location-support system. *Sixth Annual ACM International Conference on Mobile Computing and Networking*, 2000.

I. Rekleitis, G. Dudek, and E. Milios. Probabilistic cooperative localization and mapping in practice, *International Conference on Robotics and Automation*, 2003.

R. Smith, M. Self, and P. Cheeseman. *Estimating uncertain spatial relationships in robotics*, pages 167–193. Autonomous robot vehicles. Springer-Verlag New York, Inc., 1990. ISBN 0-387-97240-4.

J. R. Spletzer and C. J. Taylor. A bounded uncertainty approach to multi-robot localization, *International Conference on Intelligent Robots and Systems*, 2003.

S. Thrun and Y. Liu. Multi-robot slam with sparse extended information filers. In *Proceedings of the 11th International Symposium of Robotics Research (ISRR'03)*, Sienna, Italy, 2003. Springer.

G. Welch and G. Bishop. An introduction to the kalman filter. University of North Carolina at Chapel Hill, Chapel Hill, NC, 1995.

S. B. Williams, G. Dissanayake, and H. Durrant-Whyte. An efficient approach to the simultaneous localisation and mapping problem, *International Conference on Robotics and Automation*, 2002.

X. S. Zhou and S. I. Roumeliotis. Multi-robot slam with unknown initial correspondence: The robot rendezvous case, *International Conference on Intelligent Robots and Systems*, 2006.

10

Generic Real-Time Motion Controller for Differential Mobile Robots

João Monteiro[1] and Rui Rocha[1]
[1]*ISR - Institute of Systems and Robotics*
Faculty of Science and Technology
University of Coimbra, Portugal

1. Introduction

Trajectory planning and execution are well known tasks to be performed by mobile robots. Its importance aroused in the past from the need to replace humans when performing either complex or dangerous tasks, mainly in industrial environments. Nowadays, the capabilities and applications of mobile robots are far more generic; Complex tasks pose more demanding control requirements within trajectory execution, including high responsiveness and predictability. Robot soccer might be used as a testbed to enact most of these requirements on an amusing environment and evaluate solutions to important robotic problems such as control, perception, intelligent systems, etc. Therefore, the RAC[1] robotic soccer team has been used to study and validate the motion controller described within this chapter. The control project consists of two main parts: 1) the digital controller, and 2) the real-time system. This latter provides high responsiveness to the previous, resulting in a high-performance implementation of the projected algorithm.

Extensive work has been made on controlling differential mobile robots. For instance, [1] presents a generic controller wherein pose estimation is extracted from robot's kinematics, and an adaptive control module is introduced to deal with modelling errors. In [9], a Lyapunov based nonlinear controller is presented, where the influence of the control parameters is studied, without giving emphasis to modelling errors. The approach described herein brings together the simplicity of the Lyapunov mathematical laws, the adaptive control concept to deal with modelling errors, and also proper fusion of two sensory data - vision and odometry –, for robust pose estimation.

Besides addressing the control issue, this chapter also describes a real-time system to complement the developed control module, so that it is able to fulfil pre-defined timing constraints. A considerable number of approaches such as [7] and [5] evidence the need of integrating a real-time system in robot control. Also, the sensory fusion technique using a Kalman filter will be addressed in detail due to its importance.

The modularity of the described system provides easy integration with higher level software modules for intelligent perception and robot actuation.

---

[1] Robótica Académica de Coimbra, http://rac-uc.pt.vu

1.1 Previous Note to the Reader

This chapter aims to clearly explain to the reader an implementation of a generic real-time motion controller for differential mobile robots. To give an insight on how to implement the approach in practice, at some points we refer to specific hardware used to validate the approach. However, this is made in a way that the reader can resemble the mentioned aspects to his own implementations with ease. Practical results and analysis will be present to validate the approach.

2. Control Requirements

In this section the requirements of the controller are defined, as well as the target robot kinematics and trajectory planning. This will serve as base knowledge to understand the subsequent sections.

2.1 Case Study: The Differential Mobile Robot

The developed digital controller performs dynamic control of a mobile robot with two parallel traction wheels which, therefore, possesses differential confguration. Its core goal is to allow the robot to follow a desired trajectory - defined by velocity vectors - to achieve a desired position with high accuracy. This is made by eliminating pose error in every discrete instant of time.

2.1.1 Trajectory Definition and Robot Kinematics

The studied 2D path planner defines a trajectory as a time variant pose vector represented in the world, which has its own global cartesian system defined. The robot's pose vector q possesses three degrees of freedom (DOF): the position (x,y) and its heading h. The latter is assumed to be positive in counter-clockwise direction, beginning at the positive xx axis. The state q_0 is denoted as the zero pose state $(0,0,2n\pi)$, where n is an integer value. Since the robot is capable of moving in the world, the pose q is a function of time t. The integer representation of a set of points $(x(t),y(t))$ is defined as trajectory, and if the derivates $\dot{x}$ and $\dot{y}$ exist, $h(t)$ is no longer an independent variable, since

$$h(t) = tan^{-1}\frac{\dot{y}(t)}{\dot{x}(t)} \qquad (1)$$

from which one can see that the heading depends on the robot's velocity along each one of the two 2D axis. The robot's movement is controlled using its linear and angular velocities, v and w respectively, which are also functions of time t. The robot kinematics is defined by the Jacobian matrix

$$\begin{bmatrix}\dot{x}\\ \dot{y}\\ \dot{h}\end{bmatrix} = \dot{q} = Jp = \begin{bmatrix}\cos(h) & 0\\ \sin(h) & 0\\ 0 & 1\end{bmatrix}p, \qquad (2)$$

where the velocity vector p is defined by the linear and angular velocities v and w,

$$p = \begin{bmatrix}v\\ w\end{bmatrix}. \qquad (3)$$

This kinematics is common for all non-holonomic robots in which the number of controllable DOF's is less than the number of DOF's the robot possesses.

2.1.2 Pose Error

The core goal of any trajectory controller is to reduce the robot's pose error as much as possible, relative to a desired target point. Pose error q_e is defined as the transformation of the reference pose q_d to the local coordinate system of the robot with origin (x, y), wherein the current robot's heading is given by h's amplitude (Fig. 1).

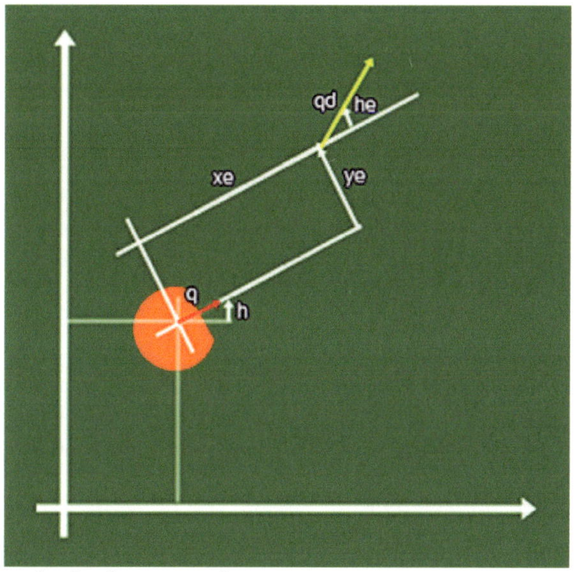

Fig. 1. Representation of robot's pose and pose error vectors.

Such transformation is the difference between q_d and q

$$q_e = \begin{bmatrix} x_e \\ y_e \\ h_e \end{bmatrix} = \begin{bmatrix} \cos(h) & \sin(h) & 0 \\ -\sin(h) & \cos(h) & 0 \\ 0 & 0 & 1 \end{bmatrix} . (q_d - q) \quad . \tag{4}$$

One can easily see that if $q_d = q$, the pose error is null, being this the ideal final state.

2.1.3 Robot Dynamic Model

Based on the Lagrange mathematical modelation of mechanical systems [1], and considering $G(q) = C(q, \ddot{q}) = 0$, the generic dynamic equations of the mobile robot can be written as

$$\begin{bmatrix} m & 0 & 0 \\ 0 & m & 0 \\ 0 & 0 & I \end{bmatrix} . \begin{bmatrix} \ddot{x} \\ \ddot{y} \\ \ddot{h} \end{bmatrix} = \frac{1}{R} . \begin{bmatrix} \cos(h) & \sin(h) \\ \sin(h) & \sin(h) \\ L & -L \end{bmatrix} . \begin{bmatrix} \tau_1 \\ \tau_2 \end{bmatrix} + \begin{bmatrix} \sin(h) \\ -\cos(h) \\ 0 \end{bmatrix} \lambda, \tag{5}$$

with τ_1 and τ_2 being the torques of left and right motors respectively, m and I the robot's mass and inertia, R the wheel radius, L the linear distance between the two wheels, and λ the Lagrange multipliers of constrained forces. The non-holonomic restriction is deduced from the above equation, and is given by

$$\dot{x} \sin(h) - \dot{y} \cos(h) = 0, \tag{6}$$

from where it is imposed that a non-holonomic mobile robot can only move in the direction normal to the axis of the driving wheels.

2.1.4 Trajectory Planning

The trajectory planning algorithm, defines intermediate target points, hereafter denoted as virtual points, which are spaced about a configurable distance, representing the intermediate desired poses between the initial instant and the target pose q_d. These poses will be used by the controller to perform pose error elimination during trajectory execution. For the iteratively placed virtual points, the desired pose is defined relative to the robot's local coordinate frame by

$$\begin{aligned} x_{d,p} &= X_{inc}.\cos(\theta_v) - Y_{inc}.\sin(\theta_v) \\ y_{d,p} &= X_{inc}.\sin(\theta_v) + Y_{inc}.\cos(\theta_v) \\ h_{d,p} &= h + \theta_v \end{aligned} \quad (7)$$

To make $x_{d,p}$ and $y_{d,p}$ consistent with a fixed coordinate frame, they need to be referenced to the world's coordinate frame. This is done using the following transformation:

$$\begin{aligned} x_{d,p,w} &= x_{d,p}.\cos(\theta_v) - y_{d,p}.\sin(\theta_v) \\ y_{d,p,w} &= x_{d,p}.\sin(\theta_v) + y_{d,p}.\cos(\theta_v) \\ h_{d,p,w} &= h_{d,p} \end{aligned} \quad (8)$$

2.2 Control Requirements Analysis and Specification

From the above sections, one can easily understand the objective of the controller, pose correction, from where a convergence of the pose error state variables to a null state is desired. To achieve this, the controller needs to deal with inevitable modeling errors which, once implemented, can impose stability problems. Therefore, in the design of the presented controller, the impact of modeling errors was carefully considered so that they can be efficiently minimized. Such errors are cumulative with $t \to \infty$, resulting in evident pose mismatches.

The developed controller is focused on the kinematic stabilization of generic mobile robots possessing differential configuration, which imposes a great challenge since no specific dynamic model is considered. Few parameters representative of the physical structure need to extracted; among them we have mass, inertia, wheel radius and distance between wheels. Despite the mentioned effort to avoid modeling errors, measurement errors can be present in the extracted parameters. Having this in mind, the concept of adaptive control is introduced, enhancing the system robustness. The controller was firstly introduced in [13], but it's further refined in this article through a real-time implementation and a more thorough experimental validation.

3. Digital Controller Design

The controller design is comprised of three parts. In the first, kinematic stabilisation is achieved using nonlinear control laws. For the second, the acceleration is used for exponential stabilization of linear and angular velocities. The uncertainty on the robot's structure parameters is compensated using an adaptive control block. Introducing suitable Lyapunov functions, stability of the system state variables is achieved. For the final part, pose estimation is made by fusing odometry and vision for robust pose feedback information by means of a Kalman filter. The latter is designed so that it is independent of the mathematical model of the robot, boosting overall performance and applicability.

3.1 Control Scheme

The developed approach is depicted in Fig. 2.

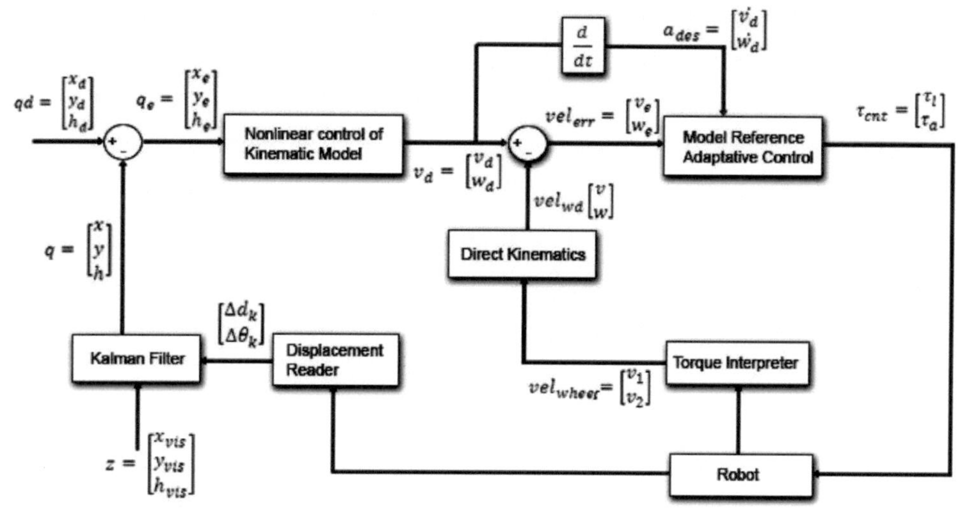

Fig. 2. Control Scheme

It is a feedback controller, in which the input state is the desired robot pose $[x_d\ y_d\ h_d]'$. At its output, proper update of the motor torques for each wheel is done to fulfill the controller's pose error elimination goal. The adaptive control block ensures that a stable condition is achieved independently of the presence of modeling errors in the robotic platform parameters. The estimation error is brought to zero in finite time. The following subsections describe the different control modules in more detail.

3.1.1 Pose Error Generator

The error dynamics is written independently of the inertial (fixed) coordinates frame by Kanayama transformation. Expanding equation 4, the following is given

$$q_e = \begin{bmatrix} x_e \\ y_e \\ h_e \end{bmatrix} = \begin{bmatrix} \cos(h) & \sin(h) & 0 \\ -\sin(h) & \cos(h) & 0 \\ 0 & 0 & 1 \end{bmatrix} \cdot \begin{bmatrix} x_d - x \\ y_d - y \\ h_d - h \end{bmatrix}, \tag{9}$$

which composes the pose error vector.

3.1.2 Nonlinear Kinematic Controller

Lyapunov based nonlinear controllers are very simple and yet very successful in kinematic stabilization. So, bringing together simplicity and functionality, the Lyapunov stability theorem proved to be of great utility for the controller presented herein. The following Lyapunov candidate function which represents, as stated in [6], the total energy[2] of the robot is considered

[2] The total sum of kinetic and potential energies.

$$V = \frac{1}{2}({x_e}^2 + {y_e}^2) + [1 - \cos{(h_e)}]. \tag{10}$$

To prove the stability condition, $\dot{V}$ needs to be obtained, and criteria must be applied. Based on [1] and [6], we have

$$\dot{V} = v_r x_e \cos(h_e) - v_d x_e + v_r \sin(h_e)\, y_e + w_r \sin(h_e) - w_d \sin(h_e) \tag{11}$$
$$\Leftrightarrow \dot{V} = v_r \cos(h_e - v_d)\, x_e + \sin(h_e)\,(v_r y_e + w_r - w_d),$$

wherein v_r and w_r are the reference velocities. v_d and w_d are chosen to make $\dot{V}$ become negative semi-definite:

$$v_d = v_r \cos(h_e) - x_e \tag{12}$$
$$w_d = w_r + v_r y_e + \sin(h_e).$$

Replacing the above expressions in (11), we obtain

$$\dot{V} = -{x_e}^2 - v_r sin^2(h_e), \tag{13}$$

from where one can easily see that $\dot{V}$ is always negative semi-definite. Considering a specific instance of the control rule at this block output, the following is given

$$v_d = v_r \cos(h_e) - K_x x_e \tag{14}$$
$$w_d = w_r + K_y v_r y_e + K_h \sin(h_e),$$

where K_x, K_h and $K_h$are positive constants. By La Salle's principle of convergence, and proposition 1 of [6], the null pose state q_0 is always an equilibrium state if the reference velocity is higher than zero ($v_r > 0$). As one can see, there are three weighting constants for the pose error, without interfering in the overall pose stability of the robot. These parameters are responsible for the influence of each component of the pose error vector on the stability process, and need to be carefully tuned.

3.1.3 Model Reference Adaptive Control

The motivation to include this module in the control mesh comes from the need of the controller to cooperate with parameter uncertainties. Based on [1], one can extract the adaptation rules for the linear velocity

$$\frac{d\theta_1}{dt} = -\epsilon_1 e \dot{v}_d \Leftrightarrow \theta_1 = \int -\epsilon_1 e \dot{v}_d \, dt \tag{15}$$
$$\frac{d\theta_2}{dt} = -\epsilon_2 e \dot{v}_d \Leftrightarrow \theta_2 = \int -\epsilon_2 e \dot{v}_d \, dt.$$

Identically, one can find similar rules for the angular velocity

$$\frac{d\theta_3}{dt} = -\epsilon_3 e' \dot{w}_d \Leftrightarrow \theta_3 = \int -\epsilon_3 e' \dot{w}_d \, dt \tag{16}$$
$$\frac{d\theta_4}{dt} = -\epsilon_4 e' \dot{w}_d \Leftrightarrow \theta_4 = \int -\epsilon_4 e' w_d \, dt$$

where v_d and w_d are the desired linear and angular velocities that constitute the output of the previous module and e and e' the respective velocity errors. The parameters $\epsilon_i, i = \{1..4\}$ are manually tuned for best performance achievement.

Practical implementation of this block is not simple, since it is necessary to determine the virtual velocity[3] before proceeding with the actual real velocities calculation, which in turn is needed to evaluate the error of the modelled parameters m and I and provide proper

[3] Velocity of the reference model.

acceleration compensation. So, for better organization and comprehension, it was important to create a schematic for the present module (Fig. 3).

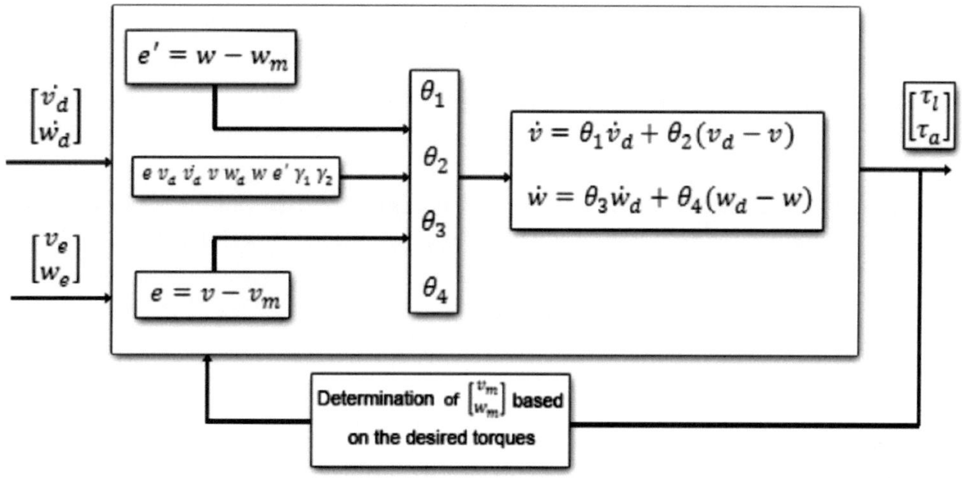

Fig. 3. Adaptive control block scheme.

From this figure, one can easily see the virtual velocity determination before evaluating the velocity error due to uncertainty on modelling parameters. The output are the linear and angular torques, which are used to easily determine the left and right wheel -- τ_1 and τ_2 -- torques,

$$\tau_l = \frac{1}{R}(\tau_1 + \tau_2) \qquad (17)$$
$$\tau_a = \frac{L}{R}(\tau_1 - \tau_2)$$

In practice, this module provides acceleration compensation to the robot. Imagine, for instance, that theoretically the robot should be navigating with a velocity of 1mps, instructed by the nonlinear control module. By theoretically we mean the robot modelled when planning the controller. This 1mps will be the velocity of the reference model, however the platform may not be navigating at that velocity due to badly modeled parameters. The adaptive control block will take notice of this difference and accelerate/decelerate the robot so that its velocity follows the reference model. Referring to Fig. 3, the reader can easily see the linear and angular acceleration equations that use the adaptation rules, which compensate the difference between the actual robot velocity relative to the reference model. The reader can think of this as a virtual (ghost) robot walking with the real platform that must be followed.

3.1.4 DC Motor Actuation Based on the Desired Torques

As we have seen from the controller scheme in Fig. 2, the output of the control mesh consists of the torque vector, but it is not explicit on how the motors are actuated in response to this torque command. The amplitude of the electric current that controls the motor power is regulated by a drive board. Such board implements a H bridge to drive each motor individually by a PWM signal. Note that this is a hardware specific section and has been put

here to give to the reader a better comprehension on how the algorithm can be implemented.
Following the testbed team's specific case, and according to the datasheet of the used DC motors, the torque-current relation is given by

$$I = k_i \tau, \tag{18}$$

where k_i is the electrical current constant, which equals 0.487. For a motor (left or right) to which a rotor torque of τ_d is desired, the necessary current that the motor must consume can be extracted as follows

$$I_d = k_i \tau_d. \tag{19}$$

Referring to the datasheet, the maximum current value that the motor should consume is 0.57A, being the ideal value 0.35A. Therefore, the I_{MAX} reference value should be between 0.35 and 0.57, letting the hardware capabilities be fully used, but avoiding persistent overload. So, a value of 0.4A is admitted to be suitable and the maximum power consumption for the robot's 12VDC motors is calculated by

$$P_{MAX} = VI_{MAX} = 12 * 0.4 = 4.8W \tag{20}$$

By assuming that the motor will not work above a maximum value of its active power of $P_{MAX} = 4.8W$, its safety is ensured and, at the same time, the robot takes advantage of its performance. It is also guaranteed that it will not overheat. Based on such value, the relation *dutyCycle-current* can be determined.

Let V_{ap} be the average voltage value applied to the motor as a consequence of a command dutyCycle of DtC, the following is given

$$V_{ap} = \frac{DtC}{100} * 12. \tag{21}$$

Therefore, if the motor consumes a maximum current of 0.4A at 12VDC, we have

$$I_{ap} = \frac{V_{ap}}{12} * 0.4. \tag{22}$$

Bringing together the two above equations, the following is given,

$$DtC = k_{ap} I_{ap}, \tag{23}$$

where $k_{ap} = \frac{1}{0.4}.100$. If the *torque-current* relation is applied, the useful *dutyCycle-torque* relation can be extracted,

$$DtC = k_{ap} k_i \tau_{ap}. \tag{24}$$

Based on this latter equation, it is possible to know by a simple calculation what value should be attributed to the duty cycle for a motor to generate a τ_{ap} torque.

3.1.5 Torque Interpreter

This module makes proper measurement of the actual velocity based on the encoder displacement during a sample time. Its output data will be used to determine the velocity error of the platform at a given time instant to feed the adaptive control module.
This module works as follows. First, a measurement of the encoder count is made and t_{samp} is waited. After this short time (the shortest possible), a new measurement is sampled and the counting is made by finding the difference of the last and actual pulse count value.
Given that the used motors possess a gear box with 1:3.71 ratio, and that the encoder is able to produce 512 pulses in a row, there are 4*412 transactions when the encoder reading is in quadrature mode (defined in the drive board). The angular increment of the encoder is

$\frac{2\pi}{2048}rad$ referred to the motor. Referring to the wheel, this results in $\frac{2\pi}{2048\times3.71}rad$. By this, the robot wheel velocity vector in rpm (rotations per minute) is

$$v_{wheel} = \begin{bmatrix} v_1 \\ v_2 \end{bmatrix}, \tag{25}$$

where $v_1 = v_2 = \frac{pulse\_count}{t_{samp}.1000} . \frac{2\pi}{2048\times3.71} . 60.$

3.1.6 Displacement Reader

Due to the fact that the implemented Kalman filter needs a specific input vector of Δd and Δh, linear and angular displacements respectively, it is necessary to determine such values for every control loop as

$$\Delta d = \frac{d_1 + d_2}{2} \; ; \; \Delta h = \frac{d_1 - d_2}{b}, \tag{26}$$

wherein b is the distance between wheels, and d_1, d_2 the distance walked by the left and right wheels respectively. This is done simply using odometry data and counting pulses between sample instants to determine d_1 and d_2. Then, knowing how many pulses it takes for a complete revolution of the wheel, i.e., $2\pi r$, we can easily determine the linear displacement of each wheel.

3.1.7 Kalman Filter for Sensory Fusion

A differential robot possessing an odometer system is equipped with an encoder in each motor. An angular displacement of α radians on the rotor corresponds to a moved distance d on the periphery of the wheel and, subsequently, to an encoder count. The distance is given by $d = k\alpha$, with $k = \frac{1}{r}$, being r the radius of the robot. If the robot's movement is assumed to be linear, the distances d_1 and d_2 walked by the left and right wheels respectively, can be transformed into linear and angular displacements through (26). For a particular sample instant k, the following is given

$$\Delta d_k = \frac{d_{1,k} + d_{2,k}}{2} \; ; \; \Delta h_k = \frac{d_{1,k} - d_{2,k}}{b}. \tag{27}$$

The robot's coordinates referenced on the world's coordinate frame can be determined by these simple relations

$$\begin{aligned} X_{k+1} &= X_k + \Delta d_k \cos\left(h_k + \frac{\Delta h}{2}\right) \\ Y_{k+1} &= Y_k + \Delta d_k \sin\left(h_k + \frac{\Delta h}{2}\right) \\ h_{k+1} &= h_k + \Delta h \end{aligned} \tag{28}$$

These coordinates constitute the state vector, and are observed by the vision's coordinate state vector z.

Note to the reader: It is important at this time to refer that we use a global vision system capable of detecting the position of the robot in the world referred to its global axis. This is composed by two non-stereo cameras placed at the top, parallel to the world, where a color code capable of identifying the robot is used so that the vision system can determine its pose vector. Therefore, we use the Kalman filter to fuse vision and odometry data to retrieve a highly accurate pose estimation for the robot in each loop. The reader should not be concerned, since Kalman filter can be used to fuse any kind of sensory data, as long as they provide the same state variables. For instance, instead of vision, the reader can fuse inertial sensory data and odometry with this technique. We will keep on mentioning vision as the sensory system used as observation in the Kalman filter. However, the reader can imagine this as a black box that retrieves an estimate of the robot's pose, i.e., a pose vector composed of $[x'\ y'\ h']$.

Such measurements can be described as a nonlinear function c of the robot coordinates, which possesses an independent noise vector v. Denoting the nonlinear equations (28) as the estimated pose vector α, and placing Δd_k and Δh_k as an input vector u_k, with associated process noise vector w_k, the robot can be modelled by the following nonlinear equations

$$x_{k+1} = \alpha(x_k, u_k, w_k, k) \qquad (29)$$
$$z_k = c(x_k, v_k, k)$$

which represent the state transition model and the observation model of the robot. The reader should note the importance of these equations, since they reveal the basis of sensor fusion with a Kalman filter, in this case, the Extended Kalman Filter (EKF), since we are dealing with a nonlinear system.

Note to the reader: Real systems are often nonlinear. This means that its state variables have a nonlinear behaviour in time. In our case, the position of the robot cannot be modelled by a linear equation, since its pose can possess unpredicted values.

The first, on equation (29) is the state transition model. The prediction stage of the Kalman filter relies only on odometry data to estimate the actual state, based on the walked amount between the actual and last iteration (determined by odometry). u_k is often denoted as the control input, which in our case, is simply the module of the distance the robot walked between two sample instants. It is important at this point to take in account the encoder resolution: it should be higher than the controller's sampling ratio so that between different control instants we can have a measurement of the distance walked by the robot as explained in section 3.1.5. So basically what the state transition model does, is to perform a naive prediction of the actual pose of the robot based only on odometry data. This prediction will next be refined by the correction phase of the filter, where the observation data is used: the vision system in our case. w_k is denoted as the state transition model associated error. Here we define the variance (or associated uncertainty) to the state variables determined by the state transition model. This vector is set up when projecting the stochastic model of the Kalman filter, and the variance may be adjusted using system simulations. If no correlation between the state variables exists, the covariance matrix of the state transition model noise Q_k will be diagonal, having w_k in its diagonal. The same goes for the noise vector associated with the observation model v_k. This vector represents the uncertainty associated to the observation model, (in our case to the vision system). Its covariance matrix is R_k and again, it will be a diagonal matrix with v_k as its diagonal values if the state variables are uncorrelated, which is the case for the mobile robot's pose estimation.

Note to the reader: The reader can easily imagine the noise by visualizing a Gaussian probabilistic density function for each state variable. If more uncertainty is given to a certain state variable, we are disbelieving that the process can provide a good estimation, hence raising the bell shape size of the Gaussian distribution in turn of the mean value, which is normally zero for white noise. We are then allowing values farther than zero for the noise, which will affect our estimated state variables.

With this being said, it is recommended to simulate the system using MATLAB for instance, so that proper tuning of the state variables can be achieved.

So, in our case $w_k \sim N(0, Q_k)$ and $v_k \sim N(0, R_k)$, being both non correlated, i.e., $E[w_l v_l^T] = 0$, where l is the number of state variables. The Extended Kalman Filter can then be constructed, using the odometry-based system model [8]

$$\begin{aligned} \hat{x}_{k+1} &= \alpha(x_k, u_k, w_k, k) \\ P_{k+1} &= A_k P_k A_k^T + Q_k, \\ K_k &= P_k C_k^T [C_k P_k C_k^T + R_k]^{-1} \\ \hat{x}_{k+1} &= \hat{x}_k + K_k [z_k - C_k \hat{x}_k] \\ P_{k+1} &= [I - K_k C_k] P_k. \end{aligned} \tag{30}$$

The filter was simulated using different noise vectors for both state transition and observation model. A DC motor transfer function was used to simulate both models (Fig. 4). This naïve simulation allowed to make concrete conclusions.

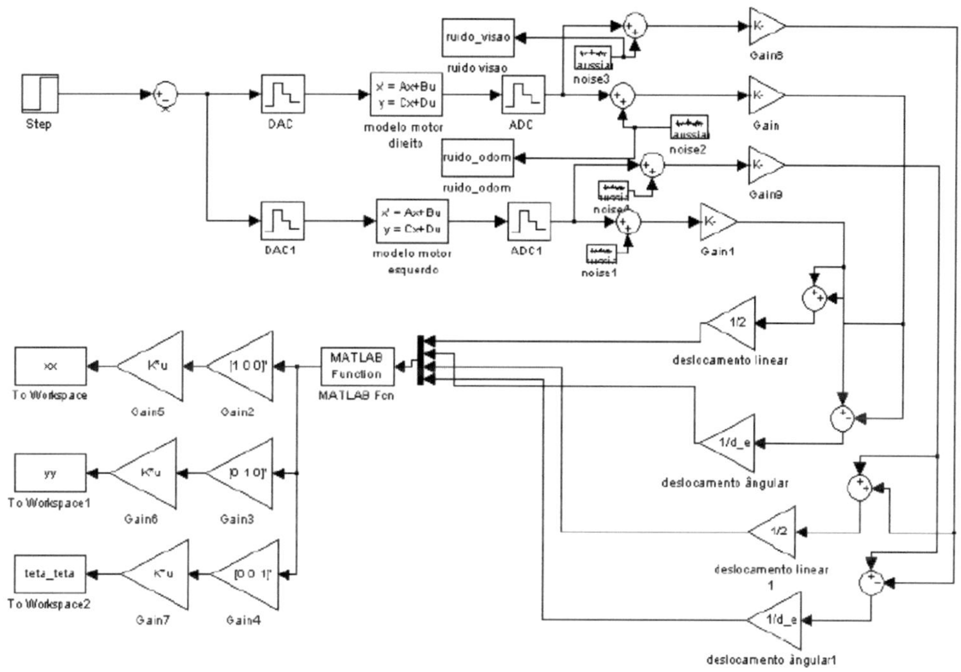

Fig. 4. Kalman filter simulation using same source for both models.

When a voltage is present at the input of each motor transfer function, a displacement is produced at its output, allowing us to simulate the robot movement within our system. In this case, and because there is intentionally no feedback loop (evident by the disconnected entrance placed before the DAC's) as the objective is to simply evaluate the filter's performance under different noise conditions for both the state transition model and for the observation model, the output will rise indefinitely in a linear form, except at the start phase of the motor's rotation movement. Observations of the filter's behavior in the present of both vision and odometry noise, and subsequent analysis of the estimated state variables were made. for this, specific situations were imposed.

3.1.7.1 Simulation 1 – robot in x=0, y=0, h=0, $\sigma_{vis}^2 = 1, \sigma_{odo}^2 = 1$

For this case, the displacement made by the robot is indefinitely linear along the *xx* axis, being the displacement over *yy* and the robot's heading equal to zero. Fig. 5 (left) shows this situation, being the blue slope the displacement over *xx*, the red slope the displacement over *yy* and the green slope, the robot's heading.

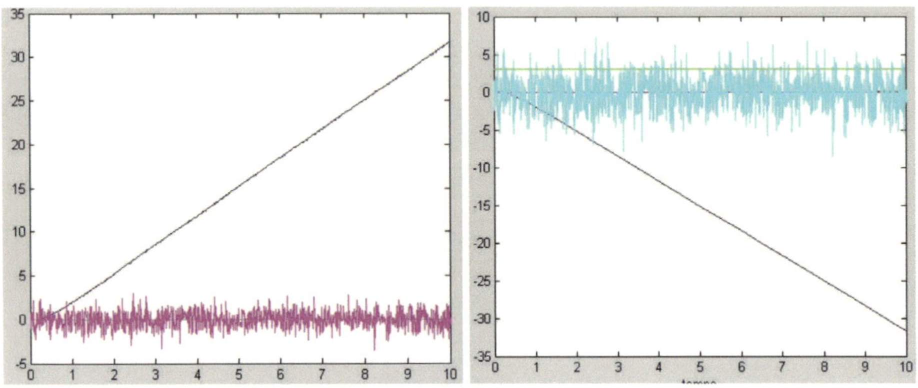

Fig. 5. Filter simulation case 1 (left) and 2 (right).

The magenta slope represents the vision noise. As one can see, the filter possesses sensitivity to the vision noise contrary to what happens with odometry noise as we will further see.

3.1.7.2 Simulation 2 – robot in x=0, y=0, h=3.14, $\sigma_{vis}^2 = 0, \sigma_{odo}^2 = 6$

In this case, the filter's robustness to odometry errors is evidenced. Note that, because the robot is oriented by π radians, letting *xx* axis be the reference of the angle, the robot will move along the negative side of this axis. Fig. 5 (right) shows this situation, having a cyan slope that represents the odometry noise and its exaggerated variance. High belief is put to the observation model, therefore, s it can be seen from the figure, in every sample instant the noise does not affect the output.

3.1.7.3 Simulation 2 – robot in x=0, y=0, h=3.14/2, $\sigma_{vis}^2 = 10, \sigma_{odo}^2 = 6$

In this case, the vision noise was drastically augmented, revealing some fragility of the filter (Fig. 6). However, this situation will not happen in practice since calibration makes the vision noise drop to very low values. One must however be aware of this fact when choosing the sensory system for observation.

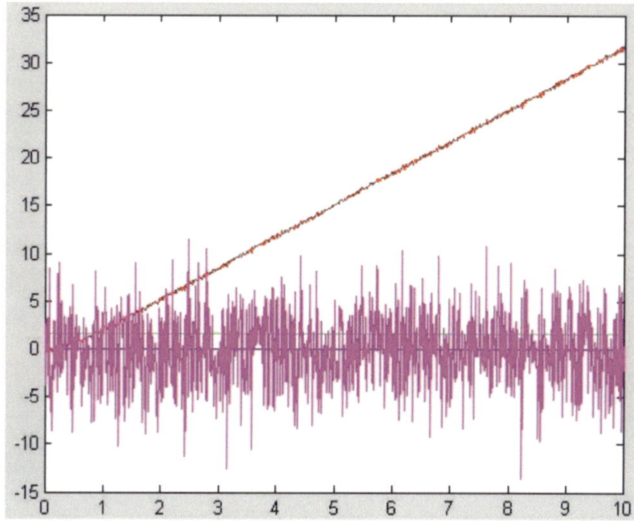

Fig. 6. Filter simulation case 3. Note the effect of observation error on the pose estimation.

3.2 Preliminary Results

The previously described control algorithm was implemented in its preliminary version 1 (without the real-time system for now), and promising results were achieved. It accepts linear velocity vectors as input commands (`setVelocities`) composed of a module and angle$[v_f\ \theta_f]$ in [m/s rad], as well as angular velocity commands in rad/s. A simple test application that sends these commands was used to test the controller, and a visualizer application was made for one to see and record what is really happening in the world. It is basically a virtual representation of the reality happening in real-time. This latter registers all the robot movements as white dots, represents the virtual point trajectory (calculated as shown in section 2.1.4), and also the initial state of the robot. The goal is that the robot autonomously follows the predefined trajectory of virtual points. All the data is sent from the robot to the visualizer using a TCP/IP wireless network.

The test tool is also able to send target desired poses, which we will call from now on as `setPoints`. This command is interpreted by the control module in its input stage, and a velocity vector is composed that allows the robot reach the desired point.

Note to the reader: Recall that we are focused on the control module, so no obstacle avoidance is taken in account here. Our experiments occurred in a world free of objects. This task should be performed by higher level software modules capable of perceiving the environment and sending velocity commands to the robot so it can then be instructed with movements that will result in obstacle avoidance. Our work is focused on the motion control only.

3.2.1 `setPoint` command to (0,0), the center of the world

For this test, we send a `setPosition` command to (0, 0). In figure 7 (left), a screenshot of the previously referred visualizer tool developed for the controller module is shown for this particular case. The robot accurately goes to the defined `setPoint`, possessing a yy axis precision error of 1 to 2 centimeters maximum. This precision error exists because of two main causes: the first is due to the backward force exerted by the energy cable that feeds the robot in test environment; the second - and main - reason is the defined tuning parameters for the influence of the robot's error over the *yy* coordinateK_y. Tuning for near-zero error is possible but leads to a very hard control scheme in the presence of physical disturbances, making the controller produce high overshoots for compensation. Since our robot is destined to walk on a world where collisions with other robots may be present, the revealed accuracy perfectly suits for the team's needs. For collision-free applications, where minimum physical errors exist, and depending on the world's physical available space, the controller can be made harder by rising not only K_y but also K_x and K_h (please recall section 3.1.2).

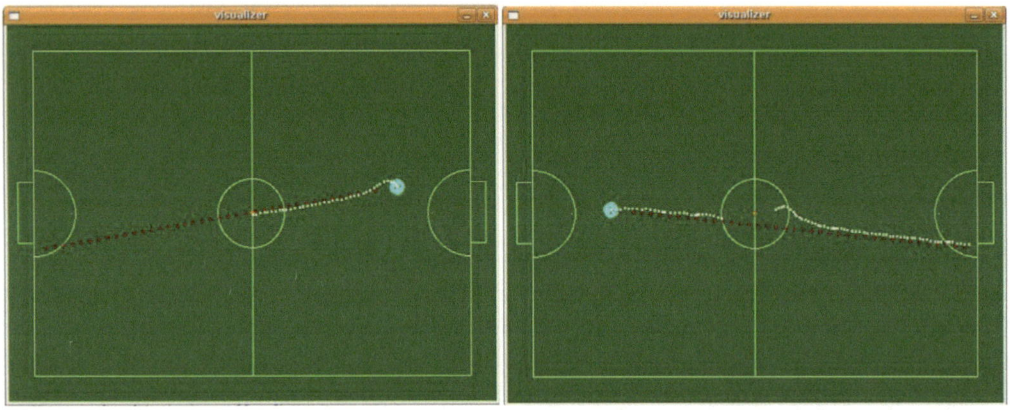

Fig. 7. `setpoint` to (0,0) (left) and `setVelocity` with v=0.3mps and desired angle = 0 (right).

3.2.2 `setVelocity` command with v = 0.3mps and desired velocity angle = 0 rad

For this test, the robot was subjected to extreme noise conditions. Referring to fig. 7 (right), it can be seen that the robot is subjected to two disturbances. One was accomplished by blocking its left wheel, evident by the multiple white dots in the same place. Another was made by blocking the color code of the robot used by the vision module to identify it, so that no vision data was being received by the controller for it to estimate the actual pose during this period (odometry-only based pose estimation: Kalman bypass). Controller's robustness is then proved, since the robot follows the desired trajectory as it starts receiving vision feedback, despite its erroneous position at that time because of the -- widely already known -- odometry alone based pose estimation cumulative error for the feedback loop of the controller. Note that a bypass to the Kalman is made when no vision data is available.

3.2.3 Sequence of `setVelocity` commands with v=0.3mps and velocity angle = 1.3 rad

In this final test (Fig. 8), a sequence of `setVelocity` instructions were sent to evaluate the control module's response in the presence of new velocity instructions. This test approaches the real application target environment, the soccer game, where high dynamic handling is required for the robot. Referring to Fig. 8, a velocity vector with zero desired angle is first sent, followed by two `setVelocity`'s with the same module and 1.3rad for the desired velocity angle. Finally, the robot is halted with a halt instruction. As it can be seen, the robot accurately executes the performed commands, evidencing the software module's robustness. More tests will be shown in further sections regarding the final version of the software implementation which is capable, for instance, of raising the acceleration so that the robot can overcome obstacles. It will also possess the real-time system for greater responsiveness achievement and more predictability.

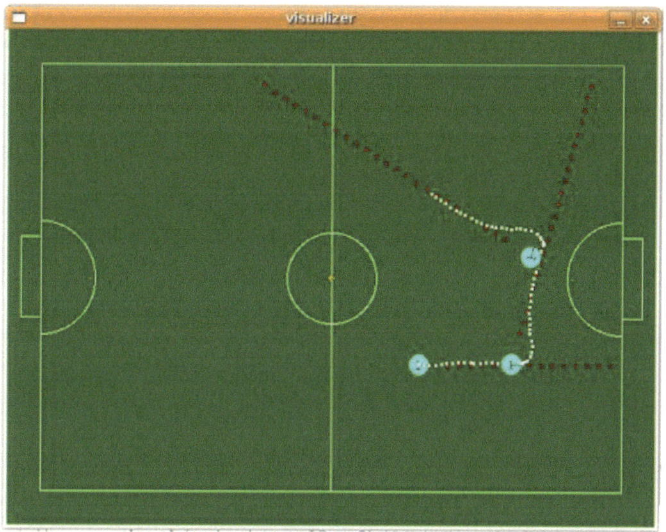

Fig. 8. Sequence of `setVelocity` commands with v = 0.3mps and velocity angle = 1.3rad.

4. Real-Time System

The modularity of the implemented controller described in the previous sections made possible the project of a suitable real-time RTAI (Real Time Application Interface) [12] based system.

Note to the reader: During the course of designing and implementing the real-time system, two articles were written [10] and [11] that may help the reader to implement an RTAI based real-time system for his own applications.

During all the control module software run-time, there are concurrent threads scheduled with the fair non-preemptive Linux scheduler which in turn is implemented in its kernel space. This imposes critical problems, since it is desired to have tasks scheduled in a premeditated fashion so that the control system can be predictable, and Linux is not able to guarantee response times for its processes. For instance, a critical situation comes when the `setTrajectory`[4] instruction is executing: the thread that calculates the distance to the `setPoint` is superimposed by the `setVelocity` thread, making the robot not realize when it is near the target. So, it is desired to have a better solution than the one implemented, which was presented above, so that full robustness can be achieved. Therefore, the desired system needs to have the following requisites,

[4] Instruction that defines various `setPoints`, i.e., target positions that the robot must achieve.

• Predictability;
• Quick response to instructions;
• Ability to choose the highest priority task within a concurrent set of tasks;
• Minimize the error of the current robot pose received from the vision system and/or odometry;
• Ensure fixed periodicity for the controller;
• Preemptivity;
• Ensure precise cycle times for periodic tasks.

Every system that deals with a high number of tasks at the same time must be predictable, i.e. such systems must always exhibit a specified behavior (e.g. a task executed with a precise periodicity and no interruptions). Or if a critical instruction is sent to be executed on the fly, the system responds to the desired order as expected, when expected. This type of behavior is completely desired for the system.

As for the pose data received from odometry and/or from the vision module, it is important to have an intelligent planned scheduling for both tasks, in order to receive the precise actual pose of the robot at any time without information delays caused by thread overlapping.

The feedback cycle needs to be periodic so that the controller executes properly. This can only be ensured by scheduling the controller task with a pre-defined period and setting a high priority to it in the preemptive real-time system. This task is classified as a hard real-time task, since its activity must be started at every period with no delays. The integration of a real-time system in the control module aims at fulfilling these requirements.

4.1 Designing the system: time line example

A real-time system must be carefully planned. We used UML (Unified Modeling Language) to aid on this complex task. This is, for sure, the most important and hard task on the process of developing a real-time system for an application. All the evolved tasks must be pointed out, and their dataflow must be clearly exposed. The importance of each task must also to be taken in account when assigning priorities to the tasks. Then, the system architect must also to evaluate if the tasks are *soft* or *hard* real-time, allowing (or not) the task to possess a delay for its activation.

We projected a system for our case, which will not be referred here due to loss of generality. However, an example of a time-line case of our system is depicted to serve as example to the reader on what is pretended. Fig. 9 shows various tasks of the controller software module and its states. The Two final are instructions which handler is implemented in the control software module. The `Controller`, `Command Interpreter` (parsing) and `System Status` are main tasks of the control software module. The sequence of instructions executed upon receiving a `setVelocity` instruction can be seen in the figure. For the tasks to run as expected their priorities, activity mode (one shot or periodic) and activation type (hard or soft) were carefully planned. For instance, the `System Status` task possesses high periodicity over all other tasks to listen for new instructions. This is visible upon receiving a HALT instruction, which makes the controller stop while running.

Note to the reader: Recall that projecting a real-time system is all about knowing the application to implement. Modularity helps a lot on this, since then we can clearly define the real-time requirements for the system. Implementing the system is in fact really easy when compared to the real-time project phase.

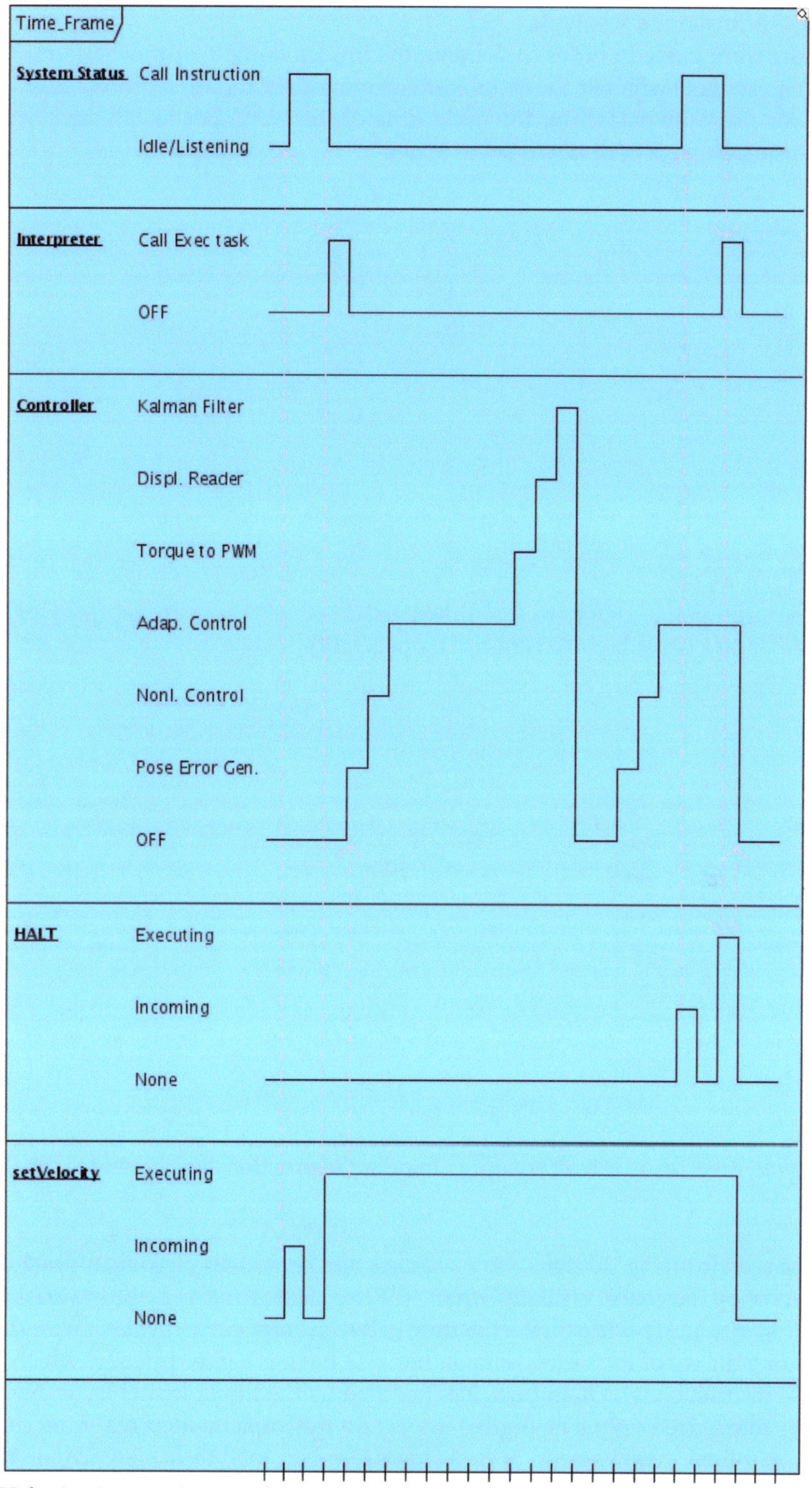

Fig. 9. setVelocity instruction receive case timeline.

4.2 Timing Performance Analysis

Various tests were made in order to validate the implemented real-time system. These tests allowed to investigate whether the tasks were meeting their timing requirements.
One test was made by extracting the cycle time of the controller task, which was set as a hard real-time task. This period was set to 52ms.

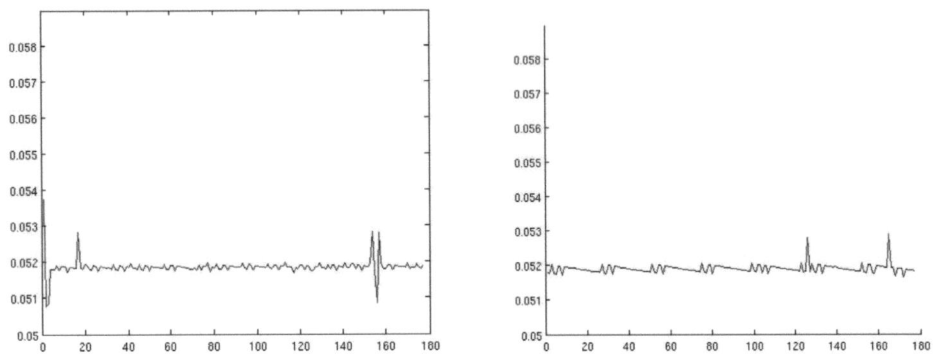

Fig. 10. Measured period of a hard real-time period task of the controller (left) and Soft real-time controller task times between control loops (right)

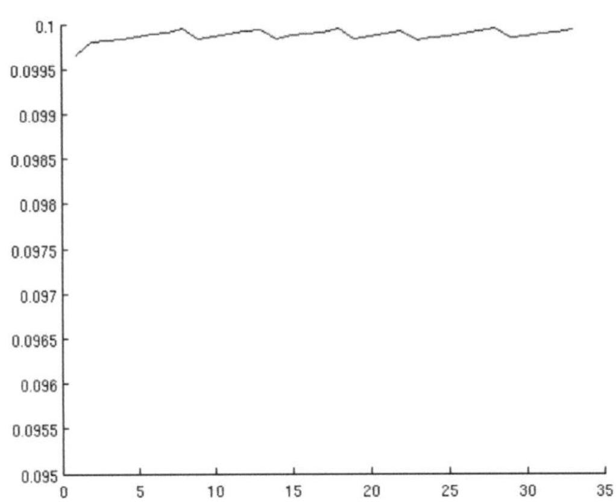

Fig. 11. Cycle time of `setTrajectory` handler instruction, implemented in the control module.

As it can be seen from Fig. 10 (left), very accurate and consistent periodic times are achieved being, in most of the cases, virtually equal to 52ms possessing very little variation. It was observed that in some few iterations, the time between consecutive cycles shifts 52ms, which was caused by delays of the vision data gather task having higher priority, which needed to call a recv() instruction which, in turn, is scheduled by the Linux kernel layer. Nevertheless, such delays introduced only a negligible impact on the implemented real-time controller. If the hard real-time requirement is not imposed for this task, few differences were surprisingly observed in run-time (Fig. 10 (left)). This leads to the conclusion that the schedulability of the system provided by RTAI is accurate and also that the system was robustly projected since there aren't observable scheduling issues during the controller execution.

Another test was made to the *setTrajectory* handler task cycle times, which in turn periodically evaluates the distance left for the robot to reach a desired target point allowing the robot to know when to stop the controller since it reached the desired target point. A period time of 100ms was set and no hard real-time imposition was made. As it can be seen from Fig. 11, the periodicity varies within less than 0.1s which confirms the robustness of the implemented real-time controller. These remarkable results allow the robot to precisely reach a desired target point and proceed to the next one with high precision. As for task activation times, the observed values possessed an average of 10ms. These times are perfectly acceptable for the application. These practical results demonstrate that the system's timing requirements are successfully met.

5. Experimental Evaluation

This section is dedicated to the analysis of the experimental results obtained by the implemented
control system with real time system totally incorporated. These tests were aimed at demonstrating about the system's integration on dynamic environments and assess its robustness. Instructions such as *setVelocity* and *setTrajectory* are sent to the robot and the pose error results obtained are analyzed.

This section also intends to analyze the influence of the controller variables K_x, K_y and K_h which need to be tuned so that a good performance is achieved. During the following set of tests, we consider that the robot is walking on a obstacle-free world, where the (x, y) pose measurement units over each axis is mm.

5.1 Analysis of the controller parameters influence: Tuning the controller

As it was introduced in section 3.1.2, there are three important weighting variables which define the controller's correction strength over each DOF of the robot, which are K_x, K_y and K_h. From a practical overview, the bigger these variables are, the more action is put to correct the trajectory (minimize the pose error) of its actuating state variable.

The following analysis will concentrate on the K_x and K_y parameter tuning, letting the correction over the heading be always the best as possible. It is important to note that this section also explains how to make proper tuning of the controller based on the pose error analysis. For the next three sections, a *setVelocity* is sent with module 0.2 mps
and angle 0 rad.

5.1.1 Very soft controller: Kx = 0.0001 ; Ky = 0.02 ; Kh = 2.4

In this case, the parameters were set for the controller to make very soft corrections over each axis. The stop condition is 16mm error over *xx*, 10mm error over *yy* and 0.5 radians for the heading. This means that a virtual point is considered to be achieved if the pose error stays bellow these values at the same iteration. Referring to figure 12 (left), it can be seen that a slightly inaccurate control is made over *yy*. The error oscillates above zero and tends to increase as time goes by. The same happens with the error over *xx* however, in this case, it can be seen that the error drops quickly every time a virtual point is achieved (at every pitch value of the blue line). As one can easily see, a more accurate tuning over *yy* is absolutely needed. Note that the blue line represents the error over *xx*, the green line, the error over *yy*, and the red line, the heading error all towards the settled virtual points (red dots in the visualizer screenshots).

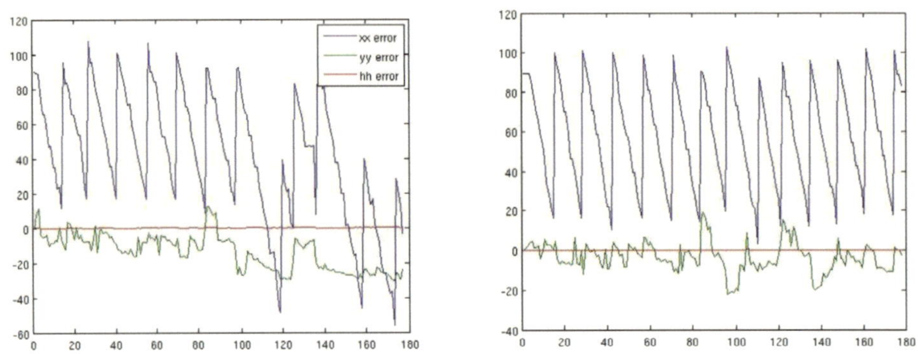

Fig. 12. `setVelocity` error for the very soft controller case (left) and `setVelocity` error for the soft controller case (right)

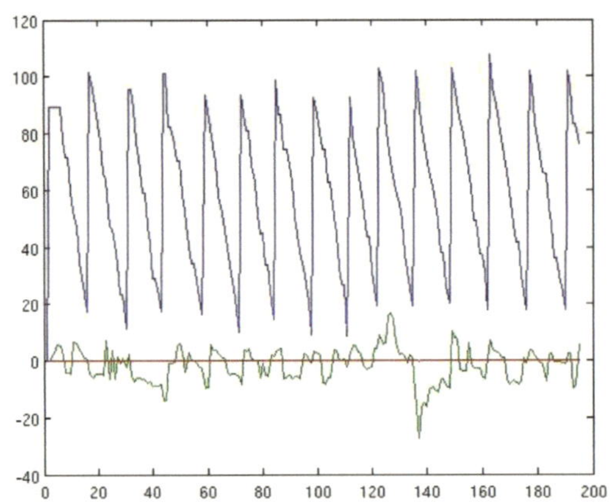

Fig. 13. *setVelocity* error for the hard controller case.

5.1.2 Soft controller: Kx = 0.0001 ; Ky = 0.09 ; Kh = 2.4

It is expected that by raising the value of K_y, the pose error results be more accurate. As it can be seen from Fig. 12 (right), the error over *yy* diminishes, rounding zero. Now, the robot follows his trajectory but still not with the desired accuracy.

5.1.3 Hard controller: Kx = 0.0001 ; Ky = 0.12 ; Kh = 2.4

Raising K_y a little more, better results are achieved. It can be seen that the robot possesses very little error during the trajectory execution, possessing a pitch of 2 cm at most. These results are remarkable, and fairly validate the implemented control system as well as its architecture, since all the projected system was conceived for the controller to possess the best performance as possible.

It is important to note that the results could be improved even more by making the controller even harder by raising the weighting variables. The drawback, however, which is high overshoot in the presence of physical disturbances, must be taken in account. By this, it must be comprehended that while tuning the controller, the overshoot of the response to physical disturbances needs to be seriously taken in account. During a soccer-game, for example, the controller can't be made too hard by any means due to the high risk of collision with other robots.

5.2 Field Experiments

A high number of experimental results with the robot on the field (see Fig. 12) were carried out in order to validate the proposed real-time controller.

Fig. 11. RAC robot soccer player robot used to validate the controller.

The following results were achieved by setting target points on the field which the mobile robot had to reach. The images represent screenshots of the developed (and already referred in section 3.2) visualization tool that depicts an on-line virtual representation of what is happening on the (real) field. It merely receives data from the robot to update the images. Recall that red dots represent the virtual points associated with the planned trajectory, white dots are the actual (real) trajectory performed by the robot (online) and the blue mark represents the real robot when starting a *setVelocity* command. The final position is not represented. As it can be seen from both Fig. 12 and 13, the robot accurately follows the self-planned virtual point paths, which validates not only the developed controller approach but also the trajectory planning capability. It is important to underline that the virtual point path plan is made accordingly with a velocity vector with no stop point defined (visible by the red dots that go farther than the desired point to achieve. The stop point is detected by the *setTrajectory* task, which is periodically checking the distance to the desired point.

The system is very accurate even in the presence of sensor noise. It is capable of reconfiguring the trajectory if it is placed elsewhere on the environment during the motion execution, due to the effective pose error elimination of the controller. The real-time system allows to attain dynamic responsiveness and task scheduling. For instance, due to the robust projected task priority system, the controller runs in parallel with the *setTrajectory* task which, in turn, checks periodically the distance left to the target point. The robot accurately switches the movement to reach the next target point as soon as the actual target point accomplishment is detected.

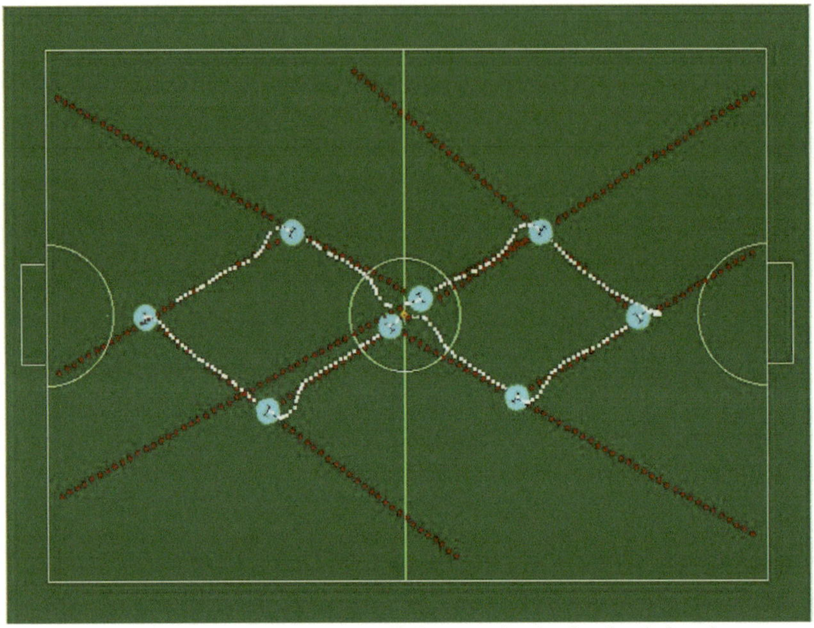

Fig. 12. Instruction `setTrajectory` - case 1: Start point is the far left robot representation and the final one is the same as the start point.

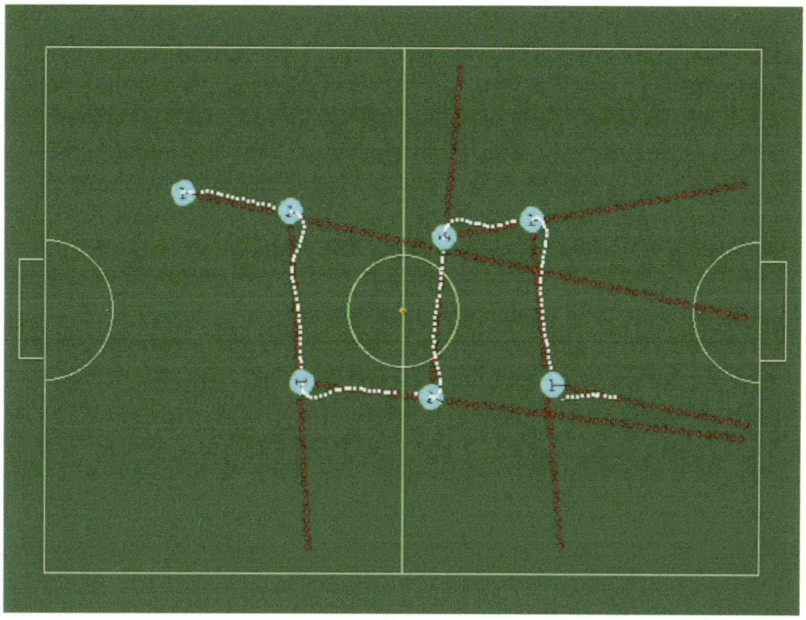

Fig. 13. Instruction `setTrajectory` - case 2: Start point is the far left robot representation and the final position is the far right white dot.

6. Conclusion

A real-time controller for pose error elimination of differential mobile robots was developed, which includes a trajectory planning algorithm. In the controller's theoretical formulation, particular emphasis was given to the design of a generic control scheme, so as to be robust against errors in the estimation of the robot's dynamic parameters. A real-time controller was designed so that specific timing constraints were also met. Experimental results obtained on a real mobile robot reveal that the presented approach is not only valid but also robust. It allows a generic differential mobile robot to correct its trajectory, leading it to converge to the desired pose.

7. References

[1] A. Gholipour, M. J. Yazdanpanah (2000). *Dynamic Tracking Control of Nonholonomic mobile robot with model reference adaptation for uncertain parameters*, University of Tehran.

[2] A. M. Bloch, M. McClamroch (1991). *Control and Stabilization of Nonholonomic Caplygin Dynamic Systems*, England

[3] A. Martinelli, R. Siegwart (1996), *Estimating the Odometry Error of a Mobile Robot During Navigation*, Autonomous Systems Lab, Swiss, Federal Institute of Technology Lausanne (EPFL)

[4] D. Wang, Guangyan Xu (2000). *Full State Tracking and Internal Dynamics of NonHolonomic Wheeled Mobile Robots*, Proceedings of the American control Conference

[5] H. Burkhard (2002). *Real time control for autonomous mobile robots*, IOS Press Amsterdam

[6] M. Vidygascar (1993). *Nonlinear Systems Analysis*, Prentice Hall, New Jersey, NJ

[7] R. Pissard-Gibollet1, K. Kapellos1, P. Rives1 Contact, J. J. Borrelly1 (1997).*Real-time programming of mobile robot actions using advanced control techniques*, Springer Berlin, Heidelberg

[8] T. Larsen, M. Bak., N. A. Andersen O. Ravn (2000).*Location Estimation for an Autonomously Guided Vehicle using an Augmented Kalman Filter to Autocalibrate the Odometry*, Technical University of Denmark

[9] Y. Kanayama, Y. Kimura, F. Miyazaki, Tetsuo Nogushi (1990). *A Stable Tracking Control Method For An Autonomous Mobile Robot*, 1990 IEEE.

[10] J. Monteiro (2008). RTAI *Installation complete guide*, https://www.rtai.org/RTAICONTRIB/RTAI_Installation_Guide.pdf

[11] J. Monteiro (2008). *Building your way through RTAI*, http://isharemyknowledge.blogspot.com/2008/03/building-your-way-through-rtai.html

[12] Real Time Application Interface, http://www.rtai.org

[13] J. Monteiro and R. Rocha (2008). *RACbot-RT: Robust Digital Control for Differential Soccer-Player Robots. In Proc. of 5th Int. Conf. on Informatics in Control, Automation and Robotics (ICINCO'08)*, Funchal, Portugal, pages 225-228.

11

Control of fuel cell systems in mobile applications

Jiri Koziorek, Bohumil Horak and Miroslav Kopriva
VSB-TU of Ostrava, Faculty of Electrical Engineering and Computer Science
Czech Republic

1. Introduction

A car is one of significant inventions in the history of the human being. With respect to ending crude oil reserves and deepening environment pollution problems, alternative power resources have been started to be found twenty years ago. The aim was and still is searching for power resources less harmful to the environment and providing further long-term carriage and transport development.
Concentrating on the land carriage, it is possible to choose from several types of alternative fuels. Significant alternatives to the current majority use of petrol and diesel oil as fuels to power transport means are hydrocarbon and hydrogen based fuels in form of compressed gases and fluids. Also solar propulsions and compressed air propulsions have been developed. Hybrid and electrical propulsions become still more and more important. The promotion of such alternatives is accompanied by a number of problems which should be addressed in relation to performance, permeability, transport distance ability and cost, as well as transport infrastructure, etc.
Alternative fuels to petrol and diesel oil used as propellants for vehicles mostly comprise:

Compressed natural gas (CNG).
Liquefied petroleum gas (LPG).
Biogas.
Bio diesel oil and fuels based on rape(-seed) oil methyl-ester.
Alcohol (ethanol and methanol) using fuels.
Hydrogen.
Electric current.

It is evident that "to burn crude oil" is too big luxury since it will be missed by time to the chemical industry which is dependent on it. Currently, compression-ignition engines can use a chemically treated fuel e.g. from rape (-seed) oil known under the name bio-diesel oil. Compared to classic diesel oil, it provides benefits in a combustion process of a thermal motor represented for example in lower quantity of harmful substances in exhaust gases. However, bio diesel oil solves the problem only partially. This is because planting of monocultures at large areas is only possible at remote world parts.
Currently, propane-butane designated as Liquefied Petroleum Gas (LPG) is the most used alternative fuel. Another variant which acquires a large development most recently is

natural gas powered transport means operation. It can be used in a minimally modified spark-ignition motor. Natural gas is a fossil fuel which releases harmful substances into the atmosphere while burning, although in much less quantity. The use of advanced exhaust gas catalysts provides meeting of emission standards for ultra low emission vehicles. They correspond to harmful emissions of an electric vehicle taking into account emission production during electric power generation. Compared to common fuels, harmful substances are produced less by 95 %. Further, Compressed Natural Gas (CNG) and Liquefied Natural Gas (LNG) can be distinguished. CNG is used for light vehicles while LNG is used for trucks and buses. Like in case of LPG, fuel tanks do not require much space. A natural gas powered vehicle can provide a shorter range and its motor output is lower if compared to the petrol or diesel oil powered vehicles. A fuel supply and pumping system is more demanding too.

The electrical vehicle is powered by an electric motor and produces minimum of harmful substances in operation. The electromotor is supplied from electrochemical accumulators. A development of electric drives in transport is currently prevented from by small one charge range, long charging time and low service time, big weight and high price of inbuilt electrochemical accumulators. It is rightly there, at a level of electrochemical accumulators, where majority environment loading by harmful substances should be found. Significant benefits comprise minimum air emissions in the electrical vehicle traffic, its silent run and low maintenance cost. Aiming to solve the problem of its limited distance reach, the electrical vehicle can be equipped by an electrical motor and combustion motor. Transport means using a hybrid propellant use electric power as a propellant for small distances and at not demanding terrains. A combustion motor is only used at demanding terrain and for long distances. In this way, the performance and distance range are optimized providing also environment friendly traffic, low noise and cost. They are rightly hybrid drive systems which may find application in transport for some transient period. The aim is such their combination that their specific benefits are used as much as possible. Currently, hybrid powered transport means fabricated in a number of car producing factories all over the world. In the future, a hybrid propulsion could become a high quality alternative provided a suitable electromotor power source, i.e. accumulator, is developed.

However, the top perspective is provided by the electromotor drive due to its suitable moment characteristics with a fuel cell current source which transfers the chemical fuel power (hydrogen or hydrocarbons) directly to the electric power. The experience acquired in natural gas powered vehicles operation serves to development of perspective hydrogen powered vehicles. Currently, hydrogen handling is problematic, and its power demands are even higher than hydrocarbon handling. It should be cooled up to the temperature of -253°C in fluidization. Unfortunately, the hydrogen production itself is much electric power demanding. Hydrogen priorities comprise that water vapor is its only burning product. However, a way of hydrogen burning in a conventional combustion motor is not long-term perspective because of its low efficiency.

At the present, the many-years research and development of fuel cells is has been prepared to their putting into a series production. Rejected fuel cells in relation to conventional electrochemical accumulators of electric power do not load the environment by heavy metals, and their service life is comparable or longer. Fuel cells produce electric power from hydrogen and oxygen.

Hydrogen can be gained by a chemical reformation from hydrocarbon fuel in unit (reformer) directly connected with the fuel cell. This presents a technical complication, efficiency decrease, increased weight, as well as other problems. However, they are not unsolvable and, therefore, currently almost all big car producing factories deal with development of fuel cells (General Motors, DaimlerChrysler, Ford, Mazda, Honda, Mitsubishi).

1.1 Hydrogen and fuel cells

A base of the power system design is to determine a suitable technology providing high efficiency and minimum surrounding environment pollution by emissions under usual conditions of the surrounding environment, and a choice of a suitable achievable power source, renewable if possible, of a high power content.

This choice is further subject to another selection run mostly comprising social and political, and economic requirements determining a success of the designed power system under competitive market conditions.

The present chapter deals with fuel cells and hydrogen, combination of a highly efficient technology and pure medium with a high power potential.

Hydrogen is colorless gas without any taste and odor. Under normal conditions, atomic hydrogen is significantly reactive (especially with oxygen and halogens). It creates compounds with all elements of the periodic table save noble gases. Despite high reactivity, it needs initialization power to the reaction. It is able to create a special type of a chemical relationship called a hydrogen bond or also hydrogen bridge, when a bound hydrogen atom indicates affinity also to other atoms which it is not bound to by a classic chemical bond. A hydrogen bond with oxygen atoms is then extraordinarily strong, which explains anomalous physical properties of water (high boiling point and melting point, etc.).

Molecular hydrogen is relatively stable and, with respect to a high value of bonding power, also little reactive. Therefore, it bonds with most elements only under increased temperature or in presence of catalysts.

Hydrogen reactions are accompanied by heat release (exothermic reactions), sometimes also by a light effect - burning. Key properties of hydrogen are presented in the Table 1.1.

General			
Name, symbol, number	Hydrogen, H, 1	Heat of vaporization	0.904 kJ mol^{-1}
Element category	Nonmetal	Heat of fusion	0.117 kJ mol^{-1}
Group, period, block	1, 1, s	Specific heat capacity	14 304 J $\cdot kg^{-1} \cdot K^{-1}$
Appearance	colorless	Electric conductivity	-----
Mass concentration in Earth´s crust	0.88 %	Thermal conductivity	0.181 5 W $m^{-1} \cdot K^{-1}$
Atomic properties:		**Miscellaneous:**	
Standard Atomic weight	1.00794(7) g mol^{-1}	Concentration of H_2 for combustion in air	4 ÷ 75 %

Atomic radius (calc.)	25 pm (53 pm)	Concentration of H_2 for explosion in air	13 ÷ 59 %
Covalent radius	37 pm	Highest temperature of combustion in air	2 318 °C (at concentration of H_2 29 %)
Van der Waals radius	120 pm	Highest temperature of combustion in O_2	Up to 3 000 °C
Physical properties		Diffusion coefficient	0.61 $cm^3 \cdot s^{-1}$
Phase	Gas	Electrode potential	0 V ($H^+ + e^- \rightarrow H$)
Crystal structure	Hexagonal	Electronegativity	2.1 (Pauling scale)
Density	0,0899 $kg \cdot m^{-3}$	Higher heating (calorific) value	2.98 $kWh \cdot m^{-3}$ = 10 728 $kJ \cdot m^{-3}$
Density of liquefied hydrogen	70,99 $kg \cdot m^{-3}$	Higher heating (calorific) value	33,3 $kWh \cdot kg^{-1}$ = 119 880 $kJ \cdot kg^{-1}$
Magnetic ordering	Diamagnetic	Ionization energies	1st: 1 312.0 $kJ \cdot mol^{-1}$

Table 1.1 Key hydrogen properties.

In the nature, hydrogen occurs as a mix of three isotopes, particularly protium (light hydrogen, 1.00794 amu) - 1H, deuterium (heavy hydrogen, 2.01363 amu) - 2H (2D) and tritium (3.01605 amu) – 3H (3T).

Elementary atomic hydrogen (H) occurs on the Earth only rarely. In such form it only occurs mostly in the star space where it amounts to 75 % of the weight, and almost 90% in relation to the quantity of atoms present in the Universe. It occurs only in higher strata in the Earth atmosphere and, due to its extraordinarily low weight, it is gradually released from the atmosphere.

On the Earth, hydrogen mostly occurs in form of two-atom molecules H2. As for compounds created by hydrogen on the Earth, most of all water is represented which in form of seas and oceans covers 2/3 of the Earth surface. Other compounds are represented by organic compounds. Hydrogen, together with carbon, oxygen and nitrogen, is included into so-called biogenic elements (elements forming basic "building stones" of all living organisms). Because of that, hydrogen can be found practically in all compounds forming the most significant raw material of the current power and organic chemistry–crude oil.

Hydrogen is also a component of any acid; in water it splits off as H+ ion and creates an oxion cation H_3O^+ which determines key properties of a material, e.g.: polymer membranes of fuel cells. Hydrogen creates compounds called hydrides in valence H^-.

Hydrogen is produced and separated in three ways:

A. Chemical - production of synthesis gas / CO conversion (steam reforming, partial oxidation of crude oil fractions, coal gasification), reaction of non rare metals with acids and hydroxides.

B. Electrochemical – electrolysis of alkalized water or water solutions of chlorides of alkali metals.

C. Physical - adsorption (PSA, TSA), diffusion (membrane technologies, metal hydrides) or fraction condensation (cryogenic technology).

In addition to production, a basic factor indicating hydrogen application is its storage and transport to a destination. Hydrogen is stored and transported in following ways:

1. In form of compressed gas (CH_2) in aluminum, carbon or composite vessels of pressure of up to 600 Bar and power loss in gas compression of up to 15 % from the power the hydrogen contains,
2. In form of fluidized gas (LH_2) at temperatures below -253°C in reservoirs with negative pressure isolations with a power loss to cooling and compression at a level of up to 30 % of power kept in hydrogen,
3. In form of metal hydride and compound with carbon (carbon absorption), more details see (Fromm, 1998),

Other methods have been still in the stage of research, particularly:

Hydrogen storage in a structure from nanofibers - much promising technology with hydrogen storage capacity of up to 70 % of the total system weight,

Hydrogen storage in a form of compound with ferrum oxides (sponge iron),

Hydrogen storage in small glass balls called microspheres, under very high pressure.

A fuel cell is an electrochemical source transforming the power bonded in chemical bonds to usable electric and thermal power. A course and power balance of this transformation mostly depends on a type of a used fuel cell which differs by used material determining its operation temperature.

A basic division of fuel cells mostly depends on a used electrolyte according to which cells are divided into cells with acid or alkali electrolyte.

On anode of a fuel cell with acid electrolyte (fluid, polymer), the hydrogen gas ionizes, releases electrons and forms H^+ ions (protons). On cathode, oxygen O_2 reacts with electrons e^- transported from the electrode and H^+ ions incoming from the electrolyte. Pure water H_2O is a product of the reaction. Processing chemical reactions can be described as follows:

Anode: $2\,H_2 \rightarrow 4\,H^+ + 4\,e^-$

Cathode: $O_2 + 4\,e^- + 4\,H^+ \rightarrow 2\,H_2O$

Generally: $2\,H_2 + O_2 \rightarrow 2\,H_2O$

In case of fuel cells with alkali (basic) electrolyte, a common reaction is the same like in case with acid electrolyte; however, a reaction on particular electrodes differs. Hydroxyl ions from the electrolyte react with hydrogen on the anode while electrons e^- and pure water H_2O are produced. On the cathode, oxygen reacts with electrons from the external electric circuit and water from the electrolyte while hydroxyl ions (OH-) are produced.

Anode: $2\,H_2 + 4\,OH^- \rightarrow 4\,H_2O + 4\,e$

Cathode: $O_2 + 4\,e^- + 2\,H_2O \rightarrow 4\,OH^-$

Generally: $2\,H_2 + O_2 \rightarrow 2\,H_2O$

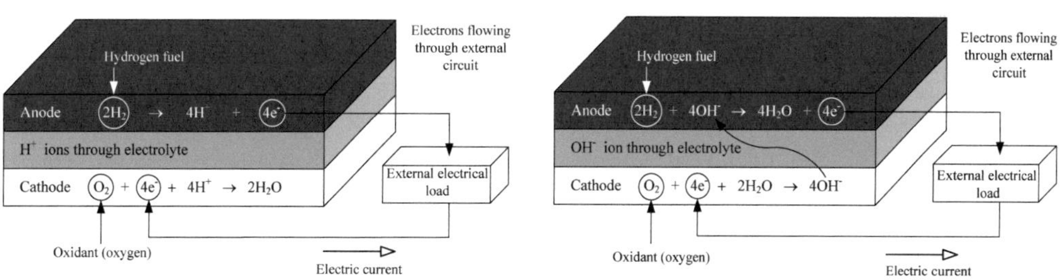

a) b)

Fig. 1. A reaction on electrodes and charge flow in a fuel cell with acid electrolyte a) and with alkali electrolyte b).

In a point of view of the operation temperature, fuel cells can be divided into high-, medium- and low temperature.
High temperature fuel cells operate under temperatures over 600 °C. Such high temperatures enable spontaneous internal reforming of light hydrocarbon fuels to hydrogen and carbon in presence of water. Reactions taking place on the anode supported by a nickel catalyst provide sufficient heat required for a steam reforming process. The internal reforming further removes a need of a separate unit for fuel processing, and enables a fuel cell to process also other than pure hydrogen fuels. These significant benefits result into the increase of the total efficiency by almost 15 %. High temperature fuel cells also produce high potential waste heat which can be used for cogeneration purposes.
High temperature fuel cells comprise Molten Carbonate Fuel Cells (MCFC) and Solid Oxide Fuel Cells (SOFC).
Low temperature and medium temperature fuel cells operate usually with temperatures lower than 220 °C. These low temperatures do not allow internal fuel reforming, due to which low temperature fuel cells require an external hydrogen source. On the other hand, they indicate a quick facility startup, and their construction materials are not so much subjected to defects. They are also much more suitable for applications in transport.
Low temperature fuel cells comprise:

Alkaline Fuel Cells (AFC),

Proton Exchange Membrane Fuel Cells (PEM FC) - Direct Methanol Fuel Cells (DMFC).

Medium -temperature fuel cells are operated under temperatures from 200°C to 600°C. Medium temperature fuel cells comprise fuel cells with the phosphoric acid based electrolyte (PAFC - Phosphoric Acid Fuel Cells).

2. PEM Fuel Cells and their characteristics

Mobile applications are a specific field requiring from driving sources particularly a quick start and high system dynamics, minimum of mobile elements (high reliability) or fluid system parts, maximum performance density and efficiency under minimum source mass, etc.
A fuel cell of the PEM type, which is a subject of the following chapter, seems to one of suitable sources.

2.1 PEM Fuel Cell

A description of the construction of a PEM fuel cell

A PEM fuel cell consists of a Membrane Electrode Assembly formed by a proton membrane, catalysts and electrodes ("sandwich structure"). It is the most important part of a fuel cell since all important chemical reactions occur there. A Proton Exchange Membrane or Polymer Electrolyte Membrane is represented by a film of the thickness usually from 50 to 200 μm, which is formed by Teflon fluoro-carbon polymers with a chain ending by a rest of sulphonic acid ($-SO_3^-$). It means its character is acid. Mostly it is sulphonated fluoro-polymer, usually fluoro-ethylene. A result of presence of these SO_3^- and H^+ ions is a strong mutual attractive force between + and - ions from each molecule. Because of that, molecules of a side chain incline to accumulation inside a total material structure. In terms of a polymer structure, hydrophyllic regions (of the group $-SO_3^-$) are made occurring within a

generally hydrophobic substance. A Proton Exchange Membrane requires sufficient humidification for its proper function, particularly approx. 20 of water molecules per one $-SO_3^-$ chain, which provides it's the conductivity of approx. 0.1 $S \cdot cm^{-1}$ (Kameš J., 2008). Membrane borders are extended with active electrode area overlaps which provide perfect MEA (Membrane Electrode Assembly) sealing against reaction gas escape.
The Proton Exchange Membrane is covered on both sides by a layer of catalysts of the thickness of units up to tens μm (currently below 0.2 $mg \cdot cm^{-2}$) to support chemical reactions. Usually platinum or nickel is used as a catalyst since they indicate a high electro-catalytic activity, chemical stability and electric conductivity. A catalyst in form of very small particles is applied on a surface of rather bigger particles of a coal powder.
Electrodes, anode and cathode are represented by porous paper, textile or felt with carbon fibers of the total thickness from 0.2 to 0.5 mm comprising PTFE (poly-tetra-fluor-ethylene), which is hydrophobic, due to which product water is expelled to the electrode surface, from where it is evaporated into the surrounding environment. Simultaneously, electrodes serve to more uniform surface density of reaction substances flow to a three phase interface (electrode/catalyst, reactant, electrolyte).

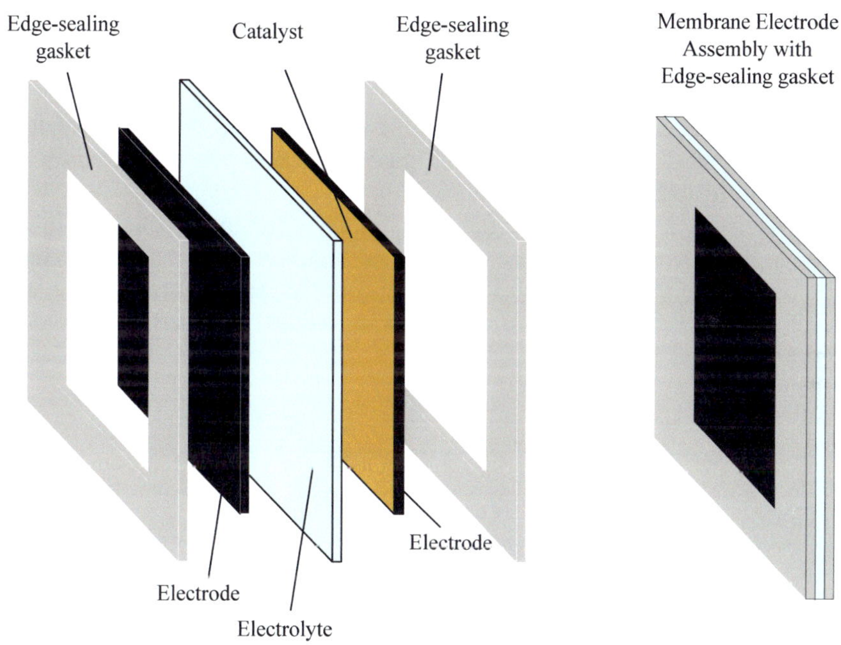

Fig. 2. A structure of an elementary fuel cell - MEA arrangement with end seals.

Bipolar Plates - these plates serving to MEA supply by reactants, discharge of products from MEA and cooling medium distribution, are usually made from graphite (carbon), stainless steel or ceramic materials. They were made by combination of unipolar plates of anode of a cell and cathode of another cell in order to reduce losses in the electric circuit inside a particular cell. On the side of electrodes, they are equipped by channels of a serpentine form providing supply by reaction gas to a membrane arrangement. The stainless steel or graphite containing PTFE is a preferred construction material, with respect to its excellent

electric conductivity, low contamination and relatively low cost. To avoid mixing of the cooling medium with reaction gases, a connection of the anode and cathode plates of two adjacent cells should be equipped by conductive sealing.
End Plates are conductive steel plates comprising inbuilt end electrodes with exit cell clamps. Their sense consists in providing a stabilized cell construction, internal tension increase in the system, power output outlet and electrode cooling.
Auxiliary operation systems and source control (Auxiliary systems) - in addition to a particular system of fuel cells, the electric power source system is equipped by a fuel gas source (hydrogen system), gas oxidation (air or oxygen system), cooling medium source (water system), wetting agents, additive units to monitor and control flows of process media, and other auxiliary process systems to control electric and thermal outputs, set of sensors of physical variables, whole power transformation process control unit. In addition, some systems comprise fuel processing reformers.
For more details see (Larminie J.,2003; Kurzweil, 2003).

2.2 PEM fuel cell polarization characteristics

Required final variable values can be achieved by a suitable construction of a fuel cell (plate/circular system, use and extension of the active catalyst surface, used construction materials, etc.), and setting suitable working points which requires a suitable mathematical physical description of a fuel cell.
This chapter describes a relationship between the efficiency and operation voltage of a fuel cell and basic physical variables of reacting gases (reaction process temperature, process substances pressure and their concentration).
The Gibbs free energy is important for fuel cells, which represents work made by movement of electrons in the external circuit. The cell does not use any work made by a change in the volume of reactants and products on the inlet and outlet from a fuel cell. The enthalpy of formation is a sum of the Gibbs free energy of formation and entropy related power. In terms of enthalpy and entropy, the Gibbs free energy of the system can be expressed as follows:

$$G = H - T \cdot S \qquad (J \cdot kg^{-1}; J \cdot kg^{-1}, K, J \cdot kg^{-1} \cdot K^{-1}),$$

Where G is the Gibbs free energy of the system ($J \cdot kg^{-1}$), H represents the system enthalpy ($J \cdot kg^{-1}$), T is a thermodynamic system temperature (K), and S means the entropy of the system ($J \cdot kg^{-1} \cdot K^{-1}$).
A following relationship can be applied for one substance mol:

$$\Delta \bar{g}_S = \Delta \bar{h}_S - T\ \Delta \bar{s}_S \qquad (J \cdot mol^{-1}; J \cdot mol^{-1}, K, J \cdot mol^{-1} \cdot K^{-1}),$$

Where $\Delta \bar{g}_S$ is a change in the molar Gibbs free energy of the system ($J \cdot mol^{-1}$), $\Delta \bar{h}_S$ represents a change in the molar enthalpy of the system ($J \cdot mol^{-1}$), T is thermodynamic system temperature (K), $\Delta \bar{s}_S$ is a change in the molar system entropy of the system ($J \cdot mol^{-1} \cdot K^{-1}$).
Values of particular variables in a range of temperature from 20°C to 1000°C are presented in the Table 2.1.
Fuel cells are not bonded by the Carnot efficiency limit as other combustion cycles. A change in the enthalpy of formation is a basic efficiency limit. If the whole power from the hydrogen

fuel (enthalpy of formation) is transformed to the electric power, then the electromotive voltage reaches the values as follows:

$$E = \frac{-\Delta \overline{h_f}}{2\,F} = 1{,}48\ V$$

for higher heating (calorific) value ($\Delta \overline{h_f}$ = - 285.84 kJ · mol-1); and = 1.25 V for lower heating value ($\Delta \overline{h_f}$ = - 241.83 kJ · mol-1).

The above values express a voltage value which could be possible to reach in case of a 100 % efficient system, depending on a choice of a related value.

The electromotive voltage of a fuel cell in the system without losses can be expressed as follows:

$$E^0 = -\frac{\Delta \overline{g_f}}{2 \cdot F}, \qquad (V;\ kJ\ mol^{-1},\ C).$$

Where E^0 is electromotive open-circuit voltage of a fuel cell (V), Δg_f is a change in the Gibbs free energy of formation of agents (kJ · mol-1), and F is a Farraday constant (F = 96, 485.341 C·mol-1).

A maximum available electric power corresponds to changes in the Gibbs free energy, so that

$$\eta_{el\,max} = \frac{-\Delta \overline{g_f}}{-\Delta \overline{h_f}} \times 100\ \%, \qquad (-;\ kJ\ mol^{-1},\ kJ\ mol^{-1}).$$

Where η_{elmax} is a maximum electric efficiency of a fuel cell (-), Δg_f is a change in the Gibbs free energy of formation of agents (reagents and reactants, kJ mol-1), and Δh_f is a change in the molar enthalpy of formation of agents (reagents and reactants, kJ mol-1). A comma above a lower case letter indicating a variable means that a respective value is expressed in unit x per 1 mol of the substance.

This efficiency limit is designated as thermo-dynamical efficiency.

Temperature (°C)	h_{fH2O}	h_{fH2}	h_{fO2}	Δh_{fH2}	s_{H2O}	s_{H2}	s_{O2}	Δs_{H2}	Δg_{fH2}	η	E^0
	kJ · mol-1			kJ · mol-1	J · mol-1 · K-1			J · mol-1 · K-1	kJ · mol-1	%	V
20	-242,0	-0,1	-0,1	-241,8	188,26	130,11	204,64	-44,17	-228,8	94,64	1,186
30	-241,7	0,1	0,1	-241,9	189,39	131,07	205,63	-44,49	-228,4	94,42	1,184
40	-241,3	0,4	0,4	-242,0	190,49	132,00	206,59	-44,81	-227,9	94,20	1,181
50	-241,0	0,7	0,7	-242,1	191,55	132,91	207,51	-45,12	-227,5	93,98	1,179
60	-240,6	1,0	1,0	-242,2	192,58	133,79	208,41	-45,42	-227,0	93,75	1,176
70	-240,3	1,3	1,3	-242,3	193,58	134,65	209,29	-45,72	-226,6	93,52	1,174
80	-240,0	1,6	1,6	-242,4	194,55	135,49	210,14	-46,00	-226,1	93,30	1,172
90	-239,6	1,9	1,9	-242,5	195,50	136,30	210,97	-46,29	-225,7	93,07	1,170
100	-239,3	2,2	2,2	-242,6	196,42	137,09	211,78	-46,56	-225,2	92,84	1,167

200	-235,8	5,1	5,3	-243,6	204,58	144,03	218,97	-48,93	-220,4	90,50	1,142
300	-232,3	8,0	8,4	-244,5	211,37	149,62	224,99	-50,75	-215,4	88,10	1,116
400	-228,7	10,9	11,6	-245,4	217,25	154,32	230,19	-52,17	-210,3	85,69	1,090
500	-224,9	13,9	14,9	-246,2	222,49	158,40	234,78	-53,30	-205,0	83,26	1,062
1 000	-204,1	29,1	32,4	-249,3	243,11	173,53	252,11	-56,47	-177,4	71,16	0,919

Table 2.1 An example of values $\Delta\bar{h}_S$, $\overline{\Delta s_S}$ a $\Delta\bar{g}_S$ for a chemical reaction between hydrogen and oxygen passing in a fuel cell of the PEM type in a range of temperatures from 20°C to 1000°C.

Not the whole fuel supplied into a fuel cell can be utilized. Some fuel passes through a fuel cell without taking part in the reaction. If this is the case, a coefficient of the fuel use, μ_f, can be defined, corresponding to a portion of electric power of a fuel cell and electric current which would be gained provided the whole fuel passed through the reaction. A suitable value of this coefficient equals 0.95.

The impact of a partial pressure of concentration and temperature of particular reacting substances to voltage in no-load running of a fuel cell is evident from, the following equation:

$$E = E^0 + \frac{R \cdot T}{2 \cdot F} \cdot \ln\left(\frac{\alpha \cdot \beta^{1/2}}{\delta} \cdot P^{1/2}\right) = E^0 + \frac{R \cdot T}{2 \cdot F} \cdot \ln\left(\frac{\alpha \cdot \beta^{1/2}}{\delta}\right) + \frac{R \cdot T}{4 \cdot F} \cdot \ln(P),$$

(V; V, kJ mol^{-1} K^{-1}; T , C mol^{-1}, bar).

Where E is electromotive voltage under particular conditions (V), E^0 is electromotive voltage at standard pressure (V), R is a molar gas constant (R = 8.314 J·mol^{-1}·K^{-1}), T is thermodynamic temperature (K), F is a Farraday constant (C·mol^{-1}), α, β and δ are constants which values depend on molar quantity and concentration of H_2, O_2 a H_2O ($P_{H2} = \alpha \cdot P$, $P_{O2} = \beta \cdot P$, $P_{H2O} = \delta \cdot P$), and P is a value of the system pressure (bar) at a supposed ambient pressure of 1 bar.

The above equation is called the Nernst equation. This value provides us a theoretical base and quantitative data for a big number of variables when designing and operating a fuel cell.

The previous paragraphs defined voltage of a fuel cell in no-load running (open circuit voltage) depending on physical parameters of reaction gases and ambient environment. In practice, the real operation voltage of a fuel cell differs much due to power losses in particular member parts.

Four major irreversibilities are defined in a fuel cell depending on their occurrence mechanism.

1. Activation losses – they are caused by a course of reactions taking place on electrode surfaces. A part of generated open circuit voltage (OCV) is lost in chemical reaction control. This voltage loss which is strongly nonlinear can be described for a hydrogen fuel cell by an equation called as a Tafel equation:

$$\Delta U_{act} = A \cdot \ln\left(\frac{i}{i_0}\right) = \frac{R \cdot T}{2 \cdot \alpha \cdot F} \cdot \ln\left(\frac{i}{i_0}\right)$$

(V; J mol^{-1} K^{-1}, K, -, C mol^{-1}, A cm^{-2}, A cm^{-2}),

Where R is a molar gas constant, T is thermodynamic temperature, a is a coefficient of charge transmission - approx. 0.5 for anode, from 0.1 to approx. 0.5 for cathode (Larminie & Dicks, 2003), F is a Farraday constant, i is a current density, and i_0 is an exchangeable current density. This exchangeable current density i_0 gives the size of current flowing from and electrode to the electrolyte, and vice versa. This parameter is decisive for control of a fuel cell on its electrodes. We aim to make this value as high as possible.

The Tafel equation applies only in case when $i > i_0$.

For a low temperature hydrogen fuel cell using the air with the ambient pressure (0.1 MPa) as an oxidizing agent, a typical value for i_0 equals 0.1 mA · cm^{-2} on a cathode, and approximately 200 mA · cm^{-2} on the anode (Larminie & Dicks, 2003).

A size of the exchangeable current density i_0 can be influenced by operation temperature increase, increase in pressure of reactants and their concentrations, increased catalysts and change in the electrode surface (area) structure.

2. A fuel passage through the electrolyte and internal current – this power loss occurs due to inability to use the whole power entering the cell. The fuel penetrates through the electrolyte to the cathode, and in a less extent, through electron conductivity of the electrolyte. The electrolyte should only transport ions through a fuel cell. Nevertheless, the passage of some fuel quantity and some number of electrons cannot be avoided and pass though a cell. This component causes significant voltage loss of OCV particularly in low temperature fuel cells.

If i_n is a value of internal current density describing passage of fuel and electrons through the electrolyte, than the equation for operation cell voltage, taking into account two above mentioned irreversibilities can be modified as follows:

$$U = E - A \cdot \ln\left(\frac{i + i_n}{i_0}\right),$$

(V; V, A · cm^{-2}, A · cm^{-2}, A · cm^{-2}).

The internal current density usually reaches units mA · cm^{-2}.

3. Resistance losses – the fuel cell voltage drop is caused by electric resistance of material of electrodes, bipolar plates, and conductive interconnections which is put in a way of the electrons flow, as well as by resistance which puts the electrolyte in a way of the ions flow. This voltage drop is directly proportional to the current density.

If the value of resistance per 1 cm^2 of active fuel cell surface is defined, designated by the symbol r and call it the resistivity (surface resistance), then the equation for the voltage drop due to resistance losses ΔU_{ohm} takes a form:

$$\Delta U_{ohm} = i \cdot r,$$

(V; mA · cm^{-2}, kΩ · cm^2)

Where i is the current density, and r is the resistivity (surface resistance) a fuel cell.

4. Mass transport and concentration losses – these power losses follow from a change in concentration of reactants on the electrodes surfaces while using the fuel. This type of losses is also designated as Nerstian. This is because of the relationship to concentrations and due to concentration of reactants modeled by the means of the Nernst equation.

Up to now, no analytic solution of a problem of voltage change modeling sufficiently functioning for all fuel cell cases exists. The voltage drop caused by concentration losses and losses caused by mass transportation ΔU_{trans} is expressed by the empiric equation (Larminie & Dicks, 2003):

$$\Delta U_{trans} = m \cdot \exp(n \cdot i), \qquad (V; V, cm^2 \cdot mA^{-1}, mA \cdot cm^{-2})$$

Where the constant m is usually approx. $3 \cdot 10^5$ V and n constant value approx. $8 \cdot 10^{-3}$ $cm^2 \cdot mA^{-1}$. This equation is applied to reach results similar to measurement results.
Giving the above mentioned voltage drops in the same equation, we can gain a complete equation describing process voltage of a fuel cell depending on the current density *i*:

$$U = E - \Delta U_{Ohm} - \Delta U_{act} - \Delta U_{trans} = E - i \cdot r - A \cdot \ln\left(\frac{i + i_n}{i_0}\right) + m \cdot \exp(n \cdot i)$$

All variables and constants stated in the equation are stated and described in already before presented equations.

3. Basic fuel cell control

3.1 A structure of a hydrogen circuit (source)

A fuel cell itself is not able of any operation. Operating it, suitable conditions for its activity should be provided - particularly supply of fuel gas, oxidizing agent, humidification, discharge of reagents from the cell, heat discharge, and other supporting processes providing a proper system function.
The characteristics of a fuel cell are often indicated in form of a polarization curve which is the cell voltage dependency on the current density. A difference between an actual cell voltage and ideal voltage depends on losses. Losses which influence a course of a polarization curve are presented in the Chapter 2.2 (Pukrushpan et al., 2004). A characteristic polarization curve course is presented in Fig. 3.

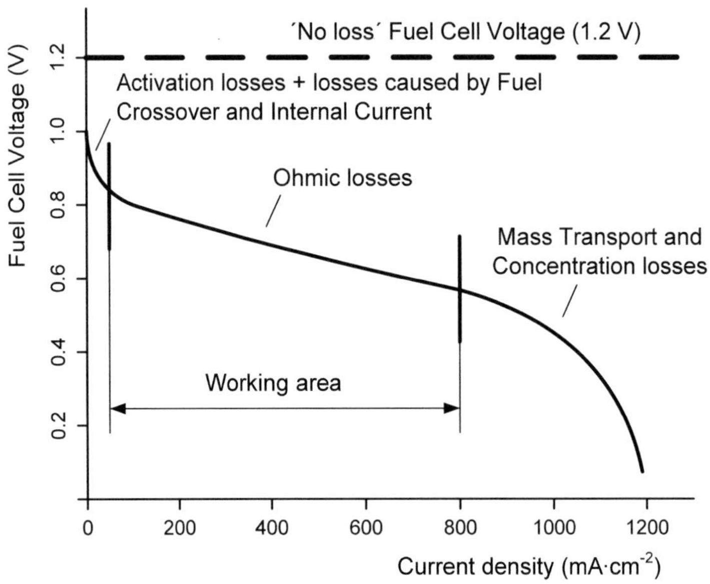

Fig. 3. A fuel cell polarization curve with mentioned main irreversibilities.

A number of factors influences a polarization curve course (See Fig. 4). One of fuel cell control system tasks is to influence the factors so that the fuel cell operation is as much efficient as possible. This particularly covers the factors as follows:

The pressure of the reaction hydrogen and air - increasing the pressure of hydrogen and reaction air usually results into the polarization curve growth.

Fuel cell temperature - increasing the temperature results into the polarization curve growth.

Reaction gases concentration - a concentration of reaction gases in the substance results into a polarization curve growth.

The increase in the reaction gas relative humidity increases a polarization curve course.

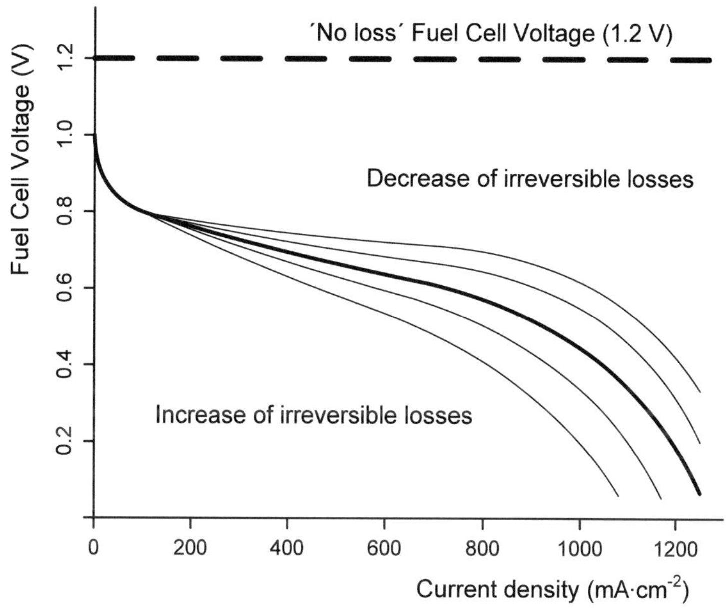

Fig. 4. The influencing of a polarization curve course.

Principal power source circuits with a fuel cell of the PEM type are described on the example of the source system with a Nexa Power Module fuel cell from the company Ballard Power Systems for purposes of the source testing. A summary chart of a hydrogen circuit is evident from Fig. 5.

3.2 Measurement and control system

A programmable automat (PLC) or embedded control system can act as a control unit. Basic fuel cell control principles are presented in Chapter 4.1. A set of sensors of electric and non-electric signals needed for a basic fuel cell system control serves a source of information for the control system. This particularly means voltage on the fuel cell, current taken from the cell, fuel cell temperature, reaction gas pressure, flow rate of reaction gases, etc.

Output block - the output block on the electric output from the fuel cell is represented by a direct current circuit breaker, DC/DC converter 24 V DC/120 V, 24 V DC/12 V DC or inverter (DC/AC converter) 24 V DC/230 V AC, followed by a single phase electrometer or voltage and current measurement (for the DC output). Parts of the AC output (interconnection with the isolated electric network or electrical power network) are reactance coils (damping of electric surges) and circuit breaker(s). The circuit breaker terminals represent a handover point with the electrification system (isolated network).

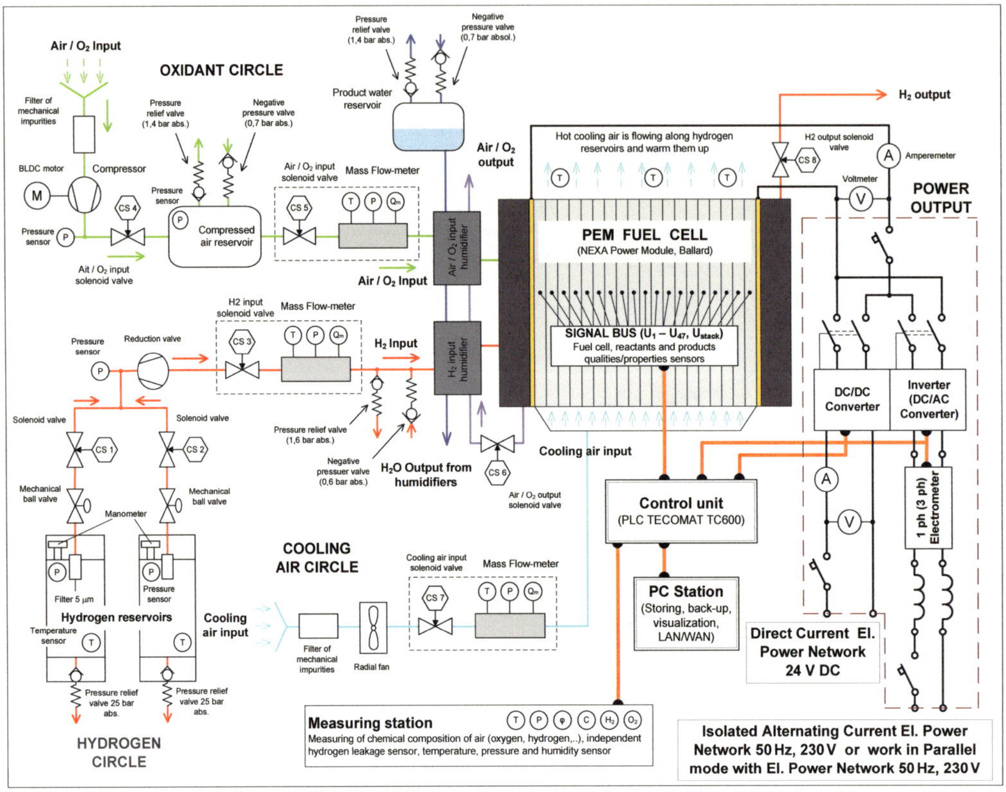

Fig. 5. A summary block diagram of a system with a fuel cell Nexa Power Module designating particular circuits and system elements.

Fuel cell signal bus - for purposes of laboratory monitoring of a source with a fuel cell, it is suitable to install a system of sensors on a fuel cell and its immediate surrounding to monitor voltage of particular elementary fuel cells and the whole stack, internal, surface and cooling air output temperature sensors, concentration of hydrogen in the nearby surrounding of the fuel cell, etc., according to required tested parameters).

A measuring station - represents a reference measurement point in the distance of approx. 5 to 10 m from a source, and its immediate vicinity, in order to find limit conditions in calculations of power balances, and in order to secure the safety of persons and equipment nearby the prototype. The measuring station is equipped by sensors of temperature, pressure and humidity of the ambient air, sensor of concentration of oxygen in the air, and sensor or detector of hydrogen escape into the surrounding environment.

A mobile PC station represented by a portable computer provides acquisition (back-up) and processing of measured data, their visualization on the intranet/internet, and interconnection of a real system with a software model.

3.3 Hydrogen circuit control

Hydrogen is supplied to the power source system from reservoirs with compressed gas or metal hydride, which is heated by the air leaving the cooling fuel cell circuit. Hydrogen reservoirs are followed by a mechanical ball and solenoid valve providing disconnection/separation of a hydrogen circuit from the gas reservoirs (tanks). This is followed by a reduction valve decreasing the hydrogen pressure from reservoirs (up to 25 bar abs.) to the operation pressure (from 1.1 to 1.5 bar abs.). Another element is most usually a mass flow-meter measuring the mass flow-rate (volume flow rate, temperature and pressure) of hydrogen passing the flow-meter (approx. 30 NL/min max.). Still before hydrogen entering the fuel cell, it passes through a humidification unit where it is saturated by water vapor as needed. A solenoid valve is situated at the exit from a fuel cell providing controlled hydrogen discharge from the fuel circuit. The fuel circle can be also equipped by a pressure sensor monitoring the hydrogen pressure in reservoirs and before the hydrogen entry into a reduction valve. Further, the surface temperature of reservoirs with a metal hydride is monitored.

3.4 Reaction air circuit control

A compressor suctions the surrounding air via a filter of mechanical impurities and drives it into a compressed air reservoir. The compressed air reservoir of the volume of approx. 10 - 20 NL serves to stabilization (uniformity) of the air/oxygen flow into a fuel cell. The reservoir is equipped by a safety/relief valve (1.5 bar abs.) and underpressure valve (0.7 bar abs.) and its input and output are blocked by a solenoid valve in order to control the pressure and volume flow of the oxidizing agent into a fuel cell. In the reach of the oxidizing agent, mass flow-meter (approx. 120 NL/min max.) and humidifier follow in order to saturate the oxidizing agent by water vapors. Water leaving the humidifier is downtaken into a product water reservoir (tank). From there, water passes through humidifiers and is discharged into the sewerage, or it is further used.

3.5 Cooling air circuit control

The cooling air is suctioned by a radial fan from the surrounding environment into a cooling circle. The cooling air passes through the filter of mechanical impurities and radial fan into a mass flow-meter equipped by a sensor of temperature and pressure (approx. 500 NL/min max.), from where it is further driven to cooling channels inside bipolar plates of a fuel cell. The exit from the fuel cell cooling air is equipped by temperature sensors. The heated cooling air is driven to hydrogen reservoirs.

4. Control of a vehicle powered by a fuel cell

The vehicle powered by hydrogen fuel cell needs an electronic control system assuring operation of its different parts. The complex electronic control is necessary already for basic operation of the vehicle, because there are a lot of subsystems that have to be coordinated and controlled. The control system assures especially following tasks:

Control of fuel cell operation - a hydrogen input valve control, a combustion products output valve control, a fuel cell fan control, coupling of produced electrical energy to an electric DC-drive system.

Control of DC-drive system - motor current control, speed control.

Processing security tasks - assuring safe operation of a fuel cell system and a drive system, processing of hydrogen detector information, temperature measuring.

Managing the driver control panel - complete interface to pilot that allows controlling the car - start/stop, speed set point, time measuring, emergency buttons and indicators.

Creating data archives with saved process variables - saving important process data to archives that can be then exported and analyzed.

Sending actual data to display panel in car - display panel in the car is the "process" visualization of the system. All important data are online displayed on it.

Communication with a PC monitoring station - control system send data and receive commands from the PC monitoring station using wireless communication system.

4.1 Basic fuel cell control

A basic fuel cell control concerns in a proper fuel cell operation in its all activity phases. For this task, most usually an electronic control system is used. The control system provides particularly (see Fig 6):

Safe start of the cell activity - fuel cell start is a sequence of activities which should be made for the cell to be transferred to a status when it supplies the electric power. This particularly means to supply the reaction air (Place 2, Fig. 6) and reaction hydrogen (Place 4, Fig. 6) into the whole volume of the fuel cell. After achieving some cell voltage level, the system enters a stage of the electric power production.

Proper cell functioning in the stage when it supplies the power to the appliance - in this stage, the cell supplies electric power to the appliance (Place 6, Fig. 6). In a point of the cell control, practically no control actions are requited, a basic control system task is to monitor statues of important variables - fuel cell voltage and current, temperature, and potentially also other ones. Based on these variables, it can be assessed whether the cell is loaded regularly or whether it is overloaded and, therefore, there is a risk of its damage.

Safe fuel cell switch-off - a process of a fuel cell switch-off contains again several actions which put the cell into a not active status. This particularly means shutting off the reaction hydrogen inlet and consumption of reaction gases which were left in the fuel cell volume (Place 8, Fig. 6). The consumption of these reaction gases will take place when the hydrogen inlet is closed however an appliance is still connected. Under such conditions, the residual hydrogen is consumed from the cell volume, and, simultaneously, cell voltage drops. In case of the voltage drop below some limit, the appliance can be disconnected, and the cell enters a state when it is switched off. In the Place 6, Fig. 6 , a situation can occur when the cell should be immediately shut off as well as the appliance should be immediately disconnected (status 9, Fig. 6). This can be done however the residual hydrogen inside the fuel cell can cause membrane damage. Therefore, the described switching off method is only suitable to apply in critical situations (e.g. when hydrogen escape is detected) and still, however, it is suitable to provide reaction gases use or removal from the cell volume.

Reaction to non-standard situations (safety functions) – this particularly means states related to the fuel cell operation which could result into system safety decrease or to its damage. This particularly means cell overheating, hydrogen escape, reaction gas pressure increase, etc.

4.2 Electric drive control

The electric drive of a car with a fuel cell is the main consumer of electric power supplied by the fuel cell. The drive control has a critical influence on the car power consumption. Two basic conceptions are applied for fuel cell driven vehicles:

A conception when a fuel cell supplies the power into some power system of a vehicle as e.g. a battery or a super capacitor. Applying this conception, a fuel cell can run in the optimal regime and need not necessarily to respond immediately to load demands. A vehicle power system designed in this way can draw the power also from other sources, it can apply power recuperation, etc.

A conception when a fuel cell supplies the power directly into the electric drive system of the vehicle. In this design, it should immediately respond to load demands. The drive unit should be controlled so that the power takeoff from the cell is sufficiently continuous and smooth and the cell is not overloaded.

A way of the drive control depends on its type. It can be a DC motor, a motor with electronic commutation, a synchronous motor, etc. A motor type should be chosen according to required properties of a vehicle. The drive usually needs to use an electronic control system. This control system can be a part of the control system for a fuel cell, or it can be a separate system.

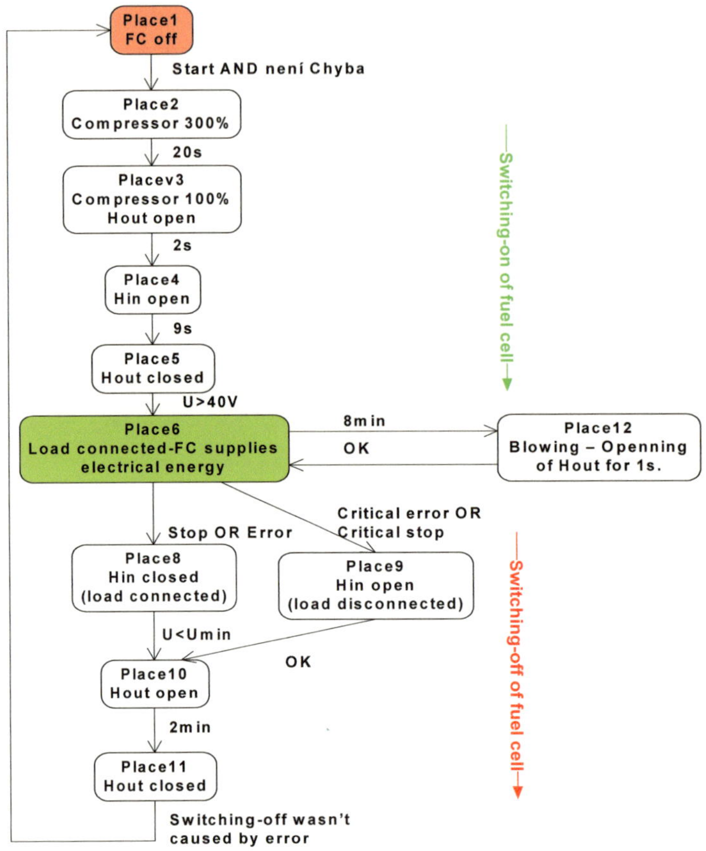

Fig. 6. Basic fuel cell control.

4.3 User-pilot interface

A fuel cell vehicle should be equipped by a user interface which enables a driver to control the vehicle. The complexity of this user interface depends on a vehicle character, however, it should contain at least primary control elements and indicators for the fuel cell system control:

Controller for the fuel cell switching on and off.

Controller for setting a required vehicle speed.

Vehicle status indicators - of a character of display or set of indicators regarding the fuel cell condition, operation mode, defects and faults, speed, electrical variables, etc.

5. An example of a real application

A team of several specialists and students of Department of Measurement and Control, VSB-Technical University of Ostrava has designed and realized a prototype of hydrogen powered car based on fuel cell technology and electrical DC drive. The car was realized according to rules of Shell Eco-Marathon competition which is focused on economization of

energy in mobile vehicles. The project is called HydrogenIX (Fig. 7), development and testing activities were realized between 2005 and today.
The project is closely related with the educational process and motivation of students for further research activity in the form of construction of a mobile system driven by electrical motor and fed from electrochemical generator with the fuel cell.
By the HydrogenIX project and related technical and technological problems, the team tries to involve the students of bachelor, master and doctoral degree on Faculty of Mechanical Engineering and Faculty of Electrical Engineering and Computer Science to problems of non-traditional power sources and their applications.

Fig. 7. The HydrogenIX car.

Car parameters:

Aerodynamic shape of the car body.
Power of the fuel cell - 1.2 kW.
2 DC motor of 150W.

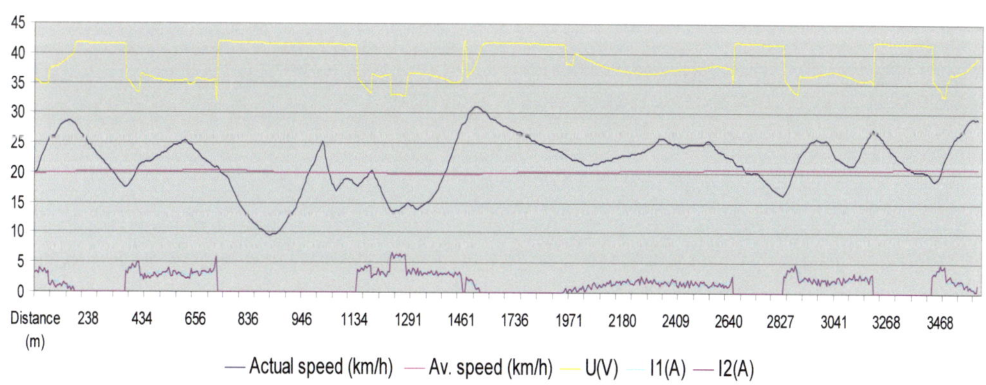

Fig. 8. A record of data of a run at the race circuit in Nogaro, France.

7. Conclusion

The top perspective is provided by an electromotor drive with a current source from a fuel cell which transfers the power contained in the fuel (hydrogen or hydrocarbon) directly to electric power. Currently, long-term verification testing has been taking place. This comprises verification of prototypes for a purpose of introduction of series and mass production. A principal fuel cell problem is represented by development of an electrolyte meeting the criteria of mass production, performance efficiency, service life and price.

8. References

Fromm E. (1998). *Kinetics of Metal-Gas Interactions at Low Temperature Hydryiding, Oxidation, Poisoning,* Springer-Verlag Berlin Heidelberg New York, ISBN 3-540-63975-6, Germany.

Kameš J. (2008). *Alternativní palivo - vodík.* Published by Czech Technical University in Prague, ISBN 978-80-254-1686-0, Prague, Czech Republic.

Kurzweil P. (2003). *Brennstoffzellen-technik, Grundlagen, Komponenten, Systeme, Anwendungen.* Vieweg, ISBN 3-528-03965-5, Wiesbaden, Germany.

Larminie J. & Dicks A. (2003). *Fuel Cell Systems Explained, Second Edition,* John Wiley & Sons Ltd., ISBN 978-0-470-84857-9, Chichester, England.

Pukrushpan, J. T.; Stefanopoulou, A. G., Peng, H. (2004). *Control of fuel cell power systems: principles, modeling, analysis and feedback design,* Springer, ISBN 1-85233-816-4, London, United Kingdom.

Modeling and Assessing of Omni-directional Robots with Three and Four Wheels

Hélder P. Oliveira, Armando J. Sousa, A. Paulo Moreira and Paulo J. Costa
Universidade do Porto, Faculdade de Engenharia
INESC-Porto – Instituto de Engenharia de Sistemas e Computadores do Porto
Portugal

1. Introduction

Robots with omni-directional locomotion are increasingly popular due to their enhanced mobility when compared with traditional robots. Their usage is more prominent in many robotic competitions where performance is critical, but can be applied in many others applications such as service robotics. Robots with omni-directional locomotion offer advantages in manoeuvrability and effectiveness. These features are gained at the expense of increased mechanical complexity and increased complexity in control. Traditional mechanical configuration for omni-directional robots are based on three and four wheels. With four motors and wheels, it is expected that the robot will have better effective floor traction (Oliveira et al., 2008), that is, less wheel slippage at the expense of more complex mechanics, more complex control and additional current consumption.
Common robotic applications require precise dynamical models in order to allow a precise locomotion in dynamical environments. Such models are also essential to study limitations of mechanical configurations and to allow future improvements of controllers and mechanical configurations.
The presented study is based on a single prototype that can have both configurations, that is, the same mechanical platform can be used with three wheels and then it can be easily disassembled and reassembled with a four wheeled configuration. A mathematical model for the motion of the robot was found using various inertial and friction parameters. The motion analysis includes both kinematical and dynamical analysis.

1.1 Context

Robots with omni-directional locomotion are holonomic and they are interesting because they allow greater manoeuvrability and efficiency at the expense of some extra complexity. One of the most frequent solutions is to use some kind of variation of the Mecanum wheels proposed by (Diegel et al., 2002) and (Salih et al., 2006).
Omni-directional wheel design is quite delicate and different wheels exhibit very different performances. Wheel construction is often application specific and the presented work uses the wheels shown in Fig. 1. These wheels are built in-house for several demanding applications. The prototype used in the experiments uses four wheels of this kind to achieve

omni-directional locomotion. A robot with four wheels is expected to have more traction than its three wheeled counterpart. In both configurations the motor plus wheel assemblies are identical to the photograph seen in Fig. 2.

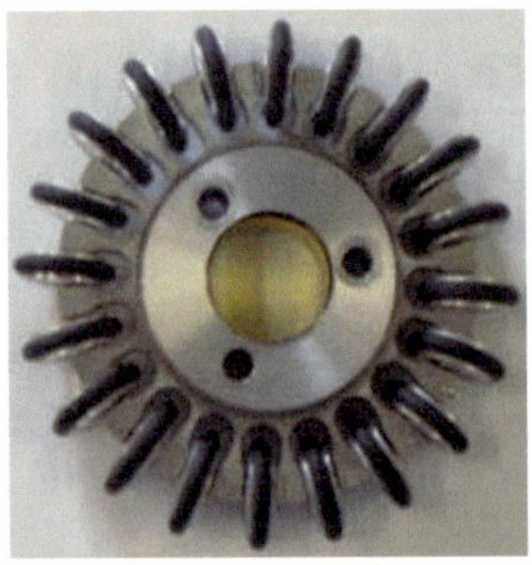

Fig. 1. 5DPO omni-directional wheel

Fig. 2. Motor and wheel

A robot with three or more motorized wheels of this kind can have almost independent tangential, normal and angular velocities (holonomic property). Dynamical models for this kind of robots are not very common due to the difficulty in modeling the several internal frictions inside the wheels, making the model somewhat specific to the type of wheel being used (Oliveira et al., 2008) and (Williams et al., 2002).

Frequent mechanical configurations for omni-directional robots are based on three and four wheels. Three wheeled systems are mechanically simpler but robots with four wheels have more acceleration with the same kind of motors. Four wheeled robots are expected to have better effective floor traction, that is, less wheel slippage - assuming that all wheels are pressed against the floor equally. Of course four wheeled robots also have a higher cost in equipment, increased energy consumption and may require some kind of suspension to distribute forces equally among the wheels.

(a) Three wheeled configuration (b) Four wheeled configuration

Fig. 3. Configurations for the prototype

In order to study and compare the models of the three and four wheeled robots, a single prototype was built that can have both configurations, that is, the same mechanical platform can be used with three wheels and then it can be disassembled and reassembled with a four wheel configuration, see Fig. 3.

Data from experimental runs is taken from overhead camera, see Fig. 4. The setup is taken from the heritage of the system described in (Costa et al., 2000) that currently features 25 frames/second, one centimeter accuracy in position (XX and YY axis) and about three degrees of accuracy in the heading of the robot.

Fig. 4. Image from overhead camera

In order to increase the performance of robots, there were some efforts on the studying their dynamical models (Campion et al., 1996), (Conceicao et al., 2006), (Khosla, 1989), (Tahmasebi et al., 2005), (Williams et al., 2002) and kinematic models (Campion et al., 1996), (Leow et al. 2002), (Loh et al. 2003), (Muir & Neuman, 1987), (Xu et al., 2005). Models are based on linear and non linear dynamical systems and the estimation of parameters has been the subject of continuing research (Conceicao et al., 2006), (Oliveira et al., 2008) and (Olsen and Petersen, 2001). Once the dynamical model is found, its parameters have to be estimated. The most common method for identification of robot parameters are based on the Least Squares Method and Instrumental Variables (Astrom & Wittenmark, 1984). However, the systems are naturally non-linear (Julier & Uhlmann, 1997), the estimation of parameters is more complex and the existing methods (Ghahramani & Roweis, 1999), (Gordon et al., 1993), (Tahmasebi et al., 2005) have to be adapted to the model's structure and noise.

2. Mechanical Configurations

Fig. 5 and Fig. 6 present the configuration of the three and four wheeled robots respectively, as well as all axis and relevant forces and velocities of the robotic system. The three wheeled system features wheels separated by 120° degrees, and the four wheeled by 90° degrees.

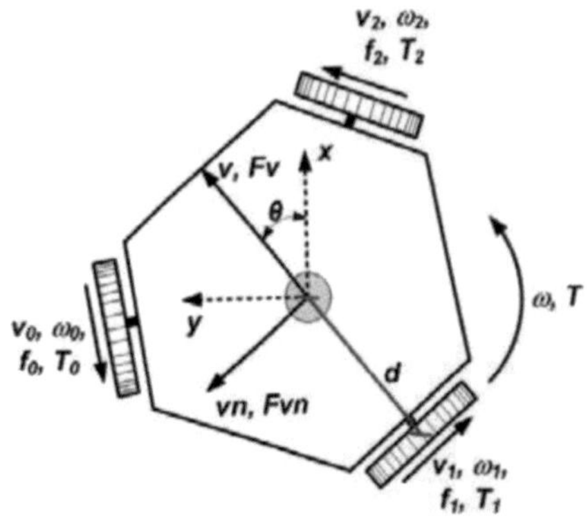

Fig. 5. Three wheeled robot

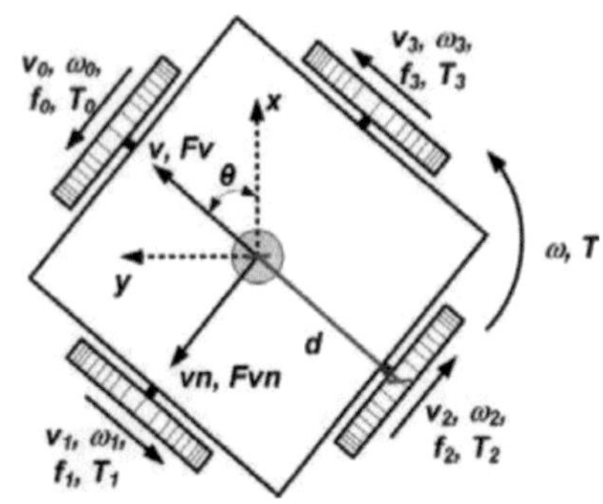

Fig. 6. Four wheeled robot

Fig. 5 and Fig. 6 show the notation used through-out this paper, detailed as follows:

- x, y, θ - Robot's position (x,y) and θ angle to the defined front of robot;
- d [m] - Distance between wheels and center robot;
- v_0, v_1, v_2, v_3 [m/s] - Wheels linear velocity;
- ω_0, ω_1, ω_2, ω_3 [rad/s] - Wheels angular velocity;
- f_0, f_1, f_2, f_3 [N] - Wheels traction force;
- T_0, T_1, T_2, T_3 [Nm] - Wheels traction torque;
- v, vn [m/s] - Robot linear velocity;
- ω [rad/s] - Robot angular velocity;
- F_v, F_{vn} [N] - Robot traction force along v and vn;
- T [Nm] - Robot torque (respects to ω).

The reader should be aware that in omni-directional robotics, the front of the robot is arbitrarily defined according to the intuitive notion of the robot mechanics. Of course, the v direction follows the front of the robot and the vn direction is orthogonal.

3. Motion Analysis and Model Determination

3.1 Kinematic model

In order to find motion models for a surface vehicle, the pose of the vehicle must be identified as (x, y, θ) and associated velocities are $v_x(t)=\frac{dx(t)}{dt}$, $v_y(t)=\frac{dy(t)}{dt}$, $\omega(t)=\frac{d\theta(t)}{dt}$.

The following text uses the notation presented in Fig. 5 and Fig. 6, where the defined "front" also defines the v direction and its orthogonal vn direction.

Equation (1) allows the transformation from linear velocities v_x and v_y on the static (world) axis to linear velocities v and vn on the robot's axis.

$$\begin{bmatrix} v(t) \\ vn(t) \\ \omega(t) \end{bmatrix} = \begin{bmatrix} \cos(\theta(t)) & \sin(\theta(t)) & 0 \\ -\sin(\theta(t)) & \cos(\theta(t)) & 0 \\ 0 & 0 & 1 \end{bmatrix} \cdot \begin{bmatrix} v_x(t) \\ v_y(t) \\ \omega(t) \end{bmatrix} \tag{1}$$

3.1.1 Three Wheeled Robot

Wheel velocities v_0, v_1 and v_2 are related with robot's velocities v, vn and ω as described by equation (2).

$$\begin{bmatrix} v_0(t) \\ v_1(t) \\ v_2(t) \end{bmatrix} = \begin{bmatrix} -\sin(\pi/3) & \cos(\pi/3) & d \\ 0 & -1 & d \\ \sin(\pi/3) & \cos(\pi/3) & d \end{bmatrix} \cdot \begin{bmatrix} v(t) \\ vn(t) \\ \omega(t) \end{bmatrix} \tag{2}$$

Applying the inverse kinematics is possible to obtain the equations that determine the robot velocities related with the wheels velocities. Solving in order of v, vn and ω, the following can be found:

$$v(t) = \left(\frac{\sqrt{3}}{3}\right) \cdot \left(v_2(t) - v_0(t)\right) \tag{3}$$

$$vn(t) = \left(\frac{1}{3}\right) \cdot \left(v_2(t) + v_0(t)\right) - \left(\frac{2}{3}\right) \cdot v_1(t) \tag{4}$$

$$\omega(t) = \left(\frac{1}{3 \cdot d}\right) \cdot \left(v_0(t) + v_1(t) + v_2(t)\right) \tag{5}$$

3.1.2 Four Wheeled Robot

The relationship between the wheels velocities v_0, v_1, v_2 and v_3, with the robot velocities v, vn and ω is described by equation (6).

$$\begin{bmatrix} v_0(t) \\ v_1(t) \\ v_2(t) \\ v_3(t) \end{bmatrix} = \begin{bmatrix} 0 & 1 & d \\ -1 & 0 & d \\ 0 & -1 & d \\ 1 & 0 & d \end{bmatrix} \cdot \begin{bmatrix} v(t) \\ vn(t) \\ \omega(t) \end{bmatrix} \tag{6}$$

It is possible to obtain the equations that determine the robot velocities related with wheels velocity but the matrix associated with equation (6) is not square. This is because the system is redundant. It can be found that:

$$v(t) = \left(\frac{1}{2}\right) \cdot \left(v_3(t) - v_1(t)\right) \tag{7}$$

$$vn(t) = \left(\frac{1}{2}\right) \cdot \left(v_0(t) - v_2(t)\right) \tag{8}$$

$$\omega(t) = \left(\frac{1}{4 \cdot d}\right) \cdot \left(v_0(t) + v_1(t) + v_2(t) + v_3(t)\right) \tag{9}$$

3.2 Dynamic

The dynamical equations relative to the accelerations can be described in the following relations:

$$M \cdot \frac{dv(t)}{dt} = \sum F_v(t) - F_{Bv}(t) - F_{Cv}(t) \tag{10}$$

$$M \cdot \frac{dvn(t)}{dt} = \sum F_{vn}(t) - F_{Bvn}(t) - F_{Cvn}(t) \tag{11}$$

$$J \cdot \frac{d\omega(t)}{dt} = \sum T(t) - T_{B\omega}(t) - T_{C\omega}(t) \tag{12}$$

where the following parameters relate to the robot are:

- M [kg] - Mass;
- J [kgm^2] - Inertia moment;
- F_{Bv}, F_{Bvn} [N] - Viscous friction forces along v and vn;
- $T_{B\omega}$ [N m] - Viscous friction torque with respect to the robot's rotation axis;
- F_{Cv}, F_{Cvn} [N] - Coulomb frictions forces along v and vn;
- $T_{C\omega}$ [Nm] - Coulomb friction torque with respect to robot's rotation axis.

Viscous friction forces are proportional to robot's velocities, see Fig. 7, and such as:

$$F_{Bv}(t) = B_v \cdot v(t) \tag{13}$$

$$F_{Bvn}(t) = B_{vn} \cdot vn(t) \tag{14}$$

$$T_{B\omega}(t) = B_\omega \cdot \omega(t) \tag{15}$$

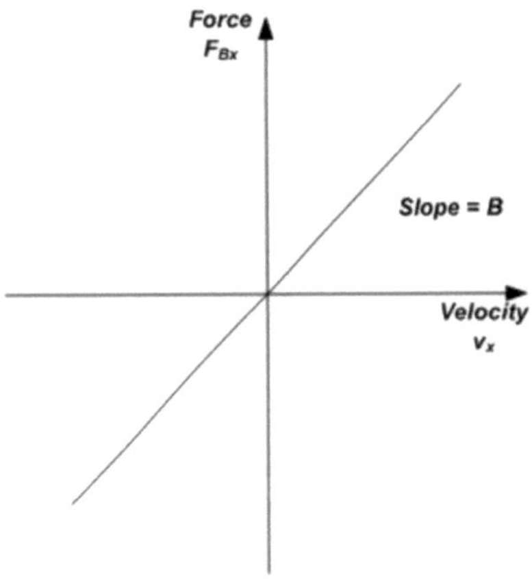

Fig. 7. Viscous friction

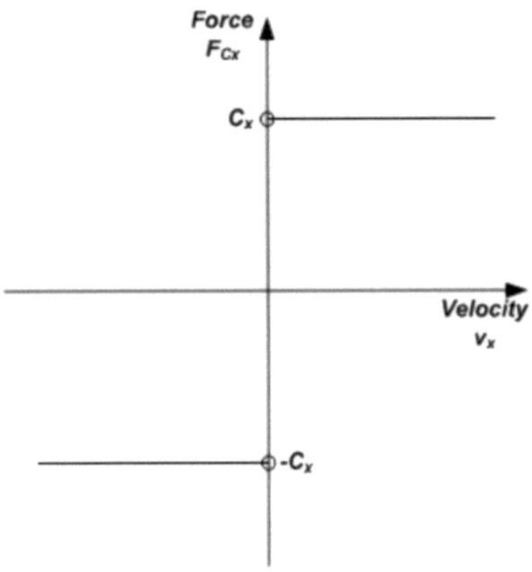

Fig. 8. Coulomb friction

where the following parameters relate to the robot as follows:

- B_v, B_{vn} [N/(m/s)] - Viscous friction coefficients for directions v and vn;
- B_ω [Nm/(rad/s)] - Viscous friction coefficient to ω.

The Coulomb friction forces are constant in amplitude, see Fig. 8.

$$F_{Cv}(t) = C_v \cdot sign(v(t)) \tag{16}$$

$$F_{Cvn}(t) = C_{vn} \cdot sign(vn(t)) \tag{17}$$

$$T_{C\omega}(t) = C_\omega \cdot sign(\omega(t)) \tag{18}$$

where the following parameters relate to the robot as follows:

- C_v, C_{vn} [N] - Coulomb friction coefficient for directions v e vn;
- C_ω [N m] - Coulomb friction coefficient for ω.

3.2.1 Three Wheeled Robot

The relationship between the traction forces and rotation torque of the robot with the traction forces on the wheels is described by the following equations:

$$\sum F_v(t) = (f_2(t) - f_0(t)) \cdot \sin\left(\frac{\pi}{3}\right) \tag{19}$$

$$\sum F_{vn}(t) = -f_1(t) + (f_2(t) + f_0(t)) \cdot \cos\left(\frac{\pi}{3}\right) \tag{20}$$

$$\sum T(t) = (f_0(t) + f_1(t) + f_2(t)) \cdot d \tag{21}$$

The traction force on each wheel is estimated by traction torque, which can be determined using the motor current, as described in the following equations:

$$f_j(t) = \frac{T_j(t)}{r} \tag{22}$$

$$T_j(t) = l \cdot K_t \cdot i_j(t) \tag{23}$$

where:

- l - Gearbox reduction;
- r [m] - Wheel radius;
- K_t [Nm/A] - Motor torque constant;
- i_j [A] - Motor current (j=motor number).

3.2.2 Four Wheeled Robot

$$\sum F_v(t) = f_3(t) - f_1(t) \tag{24}$$

$$\sum F_{vn}(t) = f_0(t) - f_2(t) \tag{25}$$

$$\sum T(t) = (f_0(t) + f_1(t) + f_2(t) + f_3(t)) \cdot d \tag{26}$$

As above, the traction force in each wheel is estimated using the wheels traction torque, which is determined by the motor current, using equations (22) and (23) where j=0,1,2,3.

3.3 Motor and Gearbox

The prototype uses brushless motors for the locomotion of the robot, photograph shown in Fig. 2. The used motors are from "Maxon Motor" (Maxon Motor, 2009), and the motor reference is "EC 45 flat 30 W". The main motor characteristics are: Nominal Power: 30W; Nominal Voltage: 12V; Nominal Current: 2.14A; Starting Current: 10A; Resistance phase to phase: 1.52 Ω; Terminal inductance phase to phase 0.56 mH; Torque constant 25.5 mNm/A; Speed constant 374 rpm/V; Mechanical time constant: 17.1 ms; Rotor inertia: 92.5 gcm^2 and a maximum efficiency of 77%.

The Gear-box Coupled to the motor is a "5:1 GS45", also from "Maxon Motor" (Maxon Motor, 2009), with a gear head of "Spur" type. The main characteristic are: Absolute Reduction: 51/10; Maximum mechanical efficiency of 90% and a mass inertia of 3.7 gcm^2.

This type of motors has been more common in the last years. The principal reason is the high performance when compared with others motors. These motors don't have mechanical switching, and this is the big difference to the common DC motors. The model for brushless motors is similar to the common DC motors, based on (Pillay & Krishnam, 1989).

$$u_j(t) = L \cdot \frac{di_j(t)}{dt} + R \cdot i_j(t) + K_v \cdot \omega_{mj}(t) \tag{27}$$

$$T_{mj}(t) = K_t \cdot i_j(t) \tag{28}$$

where:

- L [H] - Motor inductance;
- R [Ω] - Motor resistor;
- K_v [V/(rad/s)] - EMF motor constant;
- u_j [V] - Motor voltage (j=motor number);
- ω_{mj} [rad/s] - Motor angular velocity (j=motor number);
- T_{mj} [Nm] - Motor torque (j=motor number).

4. Parameter Estimation

The necessary variables to estimate the model parameters are motor current, robot position and velocity. Currents are measured by the drive electronics, position is measured by using external camera and velocities are estimated from positions.

The parameters that must be identified are the viscous friction coefficients (B_v, B_{vn}, B_ω), the Coulomb friction coefficients (C_v, C_{vn}, C_ω) and inertia moment J. The robot mass was measured, and it was 1.944 kg for the three wheeled robot and 2.340 kg for the four wheeled robot.

4.1 Experience 1 – Steady State Velocity

This method permits to identify the viscous friction coefficients B_ω and the Coulomb friction coefficients C_ω. The estimation of the coefficient according to velocity ω, was only implemented because inertia moment is unknown, and it is necessary to have an initial estimate of these coefficients. The experimental method relies on applying different voltages to the motors in order to move the robot according his rotation axis - the tests were made for positive velocities. Once reached the steady state, the robot's velocity ω and rotation torque T can be measured. The robot velocity is constant, so, the acceleration is null, and as such equation (12) can be re-written as follows:

$$\sum T(t) = B_\omega \cdot \omega(t) + C_\omega \tag{29}$$

This linear equation shows that it is possible to test different values of rotation velocities and rotation torques in multiple experiences and estimate the parameters.

4.2 Experience 2 – Null Traction Forces

This method allows for the estimation of the viscous friction coefficients (B_v, B_{vn}), the Coulomb friction coefficients (C_v, C_{vn}) and the inertia moment J. The experimental method consists in measuring the robot acceleration and velocity when the traction forces were null. The motor connectors were disconnected and with a manual movement starting from a stable position, the robot was pushed through the directions v, vn and rotated according to his rotation axis. During the subsequent deceleration, velocity and acceleration were measured. Because the traction forces were null during the deceleration equations (10), (11) and (12) can be re-written as follows:

$$\frac{dv(t)}{dt} = -\frac{B_v}{M} \cdot v(t) - \frac{C_v}{M} \tag{30}$$

$$\frac{dvn(t)}{dt} = -\frac{B_{vn}}{M} \cdot vn(t) - \frac{C_{vn}}{M} \tag{31}$$

$$\frac{d\omega(t)}{dt} = -\frac{B_\omega}{J} \cdot v(t) - \frac{C_\omega}{J} \tag{32}$$

These equations are also a linear relation and estimation of all parameters is easier.
The inertia moment J is estimated using the values obtained in the previous section. To do this, equation (32) must be solved in order of J:

$$J = -\frac{\omega(t)}{d\omega(t)/dt} \cdot B_\omega - \frac{1}{d\omega(t)/dt} \cdot C_\omega \tag{33}$$

4.3 DC Motor Parameters

The previous electrical motor model (equation (27)) includes an electrical pole and a much slower, dominant mechanical pole - thus making inductance L value negligible. To determinate the relevant parameters K_v and R, a constant voltage is applied to the motor. Under steady state condition, the motor's current and the robot's angular velocity are measured. The tests are repeated several times for the same voltage, changing the operation point of the motor, by changing the friction on the motor axis.
In steady state, the inductance L disappears of the equation (27), being rewritten as follows:

$$u_j(t) = R \cdot i_j(t) + K_v \cdot \omega_{mj}(t) \tag{34}$$

As seen in equation (35), by dividing (34) by $i_j(t)$, a linear relation is obtained and thus estimation is possible.

$$\frac{u_j(t)}{i_j(t)} = R + K_v \cdot \frac{\omega_{mj}(t)}{i_j(t)} \tag{35}$$

5. Results

5.1 Robot Model

By combining previously mentioned equations, it is possible to show that model equations can be rearranged into a variation of the state space that can be described as:

$$\frac{dx(t)}{dt} = A \cdot x(t) + B \cdot u(t) + K \cdot sign(x) \tag{36}$$

$$x(t)=\begin{bmatrix}v(t) & vn(t) & \omega(t)\end{bmatrix}^T \tag{37}$$

This formulation is interesting because it shows exactly which part of the system is non-linear.

5.1.1 Three Wheeled

Using equations on section 3.2, (19) to (23) and (34), the equations for the three wheeled robot model are:

$$A = diag(A_{11}, A_{22}, A_{33})$$

$$A_{11} = -\frac{3\cdot K_t^2\cdot l^2}{2\cdot r^2\cdot R\cdot M}-\frac{B_v}{M} \qquad A_{22} = -\frac{3\cdot K_t^2\cdot l^2}{2\cdot r^2\cdot R\cdot M}-\frac{B_{vn}}{M} \qquad A_{33} = -\frac{3\cdot d^2\cdot K_t^2\cdot l^2}{r^2\cdot R\cdot J}-\frac{B_\omega}{J} \tag{38}$$

$$B=\frac{l\cdot K_t}{r\cdot R}\begin{bmatrix} -\frac{\sqrt{3}}{2\cdot M} & 0 & \frac{\sqrt{3}}{2\cdot M} \\ \frac{1}{2\cdot M} & \frac{1}{M} & \frac{1}{2\cdot M} \\ \frac{d}{J} & \frac{d}{J} & \frac{d}{J} \end{bmatrix} \tag{39}$$

$$K = diag\left(-\frac{C_v}{M}, -\frac{C_{vn}}{M}, -\frac{C_\omega}{J}\right) \tag{40}$$

5.1.2 Four Wheeled

Using equations on section 3.2, (22) to (26) and (34), we get the following equations to the four wheeled robot model.

$$A = diag(A_{11}, A_{22}, A_{33})$$

$$A_{11} = -\frac{2\cdot K_t^2\cdot l^2}{r^2\cdot R\cdot M}-\frac{B_v}{M} \qquad A_{22} = -\frac{2\cdot K_t^2\cdot l^2}{r^2\cdot R\cdot M}-\frac{B_{vn}}{M} \qquad A_{33} = -\frac{4\cdot d^2\cdot K_t^2\cdot l^2}{r^2\cdot R\cdot J}-\frac{B_\omega}{J} \tag{41}$$

$$B=\frac{l\cdot K_t}{r\cdot R}\begin{bmatrix} 0 & -\frac{1}{M} & 0 & \frac{1}{M} \\ \frac{1}{M} & 0 & -\frac{1}{M} & 0 \\ \frac{d}{J} & \frac{d}{J} & \frac{d}{J} & \frac{d}{J} \end{bmatrix} \tag{42}$$

$$K = diag\left(-\frac{C_v}{M}, -\frac{C_{vn}}{M}, -\frac{C_\omega}{J}\right) \tag{43}$$

5.2 Experimental Data for Robot Model

5.2.1 Three Wheeled

Experience 1 was conducted using an input signal corresponding to a ramped up step, see Fig. 9. This way wheel sleeping was avoided, that is, wheel - traction problems don't exist.

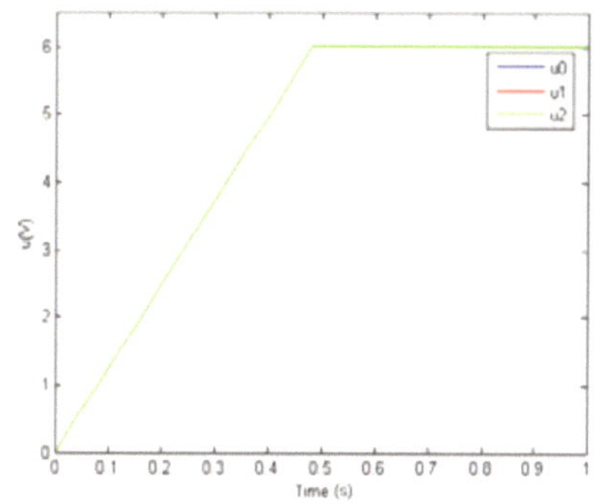

Fig. 9. Input voltage for the three wheeled robot

Table 1 and Fig. 10 shown the results for the three wheeled robot with experience 1.

u(V)	ω(rad/s)	T(Nm)
2	3.7062	0.1324
4	8.681	0.2319
6	13.4305	0.2789
8	18.3492	0.2972
10	21.8234	0.3662
12	26.2269	0.3971

Table 1. Experience 1 - Results for the three wheeled configuration

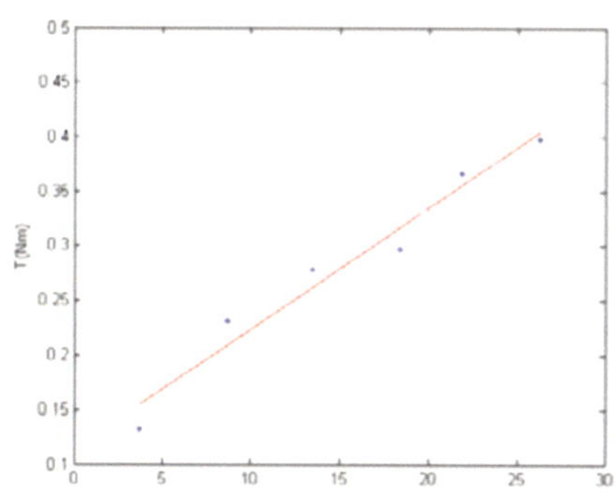

Fig. 10. Experience 1 - Results for the three wheeled configuration

The plots shown in Fig. 11 present the results obtained with experience 2.

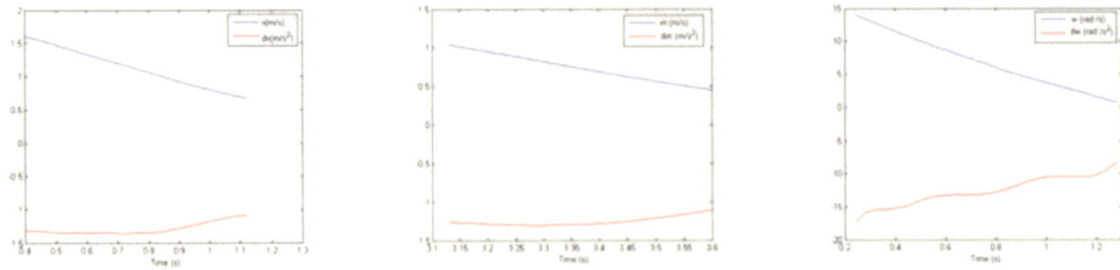

(a) Direction v (b) Direction vn (c) Rotation axis

Fig. 11. Experience 2 – Velocity and acceleration for the three wheeled robot

5.2.2 Four Wheeled

Experience 1 was conducted using an input signal corresponding to a ramped up step, similar to the chart plotted in Fig. 9. This way wheel sleeping was avoided, that is, wheel - traction problems don't exist.

Table 2 and Fig. 12 show the results for the four wheeled robot with experience 1.

u(V)	ω(rad/s)	T(Nm)
2	3.7269	0.1491
4	8.7519	0.2645
6	13.519	0.3257
8	18.357	0.3352
10	22.4065	0.3966
12	27.1416	0.4513

Table 2. Experience 1 - Results for the four wheeled configuration

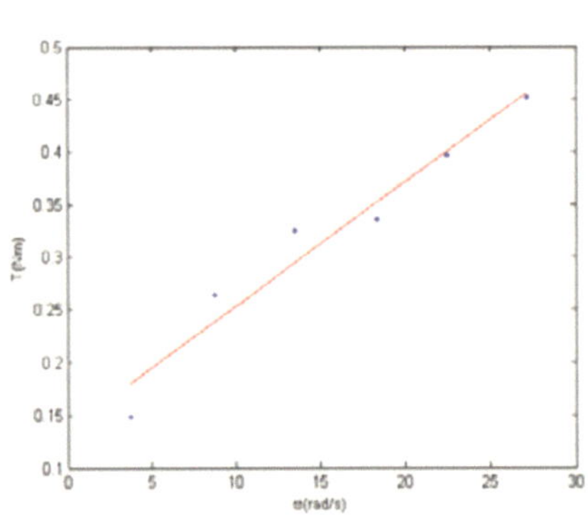

Fig. 12. Experience 1 - Results for the four wheeled configuration

The results obtained with experience 2 are shown in Fig. 13.

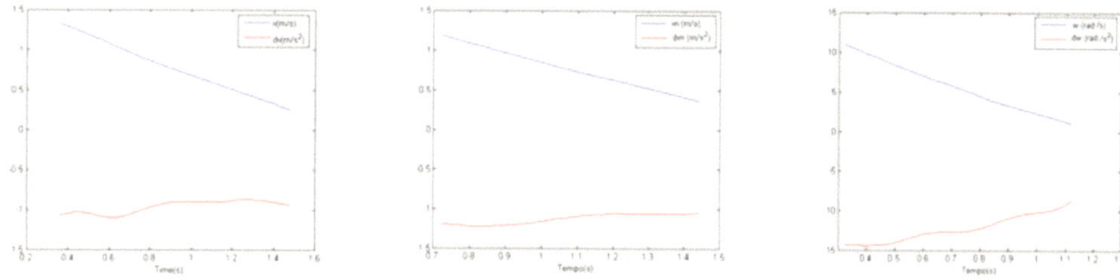

(a) Direction v (b) Direction vn (c) Rotation axis

Fig. 13. Experience 2 – Velocity and acceleration for the four wheeled configuration

5.2.3 Motor

The motor model was presented earlier in equation (34).

Experimental tests to the four motors were made to estimate the value of resistor R and the constant K_v. The numerical value of the torque constant K_t is identical to the EMF motor constant K_v.

Table 3 and Fig. 14 plots experimental runs regarding motor 0. Other motors follow similar behaviour.

u_0 (V)	i_0 (A)	ω_{m0} (rad/s)	ω_{m0}/i_0 ((rad/s)/A)	u_0/i_0 (Ω)
12	0.151	440.9	2910.1	79.2
12	0.831	345.7	415.7	14.4
12	0.425	402.1	945.0	28.2
12	0.456	397.0	870.3	26.3
12	0.715	360.5	503.6	16.7
12	0.641	371.5	579.6	18.7
12	1.164	301.0	258.4	10.3
12	0.272	423.7	1553.8	44.0
12	0.515	389.0	755.2	23.2
12	1.454	266.3	183.1	8.2

Table 3. Experimental tests with motor 0

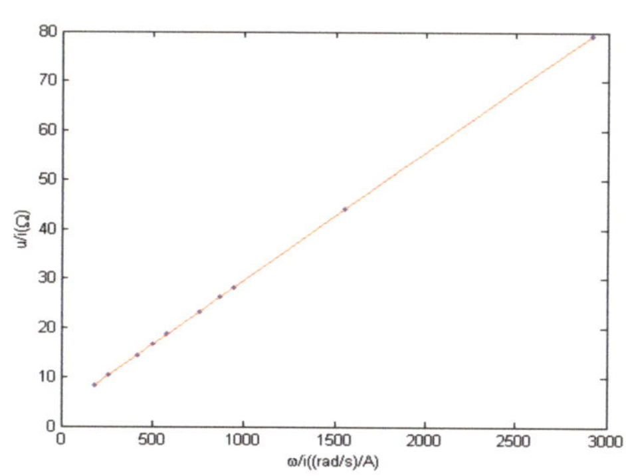

Fig. 14. Experimental tests for motor 0

5.3 Numerical Results

Table 4 presents the experimental results to the friction coefficients and inertial moment.

Parameters	3 wheels	4 wheels
J (kgm^2)	0.015	0.016
B_v (N/(m/s))	0.503	0.477
B_{vn} (N/(m/s))	0.516	0.600
B_ω (Nm/(rad/s))	0.011	0.011
C_v (N)	1.906	1.873
C_{vn} (N)	2.042	2.219
C_ω (Nm)	0.113	0.135

Table 4. Friction coefficients and inertia moment

From the experimental runs from all 4 motors, the parameters found are K_v=0.0259 V/(rad/s) and R=3.7007 Ω.

To validate the model were made some experimental test on using a step voltage with a initial acceleration ramp, see Fig. 15.

(i)

(ii)

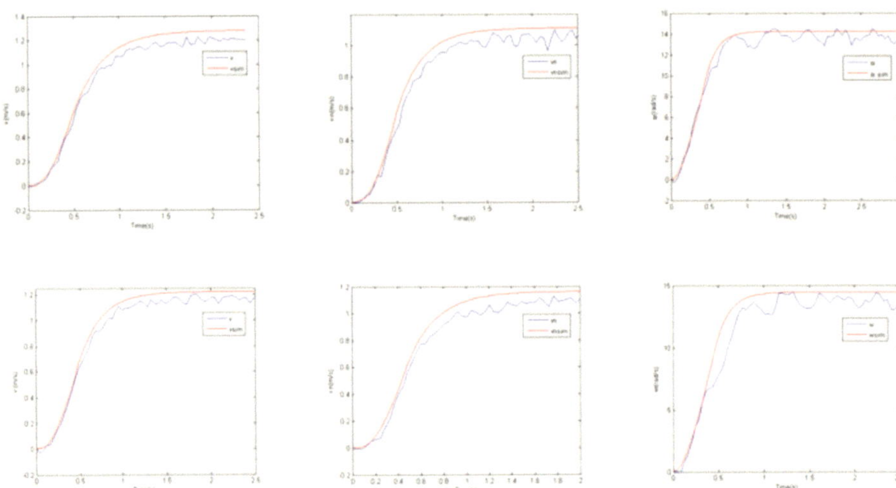

(a) Velocity along v (b) Velocity along vn (c) Angular velocity ω

Fig. 15. Experimental runs for initial model validation for the (i) three wheeled configuration and (ii) the four wheeled configuration

To evaluate the performance of the model, the real values of the velocities were compared to the estimated ones. The following numeric parameters are used to evaluate the adequacy of the model:

$$e_{\max} = \max\left|v(k) - v_{sim}(k)\right| \tag{44}$$

$$\bar{e} = \frac{1}{N} \cdot \sum_{i=1}^{N} e_i \tag{45}$$

$$\sigma = \sqrt{\frac{1}{N} \cdot \sum_{i=1}^{N} (e_i - \bar{e})^2} \tag{46}$$

where e_i represents the error in each instant.
The analysis of the errors of the experimental run shown in Fig. 15, in accordance to equations (44), (45) and (46) are analyzed later in Table 7.

5.4 Sensitivity Analysis

To understand which model parameters have more influence on the robot's dynamics, a comparison was made between the matrices of the models.
The model equation (36) is a sum of fractions. Analyzing the contribution of each parcel and of the variable portion within each fraction, a sensitivity analysis is performed, one estimated parameter at a time.

1. Matrix A, robot moving along v direction;
   - Three wheeled configuration:

$$\left(\frac{3 \cdot K_t^2 \cdot l^2}{2 \cdot r^2 \cdot R \cdot M}\right) = \frac{K_{a1}}{R} = 3.3110 \qquad \frac{B_v}{M} = K_{a2} \cdot B_v = 0.3245$$

   - Four wheeled configuration:

$$\left(\frac{2 \cdot K_t^2 \cdot l^2}{r^2 \cdot R \cdot M}\right) = \frac{K_{a1}}{R} = 3.6676 \qquad \frac{B_v}{M} = K_{a2} \cdot B_v = 0.2041$$

2. Matrix B and K, robot moving along v direction with constant voltage motor equal to 6V;
   - Three wheeled configuration:

$$\left(\frac{\sqrt{3} \cdot K_t \cdot l}{2 \cdot r \cdot R \cdot M}\right) \cdot 12 = \frac{K_b}{R} \cdot 12 = 5.7570 \qquad \frac{C_v}{M} = K_k \cdot C_v = 0.8728$$

   - Four wheeled configuration:

$$\left(\frac{K_t \cdot l}{r \cdot R \cdot M}\right) \cdot 12 = \frac{K_b}{R} \cdot 12 = 5.5227 \qquad \frac{C_v}{M} = K_k \cdot C_v = 0.7879$$

The same kind of analysis could be taken further by analyzing other velocities (vn and ω). Conclusions reaffirm that motor parameters have more influence in the dynamics than friction coefficients. This means that it is very important to have an accurate estimation of the motor parameters. Some additional experiences were designed to improve accuracy. The method used previously does not offer sufficient accuracy to the estimation of R. This parameter R is not a physical parameter and includes a portion of the non-linearity of the H bridge powering the circuit that, in turn, feeds 3 rapidly switching phases of the brushless motors used. In conclusion, additional accuracy in estimating R is needed.

5.5 Experience 3 – Parameter Estimation Improvement

The parameter improving experience was made using a step voltage with an initial acceleration ramp.
As seen in 5.1 the model was defined by the equation (36) and we can improve the quality of the estimation by using the Least Squares method (Gelb et al., 1974). The system model equation can be rewritten as:

$$y = \Theta_1 \cdot x_1 + \Theta_2 \cdot x_2 + \Theta_3 \cdot x_3 \qquad (47)$$

where: x_1=x(t); x_2=u(t); x_3=1; y=dx(t)/dt.
The parameters Θ are estimated using:

$$\Theta = \left(x^T \cdot x\right)^{-1} \cdot x^T \cdot y \qquad (48)$$

$$x = \left[x_1(1)\cdots x_1(n) \quad x_2(1)\cdots x_2(n) \quad x_3(1)\cdots x_3(n)\right]^T \qquad (49)$$

Estimated parameters can be skewed and for this reason instrumental variables are used to minimize the error, with vector of states defined as:

$$z = \left[\bar{x}_1(1)\cdots \bar{x}_1(n) \quad x_2(1)\cdots x_2(n) \quad x_3(1)\cdots x_3(n)\right]^T \qquad (50)$$

The parameters Θ are now calculated by:

$$\Theta = \left(z^T \cdot x\right)^{-1} \cdot z^T \cdot y \qquad (51)$$

Three experiments were made for each configuration of three and four wheels, along v, vn and ω. For the v and vn experiments values C_v and C_{vn} are kept from previous analysis. For the ω experiment, the value of the R parameter used is the already improved version from previous v and vn experimental runs of the current section.
The numerical value of R for each motor was estimated for each motor and then averaged to find R=4.3111 Ω. The results are present on followings tables. Table 5 shows values estimated by the experiment mentioned in this section.

Parameters	3 wheels	4 wheels
J (kgm^2)	0.0187	0.0288
B_v (N/(m/s))	0.5134	0.5181
B_{vn} (N/(m/s))	0.4571	0.7518
B_ω (Nm/(rad/s))	0.0150	0.0165
C_ω (Nm)	0.0812	0.1411

Table 5. Parameters estimated after experience 3

The final values for friction and inertial coefficients are averaged with results from all 3 experimental methods and the numerical values found are presented in Table 6.

Parameters	3 wheels	4 wheels
d (m)	0.089	
r (m)	0.0325	
l	5:1	
K_v (V/(rad/s))	0.0259	

R (Ω)	4.3111	
M (kg)	1.944	2.34
J (kgm^2)	0.0169	0.0228
B_v (N/(m/s))	0.5082	0.4978
B_{vn} (N/(m/s))	0.4870	0.6763
B_ω (Nm/(rad/s))	0.0130	0.0141
C_v (N)	1.9068	1.8738
C_{vn} (N)	2.0423	2.2198
C_ω (Nm)	0.0971	0.1385

Table 6. Parameters of dynamical models

5.6 Model Validation Experiences

The models were validated with experimental tests on using a step voltage with an initial acceleration ramp. Fig. 16 show plots for the experimental runs.

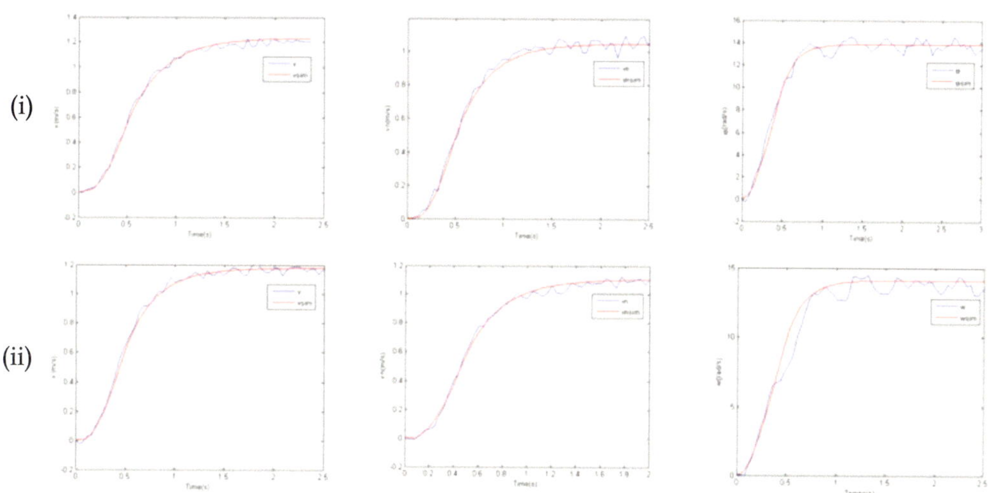

(a) Velocity along v (b) Velocity along vn (c) Angular velocity ω

Fig. 16. Experimental runs for final model validation for the (i) three wheeled configuration and (ii) four wheeled configuration

The analysis of the errors of the experimental runs shown in Fig. 16, in accordance to equations (44), (45) and (46) are presented in Table 7.

Configuration	Vels.	Initial			Final		
		e_{max}	$\bar{e}$	σ	e_{max}	$\bar{e}$	σ
Three wheeled	v	0.122	-0.059	0.034	0.067	-0.003	0.027
	vn	0.124	-0.055	0.032	0.051	0.016	0.023
	ω	1.699	-0.046	0.546	1.152	0.017	0.538
Four wheeled	v	0.096	-0.052	0.026	0.061	0.008	0.024
	vn	0.135	-0.070	0.034	0.068	-0.009	0.026
	ω	3.345	-0.919	0.852	2.504	-0.452	0.732

Table 7. Error analysis for initial and final estimated parameters (as shown I nFig. 15 and 16)

Fig. 17 shows the fit of the error of the initial experimental runs shown in Fig. 15 when compared to the fit of the errors for the final estimated parameters, as shown previously in Fig. 16. Clearly, the mean of the fits is closer to zero in the final parameters thus producing adequate model performance.

(i)

(ii)

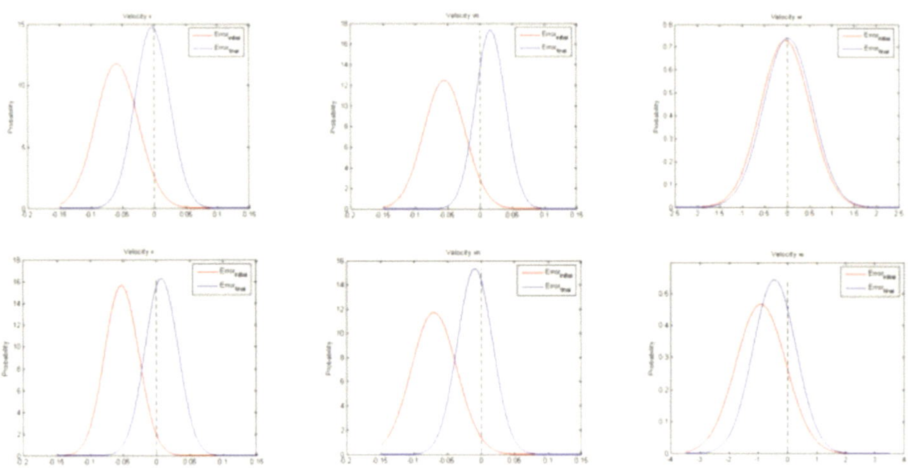

(a) Velocity along v (b) Velocity along vn (c) Angular velocity ω

Fig. 17. Comparison error fits for initial and final estimated parameters for the (i) three wheeled robot and the (ii) four wheeled

5.7 Assessment and Comparison of the Found Models

5.7.1 Preferential Directions

The models found, shown in section 5.1, prove that each mechanical configuration has different preferential direction at which the robot has maximum velocities, assuming $\omega=0$. This fact can easily be ascertained by analysing the plots shown in Fig. 18 that occur periodically over $\pi/3$ for the three wheeled and periodically over $\pi/2$ for the four wheeled case.

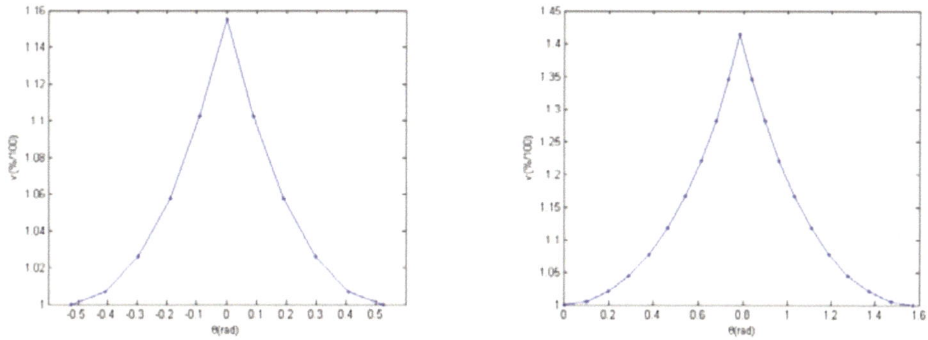

(a) Three wheeled configuration (b) Four wheeled configuration

Fig. 18. Normalized velocity variation over θ

The identified preferential directions are graphically identified in Fig. 19.

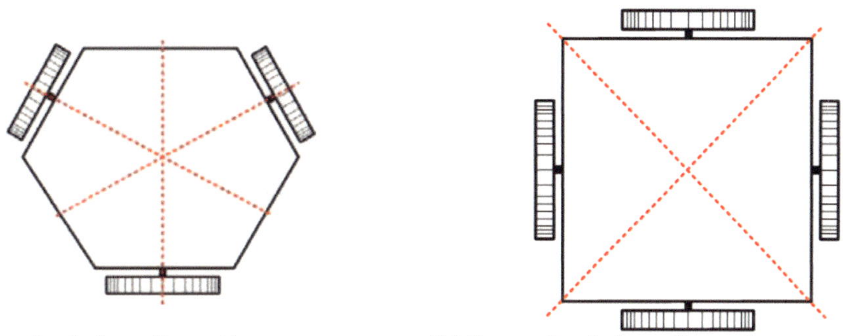

(a) Three wheeled configuration (b) Four wheeled configuration

Fig. 19. Preferential directions

5.7.2 Current Consumption

By observing Fig. 20 and Fig. 18, an interesting conclusion is revealed: on the three wheeled configuration the maximum velocity needs 8% less of the current than the current at the minimum velocity while the four wheeled configuration needs 90% more current when compared to the current needed for minimum velocity.

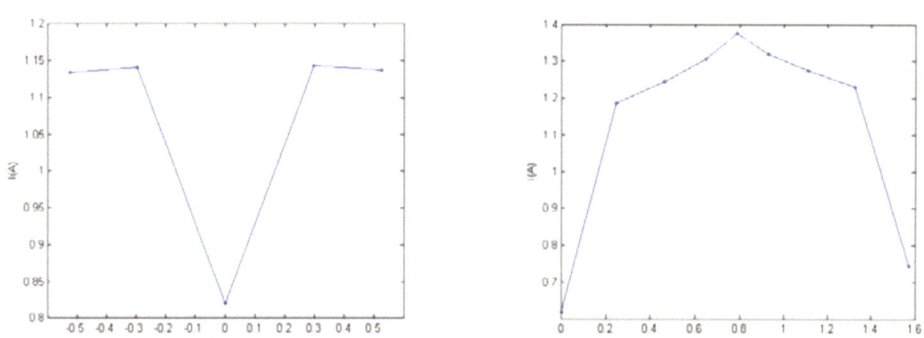

(a) Three wheeled configuration (b) Four wheeled configuration

Fig. 20. Current variation

5.7.3 Maximum Velocities

The plots shown in Fig. 21a) clearly show that the maximum velocity of the four wheeled configuration is larger than its three wheeled counterpart - the values are 3.58 m/s and 2.86 m/s respectively (at preferential directions). It thus seems natural that the accelerations shown in figure Fig. 21b) are also larger for the four wheeled configuration at the expense of a much larger current consumption, shown in Fig. 21c). This latest plots shows that roughly twice as much current is necessary to produce the previously mentioned 25% velocity increase. Fig. 21c) also shows that the sum of all the motor currents inside the robot with the four wheel configuration is about 11 A which is somewhat high and may be large enough to exceed the limits of the linearity of the system.

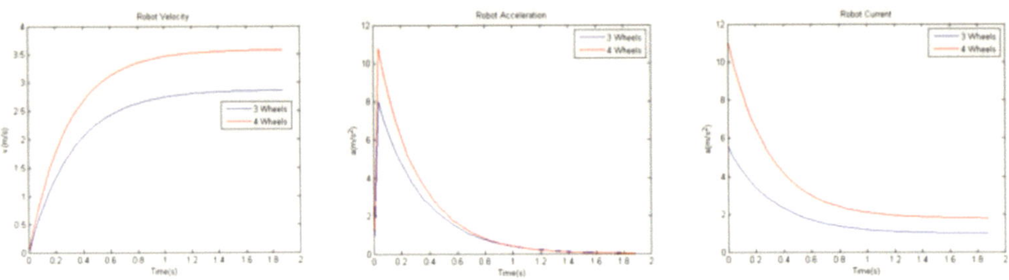

a) Velocities b) Accelerations c) Overall Currents

Fig. 21. Comparison of maximum velocities, associated accelerations and total currents

As explained earlier, the maximum speed varies with direction of the movement so it is interesting to observe the plot shown in Fig. 22 that comparatively predicts maximum speeds over the angle θ of the robot for both robotic configurations. It can be seen that the four wheeled configuration is frequently, but not always, faster than the three wheeled version. The three wheeled configuration has a smoother velocity profile whilst the four wheeled configuration has a large velocity gap in the preferential direction.

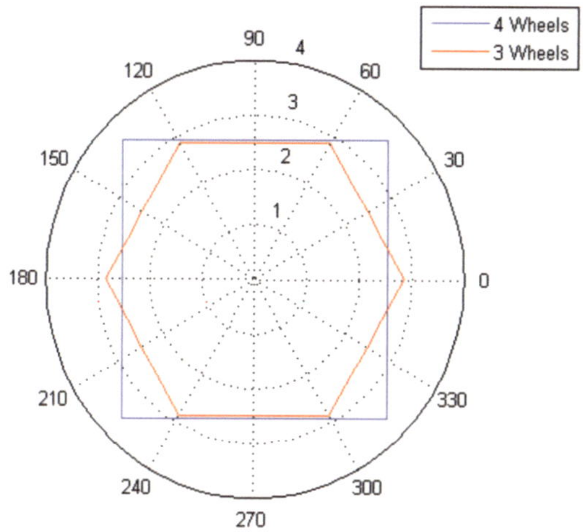

Fig. 22. Comparison of the velocity variation from the two platforms

6. Conclusions

This paper presents models for mobile omni-directional robots with three and four wheels. The derived model is non-linear but maintains some similarities with linear state space equations. Friction coefficients are most likely dependent on robot and wheels construction and also on the weight of the robot. The model is derived assuming no wheel slip as is interesting in most standard robotic applications.

A prototype that can have either three or 4 omni-directional wheels was used to validate the presented model. The test ground was smooth and carpeted. Experience data was gathered by overhead camera capable of determining position and orientation of the robot with good accuracy.

Experiences were made to estimate the parameters of the model for the prototypes. The accuracy of the presented model is discussed and the need for additional experiences is proved. The initial estimation method used two experiences to find all parameters but a third experience is needed to improve the accuracy of the most important model parameters. Sensitivity analysis shows that the most important model parameters concern motor constants.

Observing estimated model parameters, the four wheel robot has higher friction coefficients in the vn direction when compared to the v direction. This means of course higher maximum velocity for movement along v axis and higher power consumption for movements along the vn direction. This difference in performance points to the need of mechanical suspension to ensure even wheel pressure on the ground.

The found model was shown to be adequate for the prototypes in the several shown experimental runs.

From the derived models, some interesting conclusions were found. Firstly, it was proven that an omni-directional robot does not drive all directions with the same maximum velocity. As such, preferential directions for maximum velocities were found for each configuration - these preferential directions are periodic and dependent on the wheel configuration. Another conclusion is that, as expected, the four wheeled configuration has a much, much larger current consumption when compared to the three wheeled version. It is interesting to note that the three wheeled configuration has smaller current consumption at its maximum velocity than that at its minimum velocity. Also interesting is the comparative plot of maximum velocities producible by the robot with each configuration for a given direction of movement. The three wheeled robot has small velocity gap at different directions of movement while important differences exist for the four wheeled version. It was also proven that, for a large portion of directions, the four wheeled version at maximum speed is faster, but not always! There is small portion of movement directions where the maximum velocity of the three wheeled configuration is superior to that of the four wheeled counterpart.

7. Future Work

The work presented is part of a larger study. Future work will include further tests with different prototypes including prototypes with suspension. The model can also be enlarged to include the limits for slippage and movement with controlled slip for the purpose of studying traction problems. Dynamical models estimated in this work can be used to study the limitations of the mechanical configuration and allow for future enhancements both at controller and mechanical configuration level. This study will enable effective full comparison of three and four wheeled systems.

8. References

Astrom, K. & Wittenmark, B. (1984), *Computer Controlled System – Theory and Design,* Prentice-Hall, Information and System Sciences Series, 1984.

Campion, G.; Bastin, G. & Dandrea-Novel, B. (1996), *Structural properties and classification of kinematic and dynamic models of wheeled mobile robots,* IEEE Transactions on Robotics and Automation, vol. 12(1), pp. 47-62, 1996.

Conceicao, A.; Moreira, A. & Costa, P. (2006), *Model Identification of a Four Wheeled Omni-Directional Mobile Robot,* Proceedings of the 7th Portuguese Conference on Automatic Control, Instituto Superior Tecnico, Lisboa, Portugal, 2006.

Costa, P.; Marques, P.; Moreira, A.; Sousa, A. & Costa, P. (2000), *Tracking and Identifying in Real Time the Robots of a F-180 Team,* in Manuela Veloso, Enrico Pagello and Hiroaki Kitano, editors. RoboCup-99: Robot Soccer World Cup III. Springer, LNAI, pp. 286-291, 2000.

Diegel, O.; Badve, A.; Bright, G.; Potgieter, J. & Tlale, S. (2002), *Improved Mecanum Wheel Design for Omni-directional Robots,* Proceedings of the Australasian Conference on Robotics and Automation, Auckland, 2002.

Gelb, A.; Kasper, J.; Nash, R.; Price, C. & Sutherland, A. (1974), *Applied Optimal Estimation*, MIT Press, 1974.

Ghaharamani, Z. & Roweis, S. (1999), *Learning Nonlinear Dynamical systems using an EM Algorithm,* In M. S. Kearns, S. A. Solla, D. A. Cohn, (eds) Advances in Neural Information Processing Systems, Cambridge, MA: MIT Press, vol. 11, 1999.

Gordon, N.; Salmond, D. & Smith, A. (1993), *Novel approach to nonlinear/non-Gaussian Bayesian state estimation*, IEE Proceedings-F on Radar and Signal Processing, vol. 140(2), pp. 107-113, 1993.

Julier, S. & Uhlmann, J. (1997), *A New Extension of the Kalman Filter to Nonlinear Systems*, Int. Symp. Aerospace/Defense Sensing, Simul. and Controls, Orlando, FL, 1997.

Khosla, P. (1989), *Categorization of parameters in the dynamic robot model*, IEEE Transactions on Robotics and Automation, vol. 5(3), pp. 261-268, 1989.

Leow, Y.; Low, K. & Loh, W. (2002), *Kinematic Modelling and Analysis of Mobile Robots with Omni-Directional Wheels*, Proceedings of the Seventh International Conference on Control, Automation, Robotics And Vision, Singapore, 2002.

Loh, W.; Low, K. & Leow, Y. (2003), *Mechatronics design and kinematic modelling of a singularityless omni-directional wheeled mobile robot*, Proceedings of the IEEE International Conference on Robotics and Automation, vol. 3, pp. 3237-3242, 2003.

Maxon Motor (2009), *Maxon Motor Catalogue*, http://www.maxonmotor.com (visited April 2009)

Muir, P. & Neuman, C. (1987), *Kinematic modeling for feedback control of an omnidirectional wheeled mobile robot*, Proceedings of the IEEE International Conference on Robotics and Automation, vol. 4, pp. 1772-1778, 1987.

Oliveira, H.; Sousa, A.; Moreira, A. & Costa, P. (2008), *Dynamical models for omni-directional robots with 3 and 4 wheels*, Proceedings of the 5th International Conference on Informatics in Control, Automation and Robotics, pp. 189-196, 2008.

Olsen, M. & Petersen, H. (2001), *A new method for estimating parameters of a dynamic robot model*, IEEE Transactions on Robotics and Automation, vol. 17(1), pp. 95-100, 2001.

Pillay, P. & Krishnan, R. (1989), *Modeling, Simulation, and Analysis of Permanent-Magnet Motor Drives, Part 11: The Brushless DC Motor Drive*, IEEE transactions on Industry applications, vol. 25(2), pp. 274-279, 1989.

Salih, J.; Rizon, M.; Yaacob, S. ; Adom A.; & Mamat, M. (2006), *Designing Omni-Directional Mobile Robot with Mecanum Wheel*, American Journal of Applied Sciences, vol. 3(5), pp. 1831-1835, 2006.

Tahmasebi, A.; Taati, B.; Mobasser, F.; & Hashtrudi-Zaad, K. (2005), *Dynamic parameter identification and analysis of a PHANToM haptic device*, Proceedings of the IEEE Conference on Control Applications, pp. 1251-1256, 2005.

Williams, R.; Carter, B.; Gallina, P. & Rosati, G. (2002), *Dynamic model with slip for wheeled omnidirectional robots*, IEEE Transactions on Robotics and Automation, vol.18(3), pp. 285-293, 2002.

Xu, J.; Zhang, M. & Zhang, J. (2005), *Kinematic model identification of autonomous mobile robot using dynamical recurrent neural networks*, Proceedings of the IEEE International Conference Mechatronics and Automation, vol. 3, pp. 1447-1450, 2005.

13

HUNTER – HYBRID UNIFIED TRACKING ENVIRONMENT

Aislan Gomide Foina and Francisco Javier Ramirez-Fernandez
Universidade de São Paulo
Brazil

1. Introduction

Pervasive computing is an emerging paradigm used nowadays for very distinct applications. It is a model of human-computer interaction in which computer devices are embedded in an environment, exchanging information between themselves and the human preset as well. It has been called **Everyware** (Greenfied, 2006). Some of its application can go from smart homes for elderly people, to automatic toll billing system for vehicles (Xu, 2008).
In many cases, pervasive applications make use of standard technology called Radio Frequency Identification - RFID. RFID is an identification technology similar to barcode, but instead of using laser beams and optical cameras to detect a printed code in a label, it uses radio waves to communicate with a microchip built-in this same label and gets its identification wirelessly. Its application can vary as much as the pervasive computing application.
In this book chapter, a case study will be explored using pervasive computing principles and RFID technology to supervise a bulk cargo unloading process in an automated way to reduce fraud attempts and to increase the efficiency of the unloading place.
Bulk cargo is a very common transport modality for ship commodities. Large bulk wagons and bulk ships can convey hundreds of thousand tons of loads. Thus, this modality is used mostly by railway and port authorities.
A regular bulk load process consists of three tasks: 1) measurement of the weight without the product - empty weight; 2) the load of the product in the transporter; and 3) the measurement of the weight with the product loaded - full weight. The unload process is similar to the load processes, but in the reverse order, as seen in Fig. 1: 1) Full weight measurement; 2) cargo unload; and 3) empty weight measurement. The difference between the first and the last weight measurements is the weight of the product loaded or dropped.
There are two major types of trucks that convey bulk cargo: dump truck and bulk truck. The first type has a hydraulic system to tip the trailer, dumping the product by the action of gravity. The second type does not have this mechanism, so it depends on a hydraulic tipping bridge to tip the whole truck. Fig. 2. shows both types of truck unloading the cargo.
A great problem faced by the companies that work with this modality of cargo is theft. A common theft reported by the company studied is trailer exchange. A truck full of a product arrives at the unload area and it performs the full weight task normally. After that, this

truck goes to an outside area, exchanges the full trailer for an empty one, and the driver defrauds the documents. Later, it goes back to do the empty weight task and leaves. In brief, the company pays for the non-received product, and the truck driver sells this cargo to another company, getting the money two times for only one cargo. This theft is called as full load theft and, according to the company, it represents 90% of the total loss of the company with stolen cargo.

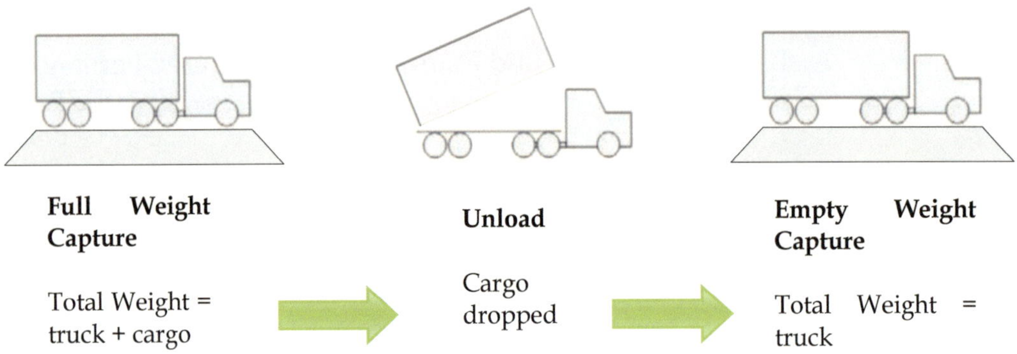

Fig. 1. The three tasks of the unload process.

Fig. 2. Most common types of truck: a) Dump truck; b) Bulk truck tipped by a hydraulic tipping bridge.

This theft is possible due to the large size of the port area, where many companies are located together. Another fact that collaborates with theft is the distance between the warehouses and the 3rd part weighting scales, seen in Fig. 3. There are two weighting scales external to the port area that serve the company when the number of trucks in the unload area is too high. Since the trucks have to move away from the port area to be weighed, it is easy for a trailer exchange to precede, for instance.

Fig. 3. The company area, and its distance to each external weighting scale.

The company unload area is 1 km x 0.5 km long, and it has 6 warehouses, seen in Fig. 4. Each of them can stock up to 30,000 tons of cargo; it is enough to completely load a bulk carrier. The company works with different types of products, so each type is stocked in one warehouse. Next to the warehouses, there is the company weighting scale. This scale has two weighting bridges, so it can weigh two trucks at the same time. Finally, the area has another small area built for administrative purposes and a datacenter building where all the servers are installed.

The solution developed and presented in this chapter, containing hardware and software to handle the trucks unload process in the port area is named **Hunter**. Its presentation starts with the introduction. Secondly the old process developed by the company studied will be presented. Later, the process and the analysis of possible technologies to be used to substitute the old process will be studied. After that, how the system was implemented is introduced, starting from the equipment and going through the software. Finally, the results, conclusion and future work are presented.

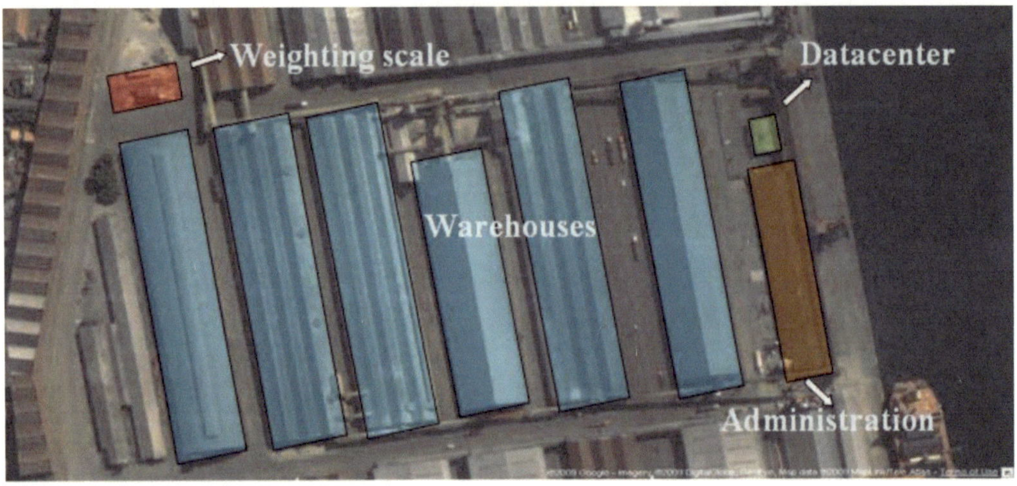

Fig. 4. Company unloading area.

2. Related Works

Nowadays, there are many works related to the use of RFID to control many types of processes. One relevant work is related to the pharmaceutical supply chain (Yue et al., 2008). The author proposes a new framework to develop a RFID system and applied it to the pharmaceutical process. The contribution of this work is the method described using the Critical Success Factor - CSF - and the "5W1H" method. Another work is related to the process control of patient discharge process (Jiao et al., 2008). A re-engineering process is described, based on the Business Process Reengineering - BRP - methodology. The focus of the system is to automate human tasks, reduce data errors and collect data in real-time. The system was tested using simulation tools. Another work describes the application of the RFID to control attendance of students in classes (Qaiser & Khan, 2006). It describes a system to track students and automatically detect their presence in the classes and measure the time that the student stayed in the classroom.

The postal service is a highly complex logistic company, and there is a relevant work describing the use of RFID for this type of service (Park et al., 2007). Here the integration of the managing system with PDAs is presented, as well as the description of the messages exchanged by the PDA and the managing system. The author creates a real test bed to

measure the operationality of the system, containing RFID readers, PDAs and all the other equipment necessary to validate it.
Finally, another work describes only the system to handle the RFID events in a process control application (Hu et al., 2008). Many software approaches can be used to implement a RFID middleware and an event handling system. In this paper, the author used a hierarchy system with three layers using Petri nets.
All the works mentioned present software models, process approaches or methodologies to assist real RFID implementation. The difference is the fact that the Hunter system is the result of a real implementation case, described in this chapter, which started in 2005 and has been used by the company for over 3 years.

2.1 Old Process

A good related work reference is the solution that the company adopted previously to control its process. All the solution was based on barcode labels, barcode readers, a paper document called *Minuta*, from the Portuguese word for protocol, and the employees in charge of the unload process. This protocol has an ID, and it is a primary key for a management system called SSA, the function of which is to handle all the trucks unload. SSA is basically a client-server application that interacts with barcode readers and weighting scales to automatically capture the values without manually input.
The company defined two positions responsible for the unload operation: security inspector, or simply **inspector**, and the logistic supervisor, or simply **supervisor**. The inspector handles the truck information and the supervisor is responsible for the cargo information. They are both preset at all places in the unload area.
The *Minuta* is a piece of paper with all the information about the truck, the cargo and the service order, seen in Fig. 5. It will be named **protocol** to keep the text more readable.
When the truck arrives at the weighting scale, the driver gives the protocol to the inspector. The latter compares both the license plates of the tractor unit and trailer, the number of wheels and spare wheels, the departure place and some other information about the truck. If all the data are in accordance with details of truck, the inspector gives the protocol back to the driver and allows the truck to get inside the weighting scale to get the full weight measurement. Inside the weighting scale, the supervisor gets the protocol to consult the service order and the protocol id in the SSA, checking if this protocol is not fake or duplicated. If the system returns that it is a valid protocol, the supervisor captures the truck full weight and inserts it into the SSA database. After the weighing, the supervisor writes the current time in the protocol, gives it back to the driver and forwards the truck to one of the unloading places in the area. The selection of the unloading place is based on the type of the truck, type of the cargo and line size of the unloading places.
If any of the validation steps in the task of getting the full weight detects any discrepancy from the protocol, the truck is forwarded to the waiting area and the divergent data is checked with the location where the truck got the cargo. Some common problems detected are mostly related to the wrong typing of the license plate.
After the truck leaves the weighting scale, it proceeds to the line of the unload place previously assigned. Arriving in the line, the unloading area inspector gets the protocol again and double checks all the same truck information checked in the weighting scale. If all the information is correct, the inspector pastes a barcode label with his unique id merged with an incremental number, gives the protocol back to the driver and the truck gets inside

the designated unloading place. There, the supervisor gets the protocol again and compares the type of cargo written in the protocol with the cargo inside the trailer. If all the data are in accordance with that information, the supervisor authorizes the beginning of the cargo unload. After the end of the unloading, the operator pastes another barcode label on the protocol with his unique id and another incremental number, writes the current time information on it and gives the protocol back to the driver.

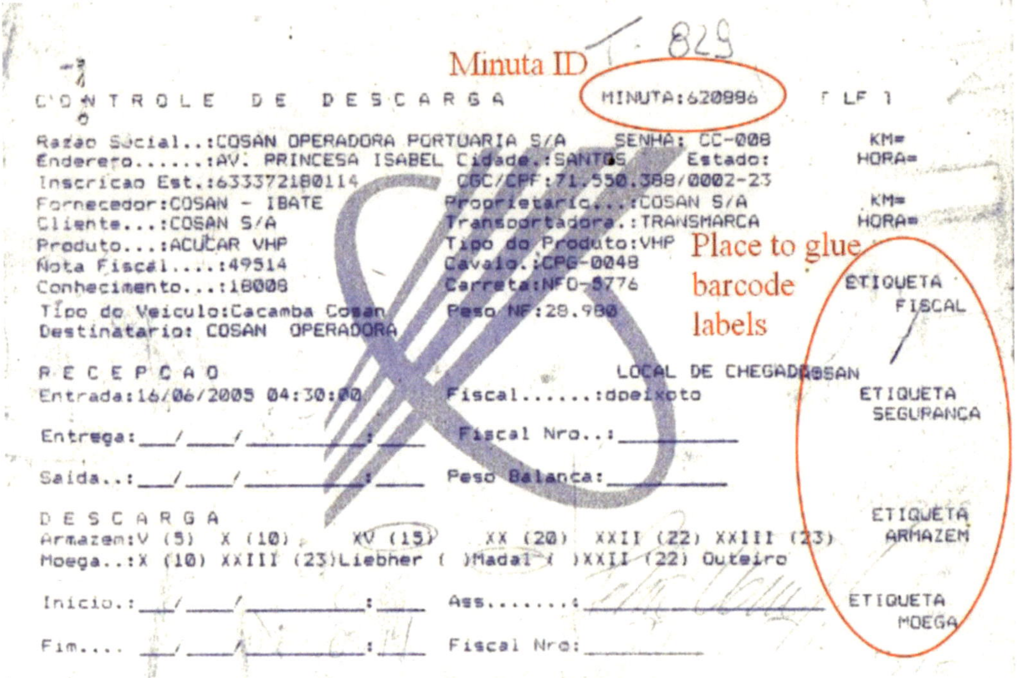

Minuta ID

CONTROLE DE DESCARGA MINUTA:620896 LF 1

Razao Social..:COSAN OPERADORA PORTUARIA S/A SENHA: CC-008 KM=
Endereco......:AV. PRINCESA ISABEL Cidade.:SANTOS Estado: HORA=
Inscricao Est.:633372180114 CGC/CPF:71.550.388/0002-23
Fornecedor:COSAN - IBATE Proprietario...:COSAN S/A KM=
Cliente...:COSAN S/A Transportadora.:TRANSMARCA HORA=
Produto...:ACUCAR VHP Tipo do Produto:VHP
Nota Fiscal....:49514 Cavalo.:CPG-0048
Conhecimento...:18008 Carreta:NFO-8776
Tipo do Veiculo:Cacamba Cosan Peso NF:28.980
Destinatario: COSAN OPERADORA

Place to glue barcode labels

RECEPCAO LOCAL DE CHEGADA COSAN
Entrada:16/06/2005 04:30:00 Fiscal......:dpeixoto

Entrega: / / : Fiscal Nro..:

Saida..: / / : Peso Balanca:

DESCARGA
Armazem:V (5) X (10) XV (15) XX (20) XXII (22) XXIII (23)
Moega..:X (10) XXIII (23)Liebher ( )Madal ( )XXII (22) Outeiro

Inicio.: / / : Ass.......:

Fim.... / / : Fiscal Nro:

ETIQUETA FISCAL
ETIQUETA SEGURANCA
ETIQUETA ARMAZEM
ETIQUETA MOEGA

Fig. 5. A scanned copy of a *Minuta*, showing where its ID number is and where the barcode labels are pasted.

Here, as with the full weight procedure, if any of the validation steps detects a mismatch, the truck is forwarded to the waiting area. It keeps there until the point of discrepancy is solved. In most cases, the driver moves the truck to the wrong line, and it is detected by the inspector that forwards it to the correct line. If, for any reason, the truck conducted a partial cargo unload, it will be forwarded to another unloading place to repeat the procedure after the problem is solved.

When the truck is released, it has to go to the weighting scale again to proceed to the empty weight task. Arriving there, the truck repeats the same procedure conducted during the full weight task, with small differences. The driver gives the protocol to the inspector, who checks the data again and releases the truck to get inside the weighting scale. Inside it, the driver gives the protocol to the weighting scale operator who reads the barcodes using a barcode reader controlled by the SSA. If all the barcodes are accordance with previous information, the operator captures the empty weight and writes the current time information in the protocol. At this moment, the total cargo weight is calculated. If the difference between the calculated weight and the protocol declared weight is more than 5%,

the driver has the option to go back to any unloading place to proceed the scrapping and repeat the whole unload task. This weight difference happens because sometimes the product gets glued in the trailer wall. If the weight is within this tolerance, the driver gets his payment approved.
At the end of the day, all the protocols data are entered in the SSA manually. The information about the cargo weight and the unload place allows the software to calculate the amount of product unloaded for each type of product and how full the warehouses are. Finally, the timing data recorded in the protocol provides statistic information to measure the performance of the working team in each unloading place.
This solution has a weakness; it is based mostly on human intervention. It is totally dependent on the inspector and on the supervisor to keep the system trustable. Hence, an easy way to compromise the whole system is subverting these two people. For instance, a corrupt driver bribes both inspector and supervisor of a specific unloading place. Each person gives one of their barcode labels to the driver who pastes the labels on the protocol and fulfills the time fields with fake information. After that, the driver sells the cargo to a second company and unloads it there. Then, the driver goes to the weighting scale of the first company, with the trailer empty and the protocol fraudulently fulfilled with the barcode labels pasted. Finally, the driver performs the entire empty weight task without any problem and gets his payment. Thus, if the overall bribe paid to the inspector and supervisor is less than the amount paid by the second company for the cargo, the driver will make money with the fraud.

3. Study on the Process and System Proposal

As seen in the previous section, the solution adopted by the company is highly human dependent and has its weak points. The main issue is that the system has to be the least human dependent as possible. Therefore, to attain this objective, some pervasive approaches were studied and adopted.
So as to make the system less human dependent, it is necessary to remove the point of decision from the inspectors and operators. One critical point, for instance, is the truck entrance being informed by the inspectors, and the supervisor informing when the truck finishes the cargo unload. A solution to get this data automatically is to use a tracking device to monitor the truck in the unload area and to create virtual zones to detect which unload place the truck goes to, how long it takes to go from one place to the other and how many minutes it parks in each place.
Some technologies can be used to track vehicles; three were selected to be compared in this project based on the availability of the market and its cost: a) Global Positioning System - GPS; b) Real-time Location System - RTLS; and c) Radio Frequency Identification - RFID (Guillemette et al., 2008). They will be explained and compared one by one further on.
Besides the system tracking functionality, some other features need to be added to the solution in order to meet the company needs. The license plate validation in each task of the process is one example of functionality the system must have. This can be done by using cameras and one optical character recognition - OCR - software. Yet, the company refused this technology arguing that the heavy dust present in the environment when the trucks unload the cargo will block the camera, compromising the capture of the license plate image. Thus, the decision was to keep the job of the security and operators just giving them

a PDA to only conduct some of the confirmations they did in the old system. In this way, when the truck gets inside the unload place, the PDA software automatically presents the truck information and license plates, together with a confirm and refuse buttons. To increase the security of the solution, digital cameras were attached to the PDAs. These cameras are used to take pictures of the license plates and of the truck while it is unloading for further auditing. Since there is still dust, it is not a problem because the person who is handling the PDA can quickly clean it with his hands.

3.1 GPS

The first technology studied was the use of a GPS-based device to get the absolute latitude and longitude of the trucks along the unloading area. This device will be fixed in the trailer the moment the truck arrives at the weighting scale to capture its full weight. This GPS device will acquire the absolute position every second and transmit it via Wi-Fi or any other non-standard radio technology.

The problem in the use of this GPS device is related to the loss of satellite link when the truck gets into a closed environment. Since the GPS technology needs a clear vision of the sky, the GPS device will not be able to update its position when the truck gets into the warehouse and the unloading places (Jardak & Samama, 2009).

A project approach to solve this problem is the use of a differential GPS - DGPS - installed inside the unloading places.

This first technology, using the GPS device attached to the truck trailer, will demand a full Wi-Fi coverage of the unloading area, a stock of GPS devices enough to be attached to all trucks in the unloading area at the peak time, and some DGPS equipment to be installed inside the unloading places to correct the positioning.

This option has the advantage of using a reliable technology such as GPS maintained by the American Department of Defense.

3.2 RTLS

The second option uses an emerging technology used to track objects inside closed areas called RTLS. This technology works based on a similar GPS principle called antenna triangulation.

To implement this approach, every square meter of the unloading area needs to be covered by at least three RTLS readers, which, in this case, will use Wi-Fi access points. In addition, one RTLS tag will be attached to each trailer, similarly to the first approach.

Briefly, this approach will need the same number of trailers attached to equipment as in the GPS approach, and the triple number of Wi-Fi APs.

3.3 RFID

Finally, the last option uses the RFID technology. Present RFID systems consist basically of four components: an electronic tag, a reader, an antenna, and application software to process the data. When a tag approaches the antenna, the latter sends a signal to the reader with the tag identification. The reader receives the signal, and, through a RS232 interface or an Ethernet port, the information is sent to a computer executing the software. This software is normally a middleware application that processes the data packets and sends them to an end-user application or a database (Kim & Kim, 2006).

RFID tags can be active or passive. The active ones are self fed by an internal battery and the passive ones are fed with the energy from electromagnetic waves sent by the reader. Regarding passive tags, the reading ranges vary between 5 cm and 10 m. For the active tags, since they have an internal battery, the reading range can reach distances in the order of 200 meters. Active tags can be connected to temperature, tamper and movement sensors, among others. For this reason, the active tags technology was chosen for truck access control in warehouses and unloading area. In this case, the reader studied operates in the UHF frequency range, at 433MHz to transmit data and at 915MHz to receive data from the tag, with 80-meter reading range.

The implementation of this technology will demand one RFID tag attached to each truck, as in both approaches already mentioned. Each unload area entrance and exit will need a RFID reader installed, as well as the weighting scale.

3.4 Best approach

In order to choose the best technology, all of them were inserted into a table to facilitate the comparison, seen in Table 1.

The criterion chosen was the final cost of the whole solution. In this context, the RFID was the best technology.

Approach	Trailer Attached Device Cost	N° of Wi-Fi APs	Cost in Network Infrastructure	Additional Equipment
GPS	Very Expensive	Regular	Regular	DGPS
RTLS	Regular	Large	Great	Wireless Switch
RFID	Regular	Small	Small	RFID Readers

Table 1. Comparison between the different tracking technologies.

4. Implementation and Implementation

The solution containing the RFID readers, RFID tags, PDAs, Wi-Fi access points, infrastructure and the software that handles all the data between these devices was named **Hunter**, an acronym for Hybrid Unified Tracking Environment. Hunter implementation was divided into two fronts: One developing the software and the second handling the network infrastructure and the installation of the equipment.

4.1 Infrastructure

To cover all the entrances and exits of every unload place and weighting scale, 32 readers were necessary. Due the different range of distance between each equipment, different technologies were used to connect them all. The warehouses equipment was at a maximum distance of 1 km from the server, so it was not possible to use Ethernet cables. Since the pipes infrastructure was already available, it was connected using optical fiber. The weighting scales were farther than the warehouses, and these scales were in an area outside the port, in the middle of the city. Based on that, a fiber link to connect both places would be very expensive. Thus, the choice was to use a point-to-point radio link to connect the weighting scale into the equipment network.

4.2 The Padlock Tag

Working as an electronic seal, a tamper sensor was attached to the tag. This sensor accuses "ok" status if its terminal is placed in short circuit and "violated" if the terminal is in open circuit. This feature allowed the development of a tag model that could be fixed at a truck dump-cart. If a tag removal attempt occurs, the opening sensor will detect it and send a violation signal to the nearest reader. Several readers were installed at the unloading area, in case any anomaly occurred within the normal procedures (for example, the truck enters some unloading area and does not unload or violates the tag) an alarm will be presented in the weighting scale operator's screen when the truck arrives there, or in the PDA of the unload operators.

The main difficulty found in the tag design was the great difference between trucks used by the company which made it difficult to develop a single tag that could be fixed in each of them. To solve the problem, the tag was designed be similar to a padlock that can be fixed to any bar up to two inches thick. Therefore, it was possible to fix a tag in multiple parts of the truck. The violation detection of a tag is made by a small switch; when the lock is closed, this switch is pressed and when the lock is opened, the switch is released. This switch was connected by two wires to the tag circuit violation terminals sensor. The padlock RFID tag can be seen in Fig. 6 and Fig. 7.

Fig. 6. Active tag in a padlock shape with opening detector disassembled.

4.3 Equipment Setup

The rule adopted to install readers was: a reader was installed at each entrance and exit. For example, at the weighting scale platform, two readers were installed, one at the entrance and another one at the exit. At the unloading warehouses, one reader was installed at each entrance/exit gate. All readers were connected by Unshielded Twisted Pair (UTP) cables Cat 5e. Warehouses and weighting scales were connected by multimode optical fiber cables, thus building a 1 GBPS local area network between all devices and the server which executes the managing application.

In sum, a reader was installed at each truck entrance/exit and an Ethernet point was installed near each reader to put it into the equipment local area network. All readers were connected to the server which executes the managing application. Each truck passage through a reader generates a data packet to the server with information containing the tag number, electronic seal status, timestamp and the number of the reader that received the packet.

Fig. 7. a) Padlock tag assembled, opened and closed; b) One type of trailer showing the iron bar where the padlock tag is attached.

4.4 Software

The managing application, called Hunter - Hybrid Unified Trekking Environment, controls the unloading process information over the monitored vehicles. Every truck is treated by the system as an event queue to be executed. Inside the system, every truck passage through an entrance/exit is an event. Full weight, unload and empty weight are examples of events to be executed by the truck. Thus, in the system, each truck has a state within a finite state machine. Through received events by that truck, it registers exactly at which stage of unloading flow the truck is in real-time. In other words, the trucks in movement inside the area are treated as independent threads in the system, which is initiated when its tag passes the weighting scale to capture the full weight, generating the first event. When these threads are created, a task queue at the data base is created as well, with all the tasks the truck must accomplish along the process, such as: weighting scale entrance for full weight, full weight capture, weighting scale exit, unloading area entrance, PDA operator confirmation, etc. While the truck moves and generates the events, these are interpreted by the system and converted into its respective tasks, and thus they are removed from the task queue and inserted in the journal table. In case the tasks orders are not obeyed, an alarm is generated on the system operator screen.

The system graphical interface shows the number of trucks on each state of the unloading process, seen in Fig. 8. The trucks in the lines between each state are shown as well. There are thus five states presented in the interface: Full weight capture, Unload line, Unloading process, Empty weight line and Empty weight capture. Through the graphical interface, it is possible to see the trucks that are at this state and how long they are there.

The application was totally developed in Java, being a part of it monolithic and a part in form of a web-base application. It is executed by a Tomcat application server and connected to an Oracle 9i data base with features of advanced task queue. As different technologies are used, the managing application was developed with four great modules, presented in Fig. 9 (Chen et al., 2003).

The first system layer, called Device Management Subsystem (DMS) is the layer responsible for the connection between the physical device (RFID readers, PDAs, etc) and the system engine for process control, times and alarms, here called Core. Inside the DMS, there is an application part (DMSApp) and a servlet part (DMSServer). The application keeps waiting for a package at the serial port or at the Ethernet connections, and when it receives data sent by any device, it interprets the information and generates an extensible markup language - XML - message with the content that is transmitted to the DMSServer. When The DMSServer receives this XML, it consults the database table where the information about the devices positioning is. After that, it generates a XML of a higher level to be sent to Core, with information on the name of the place the tag was read and with details of the truck which has the tag. DMS is a middleware, concurrently connecting different devices in the system which allows new devices to be connected to the system without any change in the source code of the application.

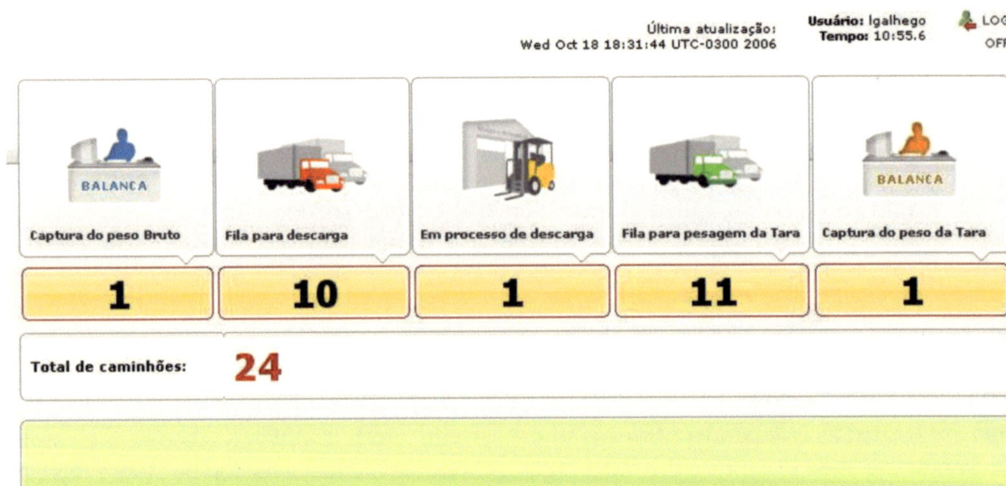

Fig. 8. Hunter screenshot showing the number of trucks in each state.

A difficulty was found in the calibration of the power emitted by the RFID equipment to obtain readings of all tags. As explained before, the dump-cart material influence and the area where the tag is placed significantly modify the reading range. Since there are trucks with wooden and steel dump-carts, the calibration was made for the worst case scenario, increasing the emitted power beyond necessary. This increase brought a few problems, such as continuous readings and anticipated readings. For example, depending on where the tag was placed in the truck, the weighting scale exit reader could read the tag constantly during its weighting before it left and, in other cases, the scale entrance reader kept reading during its weighting. To avoid this problem, a series of filters was inserted at the DMS so the real information would arrive clearly to Core. For these cases in which the entrance reader constantly reads the truck tag, an anti-repetition filter was developed, sending only the first entrance reading and ignoring the following at a certain interval of time. In the case of exit reader, a delayer filter was developed, so it would only send the last exit reading, ignoring the first ones. Therefore, the filters prevent the same entrance event from being sent twice and also prevent the exit event from being sent prematurely.

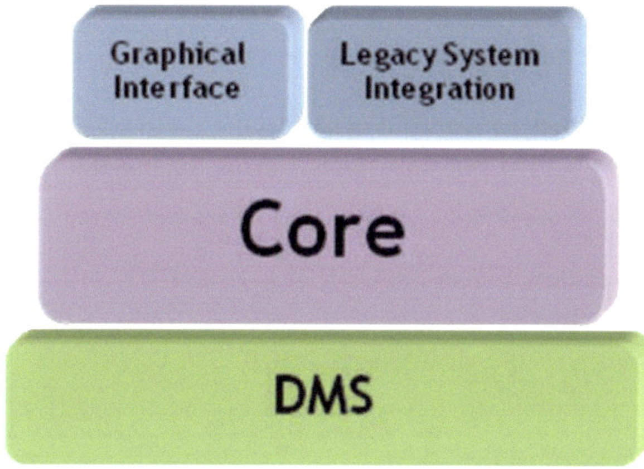

Fig. 9. Managing applicative architecture

The Core has a servlet that receives the XMLs from DMSServer and inserts them into the data base. When this insertion is made, the table where the information is stored has a trigger that initiates all the data analysis process, such as the alarms that can be generated due to this event or updates on the truck state in the unloading process. Connected to Core, there are two other modules, the graphical interface and the legacy integration systems. The first one is responsible for presenting information on the screen, such as alarms, navigation maps and reports generation. The second module is responsible for sending information to the company legacy system, updating the warehouse free space, linking timing information and alarms generated to the company corporative systems.

All the communication between system modules is made through XML messages and all the system graphical interfaces are web-based, allowing their access from any computer without the need of specific application installation, just a conventional web browser. The integration of the managing application with the company corporative systems is made through recorded text files containing the XML messages in a shared directory.

5. Problems and Solutions

The place where the tag is attached on each truck may vary depending on the truck dump-cart, and its material can vary between wood and steel, as well. These variations can influence the sensitivity of the tags, causing undesired tag readings outside the calibrated reader reading range (Foina et al., 2006). These readings happen when the truck is crossing the warehouse entrance, causing unexpected alarms and making the system call the security team unnecessarily. To solve it, two resources were used.

The first is the use of a directive antenna in the place of the omnidirecional antenna normally used in the active tags readers. The directive antenna used in the experiments was a patch antenna with the characteristics presented in Table 2.

This antenna will amplify the signal received in its directive lobule, sending a stronger signal to the reader in this case. However, if the signal is received outside its directive lobule, the reader will receive a weak signal (Balanis, 1997).
The second resource used is a feature of the active readers that allow measuring the signal strength received by the reader. This feature is called RSSI - Received Signal Strength Indication. In the equipment chosen, the RSSI is a rate from 0 to 255, where a higher value means a stronger signal. Plotting a graphic and placing the reader in mark 0 of the distance axis, it is possible to see the RSSI variation related to the tag position.

Specification	Value
E-Plane Beamwidth	65 degrees at 3 dB
Frequency Range	902-928 MHz
Gain	7 dBi
H-Plane Beamwidth	65 degrees at 3 dB
Impedance	50 Ohms
Maximum Input Power	1 Watt
Polarization	Circular
VSWR	1.5:1

Table 2. Directive antenna specification

Fig. 10 and Fig. 11 show the RSSI rate for each position of the tag in relation to the reader: the first using an omnidirectional antenna and the second using the directive antenna. The distances vary from -6 m to 4 m. The reader is positioned in mark 0 in the scale of both graphics. In Fig. 10, the directive antenna is placed in position -1 m. The difference in the curves of Fig. 10 and Fig. 11 are noticeable. Fig. 10 shows that the RSSI index slowly increases with the position until the tag reaches the reader. After that, it starts decreasing symmetrically. In Fig. 11, the RSSI rate increases rapidly when the tag moves to the front of the directive antenna, drawing an asymmetric curve.
The RSSI rate can present some oscillations due to signal reflections and noise (Zhao, 2006). For this reason, the middleware can analyze the RSSI by averaging a certain number of values using a five-period moving average. This filters out rapid variations of the signal to be presented, so each figure has two curves; the first is the signal received and the second is the averaged values.
The solution for the undesired tag readings errors is the joint use of the RSSI measurement and a directive patch antenna for the reader. In this case, RSSI will be used to inform the software if the tag is behind or in front of the antenna. The directive antenna was installed behind the reader, facing the warehouse door, approximately 1 meter away (Fig. 12). When the tag is close to the door, but behind the antenna, the RSSI index will be very low because the tag is not in the main lobule of the antenna. When the tag is positioned between the door and the antenna, the RSSI index will be much higher than if measured in the same position with an omnidirectional antenna. This enables the middleware to make the decision of whether or not to grant access into the warehouse.

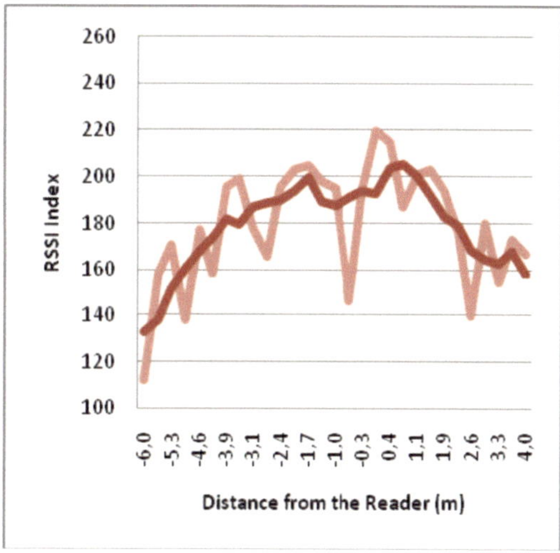

Fig. 10. RSSI changes based on the distance with an omnidirectional antenna.

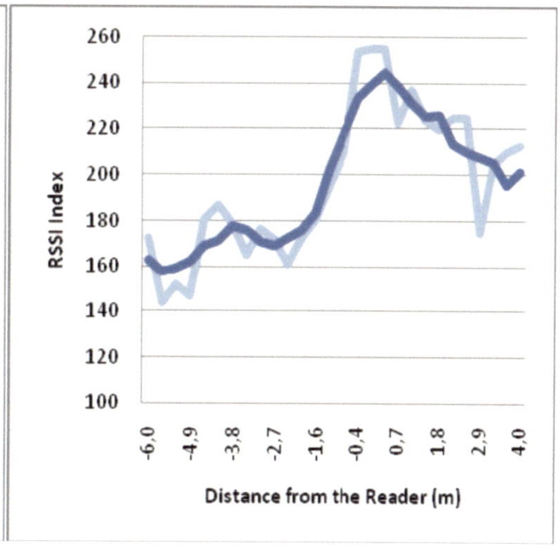

Fig. 11. The changes in the RSSI rate using a directive antenna 1 meter away from the reader.

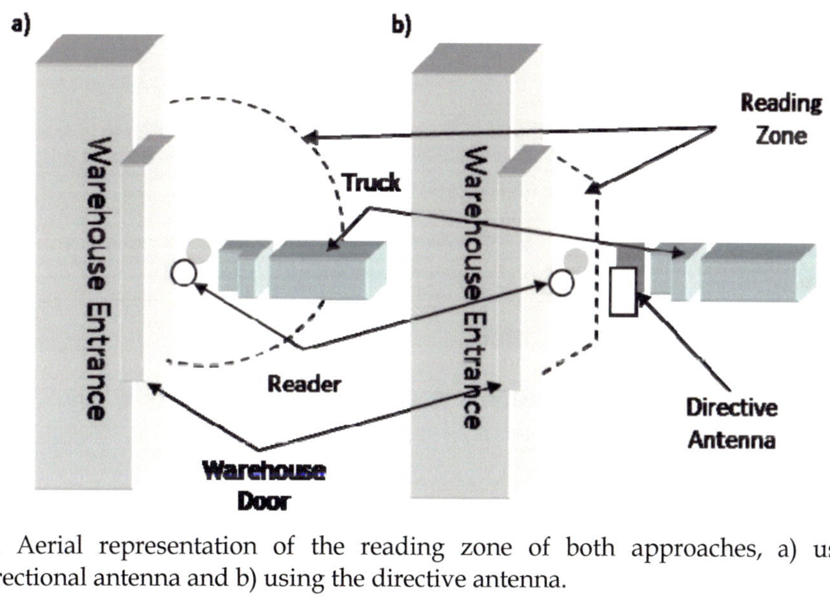

Fig. 12. Aerial representation of the reading zone of both approaches, a) using the omnidirectional antenna and b) using the directive antenna.

6. Results and Conclusion

The use of the system allows the company to have a tighter control of its operation due the real-time information. In this way, the company logistic department can keep an almost constant number of trucks in the unload area, avoiding long lines and trucks shortage, as seen in Fig. 13.

Despite the difficulties in installing a wireless network and an optical fiber network in old warehouses, and placing sensitive radio frequency equipment exposed to weather hazards, it was possible to successfully install all the equipment and to calibrate it. The PDA and server applications behaved correctly when up to 200 trucks were simultaneously circulating in the area.

With the additional information generated by the system for the logistic department about the weight, unload time statistics, and the number of trucks in real time at each stage of the unloading process, it was possible to optimize the trucks line, reducing the wait in line and therefore, reducing the average time of the unload process from 50 minutes to 30 minutes. The installation of RFID padlock tags in the trucks reduced load theft by 60% in the port area and allowed the security team to find out points of vulnerability in the previous system and to detect most of the truck drivers corrupted by the theft group.

In conclusion, Hunter showed the expected results after one year of close observation. With the reduction in the company yearly loss, the project payback occurred in less than one year.

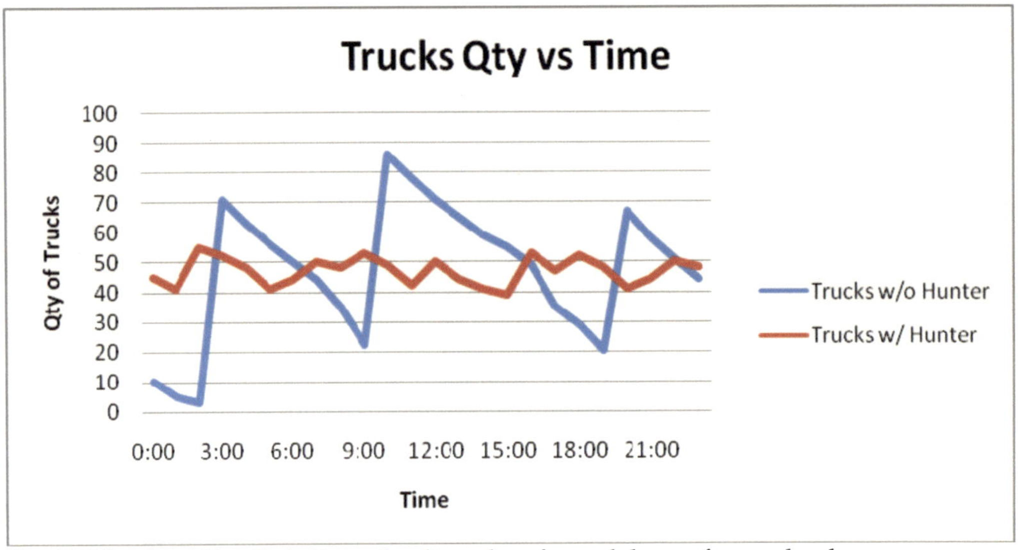

Fig. 13. Number of trucks in the unloading place for each hour of a regular day.

7. Future works

The Hunter system can be applied to many different applications and processes. For instance, it can be used with passive tags to control automated manufacture lines, supply chain pallets and forklifts. The application will control if the device goes through all the phases in the manufacturing line, from the beginning to the expedition. It can control who the forklift operator was and how long he took to move the products from one place to other. The product will thus be tracked, supplying information about the storing and movement time, and the forklifts will be tracked as well, supplying information about the efficiency of the forklift operator and generating alarms if he does something wrong. Therefore, as future work, Hunter will be tested in a different process to check its flexibility.

8. References

Almanza-Ojeda, D. L.; Hernandez-Gutierrez, A.; Ibarra-Manzano, M. A.; (2006). Design and implementation of a vehicular access control using RFID, *Multiconference on Electronics and Photonics*, pp. 223 - 225. 2006.

Balanis, C. A.; (1997). *Antenna Theory: Analysis and Design*, John Wiley & Sons, USA, second edition, pp. 86-88.

Blythe, P.; (1999). RFID for road tolling, road-use pricing and vehicle access control. *IEE Colloquium on RFID Technology*, Ref. No. 1999/123, pp. 8/1 - 816, 25 Oct. 1999.

Chen, S.; Gulati, S.; Hamid, S.; Huang, X.; Luo, L.; Morisseau-Leroy, N.; Powell, M. D.; Zhan, C.; Zhang, C; (2003). A three-tier System Architecture Design and Development for Hurricane Occurrence Simulation. *Proceedings of Information Technology: Research and Education*, pp. 113 - 117, 11-13 Aug. 2003.

Foster, P.R.; Burberry, R.A.; (1999). Antenna problems in RFID systems. *IEE Colloquium on RFID Technology*, Ref. No. 1999/123, pp. 3/1 - 3/5, 25 Oct. 1999.

Foina, A. G.; Barbin, S. E.; Ramirez Fernandez, F. J.; (2007). A New Approach for Vehicle Access Control using Active RFID Tags. *Proceedings of Microwave and Optoelectronics Conference*, pp 90-93, 29 Out. 2007.

Greenfield, A (2006); *Everyware: the dawning age of ubiquitous computing*. New Rider. ISBN 0321384016.

Guillemette, M. G.; Fontaine, I.; Caron, C. (2008). Hybrid RFID-GPS Real-Time Location System for Human Resources: Development, Impacts and Perspectives. *Proceedings of the 41st Annual Hawaii International Conference on System Sciences*, pp. 406-406, Jan. 2008

Jardak, N.; Samama, N. (2009). Indoor Positioning Based on GPS-Repeaters: Performance Enhancement using an Open Code Loop Architecture. *IEEE Transactions on Aerospace and Electronic Systems*, vol. 45, no. 1, pp. 347-359, Jan. 2009.

Jiao, Y.-Y.; Li, K.; Jiao, R.J. (2008). A case study of hospital patient discharge process re-engineering using RFID. *4th IEEE International Conference on Management of Innovation and Technology*, pp. 1342-1347, Sept. 2008.

Hu, W.; Ye, W.; Huang, Y.; Zhang, S. (2008). Complex Event Processing in RFID Middleware: A Three Layer Perspective. *Third International Conference on Convergence and Hybrid Information Technology*, vol. 1, pp. 1121-1125, Nov. 2008.

Kim, T.; Kim, H.; (2006). Access Control for Middleware in RFID Systems. *The 8th International Conference on Advanced Communication Technology*, vol. 2, pp. 1020 - 1022, 20-22 Feb. 2006.

Ni, L. M.; Liu, Y.; Lau, Y. C.; (2003). LANDMARC: Indoor Location Sensing Using Active RFID. *In IEEE International Conference in Pervasive Computing and Communications*, pp. 407-415, 2003.

Park, J.-H.; Park, J.-H.; Lee, B.-H. (2007). RFID Application System for Postal Logistics. *Management of Engineering and Technology*, pp. 2345-2352, Aug. 2007.

Qaiser, A.; Khan, S.A. (2006). Automation of Time and Attendance using RFID Systems. *International Conference on Emerging Technologies*, pp. 60-63, Nov. 2006.

Raza, N.; Bradshaw, V.; Hague, M.; 1999. Applications of RFID technology. *In IEE Colloquium on*, pp. 1/1 - 1/5, 25 October, 1999.

Tuttle, J.R.; (1997). Traditional and emerging technologies and applications in the radio frequency identification industry. *Symposium on Radio Frequency Integrated Circuits (RFIC)*, pp. 5 – 8, 8-11 Jun. 1997.

Xu, G. (2008). The Research and Application of RFID Technologies in Highway's Electronic Toll Collection System. *Wireless Communications, Networking and Mobile Computing 2008*, pp. 1-4, Oct. 2008.

Yue, D.; Wu, X.; Bai, J. (2008). RFID Application Framework for pharmaceutical supply chain. *IEEE International Conference on Service Operations and Logistics, and Informatics*, vol.1, pp. 1125-1130, Oct. 2008.

Zhao, J.; Zhang, Y.; Ye, M.; (2006). Research on the Received Signal Strength Indication Location Algorithm for RFID System. *International Symposium on Communications and Information Technologies*, pp. 881 – 885, Oct. 2006.

14

Cooperation Control in Distributed Population-based Algorithms using a Multi-agent Approach Application to a real-life Vehicle Routing Problem

Kamel Belkhelladi, Pierre Chauvet and Arnaud Schaal
LISA, Université d'Angers
IMA, Université Catholique de l'Ouest
France

1. Introduction

Combinatorial optimization problems in transportation, logistic management and other areas involve finding optimal solutions from a discrete set of feasible solutions. However, even with the advent of new computer technologies and parallel processing, many of these problems could not be solved to optimality in reasonable computation times, due to their inner nature or to their size. Moreover, reaching optimal solutions is meaningless in many practical situations, since we are often dealing with rough simplifications of reality and the available data is not precise. The goal of approximate algorithms (or heuristics) is to quickly produce good approximate solutions, without necessarily providing any guarantee of solution quality.

Metaheuristics are general high-level procedures that coordinate simple heuristics and rules to find good (often optimal) approximate solutions to computationally difficult combinatorial optimization problems. Among them, are simulated annealing (Metropolis et al., 1953), tabu search (Glover, 1989; Glover, 1990), Greedy Randomized Adaptive Search Procedure (GRASP) (Feo & Resende, 1989), genetic algorithms (Holland, 1975), scatter search (Laguna & Martí, 2003), Variable Neighbourhood Search (VNS) (Mladenović & Hansen, 1997), ant colonies (Dorigo & Di Caro, 1999), and others. They are based on distinct paradigms and offer different mechanisms to escape from locally optimal solutions, contrarily to greedy algorithms or local search methods. Metaheuristics are among the most effective solution strategies for solving combinatorial optimization problems in practice and they have been applied to a very large variety of areas and situations. The customization (or instantiation) of a metaheuristic to a given problem yields a heuristic to the latter.

Metaheuristics offer a wide range of possibilities for effective parallel algorithms running in much smaller computation times, but requiring efficient implementations. (Cung et al., 2002) showed that parallel implementations of metaheuristics appear quite naturally as an effective approach to speedup the search for good solutions to optimization problems. They allow solving larger problems and finding better solutions with respect to their sequential counterparts. They also lead to more robust algorithms and this is often reported as the main advantage obtained with parallel implementations of metaheuristics: they are less dependent on parameter tuning and their success is not limited to few or small classes of problems.

Distributed computing refers to computing that involves multiple loosely coupled processors working together to solve an overall problem (Steen & Tanenbaum, 2002). Distributed computing offers a natural approach to solving complex data and computation intensive problems that arise in industry and control. Many approaches to distributed computing have been developed over the past decades. These include Socket Programming, Remote Procedure Calls (RPC), object-oriented DCE, DCOM, CORBA, Java RMI, and Message-Oriented Middleware (MOM) (Steen & Tanenbaum, 2002). These approaches have their own advantages and disadvantages (Steen & Tanenbaum, 2002). Some major limitations with respect to the design of robust, networked applications that involve many autonomous, heterogeneous entities are: (1). Interactions among participating entities, which are fixed a priori through explicitly coded instructions by the application developer. As a result, they lack run-time adaptive behaviour. (2). Ongoing interaction that requires ongoing communication, making it unsuitable for applications that have to operate in environments where maintaining continuous communication is expensive or unfeasible and connections are unreliable. These considerations have motivated the development of approaches to distributed computing based on agents which provide ways of maintaining ongoing interaction without ongoing communication (White, 1997). Multi-agent systems (MAS) and related technologies offer an attractive paradigm for designing distributed networked applications that involve many relatively autonomous, heterogeneous entities (Honavar et al., 1998).

In recent years, the number of strategies proposed to solve complex optimization problems has increased considerably. A new set of problem and model-independent nature inspired heuristic optimization algorithms were proposed by researchers to overcome drawbacks of the classical optimization procedures. These techniques are efficient and flexible. They can be modified and/or adapted to suit specific problem requirements (see Figure 1). Research on these techniques is still continuing all around the globe.

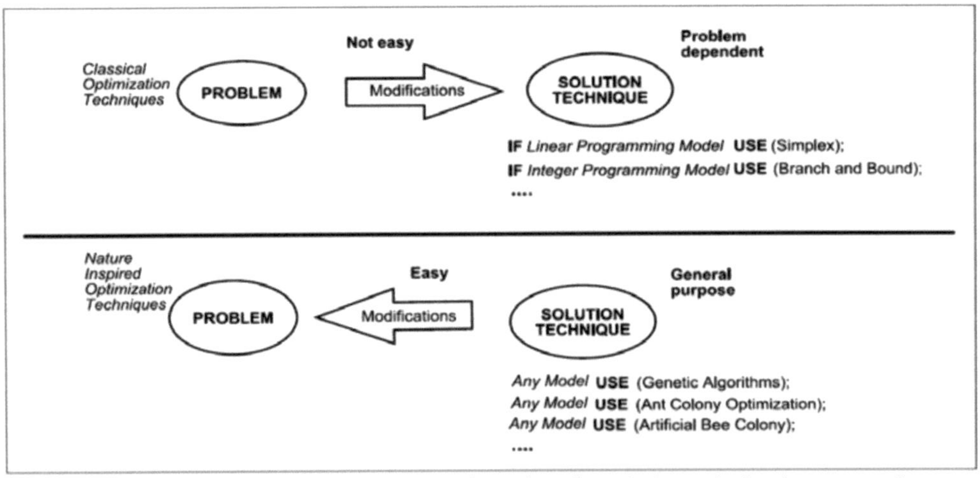

Fig. 1. A pictorial comparison of classical and modern heuristic optimisation strategies

(Adapted from (Baykasoğlu, 2001))

Population-based strategies are very frequently used in metaheuristics (Blum & Roli, 2003; Krasnogor & Smith, 2005). In addition, the complexity of these methods is even higher because they are time-consuming. In this context, the use of parallel and distributed strategies becomes a powerful tool to obtain satisfactory solutions in acceptable runtimes. Though parallel computers can dramatically speed up algorithms, they are not easily available and it is expensive to upgrade their processing power and memory. The expensive parallel computers cannot be expected especially in company-based computing environments. A promising alternative without expensive hardware facilities is to construct a parallel framework on a set of networked personal computers.

In this chapter, we propose a scalable framework for the development of parallel/distributed population-based algorithms using mobile agents launched on different hosts of available networked PCs and cooperating among them to solve large combinatorial problems efficiently. The mobile agent execution environment used to realize our framework is based on the JADE technology (Bellifemine et al., 2001). In fact, parallel/distributed computing applied to problem solving means that several processes work simultaneously on several processors with the common goal of solving a given problem instance. One then has to determine how the global problem solving process is controlled and how information is exchanged among the various processes. Hence, we define a new information exchange strategy based on dynamic migration window methods (Kim, 2002) that control the size and frequency of exchange and a selective migration model (Eldos, 2006) for the choice of information to be exchanged. The proposed framework has been experimented on an extended set of Capacitated Arc Routing Problem (CARP) and a real-life application (Underground Waste Collection Optimization Problem (UWCOP)). Several experiments are carried out on different computer networks of different sizes. Results obtained show the advantages and efficiency of our approach.

The remaining sections of this chapter are organized as follows. In section 2, we overview fundamentals of parallel/distributed population-based algorithms and more specifically parallel/distributed Evolutionary Algorithms models and the potential of multi-agent systems (MAS). Section 3 presents our mobile agent-based computing approach. The MAF-DISTA implementation is described in section 4. In section 5, an NP-hard problem, the Underground Waste Collection Optimization Problem (UWCOP), which we use as a test bed for this study, is presented. Section 6 presents a MAF-DISTA implementation for UWCOP. In section 7, different experiments and results are summarized and discussed. The remaining section presents the conclusion and outlines areas for future research.

2. Background

2.1 An overview on Parallel and Distributed population-based Algorithms

Population-based methods are inspired by natural evolution processes. They handle a population of individuals that evolves with information exchange procedures. Population-based algorithms need constructive method to generate the initial population, and any local search technique can be used to improve each solution in the population. But population-based methods combine good solutions in order to possibly get better ones. The basic idea is that good solutions often share parts with optimal solutions (Hertz & Kobler, 2000).

Parallel and distributed computing may be considered as a mechanism to speed up the search process when solving large optimization problems. Furthermore, the simultaneous use of parallelism and cooperation allows the improvement of the quality of non-dominated solution sets. Evolutionary Algorithms (EAs) are population-based algorithms and very suitable for parallelization, due to the fact that mutation and evaluation can be performed independently on each individual of the population. However, in the selection and crossover processes it is necessary to compare or combine individuals which may be handled on different processors. This implies communications among processors. Parallelization of EAs has been properly studied in the last decade (Cantü-Paz, 2000; Veldhuizen et al., 2003). Most important parallel models are:

- The *master-worker* model, which allows keeping the sequentiality of the original algorithm. The *master* centralizes the population and manages the selection and replacement steps. It sends subpopulations to *workers* that execute recombination and evaluation tasks. The *workers* return newly evaluated solutions back to the *master*. This approach is efficient when the cost of generating and evaluating new solutions is high. In some cases the island model improves the quality of the solutions reached because the islands help to maintain diversity.
- The *island* or *coarse-grain* model, which divides the entire population into several subpopulations distributed among different processors. Each processor is responsible for the evolution of one subpopulation. It executes all the steps of the metaheuristic, and occasionally individuals migrate among islands. The goal is to obtain solutions for at least the same quality than the *master-worker* model, but in a shorter runtime.
- The *diffusion* or *fine-grain* model, which works with a population. Each processor is responsible for either a single individual or at most a small number. The difference with respect to the island paradigm is that the diffusion-based scheme requires a neighbourhood structure of processors to perform the recombination and selection.

2.2 Potential of multi-agent systems

Multi-agent systems (MAS) have proven to be an effective paradigm in a number of distributed networked applications that require information integration from multiple heterogeneous autonomous entities (Caragea et al., 2001). A multi-agent system consisting of multiple agents can take advantage of computational resources and capabilities that are distributed across a network of interconnected entities. An agent-based approach allows the creation of systems that are flexible, robust, and can adapt to the environment. This is especially helpful when components of the system are not known in advance, change over time, and are highly heterogeneous. Agent-based computing offers the ability to decentralize computing solutions by incorporating autonomy and intelligence into cooperative, distributed applications. Each agent perceives (the state of its environment), infers (updates its internal knowledge according to the newly received perceptions), decides (on an action), and acts (to change the state of the environment). Agent-oriented programming (AOP) is the software paradigm used to facilitate agent-based computing and extends from object-oriented programming (OOP) by replacing the notions of class and inheritance with the notions of roles and messages, respectively (Lind, 2001). If one wants to develop useful and applicable multi-agent system development tools, several characteristics are essential in practice. Among these characteristics, one can cite:

- Simple support for the deployment of inter-computer and distributed applications.

• Significant reduction of the effort required implementing a MAS, especially when considering the important amount of programming usually involved.
• Effective and efficient abstraction in order to allow people who are less experienced with agent-oriented programming to relatively easily create MAS without in-depth knowledge of all implementation details.
• Sufficient latitude to most experienced users, who then can get access to and interact directly with various system components; the tool should add an abstraction level to programming without becoming an obstacle for programmers.
• Implementation (code) is easily extensible.
• Developers do not have to worry about system communications implementation and associated protocols used to transfer messages among agents.
• Availability of a debug utility, of a user interface facilitating development, and of an automatic source code generator.
• The environment is supported by a suitable documentation.

(Garneau & Delisle, 2003) have done a rigorous study to evaluate and compare several MAS environments in order to identify tools' and environments' strong points and weak points in the aim of determining what a complete development environment should look like. Table 1 summarizes some results obtained by (Garneau & Delisle, 2003). Eight agent-oriented programming tools have been chosen because of their popularity and their relevance: Jade (Bellifemine et al., 2001), Zeus (Lee et al., 1998), Madkit (Gutknecht & Ferber, 2001), AgentBuilder (AgentBuilder, 2004), Jack (Busetta et al., 1999), Decaf (Graham et al., 2000), JafMas/Jive (Chauhan, 1997) and AgentTool (Deloach & Wood, 2000).

Mobile agents are a convenient paradigm for distributed computing (Bic et al., 1996) (Chess et al., 1996). The agent specifies when and where to migrate, and the system handles the transmission. This makes mobile agents easier to use than low-level facilities in which the programmer must explicitly handle communication, but more flexible and powerful than schemes such as process migration in which the system decides when to move a program based on a small set of fixed criteria. Mobile agents allow a distributed application to be written as a single program.

As we are interested in JADE platform, we report a short description in the next subsection.

2.2.1 JADE

Java Agent DEvelopment Framework (JADE) middleware (JADE-Tilab, 2009) is a Java agent platform which provides a fully distributed system, where agents can reside in different hosts, an efficient transport mechanism with asynchronous messages, and internal (within a single JADE middleware platform instance) agent mobility support. An important feature of JADE is support for ontologies and content languages used in the agent messages. Each JADE agent runs as an independent thread and executes behaviours. Behaviours are tasks that can be dynamically inserted into and scheduled by an agent.

JADE is designed to ease the development of multi-agent systems by providing standard agent services and methods. It also enables interoperability among different agent platforms since it complies with *FIPA* standards. The Jade platform is composed of one *main-container* and several *containers* that can be launched on one or more hosts distributed across the network. Each *container* can run zero, one or multiple agents. The following agents and services must be present in the *main-container* at any times: the AMS (Agent Management System), the DF (Directory Facilitator) and the MTS (Message Transport System) which is also referred to as the ACC (Agent Communication Channel). The JADE middleware architecture is depicted below in figure 2.

Evaluation criteria	JADE	DECAF	AgentBuilder	Zeus	JASMAF/JIVE	Jack	AgentTool	Madkit
Methodology	0	0	4	4	3	0	3	3
Learning facility	0	3	1	1	1	0	3	2
Step transition	0	0	3	3	2	0	3	2
Tool flexibility	3	0	1	1	2	3	0	3
Communication	4	2	4	4	2	3	2	3
Debug utility	3	2	4	4	1	0	2	4
Development support	0	2	4	4	2	1	4	2
Implementation support	0	0	4	4	2	1	2	1
MAS management	4	0	3	3	0	0	1	4
Implementation simplicity	2	3	2	2	1	2	3	1
Code generation	0	0	1	3	1	0	1	0
Code extensibility	4	1	1	2	1	4	0	3
Deployment	4	1	2	2	1	2	1	3
Documentation	3	1	4	4	1	3	1	3
Total	**27**	**15**	**38**	**41**	**20**	**19**	**26**	**34**

Table 1. Grid results

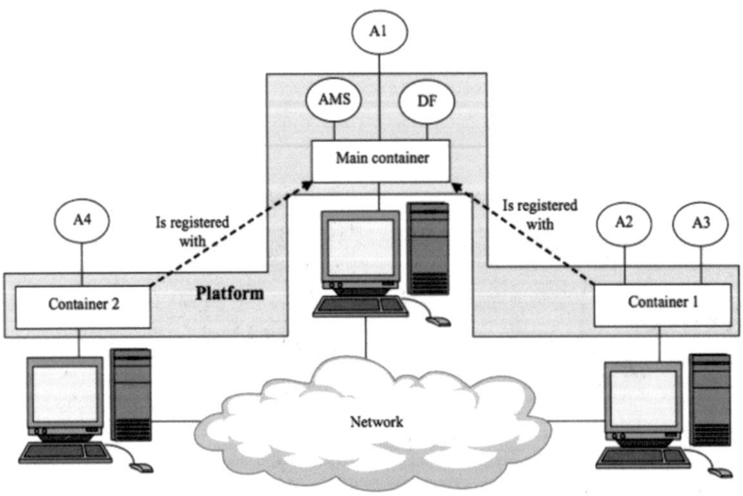

Fig. 2. The JADE agent middleware architecture (WIDE, 2009)

3. A Mobile Agent framework as a tool for implementing parallel Population-based algorithms (MAF-DISTA)

Implementing parallel EAs involves expensive parallel computers that normally cannot be expected in a company computing environment. A promising alternative without expensive hardware facilities is to construct a parallel computational framework on a set of networked computers. There are some possible ways to manage the operation of such a distributed computational framework, for example a *client-server* technique or an agent-based approach. *Client-server* is the most common paradigm of distributed computing at present. However, in this paradigm all components are stationary with respect to execution. This makes it unsuitable for the goal here no-specific computational resources are preserved and dedicated to the parallel EAs so that the EC-code needs to be moved occasionally to the available machines. Under such circumstances, mobile agent-based design paradigm provides a better choice to support parallel computing (Lange & Oshima, 1999; Macêdo & Silva, 2005). Mobile agents are software agents that are capable of being transmitted themselves (their program and their state) across a computer network and recommencing execution at a remote site. They have been proposed to solve problems of networked application domains, for example, electronic commerce (Dasgupta et al., 1999; Leeand et al., 2003), Network Management (Papavassiliau et al., 2002; Jiang et al., 2005), and information retrieval (Thati et al., 2001). Many different multi-agent platforms are publicly available, including D'Agent (Gray et al., 2002), Tracy (Braun & Rossak, 2005), Aglets (Lange & Oshima, 1998) and JADE (Bellifemine et al., 2001). Application developers can use them as platforms and focus on the software development issues at application level. In this work, we chose to use the JADE toolkit and take the mobile agent approach to develop an adaptive computational model to parallelize EAs. We implemented a *coarse-grain* population framework on a set of networked computers. This network of computers is similar to a DM-MIMD (Distributed Memory Multiple Instruction stream, Multiple Data stream) computer and its increasing availability makes it an ideal and practical platform. Figure 3 presents our framework architecture.

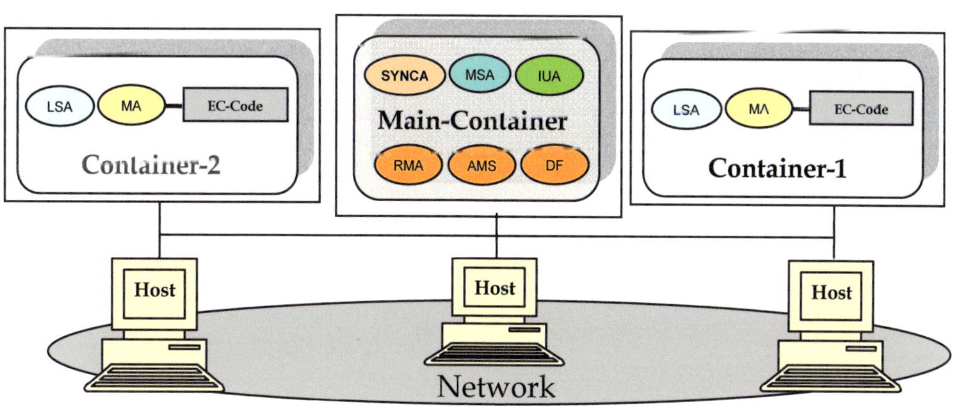

Fig. 3. MAF-DISTA architecture

We used Java mobile agents of the JADE framework, which break the barriers of heterogeneous environments. JADE is compliant with FIPA standard specifications, and agents developed with it can thus interoperate with other agents built with the same standard (O'Brien & Nicol, 1998). From the functional point of view, JADE provides the basic services necessary for distributed peer-to-peer applications in a networked environment. It supports mobility of code and execution state: an agent can stop its execution on a host, migrate to another host, and then restarts its execution. In this environment, JADE allows each agent to dynamically discover other agents and to communicate with them in a peer-to-peer manner.

4. MAF-DISTA implementation

We defined five main kinds of agents (in addition to the default agents provided by JADE for network services): the mobile agent (MA) to carry out Evolutionary Computation (EC), the main status agent (MSA) and the local status agent (LSA) to report machine status, the synchronization agent (SYNCA) to summarize the evolving progress of subpopulations and the user interface agent (UIA) to publish computing results and allow users to interact with the computing system. In the proposed framework, mobile agents have the ability to migrate from machine to machine. In our strategy, each mobile agent runs a subpopulation. Initially, a local status agent (LSA) is created for each machine in the network framework, and the main status agent (MSA) in the main host is responsible for maintaining the information of individual machine status (e.g., the CPU utility) reported by other status agents. Then, the mobile agent (MA) starts computation in the main host; it clones itself using the EC-code for each node of the prespecified n-cube model. According to the information provided by MSA, MA dispatches the duplication to each available machine. Therefore, during the period of execution if a status agent in any machine (except the main host) detects the existence of a new end-user and the CPU utility exceeds a certain *threshold*[1], the mobile agent in the same machine is informed to interrupt the execution of its corresponding EC-code, in order to release the computing resources for the end-user. The mobile agent then inquires machine availability from LSA, to resume the execution. If the mobile agent finds a machine whose CPU utility is lower than the *threshold*, it asks MSA to reserve this machine, and then carries its code and related information to that machine to continue the computation, starting from the point where it has been interrupted; otherwise it stays in the machine it has been in and waits for any available one. The implementation of the proposed approach is more detailed in the sequence diagram of figure 4.

[1] Percentage of CPU Utility

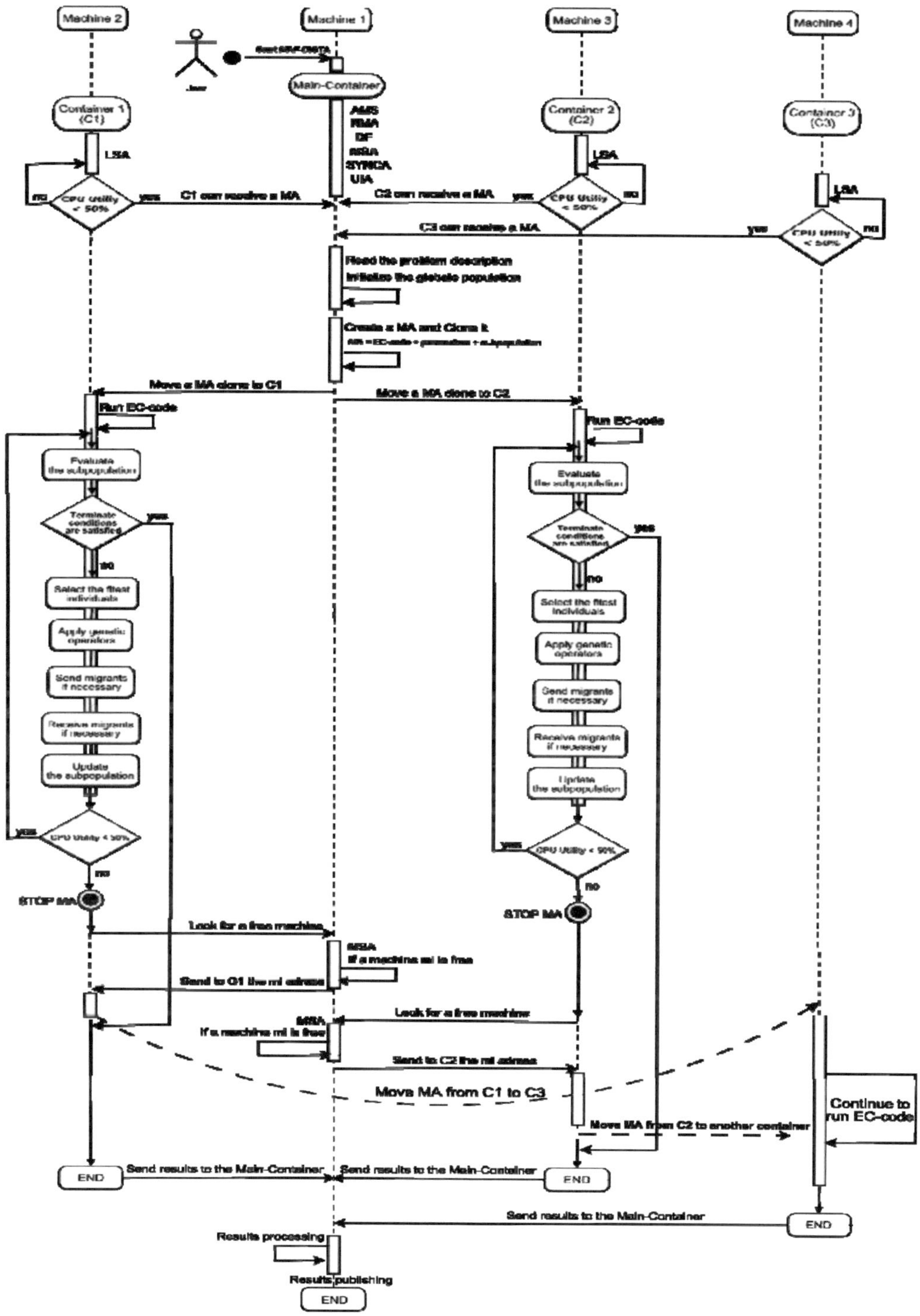

Fig. 4. MAF-DISTA Sequence Diagram

4.1 Cooperation control

The goal of maintaining genetic diversity is to prevent *premature convergence*. Separate subpopulations could have their respective genetic features. Migrations among subpopulations are necessary to help these small populations to jump out of local minima. Combinations of good genes from different species are expected to result in even better offspring. Therefore, the best individuals are the candidates to be copied to other environments. Since the clones of the global champion may dominate all subpopulations and reduce the genetic diversity, migration should be carefully directed. Figure 5 shows the aspect of our computational framework; it illustrates data and control flows of the computation within a migration interval (i.e., consecutive *k* generations). In this figure, each grey block contains a subpopulation of individuals in which evolutionary computation continues for a certain number of generations before migration happens. During this period, computation for each subpopulation is independent from others, so evolution for different subpopulations can proceed simultaneously. In our work, we developed a new information exchange strategy based on dynamic migration window methods (Kim, 2002) that control the size and frequency of exchanges and a selective migration model (Eldos, 2006) for the choice of information to be exchanged.

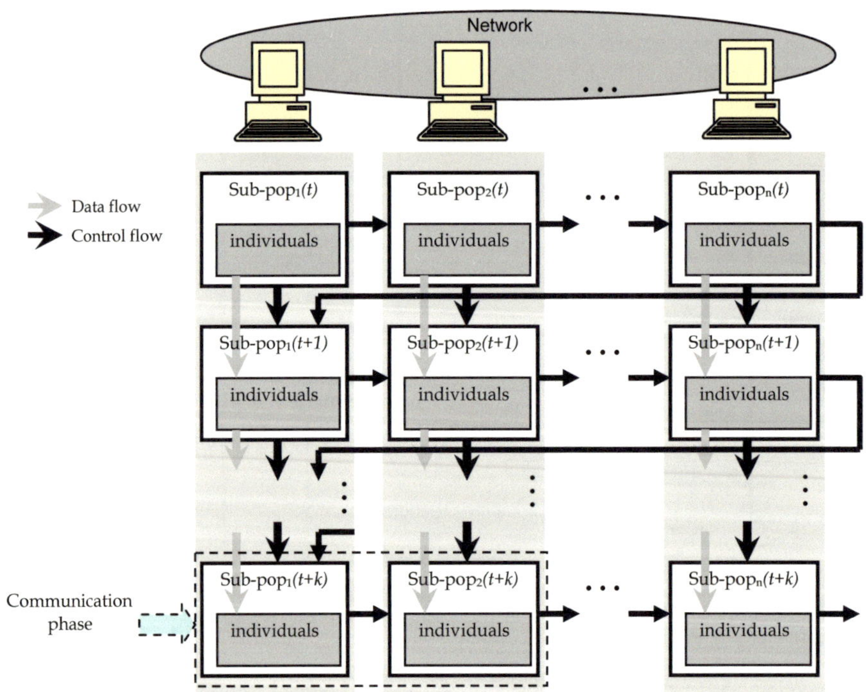

Fig. 5. MAF-DISTA cooperation control

In the proposed model, individuals are examined at both the source and destination levels, to qualify for migration. The source gives or denies a visa based on local qualification criteria and the destination grants or denies a residency based on local qualification criteria. The simplest form of qualification criteria is based on individuals' relative fitness. Every individual in a subpopulation is ranked locally; an individual qualifies for a visa at the source subpopulation if its rank is within a range, typically the middle class. In the destination subpopulation, an immigrant is accepted as a new member of the population if its rank is better than a *threshold*[2] set by the receiving subpopulation.

5. Case study in optimization problems

The proposed approach has been experimented on an industrial problem. In this section, we focused on a real-life application (Underground Waste Collection Optimization Problem).

5.1 Problem description

The underground waste storage system in Angers (France) is an innovative way of containing waste. This system consists of two main components, a *4* or *5* m^3 underground waste container with a specially designed deposit point, and a collection vehicle specifically developed and featuring a side-loading system that enables a single driver/operator to empty containers using a fully automated process. The total collection cycle for a container takes approximately 5 minutes from the moment the collecting vehicle arrives to the time it is ready to move off again. The collecting vehicle's *20* m^3 compacting container offers an equivalent load capacity of about 9.5 tons. As in other French cities, most of the population of Angers is concentrated in urban areas with a very high population density, given that people usually live in apartment buildings. As a result, the collection technique has evolved away from door-to-door collection, and the predominant system consists of collection points located throughout the city where citizens leave their refuse. Each collection point is made up of one or more underground waste containers. Different trucks collect each type of waste and transport it to its final destination (a refuse dump, an incinerator or a recycling plant). One key factor for proper operation of collection systems is the design of appropriate collection routes for each type of garbage (fraction), which may or may not be separately collected (in some cases due to technical design constraints and different treatment processes). In Angers each fraction is collected separately with a fixed periodicity (established by the municipality). Therefore, in this case the collection of each type of garbage (urban refuse, generic waste materials like paper, plastics, packaging, glass, etc.) can be considered as an independent but equivalent problem. In addition, the routes should be coherent with traffic rules, taking into account prohibited turns and traffic signals while minimizing collection costs. A container must be collected if its filling rate is more than a *threshold*[3] (20% of the container size, in our case). Some containers are "black spots" and must be collected even if their content is lower than the *threshold*. These containers are located in sensitive places, generally public structures such as markets, prisons, etc. Finally, two distinct costs per required route are considered, *deadheading* and *collecting*.

5.2 Problem modeling and formulation

UWCOP is an arc routing problem since roads are completely traversed, even if collection points are located in specific spots of the roads. In addition, the capacity of trucks is limited.

[2] Qualification criterion

[3] Percentage of container size

Thus, the basic nature of UWCOP is that of a Capacitated Arc Routing Problem (CARP) that can be defined as follows. Suppose a connected undirected graph *G=(V,E)*, deadheading costs $dc: E \to \mathbb{Z}^+$, collecting costs $cc: E \to \mathbb{Z}^+$, demands $w: E \to \mathbb{Z}^+$, a fleet of K identical vehicles with capacity *W* is based at a depot node *s*. Define $R = \{(i,j) \in E \mid w((i,j)) > 0\}$ as the set of required edges. Let *F* be a set of closed walks that start and end at the depot, where edges in a walk can be either *serviced* or *deadheaded*. Set F is a feasible CARP solution if:

- Each required edge is serviced by exactly one walk in *F*;
- The sum of demands of the serviced edges in each walk in *F* does not exceed the vehicle capacity *W*.

We want to find a solution minimizing the total costs of the walks. It can be noted that $\Sigma_{(i,j)\in R} dc(i,j) + \Sigma_{(i,j)\in R} cc(i,j)$ is a lower bound on the cost of an optimal solution, the remaining costs in a solution are the costs of the deadheaded edges.

The UWCOP can be described by the following programming problem:

Notations:

E: Set of network edges.
R: Set of network edges *(i,j)* whose demands are greater than zero.
V: Set of network nodes.
I: Set of vehicles.
W: Vehicle capacity.
K: Number of vehicles.
C: Container capacity.
Q_c: Quantity of waste contained in container *c*.
M_{ij} : Number of underground containers placed on edge *(i,j)*.
T_{cc}: A container total collection cycle.
v_c: Denotes a binary variable equal to *1* if $Q_c \geq 20\% \times C$ and 0 otherwise.
b_c: Denotes a binary variable equal to *1* if the container *c* is a black spot and 0 otherwise.
dc_{ij} : Deadheading cost of edge *(i,j)*.
cc_{ij} : Collecting cost of edge *(i,j)*.
w_{ij} : Demand of edge *(i,j)* or the total demands between nodes *i* and *j*.
x_{ijk}: Denotes a binary variable equal to *1* if the vehicle *k* traverses the edge *(i,j)* from *i* to *j* and *0* otherwise.
l_{ijk}: Denotes a binary variable equal to 1 if the edge *(i,j)* is serviced by the vehicle *k* from i to j and *0* otherwise.

$$Minimize \sum_{k\in I}\sum_{(i,j)\in E} dc_{ij}x_{ijk} + \sum_{(i,j)\in R} cc_{ij} \tag{1}$$

$$\forall i \in V, \forall k \in I: \sum_{j\in\Gamma(i)} (x_{ijk} - x_{jik}) = 0 \tag{2}$$

$$\forall (i,j) \in E, \forall k \in I: x_{ijk} \geq l_{ijk} \tag{3}$$

$$\forall (i,j) \in R: \sum_{k\in I} (l_{ijk} + l_{jik}) = 1 \tag{4}$$

$$\forall k \in I: \sum_{(i,j)\in R} l_{ijk} w_{ij} \leq W \tag{5}$$

$$\forall (i,j) \in R, cc_{ij} = \sum_{m=1}^{M_{ij}} T_{cc} \times (v_c \vee b_c) \tag{6}$$

$$u_s^k, y_s^k, x_{ijk}, l_{ijk}, v_c, b_c \in \{0,1\} \quad (7)$$

$$\forall S \neq \varnothing, S \subset \{2,..,n\}, \forall k \in I: \begin{cases} \sum_{i,j \in S} x_{ijk} - n^2 y_s^k \leq |S| - 1 \\ \sum_{i \in S} \sum_{j \in S} x_{ijk} + u_s^k \geq 1 \\ u_s^k + y_s^k \leq 1 \end{cases} \quad (8)$$

The goal (1) is to determine a set of closed walks (vehicle trips) of minimum total cost. Constraint (2) ensures that a vehicle coming into a node must leave it again. Constraint (3) ensures that a required edge is also traversed. Each required edge is serviced by one single trip (4) such that each trip starts and ends at the depot, and the total demand handled by any vehicle does not exceed W (5). A container must be collected if its filling rate is more than 20% or it is a *black spot* (6). Finally, illegal sub-trips are forbidden by constraint (8) which requires additional binary variables such as y_s^k and u_s^k .

Since the CARP is NP-hard, large scale instances must be solved in practice using heuristics. Among fast constructive methods, one can cite Path-Scanning (Golden et al., 1983), Augment-Merge (Golden & Wong, 1981) and Ulusoy's splitting technique (Ulusoy, 1985). Available metaheuristics are very recent and include tabu search methods (Hertz et al., 2000; Belenguer & Benavent, 2003), guided local search (Beullens et al., 2003) and memetic algorithms (Lacomme et al., 2001; Lacomme et al., 2004). All these heuristics algorithms can be evaluated through lower bounds (Amberg & Voß, 2002).

6. MAF-DISTA implementation for UWCOP

We conducted two series of experiments to compare the corresponding performance for evolutionary computation with and without parallelism. In these experiments, each mobile agent executes a conventional genetic algorithm. The genetic algorithm (GA) developed in this work uses several components of the genetic algorithm and the effective memetic algorithm proposed by (Lacomme et al., 2001; Lacomme et al., 2004) for the CARP. The common parts are described below:

6.1 Solution encoding

The network is coded as a symmetric digraph, in which each edge is replaced by two opposite arcs. A chromosome is an ordered list of the T tasks, in which each task may appear as one of two directions. The implicit shortest paths are assumed between successive tasks. The chromosome does not include trip delimiters and can be viewed as a *full trip* for an incapacitated vehicle. A *Split* procedure optimally partitions (subject to the sequence) the *full trip* into feasible trips. This procedure is inspired by the Ulusoy's splitting technique (Ulusoy, 1985). Usually, this technique is a CARP heuristic, which is better explained in figure 6. First, capacity is ignored to build one *full trip* τ covering all tasks (a, b, c, d and e in the figure). Second, an auxiliary graph is built, in which each arc denotes a subsequence of τ that can be done by one trip. Each arc is weighted by the cost of this trip. A shortest path in this graph shows where τ should be split into trips, and gives the cost of the corresponding solution. Third, the solution is built with one trip per arc on this path. The fitness function of the GA is the total cost of the resulting CARP solution.

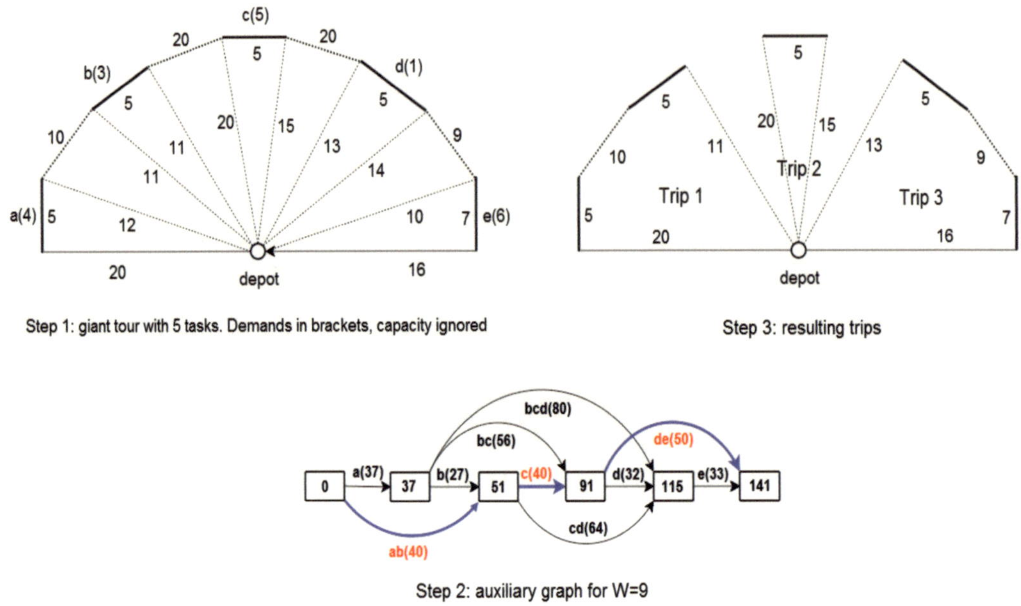

Fig. 6. Ulusoy's splitting technique

6.2 Initialization

The global population P of chromosomes is initialized with the solutions of the three CARP heuristics (PS, AM and UH) (Golden & Wong, 1981; Golden et al., 1983; Ulusoy, 1985), completed by random permutations. Clones (identical chromosomes) are forbidden.

6.3 Selection and crossover

At each iteration, two parents are selected by a biased roulette wheel (Goldberg, 1989). The crossover operator used in the algorithm consists in applying *Order Crossover (OX), Linear Order Crossover (LOX)* or *One point Crossover (X1)* operations with equal probability (see figure 7). *Order Crossover* tends to transmit the relative positions of genes rather than the absolute ones. In the *OX*, the chromosome is considered to be circular. *LOX* is a modified version of *Order Crossover*, where the chromosome is considered linear instead of circular. The *LOX* works as follows:

- Select sub-lists from parents (p_1, p_2) randomly.
- Remove $sublist_2$ from parent p_1 leaving some "holes" and then slide the holes from the extremities toward the center until they reach the cross section. Similarly, remove $sublist_1$ from parent p_2 and slide holes to cross section.
- Insert $sublist_1$ into the holes of parent p_2 to form the offspring o_1 and insert $sublist_2$ into the holes of parent p_1 to form the offspring o_2

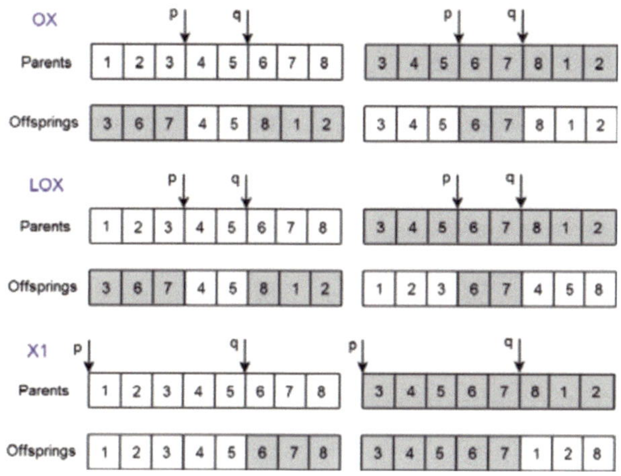

Fig. 7. Crossovers

6.4 Mutation

The mutation operator used in the algorithm consists in applying *displacement (MOVE)* or *reciprocal exchange (SWAP)* operations to each gene with equal probability (see figure 8). The *displacement* mutation selects two positions p and q in the chromosome such as $1 \le p \le T, 0 \le q \le T$, $p \ne q$ and $p \ne q-1$. Task u in rank p is moved after that of rank q *(q=0* means an insertion in the lead). *Reciprocal exchange* mutation selects two positions at random and swaps the tasks on these positions.

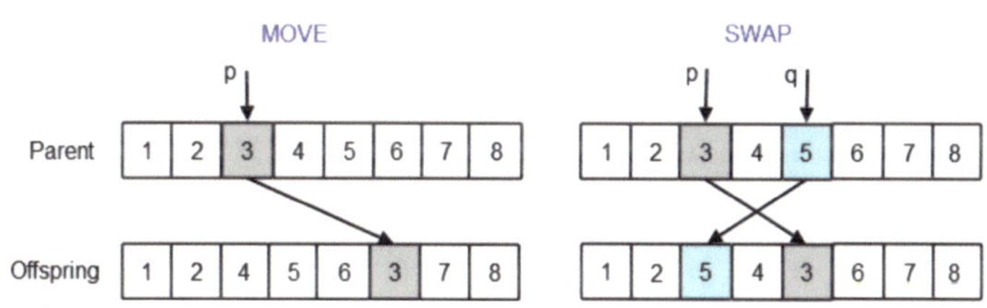

Fig. 8. Mutations

7. Experimental results and discussion

The experiment had a dual objective. On the one hand, we aimed to compare the cooperative search with the simple version and with the best performing methods proposed in the literature dealing with CARP and, thus, to validate our claim that the proposed method offers competitive performance in terms of both solution quality and computational effort. On the other hand, we also aimed to evaluate the influence of subpopulations number and how the proposed mobile agent approach could support adaptive parallelism on networked computers.

In all series of experiments, we used three crossover types (LOX, OX, X1) and two mutation types (MOVE, SWAP). Ten values for crossover rate were used ranging from 0.1 to 0.99 in increments of 0.1. Also, ten mutation rates were allowed varying from 0.1 to 0.55 in increments of 0.05. Clones (identical solutions) were forbidden in each subpopulation, to

have a better dispersion of solutions and to diminish the risk of *premature convergence*. The number of generations was fixed to 5000 for all mobile agents. The migration interval was 500 generations. Whenever migration occurred, the dynamic migration window size varied at random from 1 to θ. The θ value was generated at random within 20% of the subpopulation size. The *threshold* value was fixed to the mean fitness of the subpopulation. Furthermore, 4 runs are carried out in each experiment.
These tests were done on a standard set of undirected instances in which all edges were required. Table 2 contained 34 instances from (Belenguer & Benavent, 2003) with 24 to 50 nodes and 34 to 97 edges. All these files could be obtained at http://www.uv.es/belengue/carp.html. We modified these files by adding a deadheading and collecting costs. Thus, we applied our framework on an instance of UWCOP based on real data provided by the Waste Management and Environment Agency of Angers, France. Results are shown in table 3.
In all tables, *PB* gives the instance number and *N*, *M* the numbers of nodes and edges. *CC* gives the collecting cost. In table 2, *LBB* is a lower bound from (Belenguer & Benavent, 2003). *TS* is the result of Carpet (Hertz et al., 2000) with the parameter setting yielding the best results on average (the same setting for all instances). *BEST* gives the best solution published, generally obtained by Carpet with various parameter settings. *MA* is the solution of the competitive memetic algorithm from (Lacomme et al., 2004).
Our results are shown in the Static Parallel Algorithm (*SPA*) without individual's exchange and the distributed genetic algorithm including individual exchange strategy (*DAIE*) columns.
We compared our algorithms with the best techniques so far reported in the literature. We have selected some newer algorithms, like lower bound proposed by (Belenguer & Benavent, 2003), tabu search (Hertz et al., 2000) and the competitive memetic algorithm from (Lacomme et al., 2004). The last two techniques are known to perform very well on the CARP.

7.1 Results on Belenguer and Benavent's instances

In 90% of the 34 instances of (Belenguer & Benavent, 2003), our *DAIE* outperformed *SPA* implementation of evolutionary system by a mean advantage of 2.62% in term of solution cost. *DAIE* was very efficient: on all instances, it was at least as good as Carpet. On the 34 instances, it outperformed Carpet 17 times, improved 9 best known solutions with 6 to optimality, and reached *LBB* 27 times. The average deviation to *LBB* was roughly divided by 5 compared to Carpet, and reached 0.25%. Moreover, on the same instances, *DAIE* outperformed the Competitive Memetic Algorithm (Lacomme et al., 2004) 7 times.

PB	N	M	CC	LBB	BEST	TS	MA	SPA	DAIE
1A	24	39	195	368	368	368	368	368	368
1B	24	39	195	368	368	368	368	376	368
1C	24	39	195	430	440	440	440	445	430
2A	24	34	170	397	397	397	397	405	397
2B	24	34	170	429	429	430	429	440	429
2C	24	34	170	625	627	664	632	649	625
3A	24	35	175	256	256	256	256	257	256
3B	24	35	175	262	262	262	262	264	262
3C	24	35	175	312	313	313	313	319	312
4A	41	69	345	745	745	745	745	786	745
4B	41	69	345	757	757	761	757	775	761
4C	41	69	345	773	773	798	773	782	773
4D	41	69	345	865	886	901	886	940	881
5A	34	65	325	748	748	748	748	748	748
5B	34	65	325	771	771	773	771	773	771
5C	34	65	325	794	799	801	799	849	801
5D	34	65	325	896	906	932	906	932	902
6A	31	50	250	473	473	473	473	495	473
6B	31	50	250	481	483	491	483	495	481
6C	31	50	250	561	567	579	567	582	567
7A	40	66	330	609	609	609	609	609	609
7B	40	66	330	613	613	613	613	636	613
7C	40	66	330	663	664	673	664	673	663
8A	30	63	315	701	701	701	701	709	701
8B	30	63	315	710	710	716	710	716	710
8C	30	63	315	832	843	848	848	907	832
9A	50	92	460	783	783	783	783	830	783
9B	50	92	460	786	786	789	786	796	786
9C	50	92	460	792	792	792	792	798	792
9D	50	92	460	842	851	869	851	873	851
10A	50	97	485	913	913	913	913	915	913
10B	50	97	485	921	921	921	921	925	921
10C	50	97	485	931	931	936	931	943	931
10D	50	97	485	1009	1020	1029	1020	1045	1015

Table 2. Computational results for instances of (Belenguer & Benavent, 2003)

7.2 Results on Angers's instance

Table 3 shows the results for the file we constructed from real data provided by the Angers Waste Management and Environment Agency. On this instance, the *DAIE* outperforms the *SPA* by a mean advantage of 3% in term of solution cost.

PB	N	M	CC	SPA	DAIE
Angers1	64	67	585	920	892

Table 3. Computational results for UWCOP in Angers

7.3 The influence of subpopulation number on solution quality (Fitness)

The number of subpopulations determines the size of subpopulation because the original population is distributed among subpopulations. The greater the number of subpopulations is high, the more the size of each subpopulation is reduced (Cantü-Paz, 2000). The influence of the number of subpopulations on the quality of solution was investigated. The number of subpopulations varied from 1 to 10. The initial population in every test was the same. In fact, we conducted experiments on one subpopulation of 800 individuals, two subpopulations of 400 individuals, four subpopulations of 200 individuals, six subpopulations of 133 individuals, eight subpopulations of 100 individuals, and ten subpopulations of 80 individuals, in similar experimental conditions, as described above. Each subpopulation was executed on one computer and end-users were not allowed to access the machines used for the experiments. Figure 9 represents the variation of fitness (objective function) compared to the variation of the number of subpopulations. In this figure, one can note that the quality of the solution increases progressively as the number of subpopulations increases. Indeed, the number of subpopulations has a direct influence on the genetic diversity of this parallel model (*DAIE*). A large number of subpopulations implies a reduction of the number of individuals per subpopulation (reduction of the number of genes). As clones are forbidden, subpopulations become heterogeneous and enable interesting subspaces of solutions to be explored.

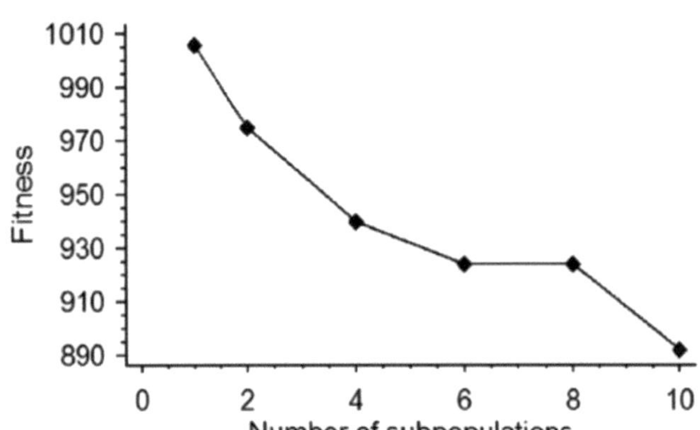

Fig. 9. Influence of subpopulation number on fitness

7.4 The influence of the number of processors on the speed-up

In the same conditions defined in the previous subsection, the influence of the number of processors on the speed-up of distributed evolutionary algorithm was considered. Figure 10 shows the computational time spent on each experiment. As can be observed, the time for running a single experiment is reduced in an almost linear manner, depending on the number of computers used. Although in the parallel model, agent communication for exchanging individuals among subpopulations needs extra computational effort, and is relatively small compared to the time for running a time-consuming evolutionary experiment.

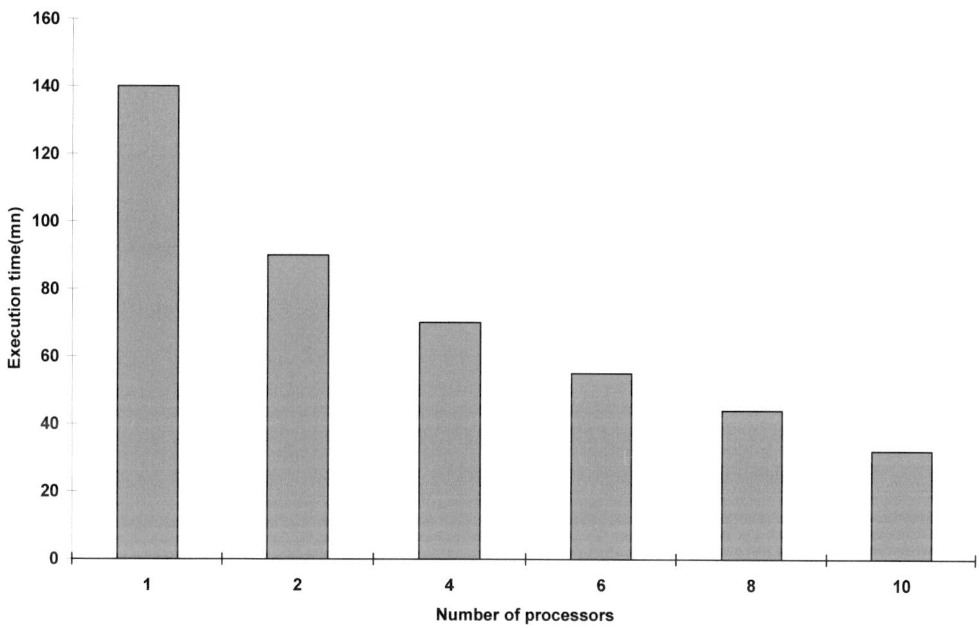

Fig. 10. Effect of the number of processors on speed-up

7.5 The influence of CPU utility on the speed-up

In this experiment, eleven hosts were connected to the JADE platform in which only one was preserved for running EC experiments. The preserved host was initiated as the main-container to enable the default agents AMS, DF, and RMA, and end-users were allowed to use the other ten computers as they usually did. Each host was a container that included a mobile agent (MA) and a local status agent (LSA).

Due to different user-accessing situations, the experiments were conducted five times in the same day. Figure 11 shows the results. As can be seen, all runs have been sped up using the proposed approach. It shows the efficiency of our mobile agent-based approach. Though the computational cost of adaptive parallelism is higher than that of static parallelism presented above, the former is nevertheless a more practical way to realize parallelism for evolutionary computation in an ordinary networked computing environment.

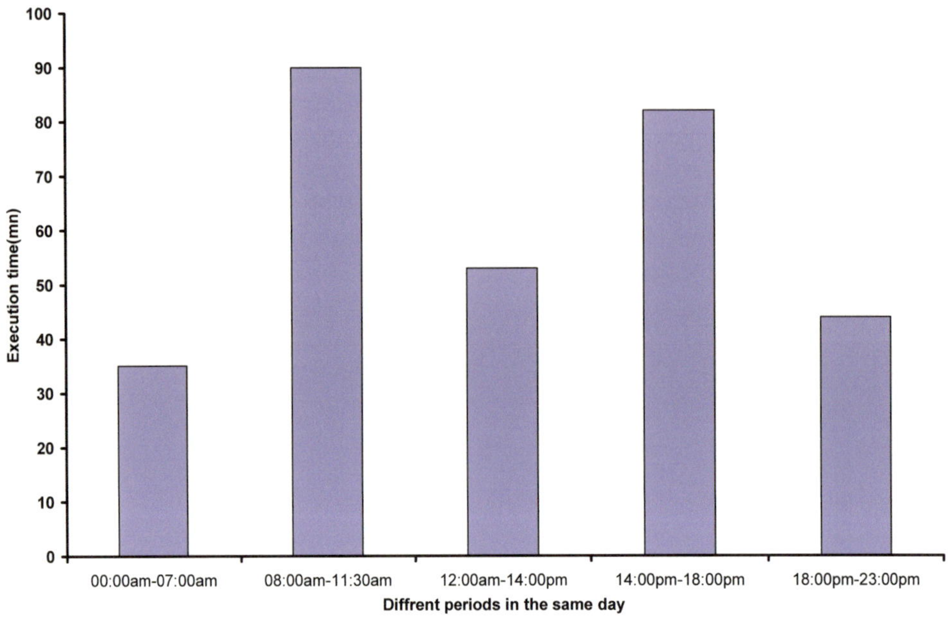

Fig. 11. Adaptive parallelism using mobile agents (influence of CPU utility on the speed-up)

8. Conclusion and future work

In this chapter, we have proposed MAF-DISTA, an efficient and scalable framework that combines the two advantages of parallelism: computational power and cooperation. We implemented an information exchange strategy based on dynamic migration window methods that control the size and frequency of migration and the selective migration model for the choice of individuals to migrate. Results confirmed the positive impact of using our individual's exchange strategy in regard to the static parallel algorithm. They also pointed out that such strategies are, at a minimum, as good as the best known methods.

Furthermore, the proposed framework was a middleware supporting development of distributed population-based techniques. Its main advantages that have been inherited from JADE, include:

- Ability to simplify the development of distributed population-based metaheuristics composed of autonomous entities that need to communicate and collaborate in order to achieve the working of the entire system.
- Interoperability - mobile agents can cooperate with other agents provided that they comply with the FIPA standard.
- Uniformity and portability - the framework provides a homogeneous set of APIs that are independent from the underlying network and Java version.
- Maximum design and Code reuse - the framework provides the user with a complete architecture design of his/her solution method. Moreover, the programmer may redo as little code as possible. It also provides a maximal conceptual separation between the solution methods and the problem to be solved. The user might therefore develop only the minimal problem-specific code.

According to these features, we can conclude that this middleware is able to be extensible.

In future work, we will implement other population-based metaheuristics on our framework such as ant colony algorithm and particle swarm algorithm. We will also test our framework on other NP-complete problems such as planning and scheduling problems.

9. Acknowledgements

Our research is supported through the municipality of Angers and the Urban Waste Collection and Environment Agency. This support is gratefully acknowledged.

10. References

AgentBuilder (2004). An Integrated Toolkit for Constructing Intelligent Sofware Agents, AgentBuilder, Reference Manual, available at http://www.agentbuilder.com, June 2004

Amberg, A. & Voß, S. (2002). A Hierarchical Relaxations Lower Bound for the Capacitated Arc Routing Problem, *Proceedings of Annual Hawaii International Conference on System Sciences*, pp. 1415 - 1424, Hawaii

Baykasoğlu, A. (2001). Goal Programming using the Multiple Objective Tabu Search. *Journal of Operational Research Society*, 52, 12, 1359-1369

Belenguer, J.M. & Benavent, E. (2003). A cutting plane algorithm for the capacitated arc routing problem. *Comput. Oper. Res.* 30, 5 (April 2003) 705-728, 0305-0548

Bellifemine, F.; Poggi, A. & Rimassa, G. (2001). Developing multi-agent systems with a FIPA-compliant agent framework. *Software: Practice and Experience, 31, 2, 103-128,* 0038-0644

Beullens, P.; Muyldermans, L.; Cattrysse, D. & Oudheusden, D.V. (2003). A guided local search heuristic for the capacitated arc routing problem. *European Journal of Operational Research,* 127, 3 (June 2003) 629-643.

Bic, L.; Fukuda, M. & Dillencourt, M. (1996). Distributed computing using autonomous objects. IEEE Computer, 18, 55-61, 1996. Aug 1996

Blum, C. & Roli, A (2003). Metaheuristics in combinatorial optimization: Overview and conceptual comparison. *ACM Computing Surveys*, 35, 3, 268-308, 0360-0300

Braun, P. & Rossak, W. (2005). Mobile agents: Basic concepts, mobility models, and the tracy toolkit, CA: Morgan Kaufmann Publishers, Los Altos, 2005.

Busetta, P.; Ronnquist, R.; Hodgson, A. & Lucas, A. (1999) Jack: Intelligent Agents Components for Intelligent Agents in Java, *AgentLink News Letter,* 2, 2-5

Cantü-Paz, E. (2000). Markov chain models of parallel genetic algorithms. *IEEE Transactions on Evolutionary Computation,* 4, 3, (September 2000) 216-226, 1089-778X

Caragea, D.; Silvescu, A. & Honavar, V. (2001). Towards a Theoretical Framework for Analysis and Synthesis of Agents That Learn from Distributed Dynamic Data Sources, In: *Emerging Neural Architectures Based on Neuroscience,* 547-559 Springer-Verlag New York, Inc., 3-540-42363-X, New York, USA

Chauhan, D. (1997). JAFMAS: A Java-Based Agent Framework for Multi-Agent Systems Development and Implementation, *Masters Thesis,* ECECS Department, University of Cincinnati

Chess, D.; Harrison, C. & Kershenbaum, A. (1996). Mobile Agents: Are They a Good Idea? *Proceedings of the Second International Workshop on Mobile Object Systems,* Jan Vitek,

Christian Tschudin (Ed.), pp. 25-47, Lecture Notes in Computer Science, Springer-Verlag, Linz, Austria, 1222, July 1996

Cung, V.-D.; Martins, S.L.; Ribeiro, C.C. & Roucairol, C. (2002). Strategies for the parallel implementation of metaheuristics, In: *Essays and Surveys in Metaheuristics,* Ribeiro, C.C. (Ed.), page numbers (263-308), Kluwer, 978-0792375203, Dordrecht

Dasgupta, P.; Narasimhan, N.; Moser, L.E. & Melliar-Smith, P.M. (1999). MAgNET: Mobile agents for networked electronic trading. *IEEE Transactions on Knowledge and Data Engineering,* 11, 4 (August 1999) 509-525, 1041-4347

DeLoach, S.A. & Wood, M. (2000). Developing Multiagent Systems with agentTool, *Proceedings of International Workshop on Agent Theories, Architectures and Languages,* pp. 46-60, 3-540-42422-9, Berlin, Germany

Dorigo, M. & Di Caro, G. (1999). The Ant Colony Optimization Meta-Heuristic, In: *New Ideas in Optimization,* 11-32, McGraw-Hill, 0-07-709506-5, New York

Eldos, T. (2006). A New Migration Model For Distributed Genetic Algorithms, *Proceedings of International Conference On Scientific Computing,* pp. 128-134, Las Vegas, NV, USA

Feo, T.A. & Resende, M. G. C. (1989). A probabilistic heuristic for a computationally difficult set covering problem. *Operations Research Letters,* 8, 2, 67–71, 0167-6377

Garneau, T. & Delisle, S. (2003). A New General, Flexible and Java-Based Software Development Tool for Multiagent Systems, *Proceedings of International Conference on Information Systems and Engineering,* pp. 22-29, Montréal (Québec, Canada), July 2003

Glover, F. (1989). Tabu Search – Part I, *ORSA Journal on Computing,* 1, 3, 190-206.

Glover, F. (1990) Tabu Search – Part II, *ORSA Journal on Computing,* 2, 1, 4-32.

Goldberg, D. E. (1989). Genetic *Algorithms in Search, Optimization and Machine Learning,* Addison-Wesley Longman Publishing Co., Inc., 0201157675, Boston, MA, USA

Golden, B.L.; DeArmon, J.S. & Baker, E.K. (1983). Computational Experiments with Algorithms for a Class of Routing Problems. *Computers and Operations Research,* 10, 1, 47-59

Golden, B.L. & Wong, R.T. (1981). Capacitated Routing Problems. *Networks,* 11, 305-315

Graham, J.R.; Mchugh, D.; Mersic, M.; McGeary, F.; Windley, M.V.; Cleaver, D. & Decker, K.S. (2000). Tools for Developing and Monitoring Agents in Distributed Multi-Agent Systems, *Proceedings of Agents Workshop on Infrastructure for Multi-Agent Systems,* pp. 12-27, 3-540-42315-X, Barcelona, Spain

Gray, R.; Cybenko, G.; Kotz, D.; Peterson, R. & Rus, D. (2002). D'Agents: Applications and performance of a mobile-agent system. *Software: Practice and Experience,* 32, 6 (2002) 543-573

Gutknecht, O. & Ferber, J. (2001). The MADKIT Agent Platform Architecture, In: *Revised Papers from the International Workshop on Infrastructure for Multi-Agent Systems,* 48-55, Springer-Verlag, 3-540-42315-X, London, UK

Hertz, A. & Kobler, D. (2000). A framework for the description of evolutionary algorithms. *European Journal of Operational Research,* 126, 1, (October 2000) 1-12, 0377-2217

Hertz, A.; Laporte, G. & Mittaz, M. (2000). A Tabu Search Heuristic for the Capacitated Arc Routing Problem. *Oper. Res.,* 48, 1, 129-135, 0030-364X

Holland, J. H. (1975). *Adaptation in Natural and Artificial Systems: An Introductory Analysis with Applications to Biology, Control, and Artificial Intelligence,* The MIT Press, 0262581116, USA

Honavar, V.; Miller, L. & Wong, J. (1998). Distributed Knowledge Networks, *Proceedings of IEEE Information Technology Conference*, pp. 87-90, 0-7803-9914-5, Syracuse, NY, USA, September 1998, IEEE Press

JADE-Tilab (2009). Java Agent Development Framework, 2009, http://jade.tilab.com

Jiang, Y.C.; Xia, Z.Y. & Zhang, S.Y. (2005). A novel defense model for dynamic topology network based on mobile agent. *Microprocessor and Microsystems*, 29, 6 (August 2005) 289-297

Kim, J. S. (2002). Distributed Genetic Algorithm with Multiple Populations Using Multi-agent, *Proceedings of the 4th International Symposium on High Performance Computing*, pp. 329-334, London, Springer-Verlag, UK

Krasnogor, N. & Smith, J. (2005). A tutorial for competent memetic algorithms: Model, taxonomy, and design issues. *IEEE Transaction on Evolutionary Computation*, 9, 5, (October 2005) 474-488, 1089-778X

Lacomme, P.; Prins, C. & Ramdane-Chérif, W. (2001). A Genetic Algorithm for the Capacitated Arc Routing Problem and Its Extensions, *Proceedings of EvoWorkshops on Applications of Evolutionary Computing*, pp. 473-483, 3-540-41920-9, Springer-Verlag, London, UK

Lacomme, P.; Prins, C. & Ramdane-Chérif, W. (2004). Competitive Memetic Algorithms for Arc Routing Problems. *Annals of Operations Research*,131, 159-185

Laguna, M. & Martí, R. (2003). *Scatter Search: Methodology and Implementations in C*, Kluwer Academic Publishers, 1-4020-7376-3, Boston

Lange, D.B. & Oshima, M. (1998). Programming and deploying Java mobile agents with aglets, In: *CA: Addison-Wesley*, Menlo Park.

Lange, D.B. & Oshima, M. (1999). Seven good reasons for mobile agents. *Communications of the ACM*, 42, 3 , (March 1999) 88-89, 0001-0782

Lee, L.C.; Ndumu, D.T. & Nwana, H.S. (1998). ZEUS: An Advanced Tool-Kit for Engineering Distributed Multi-Agent Systems, *Proceedings of Practical Application of Intelligent Agents and Multi-Agent Systems*, pp. 377-392, London, UK

Leeand, K.L.; Yun, J. S. & J. o., G. K. (2003) MoCAAS: Auction agent system using a collaborative mobile agent in electronic commerce. *Expert Systems with Applications*, 24, 2, 183-187

Lind, J. (2001). *Iterative software engineering for multiagent systems The MASSIVE Method*, Springer-Verlag New York, Inc., 3540421661, Secaucus, NJ, USA

Macêdo, R.J.A. & Assis Silva, F.M. (2005). The mobile group approach for the coordination of mobile agents. *Journal of Parallel and Distributed Computing*, 65, 3, (March 2005) 275-288, 0743-7315

Metropolis, N.; Rosenbluth, A. W.; Rosenbluth, M. N.; Teller, A. H. & Teller, E. (1953). Equation of State Calculations by Fast Computing Machines. *J. Chem. Phys.*, 21, 6, 1087-1092

Mladenović, N. & Hansen, P. (1997). Variable neighborhood search. *Computers & Operations Research*, 24, 11, November 1997, 1097-1100

O'Brien, P. & Nicol, R. (1998). FIPATowards a standard for software agents. *BT Technology Journal*, 16, 3 (1998) 51-59

Papavassiliou, S.; Puliafito, A.; Tomarchio, O. & Ye, J. (2002). Mobile agent-based approach for efficient network management and resource allocation: Framework and applications. *IEEE Journal of Selected Areas in Communication,* 20, 4 (May 2002) 858-872.

Steen, V. S. & Tanenbaum, A. (2002). Distributed Systems: Principles and Paradigms, In: *Englewood Cliffs,* NJ: Prentice Hall

Thati, P.; Chang, P. H. & Agha, G. (2001). Crawlets: Agents for high performance web search engine, *Proceedings of international conference on mobile agents,* pp. 119-134, 978-3-540-42952-4, January 2001, Springer Berlin / Heidelberg, Germany

Ulusoy, G. (1985). The feet size and mix problem for capacitated arc routing. *European Journal of Operational Research,* 22, 3, 329-337.

Veldhuizen, D. A.; Zydallys, J. B. & Lamont, G. B. (2003). Considerations in engineering parallel multiobjective evolutionary algorithms. *IEEE Transactions on Evolutionary Computation,* 7, 2, (April 2003) 144-173, 1089-778X

White, J. E. (1997). Mobile Agents, In: *Software Agents,* Bradshaw, J. (Ed.), MA: MIT Press, Cambridge

WIDE (2009). Wide user guide, http://jade.tilab.com/wade

11. Appendix - List of Symbols and Acronyms

AM	Augment-Merge
AOP	Agent Oriented Programming
CARP	Capacitated Arc Routing problem
CORBA	Common Object Request Broker Architecture
CPU	Central Processing Unit
CVRP	Capacitated Vehicle Routing Problem
DAIE	Distributed Algorithm with Individuals exchange
DCE	Distributed Computing Environment
DCOM	Distributed Component Object Model
DM-MIMD	Distributed Memory Multiple Instruction stream, Multiple Data stream
EA	Evolutionary Algorithm
EC-code	Evolutionary Computation code
FIPA	Foundation for Intelligent Physical Agents
GA	Genetic Algorithm
JADE	Java Agent DEvelopment Framework
LOX	Linear Order Crossover
LS	Local Search
LSA	Local Status Agent
MA	Mobile Agent
MAS	Multi Agent System
MOM	Message-Oriented Middleware
MSA	Main Status Agent
OOP	Object Oriented Programing
OX	Order Crossover
PS	Path Scanning
RPC	Remote Procedure Call
RMI	Remote Method Invocation
SA	Status Agent

SPA	Static Parallel Algorithm
SYNCA	SYNChronization Agent
UH	Ulusoy's Heuristic
UIA	User Interface Agent
UWCOP	Underground Waste Collection Optimization Problem
X1	One point Crossover

15

Adaptive Bio-inspired Control of Humanoid Robots – From Human Locomotion to an Artificial Biped Gait of High Performances

Aleksandar Rodić [1], Khalid Addi [2] and Georges Dalleau [2]

[1] *Mihajlo Pupin Institute, Serbia*

[2] *University of Reunion, France*

1. Introduction

Biped locomotion, in the sense of gait stability and maintenance of dynamic balance, represents one of the most complex known natural motions. Anthropomorphic locomotion is performed by a synergy of large number of body muscles. Among them, of special importance are leg muscles. Human anatomy and physical capabilities of human body determine the kinematical and dynamic performances of the gait. Stability, harmony of motion, repeatability, smoothness of trajectories, rigidity of foot landing, etc., are the attributes that characterize human gait. In that sense, performances of biped gait differs from person to person, depending on different factors such as: physical and physiological capabilities, age, professional deformation (e.g. ballerina, athletic, fashion model, etc.), some pathologic conditions but sometimes also on psychological condition of the person, etc. In this paper, the focus of research attention will be directed towards the physical capabilities (and their modulation) of human body to perform regular biped walk as a model to be used for synthesis of artificial gait for biped robots. Sophisticated synergy of muscle activities of the whole body determines characteristics of biped locomotion and its stability performances.

During walking, running or jumping, leg muscles and ligaments are being strengthened or relaxed alternately according to the physical phases of locomotion. As consequence, landing foot produces corresponding ground reaction forces through foot contact/impact to the ground, that transmit to the corresponding leg joints. In that sense, the term *leg impedance* represents a measure of how much a body structure resists motion when subjected to a given external force(s). It relates forces with corresponding velocities acting on a bio-mechanical system. The mechanical impedance is a function of the frequency of the applied force and can vary greatly over frequency. At resonance frequencies, the mechanical impedance will be lower, meaning less force is needed to cause a structure to move at a given velocity. Leg impedance has an important role in human locomotion. By modulation of leg impedance in a natural way, it is possible to produce different kinds of biped locomotion as well as to damp corresponding impact/contact forces that represent perturbation to the system and can cause in-stability. Advanced humanoid robots use electric motors in the joints to produce biped locomotion and suppress perturbing impact forces.

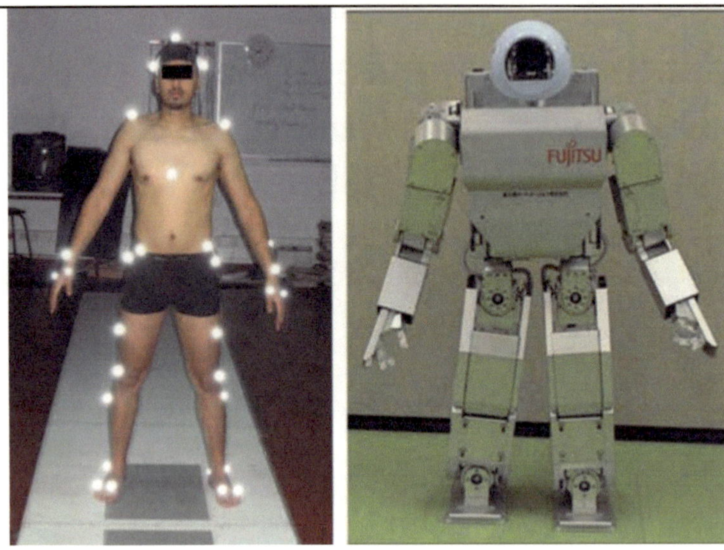

Fig. 1. Biological and artificial biped locomotion system – biped robot (Fujitsu, 2009) as an imitation of human body

2. Advances Beyond State-of-the-Art

Currently many research groups in the world are working on biped robots. The main thrust of the current research can be categorized into three parts: optimization of leg and foot trajectory, stable locomotion control, and hardware design. For optimization, reducing impact in foot landings and reducing torque and energy requirements of joint actuators are primary concerns (Rousell et al., 1998; Bruneau et al., 1998; Rostami et al., 1998). Different designs of biped robots are also explored by many research teams (Sony, 2006; Kim & Oh, 2004). However it has been regarded that walking, especially dynamic walking, is a very difficult problem to be tackled until the recent development of a few biped robots with the capability of dynamic locomotion such as for example: Wabian (Ogura et al., 2006), Asimo (Honda, 2009), Hoap-3 (Fujitsu, 2009), QRIO (Qrio, 2009), etc.
The problem of realizing a walking action in humanoid robots involves two components: generation of the basic walking pattern and the compensation required to maintain the robot's balance. Summarizing research results and practical experience of many research teams concerning implementation of different control strategies for dynamic biped locomotion, the impedance control was promoted as the most frequent strategy applied with contemporary humanoid robots. Park proposed (Park, 2001) an impedance control and corresponding switch-mode impedance modulation in particular phases of biped gait as well as in particular directions of locomotion. Both legs are controlled by impedance, where the desired impedances at the hip and swing foot are specified. The impedance parameters

(damping factors) were changed depending on the gait phase. Stiffness of the legs was assumed invariable.
Lim H-O, Setiawan and Takanishi have proposed (Lim et al., 2004; Lim et al., 2001) a position-based impedance control for biped humanoid robot locomotion. The impedance parameters of the biped leg were adjusted in real-time according to the gait phase. In order to reduce the impact/contact forces generated between the contacting foot and the ground, the damping coefficient of the impedance of the landing foot is increased largely during the first half double support phase. In the last half double support phase, the walking pattern of the leg changed by the impedance control is returned to the desired walking pattern by using a polynomial. Also, the large stiffness of the landing leg is given to increase the momentum reduced by the viscosity of the landing leg in the first half single support phase. For the stability of the biped humanoid robot, a balance control that compensates for moments generated by the biped locomotion is employed during a whole walking cycle.
The adaptive impedance control of biped robots with leg impedance modulation is proposed in the article. It controls the impedance of the swing foot as well as the hip link. The impedance control method has been successfully used for the robot manipulators (Hogan, 1986) which interact with their environment. In locomotion, a biped robot is in contact with the ground, sometimes with one foot and other times with two feet. Only in running or jumping there are phases when humans or robots have no contacts with the ground and then they have to be considered as "flyers" (Potkonjak et al., 2005). Moreover, impedance control for bipedal locomotion is similar to the control method that a human uses for his locomotion. When a human wa1ks, he does not explicitly control the trajectory of his upper body but rather controls the muscle strength of his legs that support the upper body (Park, 2001). He also rather controls the trajectory of the swing foot in order to avoid obstacles such as bumps or in order to land the foot in a safe area, for example, avoiding a pot hole. In typical human locomotion, the leg muscles are repeatedly hardened and relaxed depending on the gait phase (Leonard et al., 1995). Just before the contact of the swing foot with the ground, the leg muscle is relaxed to regulate and reduce impact, resulting in a very soft contact with the ground. Borrowing this idea from human locomotion, the authors suggest that the parameters used in the impedance control are also appropriately modulated depending on the gait phase of the biped robot. Flying phase (running or jumping) will not be considered in this paper although it represents an interesting problem how the legs prepare themselves for landing and contact with the ground.
The main contribution of this paper concerns with design of a novel, biologically-inspired, adaptive impedance control applied to biped robot locomotion. Leg impedance modulation is performed in an anthropomorphic way, implementing the biological principles of leg adaptation based on gait conditions and proprioceptive feedbacks on dynamic reactions at biped robot legs. Impedance modulation does not depend on gait phases only but also on the real gait conditions and dynamic reactions changing in real-time. The impedance parameters (leg stiffness and damping) are adjusted in real-time (continually) as humans do it. Thus, this paper covers in details how the adaptive impedance control is implemented, how to identify the variable parameters in order to achieve the high (but not optimal in the sense of energy efficiency) dynamic performances of the system. Generation of the artificial bio-inspired (anthropomorphic) gait is also concerned in the paper, using experience obtained from the capture of motion experiments with biological system. Stabile walk is enabled by implementation of a regulator based on the feedback on dynamic reactions

(Rodić, 2008). The efficiency of the control strategy proposed in the paper is verified by extensive simulation.

3. Biped Robot Model

Human body for its complex motion uses synergy of more than 600 muscles. It has more then 300 degrees of freedom (DOFs) (Vukobratović et al., 1990). Some of these particular motions are essential for the human activities (gait, work, sport, dancing, etc.) while the others give it a full mobility. In this paper, a 36 DOFs biped locomotion mechanism of the anthropomorphic structure (Fig. 2) will be considered as a mechanical representative of a human body.

Let us consider the biped robotic system of the anthropomorphic structure illustrated in Figs. 2a and 2b. Let the joints of the system be such to allow n independent motions. Let these joint motions be described by joint angles forming the vector of the generalized coordinates $q=[q_1 \cdots q_n]^T$. The terms 'joint coordinates' or 'internal coordinates' are commonly used for this vector in robotics. This set of coordinates describes completely the relative motion of the links. The basic link (such as the pelvis in this case, Fig. 2a) is allowed to perform six independent motions in 3D-space. Let the position of the basic link be defined by the three Cartesian coordinates (x, y, z) of its mass centre and the three orientation angles (φ-roll, θ-pitch and ψ-yaw), forming the vector $\underline{X}=[x\ y\ z\ \varphi\ \theta\ \psi]^T$. Now, the overall number of DOFs for the system is $N=6+n$, and the system position is defined by

$$Q=\left[\underline{X}\ q\right]=\left[x\ y\ z\ \varphi\ \theta\ \psi\ q_1 \ldots q_n\right]^T \tag{1}$$

It is assumed that each joint has an appropriate actuator. This means that each motion q_j has its own drive - the torque τ_j. Note that there is no drive associated to the basic body coordinates $\underline{X}$. The vector of the joint drives is $\tau=[\tau_1 \cdots \tau_n]^T$, and the augmented drive vector (N-dimensional) is $\mathrm{T}=[\underline{0}_6\ \tau]^T=[0 \cdots 0\ \tau_1 \cdots \tau_n]^T$.

Similarly, with human beings muscles represent biological power-trains. Pairs of muscles by their synchronized contractions and extensions move the bones of the skeleton at its joints in an articulated manner. Bearing in mind the previous mechanic assumptions, the dynamic model of the biped mechanism (humanoid) has the general form (Vukobratović et al., 1990; Park, 2001):

$$\begin{aligned} H(Q,d)\ddot{Q}+h(Q,\dot{Q},d) &= \tau+J^T(Q,d)F \\ h(Q,\dot{Q},d) &= h_{ccf}(Q,\dot{Q},d)+h_g(Q,d) \end{aligned} \tag{2}$$

or decoupled

$$\begin{aligned} H_{\underline{X},\underline{X}}\ \ddot{\underline{X}}+H_{\underline{X},q}\ \ddot{q}+h_{\underline{X}} &= \underline{0}_6+J_{\underline{X}}^T\ F \\ H_{q,\underline{X}}\ \ddot{\underline{X}}+H_{q,q}\ \ddot{q}+h_q &= \tau+J_q^T\ F \end{aligned} \tag{3}$$

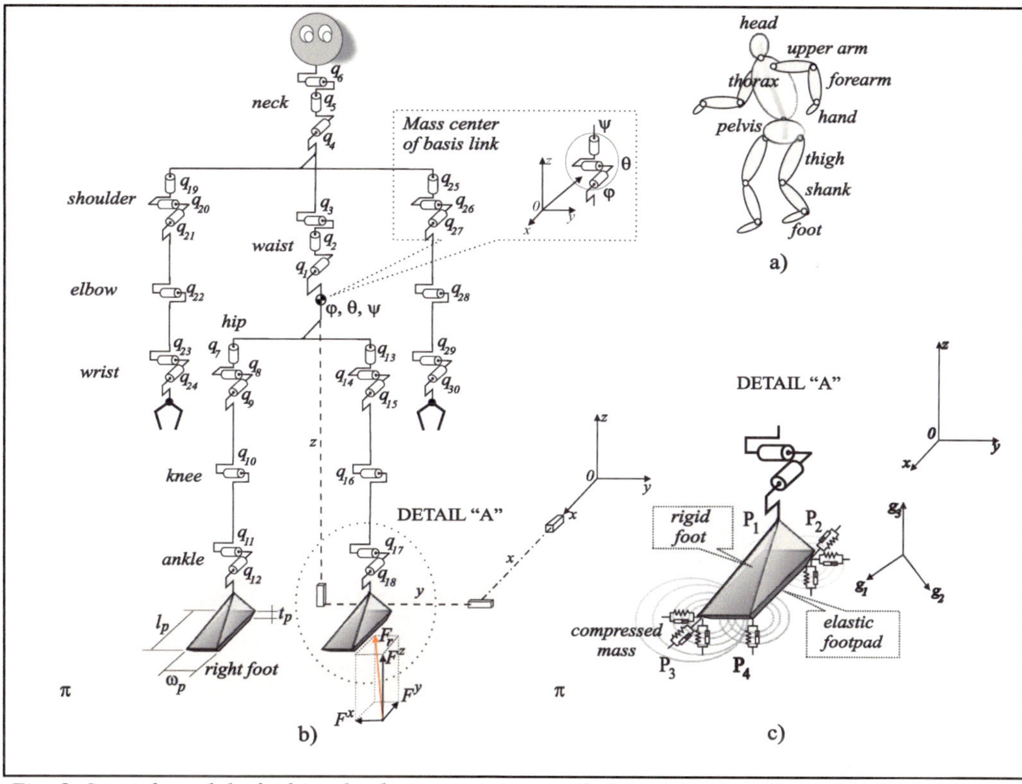

Fig. 2. Spatial model of a biped robot interacting with dynamic environement: a) Multi-body biped robot with two-link torso, b) Kinematic scheme of a 36 DOFs biped robot mechanism considered in the paper, c) 6 DOFs compliance model of a ground support

Dimensions of the inertial matrix and its sub matrices are: $H(N \times N)$, $H_{\underline{X},\underline{X}}(6\times 6)$, $H_{\underline{X},q}(6\times n)$, $H_{q,\underline{X}}(n\times 6)$, and $H_{q,q}(n\times n)$. Dimensions of the vectors containing centrifugal, Coriolis' and gravity effects are: $h(N)$, $h_{\underline{X}}(6)$, and $h_q(n)$. Vector $h(Q,d)$ consists of two vectors: the vector of centrifugal and Coriolis' forces $h_{ccf}(Q,\dot{Q},d)$ and the vector of gravity forces and torques $h_g(Q,d)$. Dimension of the vector of ground reaction, external load and disturbance forces is $F(m\times 1)$. Dimensions of the Jacobian matrix and its sub matrices are: $J(m\times N)$, $J_{\underline{X}}(m\times 6)$, $J_q(m\times n)$. Vector $d(l)$ represents a parameter vector including geometry (links' lengths, positions of the links' mass centers), as well as the corresponding dynamic parameters (links' masses, moments of inertia) of the robotic system.

3.1 Modeling of Contact Dynamics

Let us consider the link of the biped mechanism that has to establish contact with some external object. In the case considered, it is the foot that moves towards the ground, strikes it and stays in contact (e.g. walking, running or climbing the stairs). An external object may be immobile (like ground), or mobile, like part of some other dynamic system (mobile platform (Vukobratović et al., 2004), conveyer, boat, tram, etc.). To express mathematically the forthcoming contact, the motion of the considered link should be described by an appropriate set of coordinates. Since the link is a body moving in the 3D-space, it is necessary to consider six coordinates. Let this set be $\underline{g}=[g_1,\cdots,g_6]^T$ and let call them functional coordinates (Fig. 3a). Functional coordinates are introduced (Potkonjak et al., 2005) as relative ones, defining the position of the link with respect to the ground (object) to be contacted.

A consequence of the rigid link-object contact is that the link and the object perform some motions, along some axes, jointly (Fig. 3b). These are constrained (restricted) directions (e.g. g_3, g_5, Fig. 3b). Let there be *m* such directions. Relative position along these axes does not change. Along the other axes (e.g. g_1, g_2, g_4, g_6, Fig. 3b) relative displacement is possible. These are unconstrained (free) directions. In order to get a simple mathematical description of the contact, $\underline{g}$-coordinates are introduced to describe the relative position. Zero value of a coordinate indicates the contact along the corresponding axis. So, the motion of the external object (to be contacted) has to be known (or calculated from the appropriate mathematical model), and then the $\underline{g}$-frame fixed to the object (the plane π in this case, Fig.3) is introduce to describe the relative position of the link in the most proper way. Thus, in a general case, the $\underline{g}$-frame is mobile. As the link is approaching the object, some of $\underline{g}$-coordinates reduce and finally reach zero. The zero value means that the contact is established. These functional coordinates (which reduce to zero) are called restricted coordinates and they form the sub vector $\underline{g}^c$ of dimension m. The other functional coordinates are free and they form the sub vector $\underline{g}^f$ of dimension $6-m$. Now, one can write:

$$[\underline{g}^c, \underline{g}^f]^T = K\cdot\underline{g} \tag{4}$$

where K is a 6×6 matrix used to rearrange the functional coordinates (elements of the vector $\underline{g}$) and bring the restricted ones to the first positions. In order to arrive at a general algorithm, the foot motion has to be described in a general way and, once the expected contact is specified, relate this general interpretation to the appropriate $\underline{s}$-frame.

The general description of the leg motion assumes three Cartesian coordinates of a selected point of the foot link plus three orientation angles: $\underline{s}_f=[x_f\ y_f\ z_f\ \varphi_f\ \theta_f\ \psi_f]^T$, the subscript "*f*" standing for "foot link". These are absolute external coordinates defined with respect to the *Oxyz* reference coordinate system (Fig. 3a).

The relation between the foot links coordinates $\underline{s}_f$ and the leg position vector q_l is given by:

$$\underline{s}_f=\underline{s}_f(\underline{X},q_l,d) \tag{5}$$

$$\dot{\underline{s}}_f=\dot{\underline{X}}+J_l(q_l,d)\dot{q}_l \tag{6}$$

$$\ddot{\underline{s}}_f = \ddot{\underline{X}} + J_l(q_l, d)\ddot{q}_l + A_l(q_l, \dot{q}_l, d) \tag{7}$$

where q_l is a corresponding 6×1 vector of leg's (right one or left) joint coordinates, $\underline{X}$ is a 6×1 position vector of the hip (basis) link of robot mechanism defined in (1), $J_l = \frac{\partial \underline{s}_f}{\partial q_l}$ is a 6×6 Jacobian matrix of the foot link with respect to the basis link, and $A_l = \frac{\partial^2 \underline{s}_f}{\partial q_l^2} \dot{q}_l^2 = \dot{J}_l \dot{q}_l$ is a 6-dimensional adjoint vector.

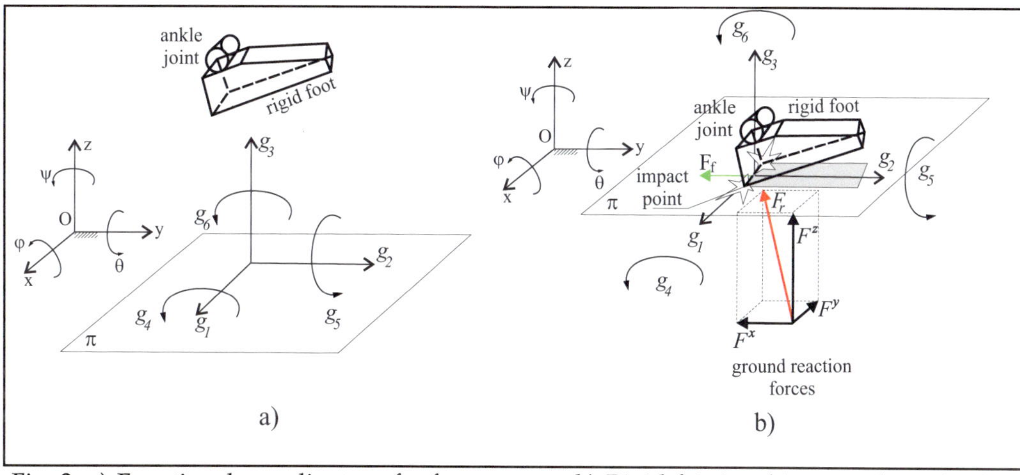

Fig. 3. a) Functional coordinates of a foot contact; b) Rigid frictional impact along the foot edge

In this paper the *compliance model* of the ground support is considered in a form of *full impedance* (inertia, damping and stiffness effects including). In that goal, a relatively complex 36 DOFs model of the biped system (Fig. 2a and 2b) is considered. Biped system considered interacts with a dynamic environment performing anthropomorphic gait. Contact dynamics is identified in a form of a 6 DOFs compliance model as presented in Fig. 2c (Park, 2001). To simulate locomotion of the biped robot, its environment should be modelled accurately so that it can provide reasonably realistic interaction forces from the ground. Typically, biped robots have elastic pads at their soles for shock absorption. These pads allow small motions of the feet on the ground. Plastic collision models such as the one proposed in (Fujimoto & Kawamura, 1995) cannot simulate contact transitions such as a foot bouncing from the ground during a foot landing. In this paper, a 6 DOFs environment model based on non-linear (Marhefka & Orin, 1996) and linear compliant models (Kraus & Kummar, 1997) is used. This model allows simulations of small movements of the feet on the ground, caused by their elastic pads and provides more realistic reaction forces.

Humans wear very often soft, elastic, tracking shoes with rubber pads on the foot sole, to ensure a comfortable and free motion (walking, running, jumping, dancing, etc.). Analogously, the biped locomotion mechanisms (humanoid robots) use elastic pads attached to the feet soles. These elastic elements damp the impact effects that are transmitted onto the structure of the system. Human/humanoid joints are especially sensitive to the dynamic loads. In the instant when the landing foot touches on the ground, a large impact force can be generated. Active control of such impact force would require robot's controllers and actuators to have very wide bandwidth and be capable of generating a large instantaneous power (Park, 2001). It is not realistic to have controllers with a wide bandwidth and huge powerful actuators, which add more weight to the biped robot. That is the reason why many biped robots are equipped with some kind of shock-absorbing elastic pads, which in turn cause small movements of the foot landing on the ground, and may destabilize the locomotion.

4. Biological Aspects of Human Gait

Aiming to the analysis of human biped gait, the extensive experiments in the capture motion laboratory have been carried out. For this purpose, a middle-age male subject, *190 cm* tall, *84 kg* weight, of normal physical constitution and functionality, played the role of an experimental biological system whose parameters are identified by direct body measurements, photometry and implementation of the anthropometry empirical relations given in (Zatsiorsky, 1990; De Leva, 1996). The following variables are captured (acquired) in the experiments: (i) body motion (joint angles), (ii) hip joint and foot sole cycloids (heel and toe trajectories) defined in the Descartes coordinate system, and (iii) ground reaction forces/torques on the footsole reduced to the coordinate system attached to the supporting surface. The mentioned measurements are used in an appropriate way to analyze the biological principles of human locomotion characteristic for a regular and stable human gait. The following parameters of biped gait are assumed as relevant: (i) *walking speed* V, (ii) *step size* s, (iii) *lifting height of the swinging foot* h_f, and (iv) *step period* T. The step period is determined from the relation $T = s / V$, under the assumption that within one step there is no variation of the forward speed V. Additionaly to capturing of the foot cycloids, the motion of the hip link is also captured. This bobbing motion is described by corresponding hip joint cycloids as well as trajectory of the hip link mass center. Forementioned cycloids are of crucial importnace for the anlysis of human gait charcateristics as well as for the synthesis of biped gait generator. For the needs of reserach in this paper, a generator of artificial biped gait was synthesized based on the experimental results obtained from the capture motion system, and the antrpopomorphic parameters of the examined biological system.

Regular biped gait consists of several characteristic gait phases that alternate periodically (Park, 2001). Hence, depending on whether the system is supported to a single or double legs, two macro-phases can be distinguished, viz.: (i) single-support phase (SSP) and (ii) double-support phase (DSP). Double-support phase has two micro-phases: (i) *weight acceptance phase* (WAP) when the heel strike happens, and (ii) *weight support phase* (WSP). Single support phase includes the *swing phase* (SP). Gait phases can be recognized also by analysis of the ground reaction forces during walking.

4.1 Foot Impact and Leg Impedance Modulation

Analysis of the foot-ground contact phenomena is essential for the research of human locomotion. The ground reaction forces (Fig. 4) that appear in contact surface of the human supporting foot, have significant dynamic influence to the body joints (Rodić, 2008): ankles, knees, hips, waist, even neck. In that sense, a foot impact is suggested to be related to pain and injury of joints, so the repetitive loads could be linked to the development of different pathologic conditions with human patients. Amplitudes of contact forces can vary significantly depending on type of locomotion (walking or jumping) as well as on gait parameters (speed, step size, lifting height of swinging leg) and ground support parameters (compliance, unevenness of surface profile, etc). With biped robots, large ground reaction forces can cause system instability as well as mechanical damages of the structure. Due to the quoted reasons, importance of leg impedance studying with human beings and biped robots is evident. A contribution to one such study will be provided in this paper.

Three time instants are of interest during a biped gait (Fig. 4): t_1 - time of heel strike to the ground (begin of the WAP phase), t_2 - time when the landing toes reach the support (begin of the WSP phase), and t_3 - time when the supporting foot leaves the supporting plane (beginning of the SP). In t_1 and t_2 time instants, the heel centre and the toe centre points reach zero speeds in the *z*-direction perpendicular to the supporting surface. In these time instants the local impacts occur producing corresponding energy dissipation. These phenomena significantly influence the stability of the system during the walk. Another critical time instant is the time when a leg leaves the ground, i.e. when the body is supported by single leg (stance phase). The way how the impact forces can be suppressed (damped) will be solved in this paper by the leg impedance adaptation. These research targets will be considered in the following sections of the paper.

By the extensive experimental measurements in the capture motion laboratory, with a complementary measuring of corresponding neuro-muscular activities with human patinets (Dalleau et al., 1998), it was discovered that an impedance of leg muscles (i.e. parameters) varies significantly in real-time during different locomotion activities such as: walking, running, jumping, climbing stairs, balancing on a mobile platform, etc.

Natural leg impedance varies not only with changing gait phases (e.g. WAP, WSP, SP) but also within duration of a single phase depending on different dynamic and kinematical gait parameters. It was proved by the experiments.

The same person changes the own leg impedance in different way depending on: forward speed of motion, lifting height of swinging leg, step size, ground support compliance, foot-ground conditions, etc. By measuring neuro-muscular activities during human locomotion it was identified that man hardens instictively leg muscles just before the foot impact in order to supress (damp) a shock to the ground suport. Since the impact period lives too short (it was measured to be just *5-30 ms*), an instantaneous modulation of the leg stiffness and damping characteristics is imposible (Dalleau et al., 2004). The damping factor of leg muscles in WAP phase is usually significantly larger (e.g. especially during fast walk and jump) than in the WSP phase. Using the proprioceptive[1] feedback as well as tactile

[1]It is the sense that indicates whether the body is moving with required effort, as well as where the various parts of the body are located in relation to each other.

information about the ground reaction forces/torques existing on the footsole, neural system commands to the leg muscles to change impedance. After the initial strengthening of leg muscles, a phase of their relaxation subsequently succeeds. The level of relaxation (decreasing stiffness and damping) depends on amplitudes of the ground reaction force F_z in the WSP phase (see the interval $t \in (t_2, t_3)$ shown in Fig. 4).

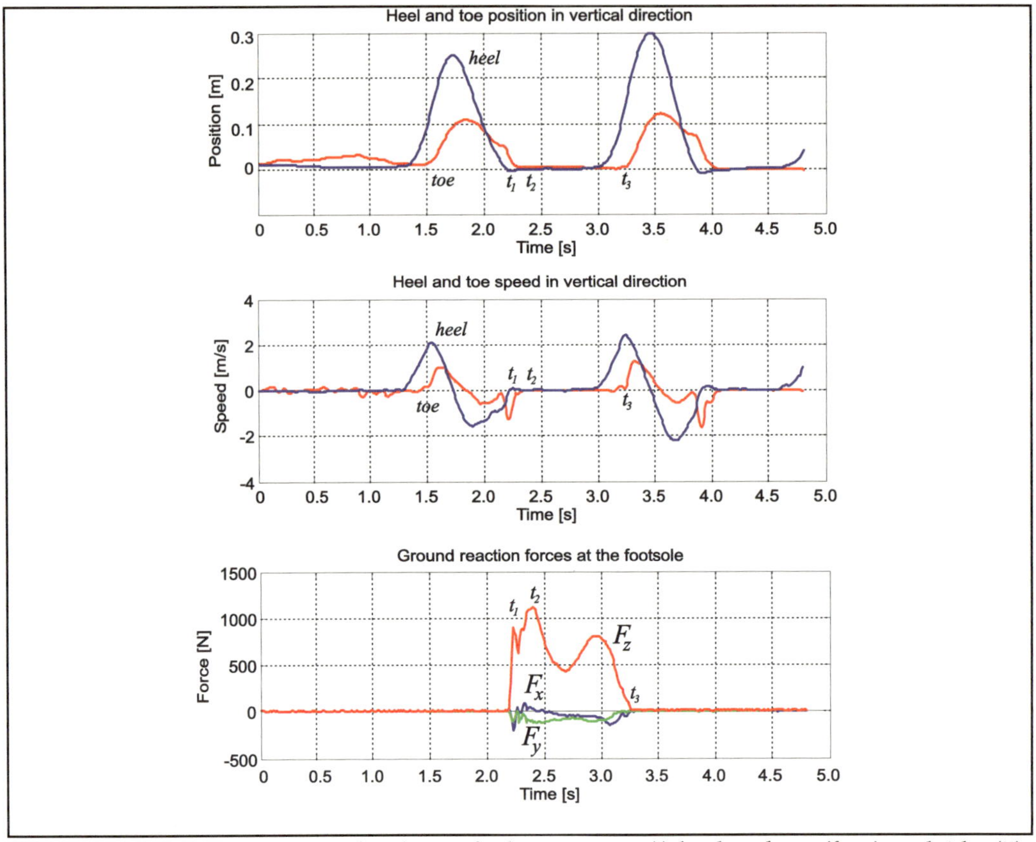

Fig. 4. Experimentaly measured indices of a human gait: (i) heel and toe (foot) cycloids, (ii) speed of the landing foot, (iii) ground reaction forces acting to the foot sole in the sagital- F_x, lateral- F_y and vertical- F_z direction

When the contact force magnitude F_z transcends the reference value corresponding to the body weight, the leg musles are consequently relaxed proportionally. Concerning the gait speed of a biped system, it was identified that the damping factor of the leg muscles is being deceresed with increasing of the speed of motion. Changes of a ground compliance (from the rigid to the soft) cause consequently changes in leg imedance, too. A soft, i.e. more compliant supporting surface (e.g. carpet, sandy terrain, etc.) needs additional muscles strengthening as well as viceversa – increasing of the ground rigidity needs relaxation of leg muscles in order to adapt to the originated changes. Forementioned biological principles, acquired from biomechanical experiments, will be elaborated in the next Sections of the paper for the synthesis of new adaptive impedance control algorithm.

5. Adaptive Control of Biped Locomotion

A good biped robot control scheme should have the following capabilities (Park, 2001). First, the desired trajectory of the swing foot should be tracked as accurately as possible in order to avoid obstacles in the way the human feet do it precisely. On the other hand, tracking of the upper body is not so much rigorous as that of the swing foot. However, maintaining a good balance and posture of the upper body is important having in mind that it carries a vision system (e.g. stereo cameras) on the head as well as robot arms should to perform different service tasks that requests posture stability. Second, footing on the ground should be safe and stable. An efficient control scheme would prevent the swing foot from being bounced from the ground during its landing on the support surface (ground). Foot bouncing from the ground, during foot landings, could cause instability in locomotion and thus should be ultimately avoided. Third, a biped robot should be also able to adapt to various conditions in the locomotion environment so that it can walk with a dynamic balance, for example, on a carpeted floor, on a compliant sandy surface, on a surface with pebbles as well as on a flat rigid floor.
The novelties presented in this paper are addressed to building of an adaptive control of biped robot mechanisms with bio-inspired, continual (real-time) leg impedance modulation. It is unique that it controls the impedance of the swing foot as well as the impedance of the hip link. The impedance control method has been successfully used for the robot manipulators (Hogan, 1986) which interact with their environment where manipulators perform their constrained motions. In biped locomotion, humanoid robot is periodically in contact with the ground, sometimes with single foot and other times with double feet. Moreover, impedance control for bipedal locomotion should to be similar to the natural method that human being uses for locomotion. When a human walks, he/she does not explicitly control the trajectory of his upper body, but rather controls the muscle strength of his legs that support the upper body. Human also rather controls the trajectory of the swing foot in order to avoid obstacles such as bumps and sills or in order to land the foot in a safe area, for example, avoiding a pot hole. In typical human locomotion, the leg muscles are repeatedly strengthen and relaxed depending on gait phases (Leonard et al., 1995). Just before the contact of the swing foot with the ground, the leg muscles are strengthen to regulate and reduce impact, resulting in a very soft contact with the ground. Borrowing this idea from human locomotion, the parameters used in the impedance control should be also appropriately modulated depending on the gait phase having in mind that the gait phase indicators can be identified in real-time as described in previous section.

5.1 Impedance Control

The impedance control law is derived under the assumption that there exist contact sensors at the feet soles and force/torque sensors at the ankles so that the controller knows when a foot reaches the ground and how much the contact force/torque is generated at the feet. Both feedbacks (sensor information) are necessary for building a regulator of dynamic balance to be described in the next section, too.
This paper proposes that the impedance of the swing foot should be specified with respect

to the ground. During the SPs and WAPs, the impedance of the tip of the swing foot depends on the entire kinematical chain, which passes through the hip and terminates at the ground contact point of the support foot of other leg. This paper also proposes a way of adaptive modulation of the set of leg impedance parameters to be used for the swing foot depending on its locomotion phase, i.e. whether it is in the SP, WSP or in the WAP. The higher damping ratio, as it will be demonstrated, is used in WAPs than that in SPs and WSP in order to absorb the impact energy in foot landings. Also, the leg in its WSPs, supporting the weight of the biped robot and propelling the hip and the upper body forward, is controlled based on the impedance model of the hip and upper body with respect to the foot on the ground. The impedance model is selected so that the hip link follows its predetermined trajectory and moves forward. It should be stressed out that the impedance of the hip link in the WSP phase depends on the entire chain up to the ground contact point or the supporting foot. The relation defining the impedance control algorithm of biped robots are presented in the text to follow.

The impedance control is applied at the level of the inertial frame rather than the joint space, as the desired behavior of the biped robot at the global coordinate is needed. For that purpose, the model of biped robot mechanism (3) is used to design the control algorithm of the overall biped mechanism including hip link and swing foot impedance control as well as posture control of the upper body and robot arms. The vector form of the integrated control algorithm has the following form:

$$\tau = H_{q,\underline{X}}\ddot{\underline{X}}^{*} + H_{q,q}\ddot{q}^{*} + h_q - J_q^T F \tag{8}$$

where, the particular variables are explained previously in Section 2, while $\ddot{\underline{X}}^{*}$ and $\ddot{q}^{*}$ represent the corresponding 6×1 and $n\times 1$ control accelerations to be determined in the text to follow. Vector $\ddot{\underline{X}}^{*}$ concerns with control of hip (basis) link motion while $\ddot{q}^{*}$ concern with biped robot joint motions including legs, trunk and arms displacements. Then, the vector of control joint accelerations $\ddot{q}^{*}$ has the following sub-vectors that correspond to the particular kinematical chains (legs, arms, and trunk) of biped mechanism.

$$\ddot{q}^{*} = \left[\ddot{q}^{*}_{tr}\ \ddot{q}^{*}_{rl}\ \ddot{q}^{*}_{ll}\ \ddot{q}^{*}_{ra}\ \ddot{q}^{*}_{la}\right]^T \tag{9}$$

where "*tr*", "*rl*", "*ll*", "*ra*" and "*la*" stands for trunk, right leg, left leg, right arm and left arm successively. Now, suppose the desired impedance of the hip (basis) link (Fig. 2a) reduced to the centre of it, is expressed in a form:

$$M_b\left(\ddot{\underline{X}} - \ddot{\underline{X}}_0\right) + B_b\left(\dot{\underline{X}} - \dot{\underline{X}}_0\right) + K_b\left(\underline{X} - \underline{X}_0\right) = \underline{0}_6 \tag{10}$$

where, M_b, B_b, K_b are the desired mass, damping ratio, and stiffness of the hip link, respectively. Impedance of the hip link in the SSP depends on the entire leg chain up to the ground contact point centre (case of foot multi-point contact) of the supporting foot. In order to achieve the desired impedance (10), acceleration of the hip link $\ddot{\underline{X}}^{*}$ should be:

$$\ddot{\underline{X}}^{*} = \ddot{\underline{X}}_0 - M_b^{-1}\left\{B_b\left(\dot{\underline{X}} - \dot{\underline{X}}_0\right) + K_b\left(\underline{X} - \underline{X}_0\right)\right\} \tag{11}$$

The obtained vector $\ddot{\underline{X}}^{*}$ is used in calculation the control law (8). Suppose that the desired impedance of the leg in the swing phases and the weight acceptance phases is expressed as:

$$M_l\left(\ddot{s}_f - \ddot{s}_f^0\right) + B_l\left(\dot{s}_f - \dot{s}_f^0\right) + K_l\left(s_f - s_f^0\right) = -\left(F_l - F_l^0\right) \tag{12}$$

where superscript "0" denotes the desired (reference) value, M_l, B_l, K_l are the desired mass, damping ratio and stiffness of the swinging leg, and F_l is the resultant ground reaction force at the considered leg (i.e. foot link). In order to make it possible to shift the load from the one leg to another one, as the second leg gradually takes the weight of the biped robot mechanism in the WAP, reference forces/torques $F_l^0 = \left[F_{l,x}^0\ F_{l,y}^0\ F_{l,z}^0\ 0\ 0\ 0\right]^T$ at the foot sole is selected. In the SPs the contact force $F_{l,z}^0$ is set to zero, while in the WAP and WSP it varies its amplitude. The vertical ground reaction force $F_{l,z}^0$ is assumed to have a trapezoidal profile, while the longitudinal $F_{l,x}^0$ and lateral $F_{l,y}^0$ components depend on the estimated friction between foot sole and support surface. The referent torques (roll, pitch and yaw) at the foot can be set to zero.

During the SP, the impedance of the tip of the swinging foot depends on the entire kinematical chain, which passes through the hip and terminates at the ground contact point of the support foot. In order to achieve the desired impedance in (12) foot acceleration should be:

$$\ddot{s}_f = \ddot{s}_f^0 - M_l^{-1}\left[B_l\left(\dot{s}_f - \dot{s}_f^0\right) + K_l\left(s_f - s_f^0\right) + \left(F_l - F_l^0\right)\right] \tag{13}$$

Having in mind the non-linear relation (7), which maps the leg coordinates q_l from joint space to the Descartes coordinates of the corresponding foot $\underline{s}_f = \left[x_f\ y_f\ z_f\ \varphi_f\ \theta_f\ \psi_f\right]^T$ in task space, then (7) can be re-written in a form:

$$\ddot{q}_l = J_l^{-1}(q_l, d)\cdot\left[\ddot{\underline{s}}_f - \ddot{\underline{X}} - \dot{J}_l(q_l, d)\dot{q}_l\right] \tag{14}$$

The explicit relation between the leg impedance and the corresponding joint coordinates is possible to be derived by combination of (13) and (14). The resulting relation has the form:

$$\ddot{q}_l = J_l^{-1}(q_l, d)\cdot\left\{\ddot{s}_f^0 - M_l^{-1}\left[B_l\left(\dot{s}_f - \dot{s}_f^0\right) + K_l\left(s_f - s_f^0\right) + \left(F_l - F_l^0\right)\right] - \ddot{\underline{X}} - \dot{J}_l(q_l, d)\dot{q}_l\right\} \tag{15}$$

The control vector of joints' accelerations for the right leg $\ddot{q}_{rl}^*$ or the left leg $\ddot{q}_{ll}^*$ (valid for SPs and WAPs), that produce the imposed leg impedance (12), is calculated from (15). In that case, $\ddot{q}_{rl}^* \equiv \ddot{q}_l$, i.e. $\ddot{q}_{ll}^* \equiv \ddot{q}_l$. Similarly, combining the relations (11) and (14), the new relation can be derived as:

$$\ddot{q}_l = J_l^{-1}(q_l, d)\cdot\left\{\ddot{\underline{s}}_f - \left[\ddot{\underline{X}}_0 - M_b^{-1}B_b\left(\dot{\underline{X}} - \dot{\underline{X}}_0\right) - M_b^{-1}K_b\left(\underline{X} - \underline{X}_0\right)\right] - \dot{J}_l(q_l, d)\dot{q}_l\right\} \tag{16}$$

The control vector of joints' accelerations for the right leg $\ddot{q}_{rl}^*$ or the left $\ddot{q}_{ll}^*$ (valid for WSPs), that produce the imposed hip link impedance (10), is calculated from (16).

Finally, when the control accelerations $\ddot{q}_{rl}^*$ and $\ddot{q}_{ll}^*$ in the vector (9) are determined (by relations (15) and (16)), providing the desired hip link and leg impedance of the biped robot, the rest of the control accelerations in the control law (8) should be determined, too. Bearing in mind that the upper body (trunk) of biped robot and robot arms have to ensure the accurate positioning (speed) and posture stability, necessary for task performance and advance vision, the control accelerations $\ddot{q}^*_{tr}$, $\ddot{q}^*_{ra}$ and $\ddot{q}^*_{la}$ can be determined as corresponding PD-regulators in a form:

$$\ddot{q}^*_{ch} = \ddot{q}^0_{ch} - K_d\left(\dot{q}_{ch} - \dot{q}^0_{ch}\right) - K_p\left(q_{ch} - q^0_{ch}\right) \tag{17}$$

where the subscript "*ch*" denotes one of the following kinematical chains: trunk "*tr*", right arm "*ra*" or left arm "*la*". K_p and K_d are corresponding positional and differential control gain matrices of dimension 6×6. The choice of the control gains K_p and K_d depends on desired characteristics of the system (stability, time of response, frequency, etc.), used actuators for biped displacements, size of the robot, etc. The structure (size), mass and inertia moments of a biped robot system determines its frequency characteristics. In that sense, the control gains in (17) could be chosen in such a way to ensure system to be stable and to avoid the resonant frequency. That is accomplished by appropriate setting of the frequency ν [*Hz*] and the relative damping factor ζ of the regulator. Bearing in mind the previously said, the control gains are calculated as to follow:

$$\begin{aligned} &K_p = diag\left\{k_p^i\right\}, \quad k_p^i = \omega_n^{i^2}, \quad i = 1,\ldots,6 \\ &K_d = diag\left\{k_d^i\right\}, \quad k_d^i = 2\zeta^i \omega_n^i, \quad i = 1,\ldots,6 \\ &\zeta^i = const, \quad \nu^i = const, \quad \omega_n^i = 2\pi\nu^i \end{aligned} \tag{18}$$

Set of parameters, including the basis (hip) link impedance (M_b, B_b, K_b) and the swing leg impedance (M_l, B_l, K_l) parameters defined in (10) and (12), is determined using relations (19). Inertial (mass) parameter matrices of the hip and leg impedance are assumed constant, and depend on the lump mass of biped mechanism. Corresponding particular damping ratio and stiffness matrices can be calculated from the relations:

$$\begin{aligned} &K_b = M_b \cdot \omega_b^2 = const, \quad B_b = M_b \cdot 2\zeta_b \cdot \omega_b = const, \\ &\omega_b = 2\pi\nu_b, \zeta_b = const, \nu_b = const \\ &K_l = M_l \cdot \omega_l^2 \cdot diag\left\{k_{l,1} \ldots k_{l,6}\right\}, \quad B_l = M_l \cdot 2\zeta_l \cdot \omega_l \cdot diag\left\{b_{l,1} \ldots b_{l,6}\right\}, \\ &\omega_l = 2\pi\nu_l, \zeta_l = const, \nu_l = const; \end{aligned} \tag{19}$$

where $k_{l,i}, b_{l,i}$ are so called the *adjustment factors* (for $i = 1,\ldots,6$ particular directions of motion) used for modulation of leg impedance parameters B_l, K_l defined in (19); ς_b, ς_l are damping factors and ν_b, ν_l are corresponding frequencies of the controller. Bearing in mind that walking conditions and gait parameters can vary significantly in real-time, impedance parameters could be varied according to the real conditions as well as according to the biological principles considered in the previous section.

5.2 Impedance Modulation

Humans adapt leg impedance parameters in real-time during a walk by changing corresponding impedance characteristics in the range from the low values (corresponding to the slack muscles) to the high values (corresponding to the stiff muscles). Magnitudes of the leg impedance parameters (B_l, K_l) depend on actual walking conditions (ground surface characteristics, compliance of the support, foot-ground contact characteristics, etc.) as well as on gait parameters (speed, step size, lifting height of foot, etc.). The leg impedance parameters can be modulated by an appropriate variation of the corresponding adjustment factors $k_{l,i}, b_{l,i}$ given in (19). The adaptive knowledge-based algorithm for leg impedance modulation was designed based on the experimental results obtained by motion capture of human gait as well as by the complementary simulation tests with the 36 DOFs model of biped robot gait. Aimed to this goal, the gait speed is varied in the range $V \in [0.5, 2.0]$ [*m/s*], step size within $s \in [0.30, 0.85]$ [*m*] and lifting height of the swing foot within the range $h_f = [0.05, 0.40]$ [*m*]. Compliance of support surface (stiffness coefficient) is varied, too. Three characteristic compliant parameters are considered as representatives: (i) stiff one $k_{z0} = 10^5$ [*N/m*], (ii) moderately compliant $k_{z0} = 4 \cdot 10^4$ [*N/m*], and (iii) compliant $k_{z0} = 10^4$ [*N/m*]. Experimentally acquired knowledge (concerning the principles of impedance modulation) enables building of the corresponding empirical algorithm for impedance parameters modulation. Flow-chart of the bio-inspired algorithm for real-time impedance modulation is presented in Fig. 5.

The procedure for identification of the adaptive (time-variable) impedance parameter values $K_l(t), B_l(t)$ is performed in two stages. First, determination of the constant-value adjustment factors $b_{l,i}^o, k_{l,i}^0$ (*i=1,…,6*) is done. Obtained constant-value factors $b_{l,i}^o, k_{l,i}^0$ can be assumed as the quasi-optimal, non-adaptive values due to the fact that they ensure the best possible performances to be achieved with implementation of constant-value adjustment factors. The identified set of adjustment factors provides a kind of satisfactory dynamic performances of biped system for variable walking conditions. Variable adjustment factors $k_{l,i}, b_{l,i}$ have to be identified in the second stage of the tuning procedure to ensure high dynamic performances of the robot. The values of the variable adjustment factors should be sought in the vicinity of the constant-value $b_{l,i}^o, k_{l,i}^0$ factors. Real-time modulation of the adjustment factors $k_{l,i}(t), b_{l,i}(t), t \in (0, \infty]$ ensures adaptive bio-inspired control of biped robot locomotion in a way similar as the humans do it. Both stages of the procedure for tuning algorithms for impedance modulation is performed beyond plenty of experimental and simulation tests seeking for the best system performances through one iterative procedure. For this purpose, different qualitative and quantitative criteria for assessment of the results, achieved by implementation of the developed adaptive impedance control algorithm, are introduced such as: accuracy of foot trajectory tracking Δs_f, posture stability in sense of tracking accuracy of the hip link cycloids Δs_h, robustness to external perturbation and environment uncertainties, criterion of the minimal peak impact force amplitudes $F_{l,z}$, minimal ground reaction force deviation from the reference values $e_F = F_{l,z} - F_{l,z}^0$, criterion of low joint torque amplitudes τ_l and energy consumptions related to the leg joints, etc. The established criteria mainly concern with stability, dynamic performances and energy issues of the control system assessment.

Concerning the first stage of the procedure for identification impedance adjustment factors, the best setup of the constant-value adjustment factors $b_{l,i}^{o} > 1, k_{l,i}^{0} > 1$ (for *i=1,…,6*) are identified for the 'moderate-case' locomotion characterized by the gait speed $V_{MC}^{ref} = 1.00\ m/s$, foot lifting height $h_{f}^{ref} = 0.15\ m$, landing speed of the swing foot $\dot{z}_{f}^{ref} = -1\ m/s$, and surface compliance of $k_{z0}^{ref} = 4 \times 10^{4}\ N/m$ stiffness coefficient. The identified constant-value adjustment factors $b_{l,i}^{o}, k_{l,i}^{0}$ enable calculation of the so called *basic impedance parameter setup* B_{l}^{0}, K_{l}^{0}, by applying the relation (19). The basic parameter setup B_{l}^{0}, K_{l}^{0} represents a reference set of parameters to be used for assessment of biped system performances obtained by implementation of the adaptive impedance control algorithm.
Impedance parameters $K_{l}(t), B_{l}(t)$ of biped legs should be changed during walk from phase to phase (SP, WAP and WSP) as well as to be varied continually in real-time within a particular phase. It was experimentally discovered that a biped robot has to change its leg impedance in the swing phase depending on: (i) landing speed $\dot{z}_{f}^{\min}$ of the swing foot, (ii) forward speed of robot V_{MC}, and (iii) estimated value of the foot-ground compliance. In that sense, two characteristic indicators are introduced: (i) rate indicator $\eta = V_{MC} / V_{MC}^{ref}$ that takes into account the relationship between the actual forward speed of biped mechanism V_{MC} and the reference speed V_{MC}^{ref} that corresponds to $V_{MC}^{ref} = 1.00\ m/s$ assumed as the moderate value; (ii) compliance indicator κ that represents a quotient $\kappa = k_{z0} / k_{z0}^{ref}$ of the foot-ground contact stiffness k_{z0} and the reference stiffness k_{z0}^{ref} that corresponds to the assumed moderately rigid surface $k_{z0}^{ref} = 4 \times 10^{4}\ N/m$; Additionally, it was explored that the damping adjustment factor $\widetilde{b}_{l}$ of a biped robot leg should to be changed according to the cubic polynomial $P(\dot{z}_{f}^{\min}) = -0.172\left|\dot{z}_{f}^{\min}\right|^{3} + 1.120\left|\dot{z}_{f}^{\min}\right|^{2} + 2.570$ identified empirically.
At the end of the SP as well as during WAP, humans instinctively strengthen leg muscles to suppress the impact. As large as the landing speed amplitude $\dot{z}_{f}$ of the foot is, the foot impact is more powerful and vice versa. Impact forces at the swing foot can be absorbed significantly by enlargement of the appropriate leg impedance parameters (e.g. damping coefficient B_{l}). During an impact, adjustment factor $\widetilde{b}_{l}$ changes its value according to the empirical rules given by the algorithm presented in Fig. 5. The magnitude of the damping adjustment factor $\widetilde{b}_{l}$ can be multi-enlarged (depending on actual walking conditions) in comparison to the value to be applied during the WSP phase. It is proportional to the landing speed quotient $\dot{z}_{f} / \dot{z}_{f}^{ref}$ (where $\dot{z}_{f}^{ref} = -1\ m/s$) as well as inverse-proportional to

the compliance indicator $\kappa = k_{z0} / k_{z0}^{ref}$. During the SP and WAP the stiffness adjustment factor of leg impedance is kept invariable, i.e. $\widetilde{k}_l = k_l^0 = const$.
During a WSP, biped robot behaves as an inverted pendulum. A body mass displacement happens in this phase, causing variation of the corresponding ground reaction forces at the foot sole. Consequently, corresponding impedance adjustment factors $k_{l,i}, b_{l,i}$ should be determined with respect to the contact force magnitudes $F_{l,z}$ as well as to their rates of changing $\dot{F}_{l,z}$. Algorithm uses force error $e_F = F_{l,z} - F_{l,z}^0$ as well as rate of force error $e_{dF} = \dot{F}_{l,z} - \dot{F}_{l,z}^0$ as the indicators necessary for stiffness adjustment factor $\widetilde{k}_l$ and damping adjustment factor $\widetilde{b}_l$ modulation (Fig. 5). Depending on the signs (less/equal/greater than zero) of e_F and e_{dF} errors, the proposed empirical algorithm calculates corresponding particular stiffness adjustment factors $\widetilde{k}_l^I$ and $\widetilde{k}_l^{II}$. The valid value takes one that satisfies criterion given by the relation $\left|\tilde{k}_l^{II} - k_l^0\right| > \left|\tilde{k}_l^I - k_l^0\right|$, (see Fig. 5). Identified stiffness adjustment factor $\widetilde{k}_l$ takes a value within the range $\tilde{k}_l \in \left[k_l^{min}, k_l^{max}\right]$. Damping ratio and stiffness factor are related each other according (19). Changes of stiffness adjustment factor $\widetilde{k}_l$, as one of leg impedance characteristics, causes consequently changes of the damping ratio factor $\widetilde{b}_l$ as presented in Fig. 5. When the amplitude of the ground reaction force $F_{l,z}$ in z-direction increases over its reference value $F_{l,z}^0$ (i.e. $e_F = F_{l,z} - F_{l,z}^0 > 0$) then the leg muscles should be relaxed. Accordingly, stiffness of robot leg mechanism should be decreased proportionally and vice versa for $e_F = F_{l,z} - F_{l,z}^0 < 0$.

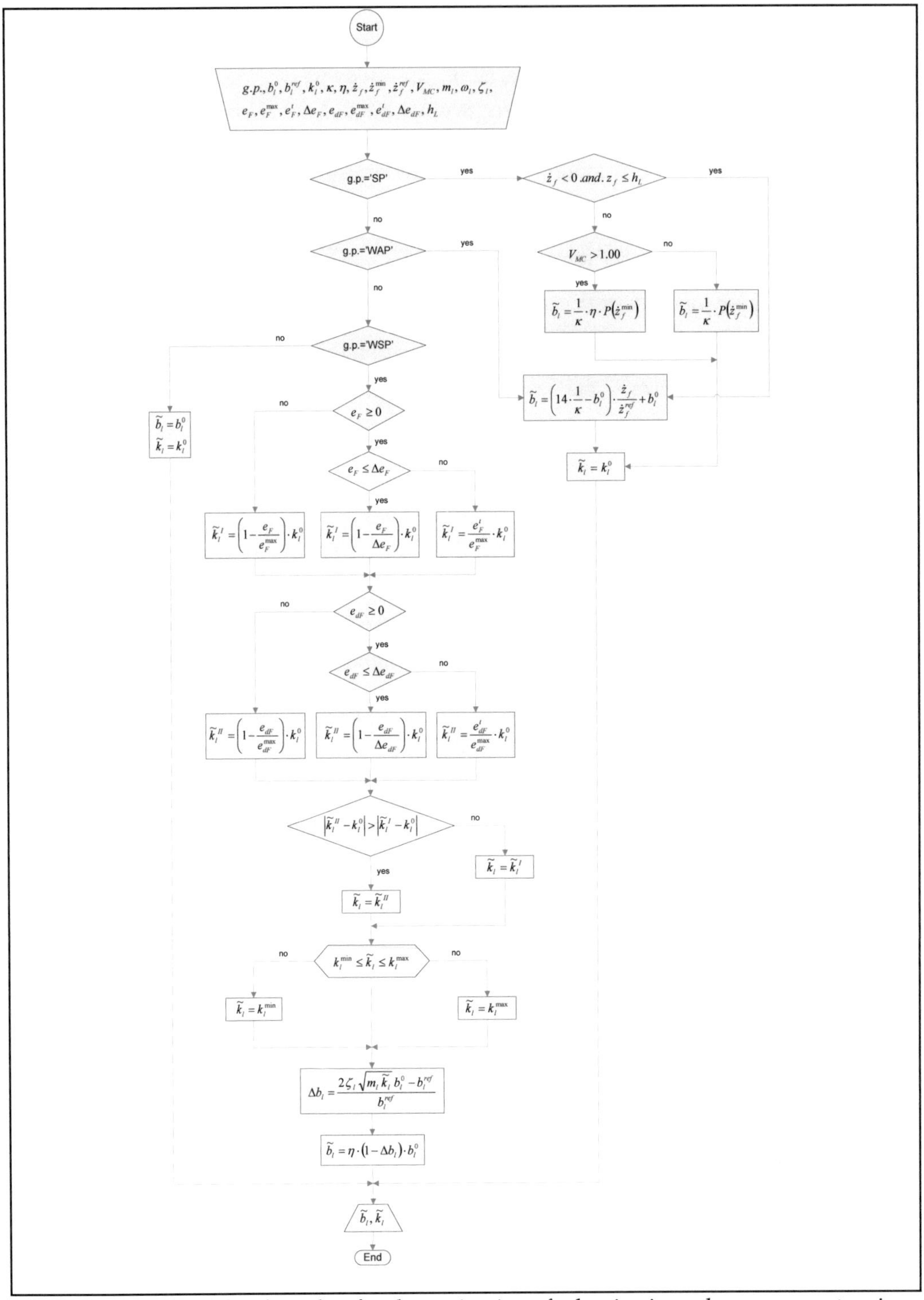

Fig. 5. Flow-chart of the algorithm for determination of adaptive impedance parameters in a way that emulates natural leg impedance modulation with human beings

The following signatures in the flow-chart diagram (Fig. 5) are used: "*g.p.*" is the acronym for gait phase; m_l represents leg mass as impedance parameter; $\zeta_l = 1.00$ is the assumed damping ratio of a leg mechanism; $\nu_l = 3.00\,Hz$ is the assumed frequency of a leg mechanism; $\omega_l = 2\pi\nu_l$ is the angular frequency of a leg mechanism; $b_l^{ref} = m_l\, 2\zeta_l\, \omega_l\, b_l^0$ is the reference damping ratio including the basic damping adjustment factor b_l^0 ; $\dot{z}_f^{\min}$ is the extreme negative landing foot speed (e.g. $\dot{z}_f^{\min} \sim -1.70\ m/s$); $h_L = 0.01 m$ is the assumed landing height threshold; $e_F^{\max}$, e_F^t are the maximal force error (assumed to be as large as body weight) and corresponding threshold of sense (assumed to be *25 %* of $e_F^{\max}$) successively; and $e_{dF}^{\max}$, e_{dF}^t are the maximal rate of contact force error (assumed to be *7000 N/s*) and the corresponding threshold of sense (assumed to be *25 %* of the $e_{dF}^{\max}$ value).
Adaptive impedance control algorithm presented in this section will be tested and verified through extensive simulation experiments under different real walking conditions. The results of evaluation of the proposed control strategy are presented in Section 7.

6. Simulation Experiments

Biped robot locomotion is simulated to enable evaluation of control system performances. For this purpose, the spatial 36 DOFs model of biped robot (presented in Fig. 2) and adaptive impedance control (defined by relations (8)-(17)) are simulated by implementation of the HRSP software toolbox (Rodić, 2009). Biped robot parameters used in simulation experiments are specified in (Rodić, 2008). Parameters of a 3D compliant model of robot environment, applied in simulation tests, are assumed as in (Park, 2001). Concerning the leg impedance parameters modulation, three qualitatively different cases are considered and evaluated by the simulation tests: (i) non-adaptive control "NA" with invariable leg impedance parameters (case without modulation of parameters), (ii) quasi adaptive "QA", switch-mode impedance modulation depending on particular gait-phases, and (iii) continual adaptive "AD" real-time impedance modulation. Concerning the first case, inertial impedance matrices defined in (10) and (12) are imposed as the constant 6×6 diagonal matrices $M_b, M_l = \Re^{6\times 6} = const$ that have the values $M_b = diag\{40\}$ and $M_l = diag\{10\}$. Particular damping ratio and stiffness coefficient of the hip link from (19) are assumed to be constant-value matrices, too. The following values $B_b = diag\{1508\}\,[Ns/m]$ and $K_b = diag\{14212\}\,[N/m]$ (when choose $\zeta_b = 1$ and $\nu_b = 3\,[Hz]$) are assumed. Their values are kept invariable in the simulation experiments. The leg impedance parameters K_l and B_l from (19) are variable in general case. They are determined on-line by use of the algorithm presented in Fig. 5. Exception of that is in the cases when choose the constant leg impedance in some of the particular simulation tests, i.e. when the adjustment factors from (19) use to have the constant unit values $b_{l,i} = 1,\ k_{l,i} = 1,\ i = 1,\ldots,6$. Then, assuming $\zeta_l = 1$ and $\nu_l = 3\,[Hz]$, the following constant-value impedance parameters are obtained to have the values $B_l = diag\{377\}\,[Ns/m]$ and $K_l = diag\{3553\}\,[N/m]$. Control gain matrices K_p and K_d of the PD regulator (17) of trajectory tracking are imposed to have the invariant values $K_p = diag\{1421\}\,[s^{-2}]$ and $K_d = diag\{76\}\,[s^{-1}]$ assuming that $\nu^i = 3.00$ [Hz] and $\zeta^i = 1.00$. Imposed gain matrices are used in simulation experiments as the invariable constant-value matrices.

In order to evaluate the control performances of the new-proposed, adaptive impedance algorithm, three different simulation examples are considered in the paper depending on type of control algorithms applied: NA, QA or AD as explained in the previous paragraph. The simulation examples are performed under the same simulation conditions such as: (i) walking on a flat surface, (ii) compliant, moderate rigid ground surface with $k_{z0} = 4\times10^4$ $[N/m]$ stiffness coefficient, and (iii) a moderate fast gait including forward gait speed of $V = 1\,[m/s]$, step size of $s = 0.7\,[m]$ and swing foot lifting height of $h_f = 0.15\,[m]$. Biped robot locomotion in the simulation examples (cases) is checked on stability, quality of dynamic performances and other relevant performance criteria such as: accuracy of trajectory tracking (hip link and foot trajectories), maximal amplitudes of ground reaction forces and joint torques, energy efficiency, anthropomorphic characteristics, etc. Simulation results obtained in three verification simulation tests are mutually compared in order to assess the quality of the applied control strategies. Simulation examples presented in Fig. 6 prove that the best dynamic performances, regarding to the smoothness of the realized ground reaction forces during a flat gait, are obtained in the Case "AD" (adaptive control). That is the example when the continual leg impedance parameter modulation was applied. The worst performances were obtained as expected in the Case "NA", where there is no adaptive control and when the impedance parameters have exclusively the constant, non-adaptive values. The Case "QA" regards to the quasi-adaptive switch-mode modulation of leg impedance parameters. It provides a moderate quality of system performances. The minimal force peak amplitudes and deviations from the referent values are in the Case "AD". The peaks are well damped by introducing of the continually modulated impedance, with an on-line modulation of the leg stiffness and damping ratio. The results presented in Fig. 6 prove that the new-proposed bio-inspired algorithm of adaptive impedance with continually modulated leg stiffness and damping ensures the best system performances with respect to other candidates - Case "NA" and Case "QA". The labels in the figure marked as "rf" and "lf" will be used in the paper to indicate the right, i.e. the left foot.

Concerning the criterion of the accuracy of trajectory tracking, the following results are obtained and described in the text to follow. In the Case "NA", when the control (impedance) parameters take the constant values then the biped robot performs trajectory tracking in a rather poor way. The better results are obtained when implement the switch-mode impedance modulation (Case "QA"). But, the best accuracy of tracking is achieved by implementation of the continually modulated impedance parameters (Case "AD") as presented in Fig. 7. In this case, the swing foot lifting height is maintained almost constant all the time at the level of $h_f = 0.15\,m$.

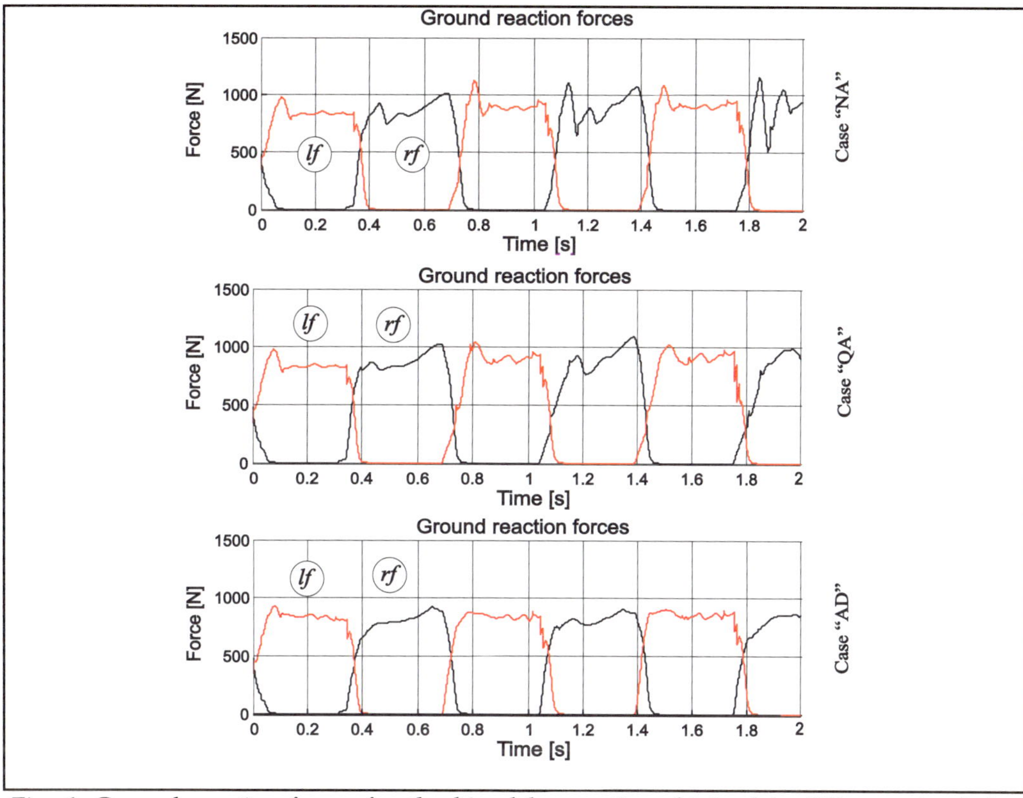

Fig. 6. Ground reaction forces for the biped locomotion obtained in different simulation examples - comparison of Cases "NA", "QA" and "AD"

The heel and toe cycloids are performed regularly and without significant deviations. That guarantees a fine locomotion and desired accuracy. Control algorithm with the continual impedance modulation of leg stiffness and damping ratio ensures that there is no foot bouncing from the ground support during locomotion (Case "AD", Fig. 7). In the Case "QA", a slight bouncing exists as presented in Fig. 7, Detail "A". Suppression of the foot bouncing is very important for the system performances, because the bouncing feet can cause system instability. Complementary to the previous results, the quality of trajectory tracking of the hip link as well as right foot of biped robot in different coordinate directions is presented in the phase-planes in Fig. 8. In both cases "QA" and "AD", a stable walking is ensured since the actual trajectories (hip and foot trajectories) converge to the referent trajectories (Fig. 8).

Although there is a certain delay in velocity tracking in the sagital direction, the foot centre tracks its nominal (referent) cycloid in a satisfactory way in the case when the continual modulation of leg impedance (Case "AD") is applied. The delay is more expressed (larger) in the case when the switch-mode modulation (Case "QA") is applied (Fig. 8). Better tracking of the foot and the hip link trajectories (in the vertical direction) are appreciable, too. The trajectory of the hip link as well as the foot centre trajectory converges to the referent cycles (Fig. 8) as locomotion proceeds. The convergence is better in the Case "AD" than in the Case "QA". That proves the advantage of the new proposed control algorithm in the sense of achievement of a better accuracy of locomotion.

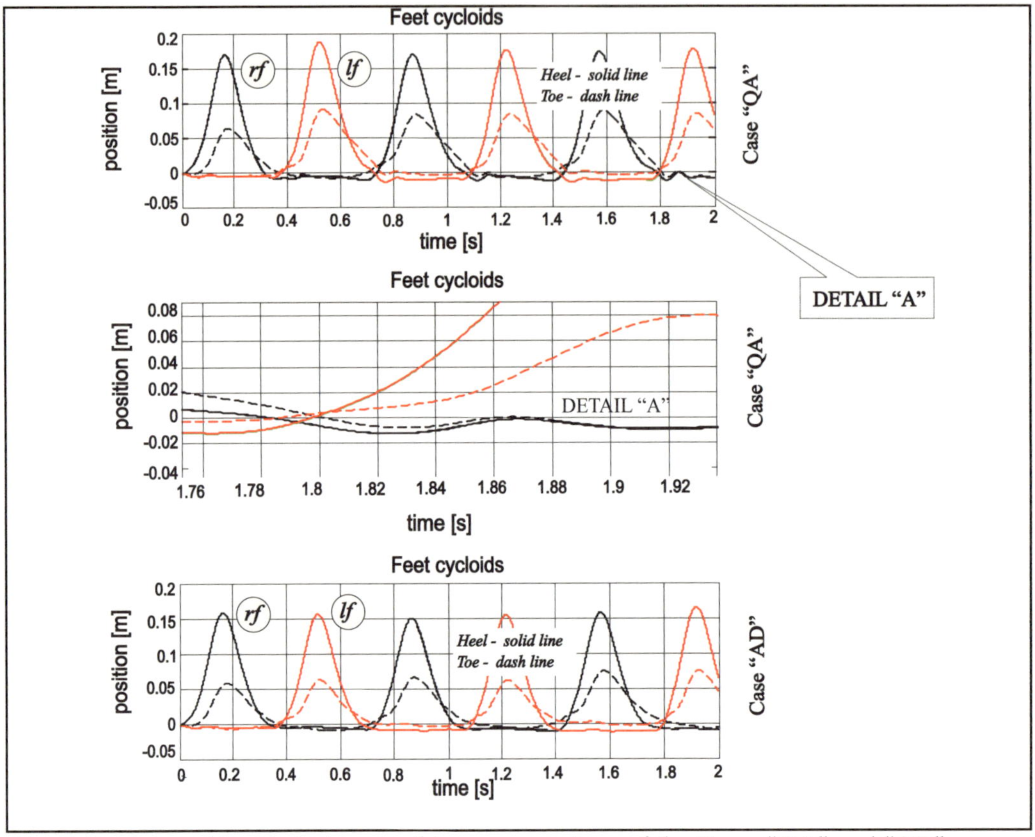

Fig. 7. Precision of foot trajectory tracking - comparison of the Cases "QA" and "AD"

The stiffness and damping ratio adjustment factors $k_{l,i}$ and $b_{l,i}$, defined in (19), are presented in Fig. 9. In the Case "QA", stiffness factor is kept constant $k_{l,i} = 1$ while the damping ratio factors (for the both legs) vary from gait phase to gait phase. In the case of continual modulation (Case "AD"), stiffness factors as well as damping ratio factors change their amplitudes as presented in Fig. 9. Variable adjustment factors provide better adaptation of the biped robot system to the variable gait as well as to different environment conditions. The hip joints as well as the knee joint endure the most effort to adapt biped gait to the actual conditions. In that sense, especially critical moments represent moments of foot impacts. Bearing in mind this fact, the variable leg impedance enables biped system to prevent enormous impact loads and serious damages of its leg joints.

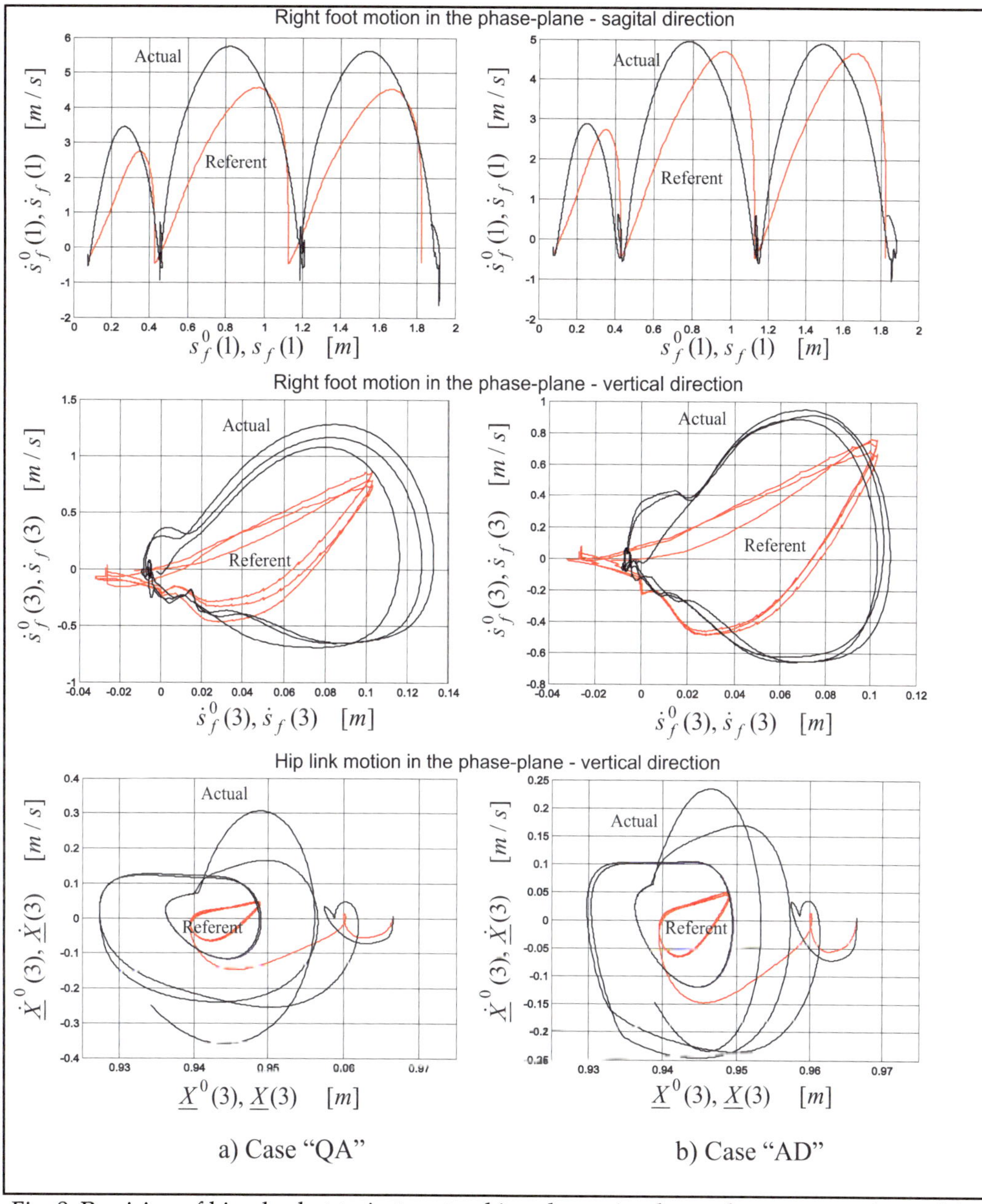

Fig. 8. Precision of biped robot trajectory tracking shown in phase-plane – comparison of the Case "QA" and Case "AD"

Performances of the adaptive impedance control with continually modulated leg impedance parameters can be validated by the analysis of some additional numerical indicators, too. The appropriate criteria indicators, related to the extreme dynamic reactions as well as to the energy efficiency, are imposed such as: (i) the relative average magnitude of the ground reaction forces' deviation η_F with respect to the referent "NA" case; (ii) extreme relative peak amplitude of dynamic reactions η_{Peak} with respect to the "NA" case, and (iii) indicator of the relative energy efficiency η_E with respect to the referent "NA" case. Systematized indicators of performance quality are shown in Tab.1. According to this table, it is evident that an adaptive control with real-time modulation of leg impedance parameters ensures significantly better characteristics than non-adaptive and quasi-adaptive cases of control.

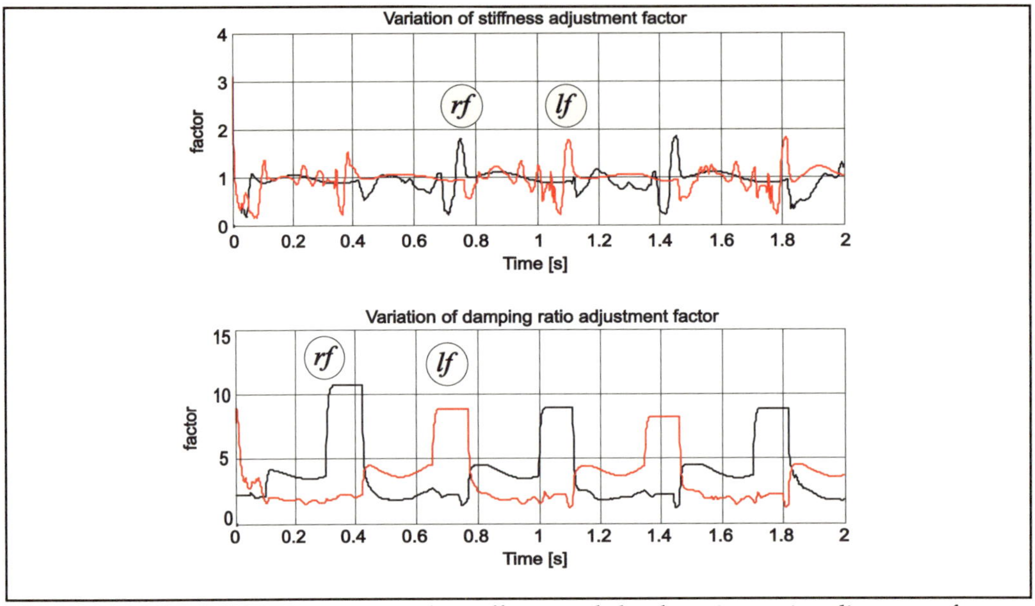

Fig. 9. Leg impedance modulation - the stiffness and the damping ratio adjustment factors for the Case "AD"

Comparative pair	η_F %	η_{Peak} %	η_E %
QA : NA	-23.82	-8.82	-1.13
AD : NA	-66.45	-47.05	-30.93

Table 1. Table of criteria indicators depicting the quality of control performances against the indices of dynamic reactions deviations, extreme payload and energy efficiency

7. Conclusion

Stable and robust walking on irregular surfaces and compliant ground support as well as walking with variable gait parameters request advanced control performances of biped robots. In general case, walking conditions are unknown and cannot be anticipated confidently in advance to be used for trajectory generation. As consequence, path generator produces biped trajectories for non-perturbed locomotion such as: flat gait, climbing stairs, spanning obstacles, etc. In the case of a perturbed locomotion, robot controller is charged to manage the dynamic performances of the system and to maintain dynamic balance. In that sense, we speak about the robustness of control structure to the gait parameters variation as well as to the external perturbations concerning uncertainties (structural and parametric) of

the ground support. Bearing in mind previous facts, promising control architecture capable to cope with the fore mentioned uncertainties is the adaptive impedance control with continually modulated impedance parameters.
Main contribution of the article is addressed to a synthesis of the bio-inspired, experimentally-based, adaptive control of biped robots. Aimed to this goal, the adaptive bio-inspired algorithm designed for real-time modulation of leg impedance parameters are proposed in the paper. Proposed control structure is robust to variation of gait parameters as well as uncertainties of the ground support structure. The proposed control algorithm was tested through the selected simulation experiments to verify the obtained control system performances. Developed control algorithm is valid and can be applied for control of any biped robot of anthropomorphic structure regardless to its size, kinematical and dynamic characteristics. It was proved through the simulation experiments that the biological principles of leg impedance modulation are valid with artificial systems such as biped robots, too.

8. References

Bruneau, O.; Ouezdou, ben F.; Wieber, P. B. (1998). Dynamic transition simulation of a walking anthrpomorphic robot, *Proceedings of IEEE International Conference on Robotics and Automation*, pp. 1392-1397, May, Leuven, Belgium

Dalleau, G.; Belli, A.; Bourdin, M.; Lacour, J-R. (1998). The spring-mass model and the energy cost of treadmill running. *European Journal on Applied Physuiology*, Springer-Verlag, Vol. 77, pp. 257-263

Dalleau, G.; Belli, A.; Bourdin, M.; Lacour, J-R. (2004). A Simple Method for Field Measurements of Leg Stiffness in Hoping, *International Journal on Sports and Medicine*, Georg Thieme Verlag Stuttgart, Vol. 25, pp. 170-176

De Leva, P. (1996). Adjustments to Zatsiorsky-Seluyanov's segment Inertia Parameters. *Journal of Biomechanics*, Vol. 29, No. 9, pp. 1223-1230

Fujimoto, Y.; Kawamura, A. (1995). Three dimensional digital simulation and autonomous walking control for eight-axis biped robot, *Proceedings of IEEE International Conference on Robotics and Automation*, pp. 2877-2884, May, Nagoya, Japan

Fujitsu HOAP-3 bipedal robot. (2009). http://www.techjapan.com/Article1037.html

Hogan, N. (1986). Impedance control: An approach to manipulation, Part I-III. *Journal of Dynamic Systems, Measurements and Control*, Vol. 107, pp. 1-24

Honda humanoid robots. (2009) http://world.honda.com/ASIMO/

Kim, J. H.; Oh, J. H. (2004). Walking Control of the Humanoid Platform KHR-1 based on Torque Feedback, *Proc. of the 2004 IEEE Int. Conf. on Robotics & Automation*, pp. 623-628, Los Angeles, USA

Kraus, P. R.; Kummar, P. R. (1997). Compliant contact models for rigid body collisions. *Proceedings of IEEE International Conference on Robotics and Automation*, April, pp. 618-632, Albuquerque, NM

Leonard, T. C.; Carik, R. L.; Oatis, C. A. (1995). *The neurophysiology of human locomotion in Gait Analysis: Theory and Application*. Eds: Mosby-Year book

Lim H-O.; Setiawan S. A.; Takanishi A. (2004). Position-based impedance control of a biped humanoid robot. *Advanced Robotics, VSP*, Volume 18, Number 4, pp. 415-435

Lim H-O; Setiawan, S. A.; Takanishi, A. (2001). Balance and impedance control for biped humanoid robot locomotion. Department of Mechanical Engineering, Waseda University, Tokyo, *Proceedings 2001 IEEE/RSJ International Conference on Intelligent Robots and Systems*, Vol. 1, pp. 494-499, ISBN: 0-7803-6612-3, October, Maui, HI, USA

Marhefka, D. W.; Orin, D. E. (1996). Simulation of contact using a non-linear damping model, *Proceedings of IEEE International Conference on Robotics and Automation*, pp. 88-102, Minneapolis, USA, April

Ogura, Y.; Aikawa, H.; Shimomura, K.; Morishima, A.; Hun-ok Lim; Takanishi, A. (2006). Development of a new humanoid robot WABIAN-2, *Proceedings* 2006 *IEEE International Conference on Robotics and Automation*, pp. 76 - 81, ICRA 2006, 15-19 May, Orlando, Florida, USA

Park, J. H. (2001). Impedance Control for Biped Robot Locomotion. *IEEE Transactions on Robotics and Automation*, Vol. 17, No. 6, pp. 870-882

Potkonjak, V.; Vukobratović, M. (2005). A Generalized Approach to Modeling Dynamics of Human and Humanoid Motion, *International Journal of Humanoid Robotics*, World Scientific Publishing Company, pp 65-80.

Qrio Sony robot. (2009). http://www.sony.net/SonyInfo/News/Press_Archive/200310/03-1001E/

Rodić, A. (2009). *Humanoid Robot Simulation Platform.* http://www.institutepupin.com/RnDProfile/ROBOTIKA/HRSP.htm

Rodić, A.; Vukobratović, M.; Addi, K.; Dalleau, G. (2008). Contribution to the modeling of non-smooth multipoint contact dynamics of biped locomotion - Theory and experiments. *Robotica*, Cambridge University Press, Vol. 26, pp. 157-175, ISSN: 0263-5747

Rostami, M.; Bessonnet, G. (1998). Impactless sagital gait of a biped robot during the single support phase, *Proceedings of IEEE International Conference on Robotics and Automation*, May, pp. 1385-1391, Leuven, Belgium

Rousell, L.; Canudas de Wit C.; Goswami, A. (1998). Generation of energy optimal complete gait cycles for biped robots. *in Proceedings of IEEE International Conference on Robotics and Automation*, May, pp. 2036-2041, Leuven, Belgium

Sony entertainment robot. (2006). *http://www.tokyodv.com/news/RoboDex2002SDR-3XSonybot.html*, (2006)

Vukobratović, M.; Borovac, B.; Surla, D.; Stokić, D. (1990). *Biped Locomotion - Dynamics, Stability, Control and Application*, Springer-Verlag, Berlin

Vukobratović, M.; Potkonjak, V.; Rodić, A. (2004). Contribution to the Dynamic Study of Humanoid Robots Interacting with Dynamic Environment, *Robotica*, Vol. 22, Issue 4, Cambridge University Press, ISSN: 02 63-5747, pp. 439-447.

Zatsiorsky, V.; Seluyanov, V.; Chugunova, L. (1990). Methods of Determining mass-inertial Characteristics of Human Body Segments. *Contemporary Problems of Biomechanics*, CRC Press, pp. 272-291

16

Dynamic-Based Simulation for Humanoid Robot Walking Using Walking Support System

Aiman Musa M. Omer[1], Hun-ok Lim[2], and Atsuo Takanishi[1]
[1]*Waseda University*
[2] *Kanagawa University*
[1,2]*Japan*

1. Introduction

With the rapid aging of society in recent times, the number of people with limb disabilities is increasing. According to the research by the Health, Labour and Welfare Ministry, Japan, there are around 1,749,000 people with limb disabilities; this accounts for more than half of the total number of disabled people (3,245,000 handicapped people)[1]. The majority of these people suffer from lower-limb disabilities. Therefore, the demands for establishing a human walking model that can be adapted to clinical medical treatment are increasing. Moreover, this model is required for facilitating the development of rehabilitation and medical welfare instruments such as walking machines for assistance or training (Fig. 1(a)). However, experiments that are carried out to estimate the effectiveness of such machines by the elderly or handicapped could result in serious bodily injury.
Many research groups have been studying biped humanoid robots in order to realize the robots that can coexist with humans and perform a variety of tasks. For examples, a research group of HONDA has developed the humanoid robots—P2, P3, and ASIMO [2]. The Japanese National Institute of Advanced Industrial Science and Technology (AIST) and Kawada Industries, Inc. have developed HRP-2P. The University of Tokyo developed H6 and H7, and the Technical University of Munich developed Johnnie. Waseda University developed the WABIAN series that realized various walking motions by using moment compensation. Korea Advanced Institute of Science and Technology (KAIST) also developed a 41-DOF humanoid robot— KHR-2 [3].
The above mentioned human-size biped robots achieved dynamic walking. If these humanoid robots can use rehabilitation or welfare instruments (as shown in Fig. 1(b)), they will be able to help in testing such instruments quantitatively. The main advantages of the human simulator can be considered to be as follows: (1) The measurement of the angle and the torque required at each joint can be measured easily and quantitatively as compared to the corresponding values in the case of a human measurement. (2) Experiments using such robots can help identify leg defects of a human from an engineering point of view. (3) A robot can replace humans as experimental subjects in various dangerous situations: experiments involving the possibility of falling, tests with incomplete prototype instruments, simulations of paralytic walks with temporarily locked joints.

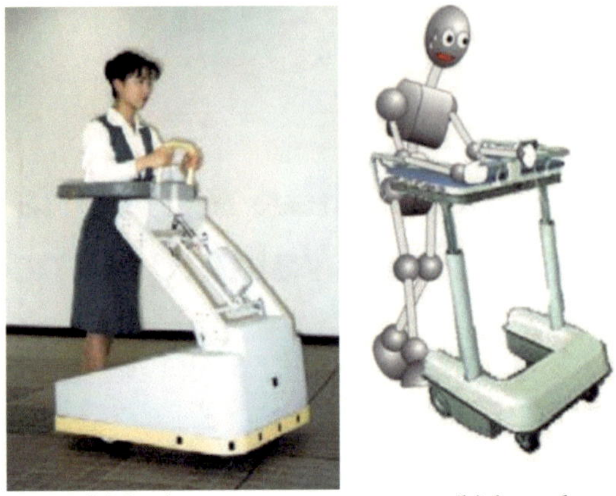

(a) by human (b) by robot

Fig. 1. Walking support system

Such experiments require a humanoid robot that enables it to closely replicate a human. However, humans have more redundant DOFs than conventional biped humanoid robots; this feature enables them to achieve various motions. Therefore, a DOF configuration that is necessary to reproduce such motions is one of the very important issues in the development of a humanoid robot [4].

The Waseda Bipedal Humanoid Robot WABIAN-2R has been developed to simulate human motion. WABIAN-2R performed human-like walking motions (Fig. 2). Moreover, WABIAN-2R achieved to perform walking motion using walk-assist machine. However, the walk-assist machine was freely rolling without activating its wheels motors. In this case, the robot faced the minimum resistance or disturbance case by the walk-assist machine. On the other hand, activating the walk-assist machine may create a large disturbance for robot due to separate control for each of them. Conducting this experiment may be highly risky.

Fig. 2.WABIAN-2R

As we develop humanoid robot to coexist in the human environment, we need to conduct many experiments such as robot walking on uneven surface, climbing the stairs, and robot interact with other machine and instruments. Doing any new type of experiment using WABIAN-2 might be risky. Therefore, we need find a safer method for initial experimental testing. Using a dynamic simulation is useful method due to some reasons such as: (a) It is safer in terms of cost and risk. (b) It is easy to monitor and view motion outputs. (c) It can show the variation cased by any external disturbances. In this paper, a dynamic simulator is described, which is able to easily simulate any new type of walking. Using the dynamic simulator, we can monitor the motion performance and output all needed data that is useful for further development. This paper is aimed to simulate the walking motions of WABIAN-2 using walk-assist machine.

2. Dynamic simulation

Dynamic simulation could be used the purpose of testing and checking the dynamic motion of a mechanical structured model. It has the advantages of saving cost and risk which are highly needed in a development of a mechanical structure. There are many simulation software have been developed for robotics application, mainly for the industrial robot applications. However, there are some software used for mobile robot simulation. For examples, RoboWorks, SD/FAST, OpenHRP, Webots, and Yobotics are used for mobile and legged robot simulation. Webots is high and advanced simulation software used in Robotics simulation. It is use for prototyping and simulation of mobile robots. It has many advanced functions and techniques. Webots is very easy to use and implement. Therefore, we choose it as simulation software for our research [5].

2.1 Modeling

In order to develop a dynamic simulation, we need to go through several steps. First is modeling where we set up the simulation environment and initial parameters. We set up a full structure of WABIAN-2, based on the specifications (size, shape, mass distribution, friction, .etc) of components of WABIAN-2 (Fig. 3).

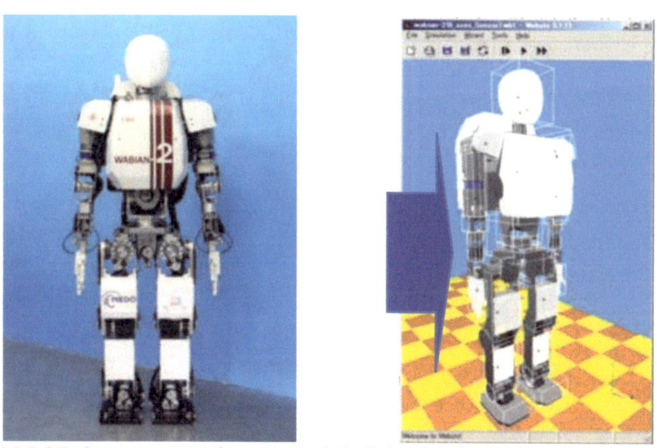

Fig. 3.WABIAN-2R Structure is been modeled in the Simulated World

2.2 Controlling

Second is controlling, which identifies simulation objects and controls the simulation procedures. The controller is some how similar to the WABIAN-2R control. It gets the input data from the CSV pattern file, and sets the position angle of each joint through inverse kinematics techniques. Moreover, the controller sets the simulate time step and the measurement of data.

2.3 Running

The program in the controller section of the simulator will run by going through the main function. There are several steps the controller will go through. First, check the pattern file and prepare to read through the lines. Then read the data from one line. The data is in terms of position and orientation of foots and hands. Using these data we calculate each joint position through inverse kinematics techniques. After that it will set all positions to its joint. The controller runs one control step of 30ms which is similar to the real robot. The controller goes through all the lines in the pattern file until it is completed in the last line.
When the simulation runs it can be viewed the simulation from different view sides. This can gives us a clear idea about the simulation performance. Moreover, most of the needed data could be measured through several functions.

3. Walking with Walking Assist Machine

WABIAN-2 performed some walking experiments using walking assist machine. The performance was conducted by leaning its arms on the walking assist machine holder. The walking assist machine moves passively without generating its own motion. The robot was able to walk and push the walking assist machine forward. The experiments were conducted with different walking styles and different heights of arm rest (Fig. 4).
The walking performance of WABIAN-2 using an active walking assist machine, expected to be unstable. The walk-assist machine has its own control system, not connected to WABIAN-2 control system. The walking assist machine moves with constant velocity in a forward direction, while the robot moves by setting its position. The robot arms may displace from its position on the arm rest of the machine which will case external forces on WABIAN-2. In order to stabilize the walking, the external force has to be minimized.

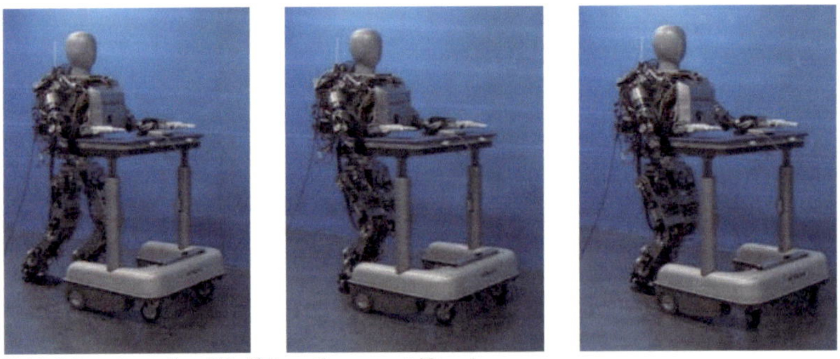

Fig. 4. WABIAN-2 using the Walking Support Device

3.1 Force Sensor

The researches conducted on the walking assist machine are focusing on the relation between the machine and the human user. One of the latest studies promote the idea of measure the forces applied by the user on the machine arm rest. They develop a force sensor in terms of a displacement sensor (Fig. 5). The force sensor is constructed by connecting two flat plates with displacement sensors in between.

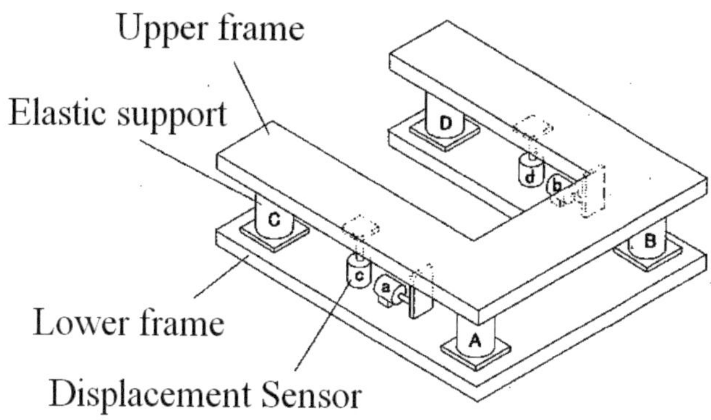

Fig. 5. Force Sensor

This force sensor is attached on top of the arm rest in the walking assist machine. It measures the forces by sensing the amount of displacement measured by the position sensors. The signals generated by the sensor are sent to the controller of the walking assist machine in order to set the velocity and direction of motion.

Measuring forces acting between upper and lower frame are determine through the amount of displacement and orientation between them. Assuming that each frame has its own coordinate system, the displacements in each axis are set as D_x, D_y, and D_z and the orientation around Y axis and Z axis are set as D_{ry}, and D_{rz} (Fig. 6).

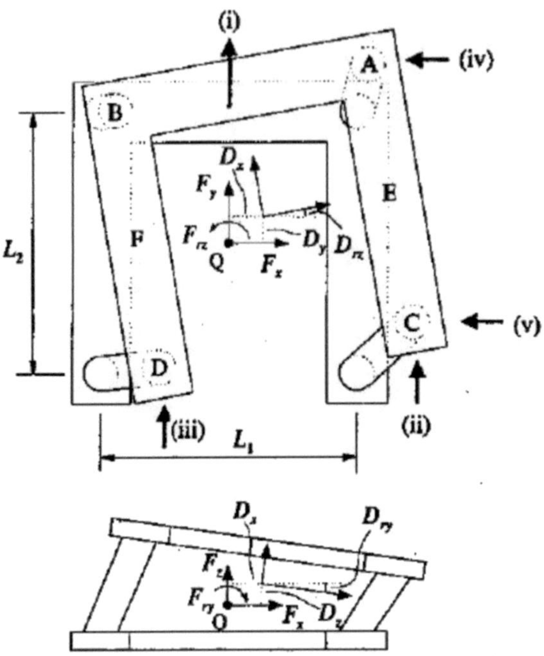

Fig. 6. Definition of Forces and Displacements

The forces are determined through the following equations:

$$F_y = 4\,C_{sx}\,D_y,$$
$$F_z = 4\,C_{sz}\,D_z \qquad (1)$$
$$F_{rz} = (L_1^2 + L_2^2)\,C_{sx}\,D_{rz}$$

Where C_{sx} and C_{sz} are the spring constant of the displacement sensors in horizontal and vertical directions. L_1 and L_2 are the distance between the elastic support in X and Y directions. The output of the displacement sensors a, b, c, and d are set as S_a, S_b, S_c, and S_d. The amounts of displacement are determined through the following equations:

$$S_a = D_y + (L_1/2)\,D_{rz}$$
$$S_b = D_y - (L_1/2)\,D_{rz}$$
$$S_c = D_z - (L_1/2)\,D_{ry} \qquad (2)$$
$$S_d = D_z + (L_1/2)\,D_{ry}$$

Obtaining the previous formulas in (1) and (2) we can define the forces measurement from the displacements as follow:

$$F_y = 2\,C_{sx}\,(S_a + S_b)$$
$$F_z = 2\,C_{sz}\,(S_c + S_d) \qquad (3)$$
$$F_{rz} = (L_1 + L_2^2/L_1)\,C_{sx}\,(S_a - S_b)$$

The forces and torque can be determined from the displacement cased in all the sensors. The amount of the spring constant of the horizontal direction (C_{sx}) is 105 kN/m and for the vertical direction (C_{sz}) is 490 kN/m. The displacement between the upper and lower frame is limit to 500mm to the sides (Right and Left) and 355mm forward and backward.

3.2 Velocity Control

The walking assist machine control system is designed and developed to adjust its speed and direction according to the force applied on the arm rest [7]. The arm rest is designed to measure the force and torques applied by the user of the machine. The controller uses those measure data as an input data to set the velocity of each motor of the machine (Fig. 7). The force f_y and the turning moment m which applied by the arm of the user is calculated in the sensor by the following equations:

$$m_z = m + s_x f_y \tag{4}$$

where m_z is the moment measured by the sensor, s_x the distance shifted from the arm position to the sensor position. The values for m_z and f_y are the input data for the controller that set the velocity of each wheel motor (Fig. 7) [7].
In this study, we have developed the control system model that controls the velocity of the walking assist machine. The system adjusts the velocity according to the force measured by the force sensor. The new adjusted velocity is based on current velocity and the displacement with WABIAN-2.

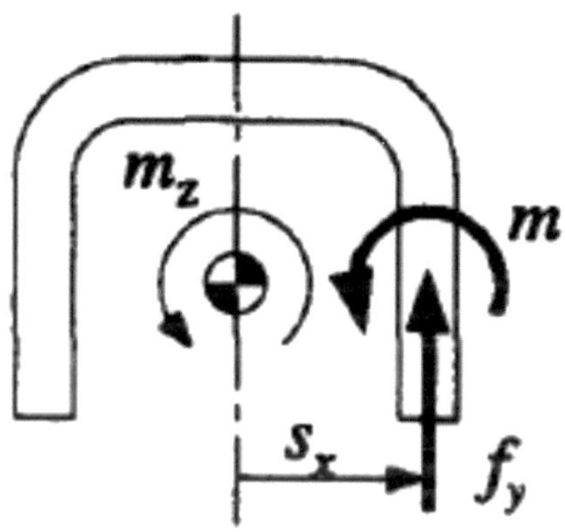

Fig. 7. Force and Moment Applied to Arm rest

Developing the equations of the modeled system, we can have the following equation:

$$F_y = ma \tag{5}$$

where m is the total mass of the walking assist machine, a is the acceleration, and F_y is the force measured by the spring. The force is the result of displacement of the spring mechanism, which can be expressed as

$$F_y = Cx \tag{6}$$

where C is the spring constant and x is the amount of displacement. Substitute equation (3) in (2), we will have

$$a = (C/m)\ x \tag{7}$$

the acceleration is the derivative of velocity. Approximately, it is equal to the difference in velocity over step, which could be express as

$$a(t) = (v\ (t + \Delta t) - v\ (t))\ /\ \Delta t \tag{8}$$

since we are dealing with discrete time, we can rearrange equation (5) to

$$a(k) = (v(k+1) - v(k))\ /\ T \tag{9}$$

where v(k) is the current velocity, v(k+1) is the next velocity, and T is the step time. Substitute equation (4) in (6), we will have

$$v\,(k+1) = (C\,T\,/\,m)\,x(k) + v(k) \qquad (10)$$

where x(k) refer to the displacement measured by the spring of the sensor. The constant value in equation (10) will be considered as the system gain. Therefore, equation (9) can be changed to

$$v\,(k+1) = G \cdot x(k) + v(k) \qquad (11)$$

where G is the control gain. The gain could be adjusted to check the response of the system according to the value set.

4. Simulation Result

Many simulations were conducted to test the walking performance of the robot using the walking support device. In the simulator the control gain of the walking device could be adjusted. The simulation result shows different response from the walking support device to the robot motion. Different control gain values of 1000, 2000, 5000, 7000, and 10000 were set to check the response of the system. Smaller gain value, like 1000 or 2000, result in slow response from the walking support device to the force applied by the robot on the arm rest. However, the robot faced some difficulties to walking with support device due to some differences in velocity between robot and the support device (see Fig. 8), or it can stop when the robot stop due to the slow response (see Fig. 9). On the other hand, when the gain was set to higher value, like 5000, 7000 or 10000, the response of the system get better by having much stable walking motion with the walking support device (see Fig. 10, Fig. 11 and Fig. 12).

The velocity set for each wheel is set according to the force applied by the robot arm on each side of the arm rest. If different amount of forces are applied in each side the velocity of each wheel is different which cases the walking support device make a turn. The device controller can set a high for velocity at the end of the robot walking due to sudden stop of the robot which cases a high force on the arm rest. This amount to high force stops to walking support device with the robot weight which is loaded on top of the arm rest (see Fig. 13, Fig. 14, Fig. 15, Fig. 16, and Fig. 17).

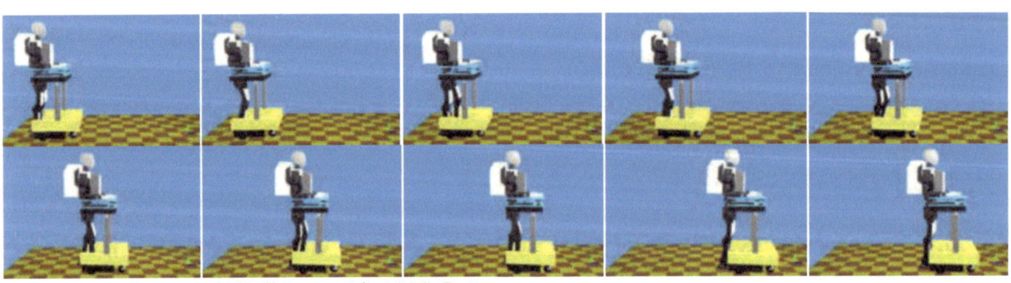

Fig. 8. Simulation of Walking with 1000 Gain

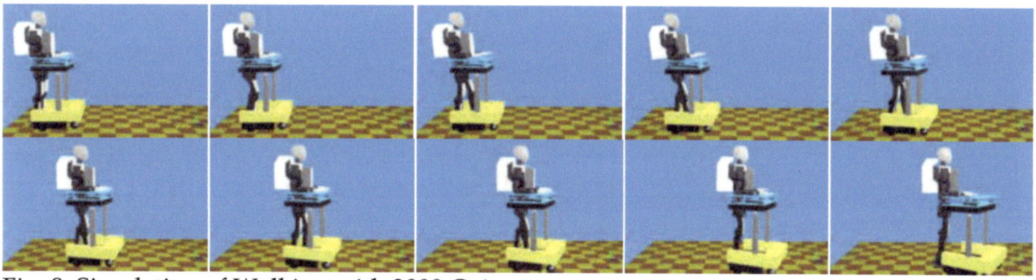

Fig. 9. Simulation of Walking with 2000 Gain

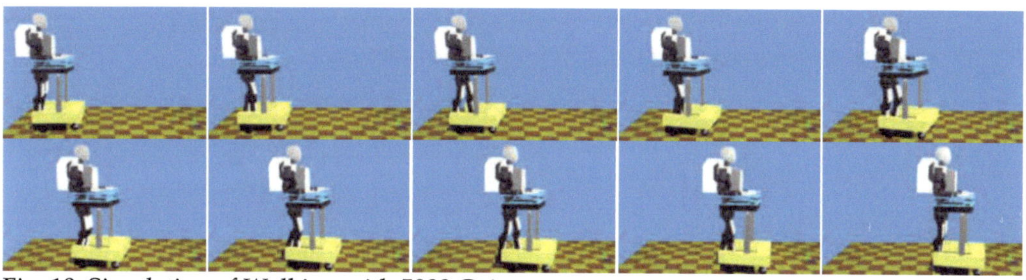

Fig. 10. Simulation of Walking with 5000 Gain

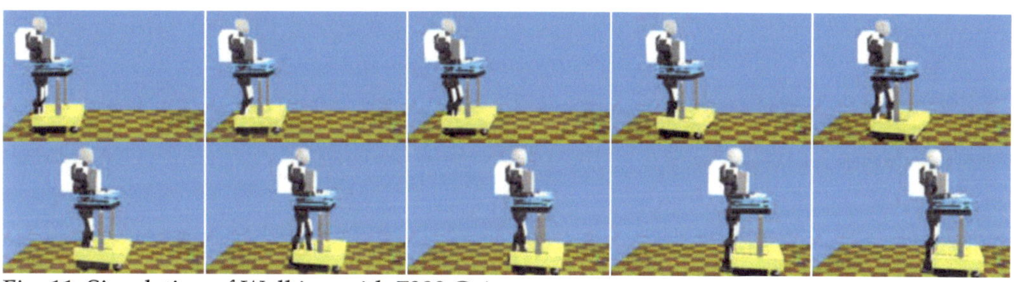

Fig. 11. Simulation of Walking with 7000 Gain

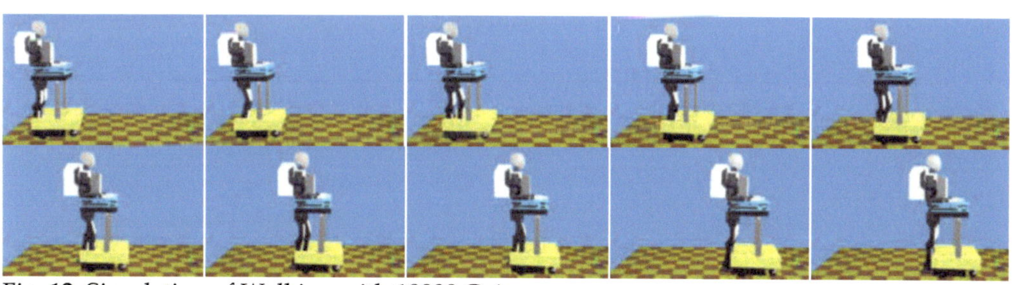

Fig. 12. Simulation of Walking with 10000 Gain

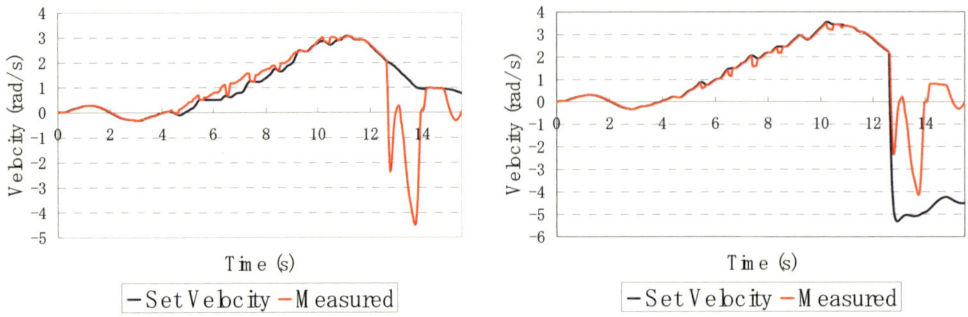

Left Wheel Right Wheel

Fig. 13. Velocity set and measured for each Wheel with 1000 Gain

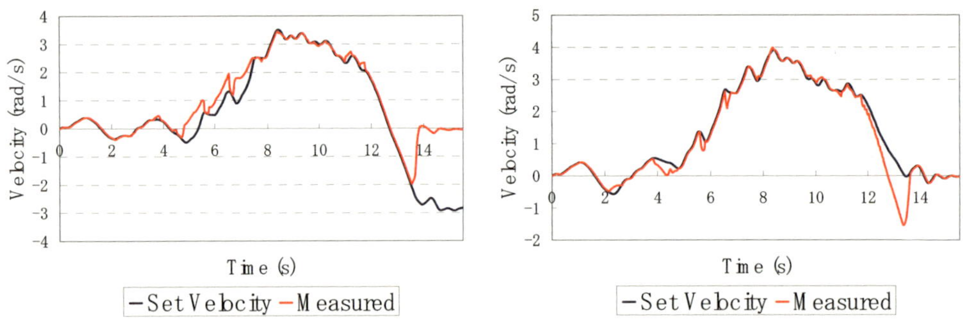

Left Wheel Right Wheel

Fig. 14. Velocity set and measured for each Wheel with 2000 Gain

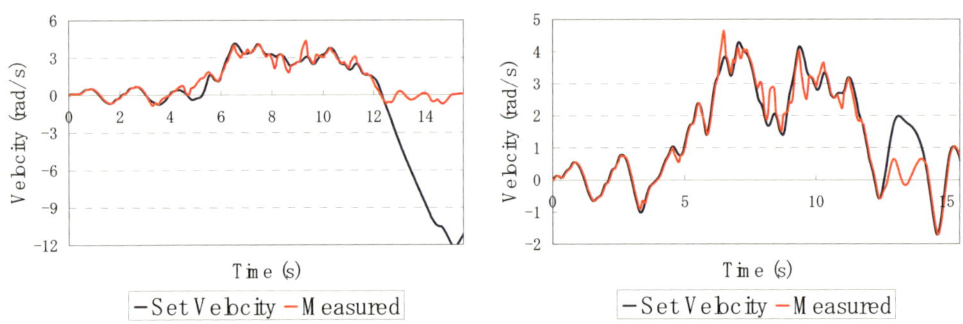

Left Wheel Right Wheel

Fig. 15. Velocity set and measured for each Wheel with 5000 Gain

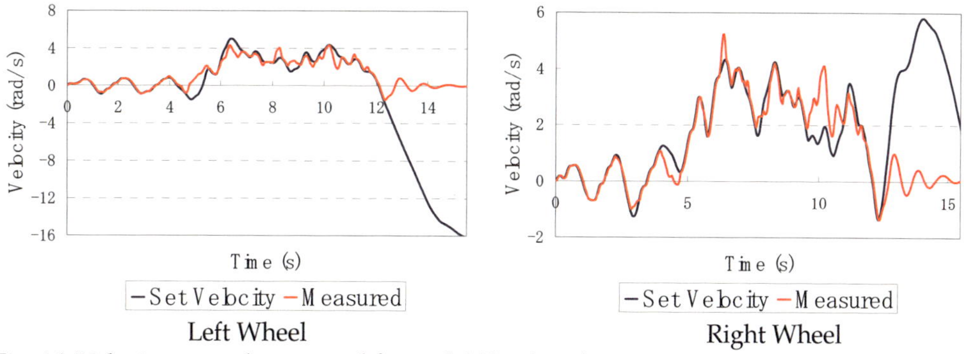

Fig. 16. Velocity set and measured for each Wheel with 7000 Gain

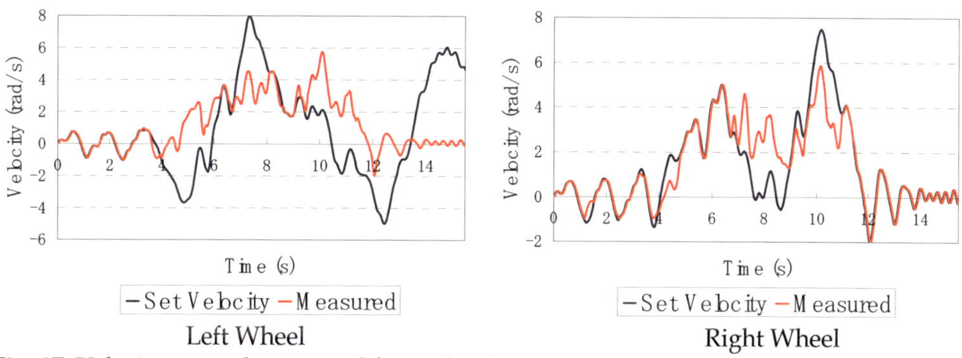

Fig. 17. Velocity set and measured for each Wheel with 10000 Gain

5. Conclusion and Futurework

This paper describes the simulation of walking by WABIAN-2R with the walking assist machine. The dynamic simulation is very important to check the motion of any new pattern generated. Using the dynamic simulation we can see the effect of the walking assist on WABIAN-2R. As expected, the walking was unstable due to the effect of external forces created from the arm rest. By using the velocity control in the control system of the simulation, the robot is able to walk stably with the walking assist machine.

In the near future, it is important to develop WABIAN-2R system to be stabilized during walking. The stabilization control will be based on Zero Moment Point. Moreover, it is necessary to develop the robot to interact with other objects and equipments. This will make the robot can interact with its surrounding environment.

6. References

Health, Labour and Welfare Ministry of Japan, http://www.mhlw.go.jp/english/wp/wphw/vol1/p2c4s2.h-tml.

Y. Sakagami, R. Watanabe, C. Aoyama, S. Matsunaga, N. Higaki, and K.Fujimura, "The intelligent ASIMO: System overview and integration," Proc. IEEE/RSJ Int. Conference on Intelligent Robots and Systems, pp.2478-2483, 2002.

Aiman Musa M. Omer, Yu Ogura, Hideki Kondo, Akitoshi Morishima, Giuseppe Carbone, Marco Ceccarelli, Hun-ok Lim, and Atsuo Takanishi. Development of A Humanoid Robot Having 2-DOF Waist and 2-DOF Trunk. Humanoid2005, Tsukuba- December 2005.

Yu Ogura, Hiroyuki Aikawa, Kazushi Shimomura, Hideki Kondo, Akitoshi Morishima, Hun-ok Lim, and Atsuo Takanishi. Development of a New Humanoid Robot WABIAN-2. Proceedings of the 2006 IEEE International Conference on Robotics and Automation Orlando, Florida - May 2006.

Webots. http://www.cyberbotics.com. Commercial Mobile Robot Simulation Software.

S. Mojon. Realization of a Physic Simulation for a Biped Robot. Semester Project at BIRG laboratory Swiss Federal Institute of Technology, Summer 2003.

S. Egawa, Y. Nemoto, M. G. Fujie, A. Koseki, S. Hattori, T. Ishii S. Egawa, Y. Nemoto, M. G. Fujie. POWER-ASSISTED WALKING SUPPORT SYSTEM WITH IMBALANCE COMPENSATION CONTROL FOR HEMIPLEGICS. Proceedings of The Rrst Joint BMES/EMBS Conference Serving Humanity, Advancing Technology o& 1&16, 99, Athn$, GA, USA.

Saku EGAWA, Ikuo TAKEUCHI, Atsushi KOSEKI, Takeshi ISHI. Force-sensing Device for Power-assisted Walking Support System. System Integration Conference, December 2002.

P. E. Klopsteg and P. D. Wilson et al., Human Limbs and Their Substitutes, New York Hafner, 1963.

F. Kanehiro, K. Fujiwara, S. Kajita, K. Yokoi, K. Kaneko, H. Hirukawa, Y. Nakamura, K. Yamane. Open architecture humanoid robotics platform. ICRA '02. IEEE International Conference on, Volume: 1, 11-15 May 2002 Robotics and Automation, 2002. Proceedings.

Philippe Sardain and Guy Bessonnet. Force Acting on a Biped Robot Center of Pressure-Zero Moment Point. IEEE TRANSACTIONS ON SYSTREMS, MAN, AND CYBERNETICS-PART A: SYSTEMS AND HUAMNS, VOL. 34, NO. 5, SEPTEMBER 2004.

17

Output Feedback Adaptive Controller Model for Perceptual Motor Control Dynamics of Human

Hirofumi Ohtsuka and Koki Shibasato
Kumamoto National College of Technology
Japan
Shigeyasu Kawaji
Kumamoto University
Japan

1. Introduction

The construction of collaborative human-machine system is being recognized as an important technology from the viewpoint of human centered assisting system development (Takahashi and Ikeura, 2006; Yamada and Utsugi, 2006). While such assisting systems aim at partial replacement of control task or an amplification of control power, those have insufficiency in order to achieve the accurate maneuvering, where human performs as a main controller in the human-machine system. For the purpose of improvement of the maneuvering performance and the response of human-machine system, authors have developed a new compensator named as "collaborater", which can support the collaborative work of human and machine (Ohtsuka et al., 2007, Ohtsuka et al., 2009). The model of human response behavior is required to design the collaborater and the collaborative assisting system, but it has been difficult to construct an accurate model of human perceptual motor control system (e.g., limb and muscle). Kleinman et al. applied optimal control theory to develop a model of human behavior in manual tracking tasks (Kleinman et al., 1970). Their model contains time delay, a representation of neuromotor dynamics, and controller remnant as limitations.

Recently, Furuta considers that the analysis of human control action is one of fundamental problems in the study of human adaptive mechatronics (Furuta et al., 2004). From such a viewpoint, in the authors' previous study, Delayed Feed-Forward (DFF) Model has been used for describing human's hand-tracking motion with visual information (Ishida and Sawada, 2003). The DFF model can realize the characteristics that the limb motion, with prediction of target position, makes the predicted value to minimize the transient error in the considering frequency range. However, for the non-cyclical target value and/or the controlled machine output, it has been resulted in that the DFF model has an insufficient reliance because of the shortage of consideration through the experimental study.

In this paper, for the upper limb motion in the hand-tracking control, a new Perceptual Motor Control Model (PMCM) is considered. Namely, the visual feedback controller is represented as the output feedback type adaptive controller stabilizing the closed loop system based on an Almost Strict Positive Real (ASPR) characteristic of the controlled system. The Parallel Feed-forward Compensator (PFC) has been introduced in order to make an ASPR augmented system (Iwai et al., 1993). And, Miall et al. have proposed a human's brain model by introduction of Smith Predictor (as forward internal model) in order to predict the consequences of actions and to overcome pure time delays of neuro-motor signal transmission associated with feedback control (Miall et al., 1993). So, taking into account of those approaches, both PFC and Smith Predictor are located into the minor feedback loop for the output feedback adaptive controller. So, the PMCM has similar structure to the cerebrum-cerebellum neuro-motor signal feedback loop. The effectiveness of the proposed PMCM is discussed through a comparison of the experiment and simulation results.

2. Output Feedback type Adaptive Control System

In this section, as a preparation for discussion about the PMCM of human, we briefly outline an output feedback adaptive control method, where the controller is designed to realize the plant output converging to reference signal.

2.1 Configuration and Controlled Plant

Let us consider the following SISO plant:

$$\begin{aligned} \dot{\mathbf{x}}(t) &= A\mathbf{x}(t) + \mathbf{b}u(t) \\ y(t) &= \mathbf{c}^T\mathbf{x}(t) + du(t) \end{aligned} \tag{1}$$

, where $\mathbf{x}$ is the nth order state vector, u and y are scalar input and output, respectively. A, $\mathbf{b}$, $\mathbf{c}$ and d are unknown matrix, vectors with appropriate dimensions, and scalar. The transfer function form of the plant Eq.(1) is expressed by

$$G(s) = \mathbf{c}^T(sI - A)^{-1}\mathbf{b} + d = \frac{N(s)}{D(s)} \tag{2}$$

, where

$$\begin{cases} N(s) = b_m s^m + b_{m-1}s^{m-1} + \cdots b_1 s + b_0 \\ D(s) = s^n + a_{n-1}s^{n-1} + \cdots a_1 s + a_0 \end{cases} \tag{3}$$

Now, we make the following assumption.

Assumption 1 The Plant Eq.(1) or Eq.(2) is ASPR(Almost Strictly Positive real). From this assumption, there exists a constant gain k_p such that the transfer function

$$G_C(s) = (1 + k_p G(s))^{-1} G(s) \tag{4}$$

is SPR(Strictly Positive Real). $G_C(s)$ Eq.(4) can be expressed by the following state space representation:

$$\begin{aligned} \dot{\mathbf{x}}(t) &= (A - k_p\bar{\mathbf{b}}\mathbf{c}^T)\mathbf{x}(t) + \bar{\mathbf{b}}v(t) \\ y(t) &= \bar{\mathbf{c}}^T\mathbf{x}(t) + \bar{d}v(t) \end{aligned} \tag{5}$$

where,

$$\bar{\mathbf{b}} = \frac{\mathbf{b}}{1 + dk_p},\ \bar{\mathbf{c}}^T = \frac{\mathbf{c}^T}{1 + dk_p},\ \bar{d} = \frac{d}{1 + dk_p} \tag{6}$$

by taking into account of that state space representation of $G(s)$ as Eq.(1).
Sufficient condition for Assumption 1 can obtained, such that (1) $N(s)$ is Hurwitz polynomial, (2) $\gamma = n - m \leq 1$, and (3) $b_m > 0$ (Kaufman et al., 1998).

In practice, it is necessary for the realization of the output feedback adaptive control system that the controlled plant must satisfy the ASPR condition in Assumption 1. Unfortunately, this condition is not satisfied by most real systems. Namely, many practical plant satisfies $d = 0$ and the relative degree of plant is larger than 1. To overcome this problem, several types of PFC (Parallel Feed-forward Compensator) have been proposed.(Z.Iwai and M.Deng, 1994; Z.Iwai and H.Ohtsuka, 1993; H.Kaufman and K.Sobel, 1998) For example, Iwai et.al. (Z.Iwai and M.Deng, 1994) have shown the following theorem giving the design procedure of PFC.

Theorem 1(Z.Iwai and M.Deng, 1994) Augmented plant $G_a(s)$:

$$G_a(s) = G(s) + G_f(s) \tag{7}$$

becomes ASPR system and the output of augmented plant $y_a(t)$ approximately equals to the plant output $y(t)$, if the transfer function of PFC $G_f(s)$ is given as

$$G_f(s) = \sum_{i=1}^{\gamma-1} \bar{\delta}^i G_i(s) \tag{8}$$

$$G_i(s) = \frac{\bar{\beta}_i n_i(s)}{d_i(s)} \tag{9}$$

where, $d_i(s)$ is n_{di} -th order monic stable polynomial, $n_i(s)$ is $m_{ni} = \{n_{di} - (\gamma - i)\}$-th order monic polynomial ($m_{ni} \geq 0$), $\bar{\delta}$ is sufficiently small positive constant, and $\bar{\beta}_i$ are coefficients of the Hurwitz polynomial:

$$R(s) = \bar{\beta}_{\gamma-1} s^{\gamma-1} + \cdots \bar{\beta}_1 s + \bar{\beta}_0 \tag{10}$$

(Proof) See the reference. (Z.Iwai and M.Deng, 1994)

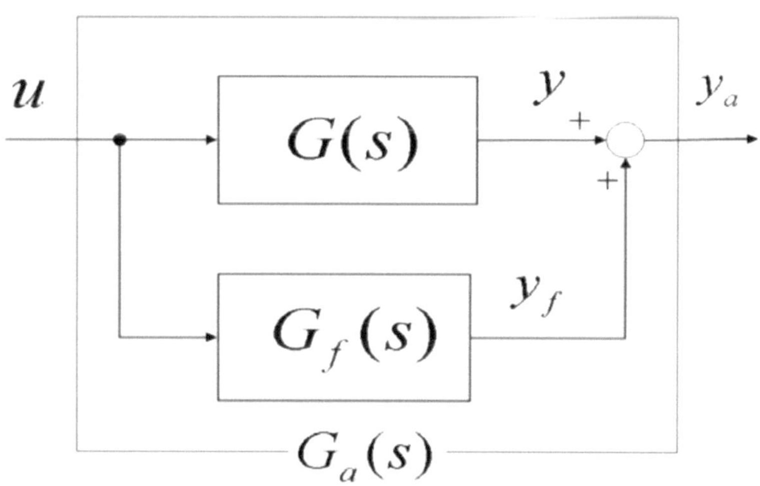

Fig. 1. Augmented Plant with PFC

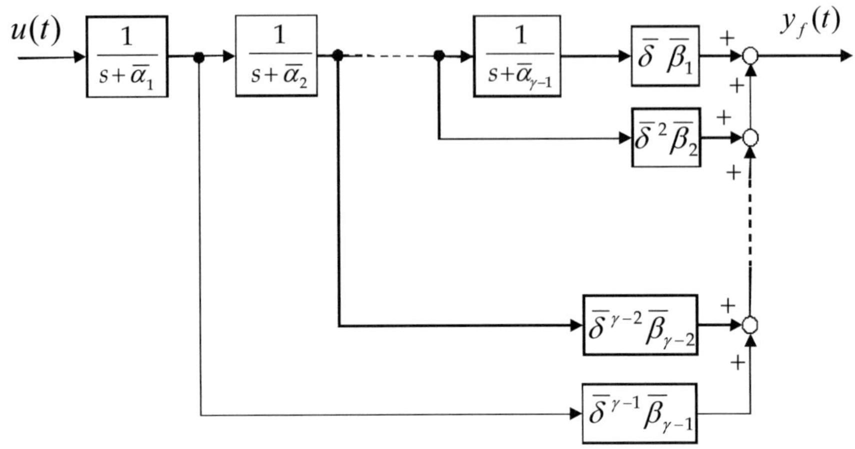

Fig. 2. Ladder Network type PFC

While Theorem 1 gives a general structure of PFC, the practical realization of PFC with simple structure is shown in Fig.1 and can be described as follows.

$$\begin{cases} n_i(s) = s^{\rho} \\ d_{\gamma-i}(s) = (s+\bar{\alpha}_0)^{\rho}\bar{d}_{\gamma-i}(s) \\ \bar{d}_{\gamma-i}(s) = (s+\bar{\alpha}_i)\bar{d}_{\gamma-i+1}(s) \\ \bar{d}_{\gamma}(s) = 1 \\ \bar{\alpha}_i > 1 \end{cases} \tag{11}$$

where $i = 1,2,\cdots,\gamma-1$. In Theorem 1, $n_i(s) = s^{\rho}$ is introduced in order to remove an offset (steady state error on plant output) caused by the addition of PFC. In the case of the step type reference signal, r is given as 1. Fig.2 shows the one of practical realization of PFC based on the Theorem 1.

2.2 Basic Adaptive Control Algorithm

Under the Assumption 1, the following adaptive algorithm:

$$u(t) = k(t)e(t) \tag{12}$$

$$\dot{k}(t) = g\, e(t)^2 \tag{13}$$

generates the control input of the plant Eq.(1), where $e(t)=r(t)-y(t)$ and g is positive constant.

2.3 Stability

The following theorem can be obtained under the Assumption 1.

Theorem 2 Suppose that the Assumption 1 is satisfied. Then, the adaptive control law Eqs.(12),(13) can achieve the output error convergence to zero, namely

$$\lim_{t\to\infty} e(t) = 0 \tag{14}$$

(Proof) Under the assumption 1, let us consider an ideal plant state vector $\mathbf{x}^*(t)$ which can satisfies $e(t)=0$, $(t \geq 0)$, then the following relationship:

$$\begin{aligned} \dot{\mathbf{x}}^*(t) &= A\mathbf{x}^*(t) \\ y^*(t) &= \mathbf{c}^T\mathbf{x}^*(t) \end{aligned} \tag{15}$$

is held by using the output feedback control:

$$u^*(t) = k^*e(t) \tag{16}$$

from Eqs.(1)and(12). Eq.(15) is called as an ideal plant. Now, suppose that the state error vector is defined as

$$\mathbf{e}_x(t) = \mathbf{x}(t) - \mathbf{x}^*(t) \tag{17}$$

from Eqs.(1) and (15), we have

$$\begin{aligned} \dot{\mathbf{e}}_x(t) &= A\mathbf{e}_x(t) + \mathbf{b}u(t) \\ e(t) &= \mathbf{c}^T\mathbf{e}_x(t) + du(t) \end{aligned} \tag{18}$$

So, rewriting the control law as

$$u(t) = \bar{k}(t)e(t) + k^*e(t) = \bar{u}(t) + k^*e(t) \tag{19}$$

$$\bar{k}(t) = k(t) - k^* \tag{20}$$

$$\bar{u}(t) = \bar{k}(t)e(t) \tag{21}$$

gives next equation from Eqs.(18) and (19).

$$e(t) = \bar{\mathbf{c}}^T\mathbf{e}_x(t) + \bar{d}\bar{u}(t) \tag{22}$$

Substitution of the above equation into Eq.(18) gives

$$\begin{aligned} \dot{\mathbf{e}}_x(t) &= A\mathbf{e}_x(t) + \mathbf{b}(\bar{u}(t) + k^*e(t)) \\ &= (A + k^*\bar{\mathbf{b}}\mathbf{c}^T)\mathbf{e}_x(t) + \mathbf{b}\frac{1}{1+dk^*}\bar{u}(t) \\ &= (A + k^*\bar{\mathbf{b}}\mathbf{c}^T)\mathbf{e}_x(t) + \bar{\mathbf{b}}\bar{u}(t) \end{aligned} \tag{23}$$

Thus, we have the following error system representation.

$$\begin{aligned} \dot{\mathbf{e}}_x(t) &= (A + k^*\bar{\mathbf{b}}\mathbf{c}^T)\mathbf{e}_x(t) + \bar{\mathbf{b}}\bar{u}(t) \\ e(t) &= \bar{\mathbf{c}}^T\mathbf{e}_x(t) + \bar{d}\bar{u}(t) \end{aligned} \tag{24}$$

Then, it follows from assumption 1 and the Kalman-Yakubovich lemma (H.Kaufman and K.Sobel, 1998) that there exist $n \times n$ positive symmetric matrices P and Q and vector l and scalar w satisfying the following equations:

$$\begin{aligned} &(A + k^*\bar{\mathbf{b}}\mathbf{c}^T)^T P + P(A + k^*\bar{\mathbf{b}}\mathbf{c}^T) = -Q - ll^T \\ &P\bar{\mathbf{b}} = \bar{\mathbf{c}} - wl \\ &2\bar{d} = w^2 \end{aligned} \tag{25}$$

Take the positive function:

$$V(t) = \mathbf{e}_x(t)^T P \mathbf{e}_x(t) + \frac{\bar{k}(t)^2}{g} \tag{26}$$

Then, because the following relationship holds

$$\dot{\bar{k}}(t) = g\, e(t)^2 \tag{27}$$

from Eqs.(13) and (20), the following equation is obtained.

$$\begin{aligned}\dot{V}(t) &= \mathbf{e}_x(t)^T \left\{(A - k^*\bar{\mathbf{b}}\mathbf{c}^T)^T P + P(A - k^*\bar{\mathbf{b}}\mathbf{c}^T)\right\}\mathbf{e}_x(t) + 2\mathbf{b}^T P\mathbf{e}_x(t)\bar{k}(t)e(t) + 2\bar{k}(t)e(t)^2 \\ &= -\mathbf{e}_x(t)^T Q\,\mathbf{e}_x(t) - \mathbf{e}_x(t)^T ll^T \mathbf{e}_x(t) + 2\mathbf{c}^T\mathbf{e}_x(t)^T\bar{k}(t)e(t) \\ &\quad - 2wl^T\mathbf{e}_x(t)\bar{k}(t)e(t)^2 + 2\bar{k}(t)e(t)^2 \\ &= -\mathbf{e}_x(t)^T Q\,\mathbf{e}_x(t) - \mathbf{e}_x(t)^T ll^T \mathbf{e}_x(t) - w^2(\bar{k}(t)e(t))^2 - 2wl^T\mathbf{e}_x(t)\bar{k}(t)e(t) \\ &= -\mathbf{e}_x(t)^T Q\,\mathbf{e}_x(t) - \left\{\mathbf{e}_x(t)^T l + w\bar{k}(t)e(t)\right\}^2 \le 0\end{aligned} \tag{28}$$

From the above relationships, we can see that $V(t)$ is the Lyapnov function and both $\mathbf{e}_x(t)$ and $\bar{k}(t)$ asymptotically converge to zeros. Namely, from (28), we obtain Eq.(14).

(End of Proof)

2.4 Modified Adaptive Adjusting Law

Furthermore, against to the input disturbance and to the un-modeled dynamics of the plant, the following modified adaptive adjusting law

$$\dot{k}(t) = -\sigma\, k(t) + g\frac{e(t)^2}{1 + \varepsilon\, e(t)^2} \tag{29}$$

can be utilized in order to maintain that the all signals in the closed loop system become uniformly ultimate bounded (UUB), where σ and ε are given as sufficiently small positive constants (Iwai et al., 1993).

However, Assumption 1 is not satisfied by most practical systems with large relative degree $\gamma > 1$. In this case, the stability of closed loop system can also be maintained while the all signals in the closed loop system are uniformly ultimately bounded (UUB). (Z.Iwai and H.Ohtsuka, 1993)

3. Neuro-motor Apparatus Model

In the brain science, the cerebellum has attracted the attention of theorists and modelers and the need for a unifying theory for the role of the cerebellum in motor control has been recognized for many years (R.C.Miall and J.F.Stein, 1993; M.Ito, 1970; D.M.Wolpert and M.Kawato, 1998; D.L.Kleinman and W.H.Levison, 1970). Specially, based on data from the control of the primate arm in visually guided tracking tasks, Miall et.al. suggested that the cerebellum acts as a Smith Predictor, which is based on internal representation of controlled object suffering with long and unavoidable feedback delays. Ito et.al. (M.Ito, 1970) also suggested that there exists the cerebrum-cerebellum neuro-motor signal feedback loop (Fig.3) and the cerebellum may form the internal model, based on physiological and clinical evidence. There are two variety of internal model, forward and inverse models (D.M.Wolpert and M.Kawato, 1998). Forward models capture the forward or causal relationship between inputs to the system, such as the arm, and the outputs. The Smith

predictor can be regarded as a kind of forward model. While we can overcome the issue for the pure time delay by using a Smith predictor, the performance of visual feedback control is mainly affected by the setting of output feedback gain.
However, conventional most of neuro motor models have fixed the feedback gain as constant. On the other hand, many control engineering researcher study about the adaptive control method based on the ability of animal to adapt itself to changes in its surroundings.
Taking into account the above-mentioned brain science researchers' suggestions, and based on the output feedback type adaptive control strategy described in above section, let us construct a new perceptual motor control model as shown in Fig.5 for the control problem as shown in Fig. 4 in which a human operator controls the machine to follow the target. In later the time delay of nervous system transmission is successfully compensated, the controlled system from a side of the output feedback controller becomes a series of three elements consisted of a first lag model with time constant τ_1 which is a model of brain dynamics, a first lag model with time constant τ_2 which is one of muscle dynamics, and a controlled machine dynamics $G_P(s)$. To construct a stable output feedback adaptive control system, the ASPR compensation must be implemented for such a series of three elements. Here, suppose that such ASPR compensator forms as PFC whose transfer function described as $F(s)$. Then, both the Smith predictor and PFC can located into the minor feedback loop for the adaptively adjusted output feedback gain k, as shown in Fig.5. Here, it eases to recognize that the structure of proposed perceptual motor control model is very similar to the cerebrum-cerebellum neuro-motor signal feedback loop model (Fig.3). Namely, we can imagine that the Smith predictor and PFC perform the role of cerebellum, which generates the forward model of controlled object.

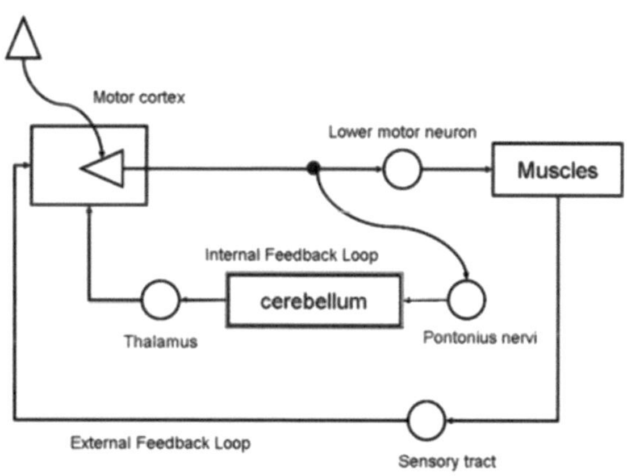

Fig. 3. Cerebrum & cerebellum (M.Ito, 1970)

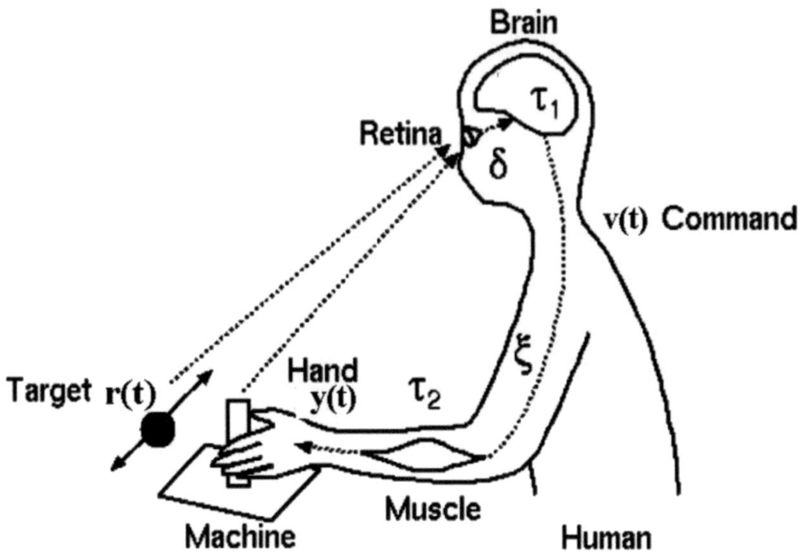

Fig. 4. Human Body Dynamics

Notation	Parameters and Variables
$r(t)$	position of the target
$y(t)$	position of the hand
$v(t)$	command signal from the brain
δ	dead time in the nervous system from the retina to the brain
ξ	dead time in the nervous system from the brain to the muscle
τ_1	time constant of the brain
τ_2	time constant of the muscle dynamics

Table 1. Parameters and variables

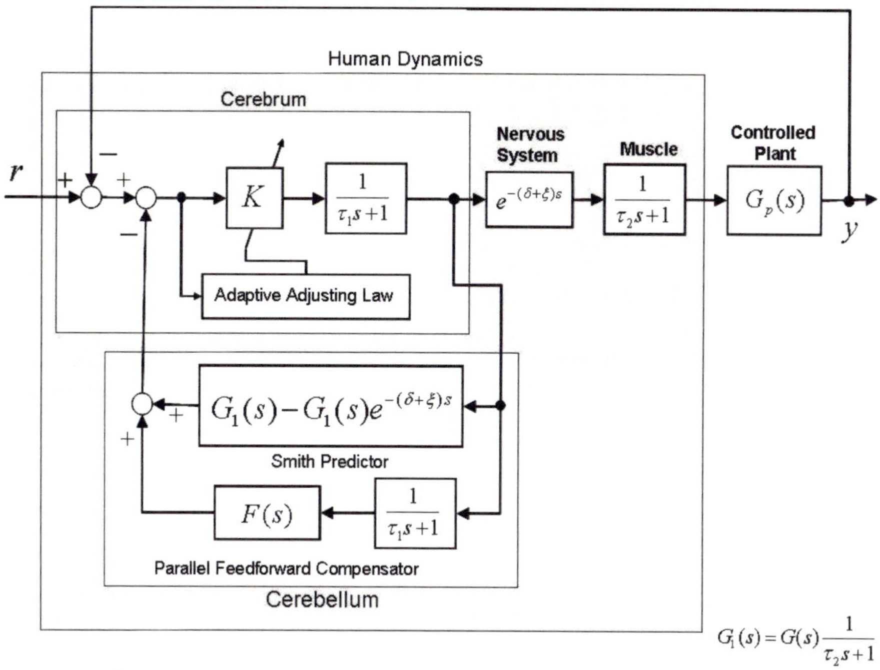

Fig. 5. Perceptual Motor Control Model

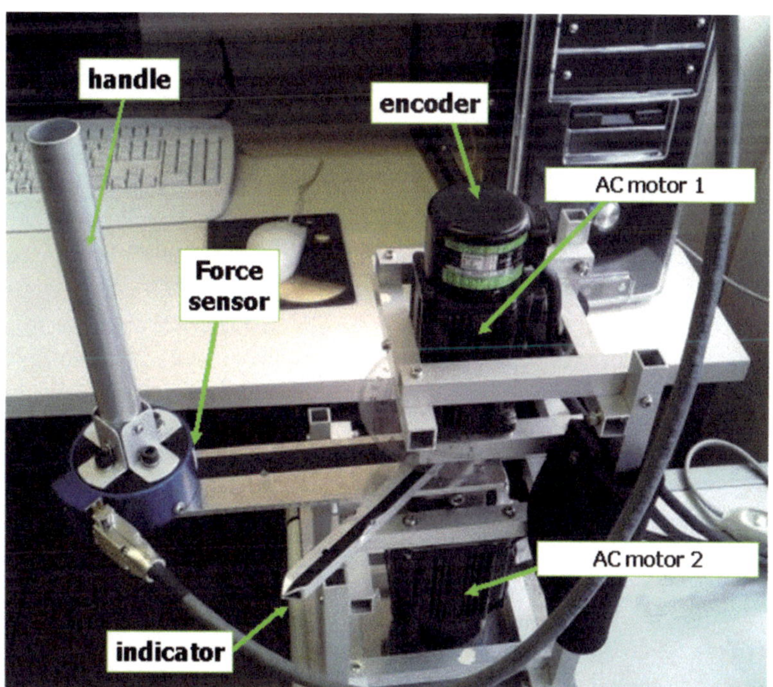

Fig. 6. Experimental Equipment

4. Experiments

Fig.6 shows the experimental equipment. An indicator shows the target position, which is driven by AC motor 1, and an operator controls a handle to follow the indicator. AC motor 2 is assembled in order to generate the assisting torque for human, while it performs as load inertia for human in this stage.

Mechanical System: From the experimental results of automatic positioning control, the transfer function of the one-link arm mechanism involving AC motor 2: $G_P(s)$ was estimated as follows.

$$G_P(s) = \frac{4213}{s(s+1)} \tag{30}$$

Human Dynamics model: Through the experimental results, the parameters of human dynamics model are estimated such that $\delta + \varsigma = 0.13\,[\mathrm{s}]$, $\tau_1 = \tau_2 = 0.03\,[\mathrm{s}]$, respectively (Saito and Nagasaki, 2002).

Perceptual Motor Control Model: In this case, the controlled system from a side of the output feedback controller, which is the above-mentioned series of three elements are given as follow.

$$G_1(s) = \frac{4213}{s(s+1)(0.03s+1)^2} \tag{31}$$

Because it has a relative order as 4 and minimum phase characteristics, PFC: $F(s)$ in Fig.5 is constructed based on Theorem 1 as follows:

$$\begin{aligned} F(s) &= \frac{f_1 s}{(\tau_1 s+1)(s+\alpha)^2} + \frac{f_2 s}{(\tau_1 s+1)(s+\alpha)} \\ &= \frac{350\,s}{(0.03\,s+1)(s+0.5)^2} + \frac{6\,s}{(0.03\,s+1)(s+0.5)} \end{aligned} \tag{32}$$

Results of Experiment and simulation: Experimental results for the target position $r(t)=30$ [degree] are shown as Fig.7 and Fig.8. And, Fig.9 and Fig.10 also shows the simulation results for the variance of design parameter g in Eq.(13). For the variance of design parameter of PFC, we can obtain the simulation results shown in Figs.11 and 12. In the simulation, the other parameters in Eq.(6) are given as $k(0) = 0$, $\sigma = 0.1$, $g = 0.009$, $\varepsilon = 0.01$. Although there exists some fluctuation in the experimental results obtained for 3 testers, we can recognize that the both responses are very similar. Because, by comparing between Fig.7 and Fig.9/Fig.11, the overshoots are almost same level and the damping ratio and the values of peak time are close resemblance.

Furthermore, comparing between Fig.8 and Fig.10/Fig.12, these signals also show a close

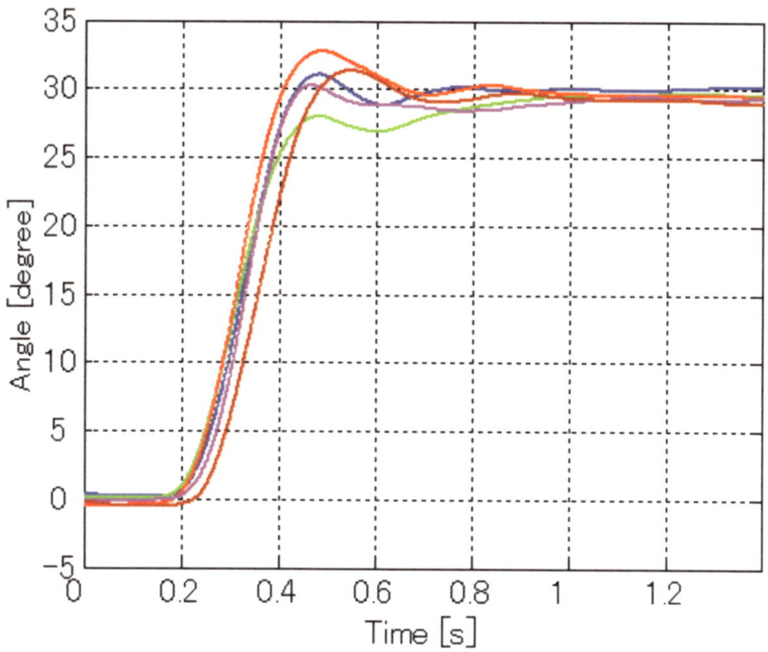

Fig. 7. Experimental Result (Output: Angle)

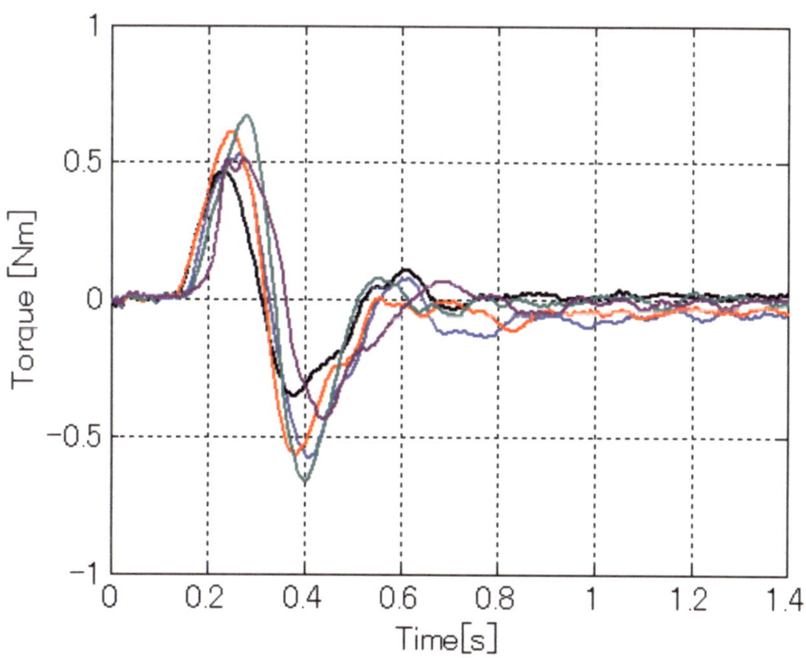

Fig. 8. Experimental Result (Input: Torque)

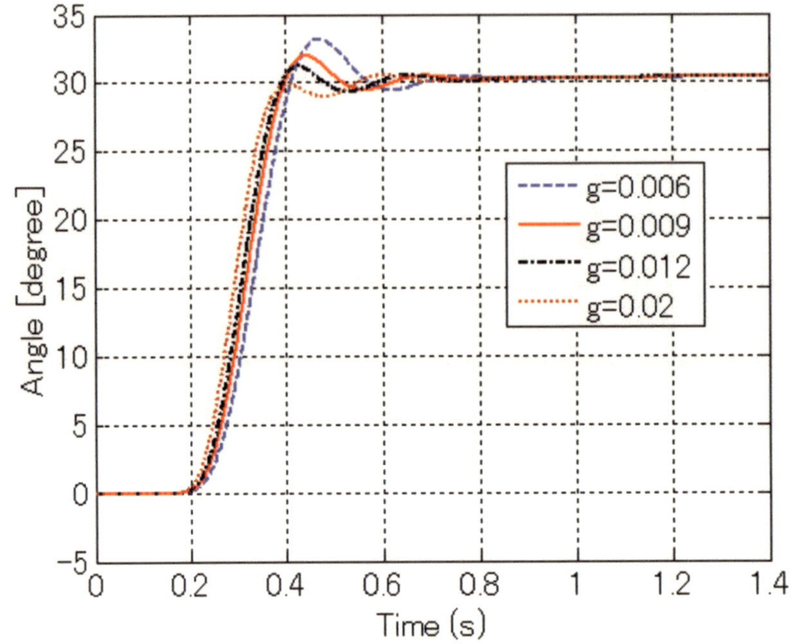

Fig. 9. Simulation Result (Output: Angle)

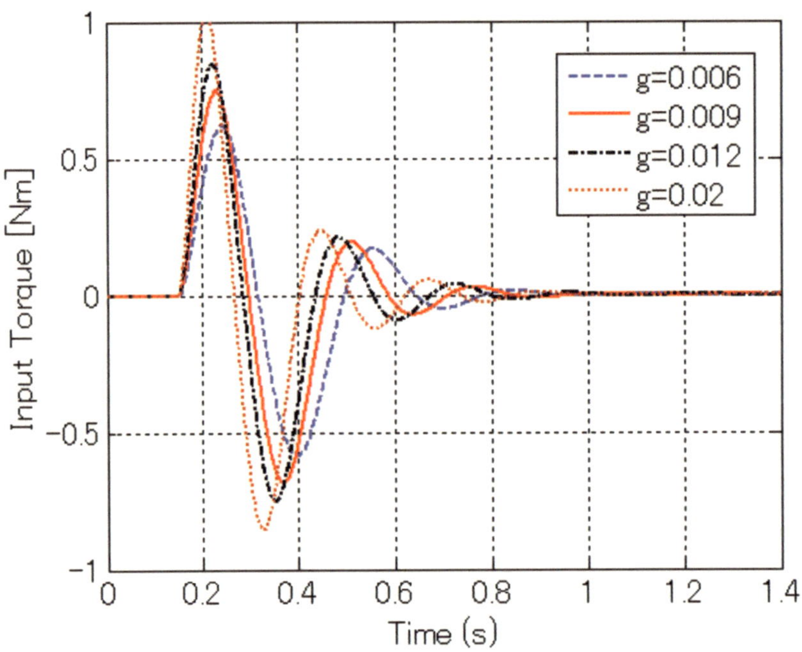

Fig. 10. Simulation Result (Input: Torque)

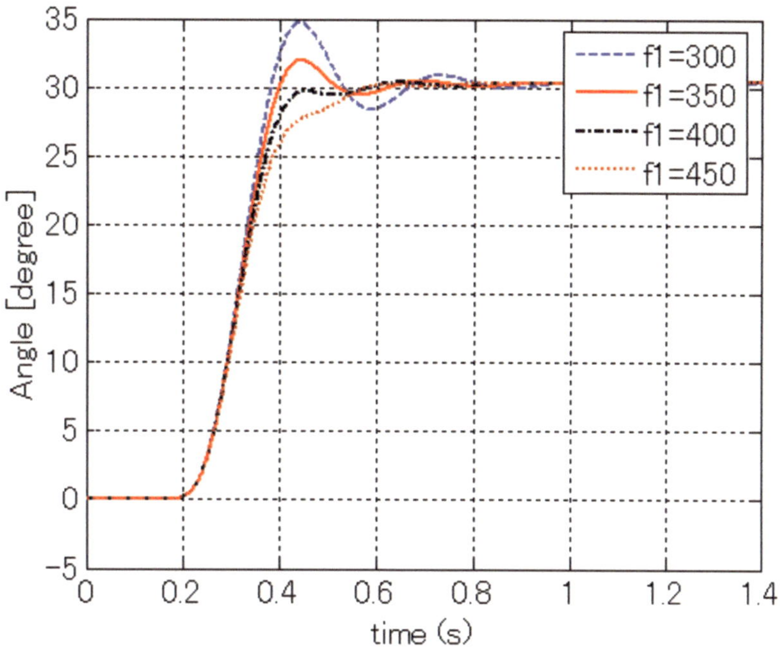

Fig. 11. Simulation Result (Output: Angle)

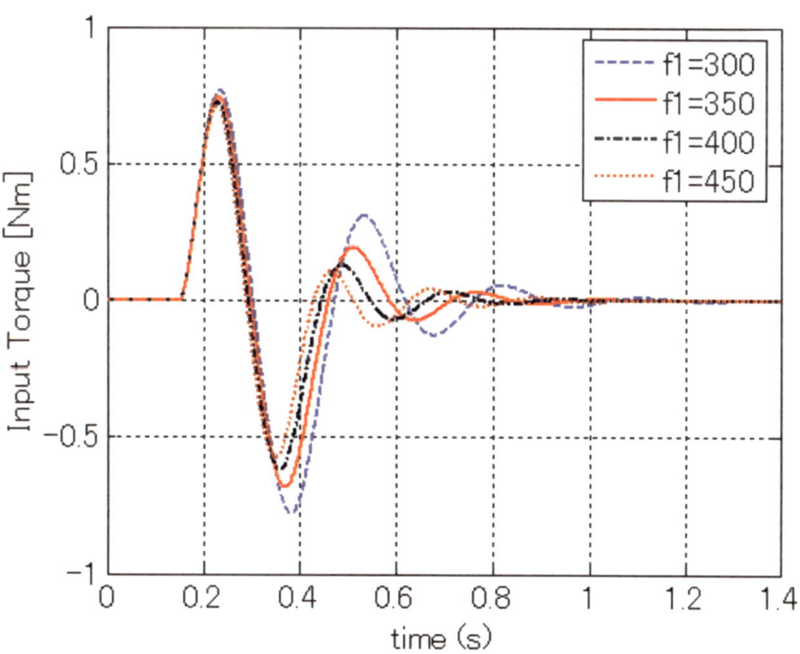

Fig. 12. Simulation Result (Input: Torque)

similarity. So, we can note that the proposed model can maintain its good performance.

Furthermore, we can set up a hypothesis such that the fluctuation in the response can be interpreted as the fluctuation of PFC parameters and/or parameter of adaptive adjusting law g.

5. Conclusions

From the point aimed at the minor feedback loop in the brain, that is, the nervous network between the cerebrum and the cerebellum performing minor feedback loop element, and a hypothesis for cerebellum generating a forward model of motor apparatus dynamics, a perceptual motor control model is discussed. The proposed method is based on output feedback type adaptive control using a ASPR characteristics of the controlled plant, which accompany with PFC. In the nervous network, there necessarily exists dead time (pure time delay) of signal transmission between cortex and lower apparatus. To overcome the influence of the feedback of the sensed signal involving time delay, the Smith predictor method is introduced. The effectiveness of proposed model are examined through the comparison between of experimental results and simulation results for one-link arm positioning control problem. And, it is confirmed that the proposed model can represent the manual control response with sufficient accuracy. Furthermore, we suggest that the fluctuation in the response can be interpreted as the fluctuation of PFC and/or adaptive adjusting law parameters. The proposed model will be utilized to design and realize an assisting system for human-machine system, that is, "Collaborater".

6. References

Arai, B. & Yokogawa, H. (2005). A novel hoist system for the disable to support independence and nursing, In: *Journal of the Japan Society of Mechanical Engineers, Vol.108, No.1038, pp.406.*

Furuta, K., Iwase, M., & Hatakeyama, S. (2004). Analysing saturating actuator in human-machine system from view of human adaptive mechatronics. In: *Proceedings of REDISCOVER 2004, Vol.1, pp.(3-1)-(3-9).*

Ibuki, S.; K. & Takeda, T. (2005). Living assistance system by communication robot for elderly people, In: *Journal of the Japan Society of Mechanical Engineers, Vol.108, No.1038, pp.392-395.*

Ishida, F. & Sawada, Y. (2003). Quantitative studies of phase lead phenomena in human perceptro-motor control system. In: *Trans. of SICE, Vol.39, No.1, pp.59-66.*

Ito, M. (1970). Neurophysiological aspects of the cerebellar motor control system, In: *International Journal of Neurology, Vol. 7, pp.162-176.*

Iwai,Z; Mizumoto, I. & Ohtsuka, H. (1993). Robust and simple adaptive control system design, In: *International Journal of Adaptive Control and Signal Processing, Vol.7, pp.163-181.*

Iwai, Z.; Mizumoto, I. & Deng, M. (1994). A parallel feedforward compensator virtually realizing almost strictly positive real plant, In: *Proc. of 33rd IEEE CDC, pp.2827-2832.*

Kaufman, H.; I.-K. & Sobel, K. (1998). *Direct Adaptive Control Algorithms Theory and Application*, Springer-Verlag, New York, 2nd edition.

Kiguchi, K. (2006). Power suits, In: *Journal of the Society of Instrument and Control Engineers, Vol.45, No.5, pp.436-439.*

Kleinman, D.L.; S. & Levison, W.H. (1970). An optimal control model of human response part i: Theory and validation, In: *Automatica, Vol.6, pp.357-369.*

Lee, S. & Sankai, Y. (2002). Power assist control for walking aid with hal-3 based on emg and impedance adjustment around knee joint, In: *Proc. of IEEE/RSJ International Conf. on Intelligent Robots and Systems, pp.1499-1504.*

Miall, R.C.; Weier, D.J.; D. & Stein,J.F. (1993). Is the cerebellum a smith predictor ? , In: *Journal of Motor Behavior, Vol.25, No.3, pp.203-216.*

Obinata, G. (2005). Special issue on mechanical technology for aged society: Its contribution to the society and itsexpectancy for the industry, In: *Journal of the Japan Society of Mechanical Engineers, Vol.108, No.1038, pp.368.*

Ohtsuka, H.; Shibasato, K. & Kawaji, S. (2007). Collaborative control of human-machine system by collaborater, In: *Trans. of The Japan Society of Mechanical Engineers, Series C, Vol.73, No.733, pp.2576-2582.*

Ohtsuka, H.; Shibasato, K. & Kawaji, S. (2009). Experimental Study of Collaborater in human-machine system, In: *IFAC Journal of Mechatronics, Vol.19, Issue 4, pp.450-456.*

Saito, H. & Nagasaki, H. (2002). *Clinical Kinesiology*, Ishiyaku Publishers, Inc., 3rd edition, ISBN 978-4-263-21134-2, Japan.

Takahashi, T. & Ikeura, R. (2006). Development of human support system, In: *Journal of the Society of Instrument and Control Engineers, Vol.45, No.5, pp.387-388.*

Vlacic, L.; M. & Harashima, F. (2001). *Intelligent Vehicle Technologies, Theory and Applications.*, Butterworth Heinemann, 1st edition, ISBN 0-7506-5093-1, Oxford.

Willems, J. & Polderman, J. (1998). *Introduction to Mathematical Systems Theory*, Springer, ISBN 978-0-387-35763-8, New York.

Wolpert, D.M.; R. & Kawato, M. (1998). Internal models in the cerebellum. In: *Trends in Cognitive Sciences, Vol.2, No.9, pp.338-347.*

Yamada, Y. & Utsugi, A. (2006). Human intention inference techniques in human machine systems and their robotic applications, In: *Journal of the Society of Instrument and Control Engineering, Vol.45, No.6, pp.407-412.*

Biomimetic approach to design and control mechatronics structure using smart materials

Nicu George Bîzdoacă[1] , Daniela Tarniță[2], Anca Petrişor[3], Ilie Diaconu[1], Dan Tarniță[4] and Elvira Bîzdoacă [5]
[1] Department of Mechatronics,University of Craiova,
[2] Faculty of Mechanics,University of Craiova,
[3] Faculty of Electromechanics,University of Craiova
[4] University of Pharmacology and Medicine of Craiova,
[5] National College Ghe. Chitu
Craiova, Romania

1. Introduction

Life's evolution for over 3 billion years resolved many of nature's challenges leading to solutions with optimal performances versus minimal resources. This is the reason that nature's inventions have inspired researcher in developing effective algorithms, methods, materials, processes, structures, tools, mechanisms, and systems.

Animal -like robots (biomimetic or biomorphic robots) make an important connection between biology and engineenng.

Biomimetics is a new multidisciplinary domain that include not only the uses of animal-like robots – biomimetic robot as tools for biologists studying animal behavior and as research frame for the study and evaluation of biological algorithms and applications of these algorithms in civil engineering, robotics, aeronautics.

The biomimetic control structures can be classified by the reaction of living subject, as follows:

- reactive control structures and algorithms
- debative control structures and algorithms
- hybrid control structures and algorithms
- behavior control structures and algorithms.

Reactive algorithms can be defined, regarding living subject reaction, as being characterized by the words : "React fast and instinctively". This kind of control is specific to reflex reactions of the living world, fast reactions that appear as reply to the information gathered from the environment that generate reactions to variable conditions like fear, opportunities, defense, attack. For such algorithms there is available a small number of internal states and representations with the advantages (fast answer time, low memory for taking decisions) and disadvantages (lack of ability to learn from these situations, implicit repetitive reaction) that goes with them. Studies regarding this kind of control were started by Schoppers 1987 and Agre and Chapman 1990 that have identified the strong dependence of this control by the environment and evolutive situations. In robotics, alternatives for this control are applicable in mobile structures that work in crowded places.

Debative algorithms can be defined by the following words: "Calculate all the chances and then act". This kind of control is an important part of artificial intelligence. In the living world, this type of control is specific to evolved beings, with a high level of planned life. For example, man is planning ahead its route, certain decisions that must be taking during its life, studies possible effects of these decisions, makes strategies. From a technological point of view, this kind of control has a complicated internal aspect, internal representations and states being extremely complex and very strong linked by predictive internal and external conditions with a minor or major level of abstract. Consuming a lot of memory and calculus, this kind of control doesn't fit, for now, to real time control, the technological structures that benefit from such control might suffer decisional blocks or longer answer times. Even the solution given by this algorithm is optimal, the problem of answering in real time makes alternatives for this control to be partly applied, less then optimal solutions being accepted.

Hybrid algorithms can be defined by the phrase "Think and act independently and simultaneous". Logical observation that living world decisions are not only reactive or debative has led to hybrid control. The advantages of reactive control - real time answers - together with the complexity and optimal solutions provided by debative control has led to a form of control that is superior from a decisional and performance point of view. The organization of control architecture consists of at least two levels: the first level - primary, decisional - is the reactive component that has priority over the debative component due to the need of fast reaction to the unexpected events; the second level is that of debative control that operates with complex situations or states, that ultimately lead to a complex action taking more time. Due to this last aspect, the debative component is secondary in importance to the reactive component. Both architectures interact with each other, being part of the same system: reactive architecture will supply situations and ways to solve these situations to the debative architecture, multiplying the universe of situations type states of the debative component, while the last one will create new hierarchic reactive members to solve real time problems. There is the need for an interface between the two levels in order to have collaboration and dialogue, interface that will lead to a hierarchy and a correspondence between members of the same or different levels. That's why this system is also called **three levels of decision system.** In robotics this system is used with success, the effort of specialists is focused on different implementations, more efficient, for a particular level, as well as for the interactions between this levels (Giralt 1983, Firby 1987, Arkin 1989, Malcolm and Smithers 1990, Gat 1998).

Behavioural algorithms can be defined by the words: "Act according with primary set of memorized situations". This type of system is an alternative to the hybrid system. Thou the hybrid system is in permanent evolution, it still needs a lot of time for the decisional level. The automatic reactions identified when the spinal nervous system is stimulated have led to the conclusion that there is a set of primary movements or acts correspondent to a particular situation. This set is activated simultaneously by internal and external factors that leads to a cumulative action (Mataric 1990). This type of architecture has a modular organization splitted in behavioral sets that allows the organization of the system on reactive states to complex situations, as well as the predictive identification of the way that bio-mimetic system responds (Rodney 1990). This response is dependant of the external stimulations and the internal states that code the anterior evolution and manifests itself by adding contribution of the limited number of behavioral entities (Rosenblatt 2000). The complexity of this approach appears in situations in which, due to internal or external conditions, are activated more behavioral modules that interact with each other and that are also influenced differently by the external and internal active stimulations at a specific moment in time. (Pirjanian 2002).

Cognitive model refers to essential aspects of the level of intelligence associated with a living or bio-mimetic system. The main models involved in this assembly are associated with **visual attention, motivation and emotions. Visual attention** is achieved in two stages (Chun 2001): first stage is a global, unselected, acquisition of visual information - **prefocus period** - and the second stage is **selective focus** that identifies a center of attention, a central frame in which the objective is found, objective that corresponds to the target image stocked in system memory.

Motivational model (Breazeal 1998) identifies all internal and external stimulations that trigger a basic behavior (movement, food, rest, mating, defense, attack). If animals are thought to have only one behavior at a certain moment in time because they receive only one primary motivational stimulation at a time, in humans this system must be extended. This extension results from numerous internal variables that are taken into account in human motivational analysis, external stimulations might be interpreted differently related to the internal states. Inside this motivational molding one must take also into account the complexity of reactions of different groups of people. These situations mustn't be looked like a sum of factors, the group reactions being, at least in most cases, a motivational reactions that neglects the individual (the survival of the group might accept the loss or disappearance of an individual or of a group of people, a fact that is practically impossible for an individual).

Emotional model is considered to be an identification system for major internal and external stimulations, as well as system to prepare the reaction response of the global system. Thus, based on low level entries and beginning initial states, the emotional model is activated in a different degree of excitation that will lead to a response of the global system correspondent to the generated states by the model, response different by the major actions with which the global system answers to emergent situations.

Fig. 1. Android robot Repliee R1 - Osaka University

The way that emotional system manifests itself is very different with every biological system: changing skin color, changing feathers arrangement, repeated movements that do not generate movement indicating fear or trying to intimidate, different sounds, changing face physiognomy. This last aspect was studied mainly in the last years, to achieve a

humanization of the technological environment that is evermore present (Pioggia 2006, Goetz 2003).
The androids made in Japan, the researches in USA, pet animals are only few examples for the evermore increasing interest for this type of research.
A promising field in practical implementation of biomimetics devices and robots is the domain of intelligent materials. Unlike classic materials, intelligent materials have physical properties that can be altered not only by the charging factors of that try, but also by different mechanisms that involve supplementary parameters like light radiation, temperature, magnetic or electric field, etc. This parameters do not have a random nature, being included in primary maths models that describe the original material. The main materials that enter this category are iron magnetic gels and intelligent fluids (magneto or electro-rheological or iron fluids), materials with memory shape (titan alloys, especially with nickel), magneto-electric materials and electro-active polymers. These materials prove their efficiency by entering in medical and industrial fields, a large number of them, due to their biocompatibility, being irreplaceable in prosthesis structures. Electro-active polymers, due to the flexibility of the activator potions, are a perfect solution for the implementation of animatronic projects. A special attention deserve the researches made by NASA, Jet Propulsion Laboratories - project Lulabot, Dept. of Science and Technology, Waseda University in Tokyo - project Humanoid Cranium, Cynthia Breazeal MIT (Cambridge, Mass.) – Kismet.

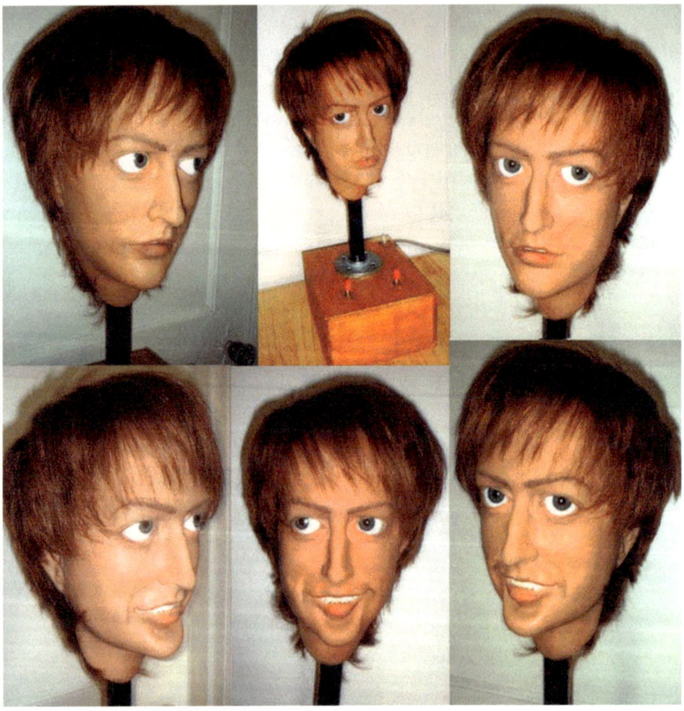

Fig. 2. Lulabot -David Hanson, NASA, JET Laboratory

Fig. 3. Humanoid Cranium - Prof. Takanishi Atsuo, Waseda University in Tokyo

Fig. 4. Robotul Kismet dezvoltat de Cynthia Breazeal, MIT (Cambridge, Mass)

2. Fundamental characteristics of shape memory alloys

The unique behavior of SMA's is based on the temperature-dependent austenite-to-martensite phase transformation on an atomic scale, which is also called thermoelastic martensitic transformation. The thermoelastic martensitic transformation causing the shape recovery is a result of the need of the crystal lattice structure to accommodate to the minimum energy state for a given temperature [Otsuka and Wayman 1998].
The shape memory metal alloys can exist in two different temperature-dependent crystal structures (phases) called martensite (lower temperature) and austenite (higher temperature or parent phase).

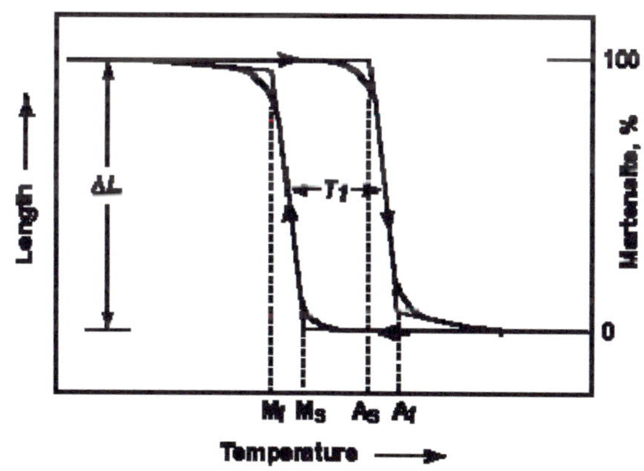

Fig. 5. Shape memory alloy phase transformation

When martensite is heated, it begins to change into austenite and the temperatures at which this phenomenon starts and finishes are called austenite start temperature (A_s) and respectively austenite finish temperature (A_f). When austenite is cooled, it begins to change into martensite and the temperatures at which this phenomenon starts and finishes are called martensite start temperature (M_s) and respectively martensite finish temperature (M_f) (Buehler et al. 1967).

Several properties of austenite and martensite shape memory alloys are notably different. Martensite is the relatively soft and easily deformed phase of shape memory alloys, which exists at lower temperatures. The molecular structure in this phase is twinned.
Austenite is the stronger phase of shape memory alloys, which exists at higher temperatures. In Austenite phase the structure is ordered, in general cubic.
The thermoelastic martensitic transformation causes the folowing properties of SMA's (Waram, 1993, Van Humbeeck, 1999, Van Humbeeck, 2001).

- *One-way shape memory effect* represents the ability of SMA to automatically recover the high temperature austenitic shape upon heating, but it is necessary to apply a force to deform the material in the low temperature martensitic state.
- *Two-way shape memory effect* or *reversible shape memory effect* represents the ability of SMA's to recover a preset shape upon heating above the transformation temperatures and to return to a certain alternate shape upon cooling.

Note that both the one-way and two-way shape memory effects can generate work only during heating (i.e. force and motion).

- *All-round shape memory effect* is a special case of the two-way shape memory effect (Shimizu et al. 1987). This effect differs from the two-way effect in the following ways:
  (I) a greater amount of shape change is possible with the all-around effect,
  (II) the high and low temperature shapes are exact inverses of each other, that is a complete reversal of curvature is possible in the case of a piece of shape memory strip.
- *Hysteresis behavior*. Due to processes which occur on an atomic scale, a temperature hysteresis occurs. In other words the austenite to martensite transformation (the "forward reaction") occurs over a lower temperature range than the martensite to austenite transformation. The difference between the transition temperatures upon heating and cooling is called hysteresis. Most SMA's have a hysteresis loop width of 10-50°C.
- *Superelasticity* can be defined as the ability of certain alloys to return to their original shape upon unloading after a substantial deformation has been applied.
- *Vibration damping capacity*. Due to the special micro structural behavior, SMA's exhibit the highest vibration damping property of all metal materials. The damping is non-linear and frequency independent, but it's sensitive to temperature variations and the antecedents of thermal cycling.

3. Design strategies for SMA elements

The first step an engineer should take when undertaking a design involving shape memory material is to clearly define the design requirements. These usually fall into one of the following interrelated areas: operating mode, mechanical considerations, transformation temperatures, force and/or motion requirements, and cyclic requirements.

3.1 Operating modes of SMA's

The most used operating modes of SMA's are:

- *Free recovery* which consists of three steps: shape memory material deformation in the martensitic condition at low temperature, deforming stress release, and heating above the A_f temperature to recover the high temperature shape. There are few practical applications of the free recovery event other than in toys and demonstrations.

▪ *Constrained recovery* is the operation mode used for couplings, fasteners, and electrical connectors.

▪ *Work production – actuators.* In this operation mode a shape memory element, such as a helical springs or a strip, works against a constant or varying force to perform work. The element therefore generates force and motion upon heating.

3.2 Mechanical considerations and design assuptions

The most successful applications of shape memory alloy components usually have all or most of the following characteristics:

- A mechanically simple design.
- The shape memory component "pops" in place and is held by other parts in the assembly.
- The shape memory component is in direct contact with a heating/cooling medium.
- A minimum force and motion requirement for the shape memory component.

The shape memory component is isolated ("decoupled") from incidental forces with high variation.

The tolerances of all the components realistically interface with the shape memory component.

3.3 Transformation temperatures

The force that a spring or a strip of any material produces at a given deflection depends linearly on the shear modulus (rigidity) of the material. SMA's exhibit a large temperature dependence on the material shear modulus, which increases from low to high temperature. Therefore, as the temperature is increased the force exerted by a shape memory element increases dramatically [Dolce, 2001]. Consequently the determination of the transformation temperatures is necessary to establish the shear modulus values at these functional temperatures for a high-quality design.

This section presents the transformation temperatures obtained for the studied SMA elements (strip and helical spring) using Thermal Analysis Methods. Ni-Ti-Cu (Raychem proprietary alloy) is the material used for the two SMA elements.

Thermal Analysis Methods comprises a group of techniques in which a physical property of a sample is measured as a function of temperature, while the sample is subjected to a controlled temperature program.

Thermogravimetric Analysis (**TGA**), **D**ifferential **T**hermal **A**nalysis (**DTA**) and **D**ifferential **S**canning **C**alorimetry (**DSC**) methods were used to determine the required parameters.

TGA is a technique which relies on samples that decompose at elevated temperatures. The **TGA** monitors changes in the mass of sample on heating.

In **DTA**, the temperature difference that develops between a sample and an inert reference material is measured, when both are subjected to identical heat-treatments. **DTA** can be used to study thermal properties and phase changes.

The related technique of DSC relies on differences in energy required to maintain the sample and reference at an identical temperature.

The **DTA** and **DSC** curves use a system with two thermocouples. One of them is placed on the sample and the other on the reference material.

In this paper, both isothermal and non-isothermal regimes combined with heating-cooling experiments, were used in order to characterize SMA test samples.
The measurements were carried out on a Perkin Elmer Thermobalance in dynamic air atmosphere, in the aluminium crucible.
The test sample's phase transitions were identified by analyzing their behavior at programmed heating up to 200°C and cooling at ambient temperature. In addition we can notice that the sample's mass does not undergo any changes at heating and cooling. In consequence, the TGA curves are ignored in further measurements.

3.3.1 SMA strip transformation temperature

The temperature-control program used for SMA strip measurements contains the following sequences:

- heating from 30°C to 160°C at 5°C/min;
- holding for 10 min at 160°C;
- cooling from 160°C to 20°C at 5 °C/min.

The measurements were carried out in dynamic air atmosphere. The results are presented in Fig. 6.
By analyzing Fig. 6 we can observe two phase transitions. The first occurs during the heating process while the second one appears during the cooling process. The details of these thermal effects are presented in Fig. 7 and Fig. 8 (reported from the DSC curve). Figure 6 shows that the determined transformation temperatures at heating (martensite to austenite) are A_s=80°C and A_f=111°C. The enthalpy of the endothermal transition process is ΔH_h = 36.8858 J/g. The temperature corresponding to maximum transformation speed is 98.79°C.
The transformation temperatures at cooling (austenite to martensite) result from Fig. 8: M_s=69°C and M_f=48.25°C. The enthalpy of the exothermal transition process is ΔH_c=-28.7792 J/g and the temperature corresponding to maximum transformation speed is 59.75°C.

3.3.2 SMA helical spring transformation temperature

The transformation temperatures of SMA helical spring are obtained by similar measurements as in the case of SMA strip, using the following temperature-control sequences:

- heating from 30°C to 100°C at 5°C/min;
- holding for 10 min at 100°C;
- cooling from 100°C to 20°C at 5 °C/min.

The form of DTA and DSC curves is similar to the ones represented in Figure 5, for 6.849 mg SMA spring sample.
The determined transformation temperatures at heating (martensite to austenite) are A_s=58.89°C and respectively A_f=67.93°C. The enthalpy of the endothermal transition process is ΔH_h=9.2 J/g and the temperature corresponding to maximum transformation speed is 60.42°C.
The transformation temperatures at cooling (austenite to martensite) are M_s=45°C and M_f=33°C, the enthalpy of the exothermal transition process is ΔH_c = -5.03 J/g and the temperature corresponding to maximum transformation speed is 39.07°C.

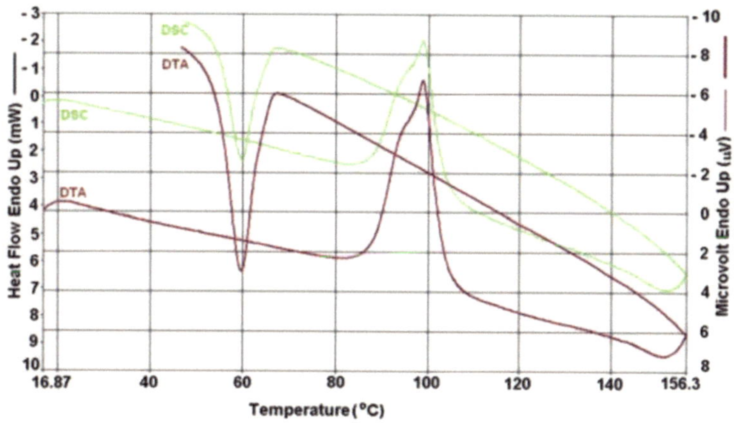

Fig. 6. DTA and DSC curves for 18.275mg SMA strip

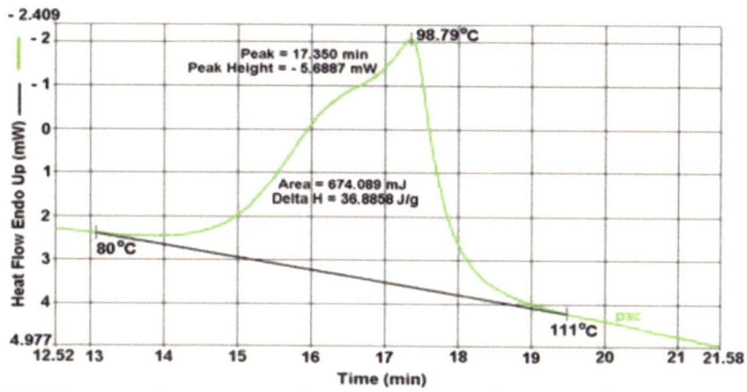

Fig. 7. Detail of DSC curve for computation transition at heating of SMA strip.

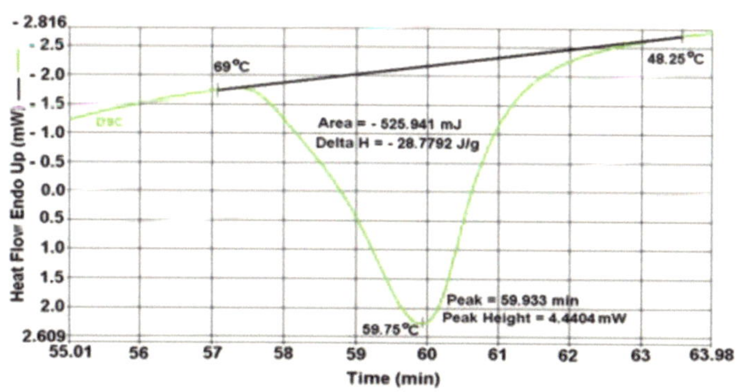

Fig. 8. DTA and DSC curves for 18.275mg SMA Detail of DSC curve for computation transition at cooling of SMA strip.

3.4 Mathematical model of constitutive behavior of shape memory alloy

A variety of mathematical models describing the constitutive behavior have been proposed over the past 15 years, which has not made it an easy for the designer to select. The frequently used SMA constitutive laws are:

- The Landau-Devonshire theory
- The mathematical model of Graesser and Cozzarelli
- The model of Stalmans, Van Humbeeck and Delaey

The Landau-Devonshire (Devonshire,1940) theory is one of the early models introduced. The free energy Ψ of SMA is a function of temperature T and strain ε with positive constants a_i:

$$\Psi(T,\varepsilon) = a_0 + a_1 T\log(T) + a_2(T - T_1)\varepsilon^2 - a_4\varepsilon^4 - a_6\varepsilon^6 \tag{1}$$

The $\sigma(T,\varepsilon)$ stress reaction of the material system, with ρ - material density, in varying the order parameterε, is proportional to the partial derivative of equation with respect to ε:

$$\sigma(T,\varepsilon) = \rho\frac{\partial\Psi(T,\varepsilon)}{\partial\varepsilon} = \rho\left[2a_2(T - T_1)\varepsilon - 4a_4\varepsilon^3 + 6a_6\varepsilon^5\right] \tag{2}$$

Unfortunately the Landau Devonshire theory can only reproduce the isothermal constitutive behavior of SMA. Neither the constant-stress transformation nor the free or constrained memory effect can be modeled.
The mathematical model of Graesser and Cozzarelli (Graesser & Cozarelli, 1994) uses for one dimensionality case only two basic equations. First equation is related to the stress rate $\dot{\sigma}$ and the second equation determinates so-called one dimensional back stress β:

$$\dot{\sigma} = E\left[\dot{\varepsilon} - |\dot{\varepsilon}|\left|\frac{\sigma-\beta}{Y}\right|^{n-1}\left(\frac{\sigma-\beta}{Y}\right)\right] \tag{3}$$

$$\beta = E\gamma\left[\varepsilon - \frac{\sigma}{E} + f_T|\varepsilon|^c \operatorname{erf}(a\varepsilon)\right] \tag{4}$$

Three model parameters are directly related to material (elasticity modulus E, the slope of inelastic region γ, the threshold stress Y) while the others have to be determined empirically (f_T controls the type and size of the hysteresis, c is assumed zero, *n* depends on influence in forming the stress strain hysteresis, a describes the transition from the linear-elastic to the inelastic region and the other way around).

The numerical stability of the model is the main advantage, but the model does not consider the difference between martensite and austenite elasticity modulus.

The model of Stalmans (Stalmans,1994), Van Humbeeck and Delaey (Delaey,1987)0 was developed to describe the change of one-dimensional composite material with embedded SMA wires. Differentiation of the global equilibrium condition gives a generalized Clausius-Clapeyron equation:

$$\left\{\frac{d\sigma_{SMA}}{dT}\right\}_{\varepsilon,v=const} = -\frac{\rho_0\Delta s}{\Delta\varepsilon_{tr\Sigma A}(\xi_V)} = C_A \tag{5}$$

The material constant Δs is the entropy change during transformation from austenite to martensite, ρ_0 is the mass density of SMA material in stress free condition and $\Delta\varepsilon_{tr\Sigma A}(\xi_V)$

is the transformation strain in dependence of the martensite volume fraction, C_A represent the gradient.
During transformation from martensite to austenite, the stress-rate is calculated during the following equation:

$$\frac{d\sigma_{SMA}}{dT} = \frac{\left(\alpha_{mar} - \alpha_{sma}\right) + \dfrac{\left(P_{sma} - E_{sma}\right)\rho_0 \Delta s}{\Delta\varepsilon_{tr\Sigma A}\left(\xi_V\right)P_{sma}E_{sma}} - \dfrac{k_{spr}\alpha_{matr} l}{Q_{matr}E_{matr} + k_{spr} l}}{\dfrac{1}{P_{sma}} - \dfrac{Q_{mar}}{Q_{matr}E_{matr} + k_{spr} l}} \tag{6}$$

The parameters implied in the last equation are:
l - the SMA matrix beam, *P* - the initial pseudo plastic strain, the cross section of the SMA wire, *Q* - pseudo elastic modulus, α - the average values of the thermal dilatation coefficient, *E* - the elasticity modulus. The subscript *matr* represent the specific parameters of matrix beam to specific load or temperature.
The stress of the matrix material and the strain temperature can also be found out. However the model can so far only describe the transformation from martensite to austenite.
Comparing the different models (Schroeder & Boller,1998) shows that each model has its own characteristic. Each of the models still lacks the one or the other of the proprieties being mentioned. A combination of models, such as done with the models of Stalmans and Brinson leading to a new model called Stalman modified can result in an improvement of the model's proprieties.

3.5 Numerical tools for modelling shape memory alloy behavior

Based a description of shape memory alloy materials, a SMA Simulink block was developed. The characteristic of material is idealized, but the approximations made are suitable for an efficient simulation. The user can indicate the start and stop martensitic and austenitic temperature and the force, momentum evolution.

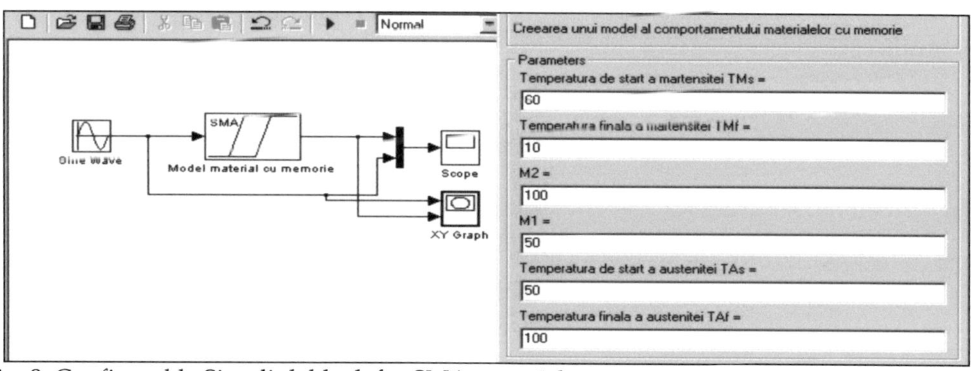

Fig. 9. Configurable Simulink block for SMA material

The numerical results respect the real comportment of the user specified shape memory alloy:

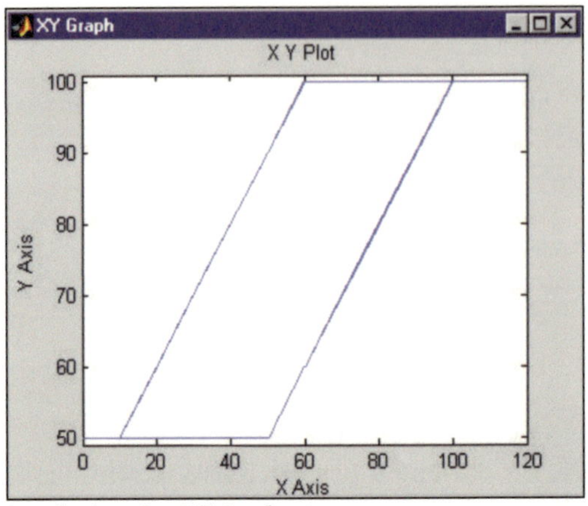

Fig. 10. The numerical simulation for Nitinol

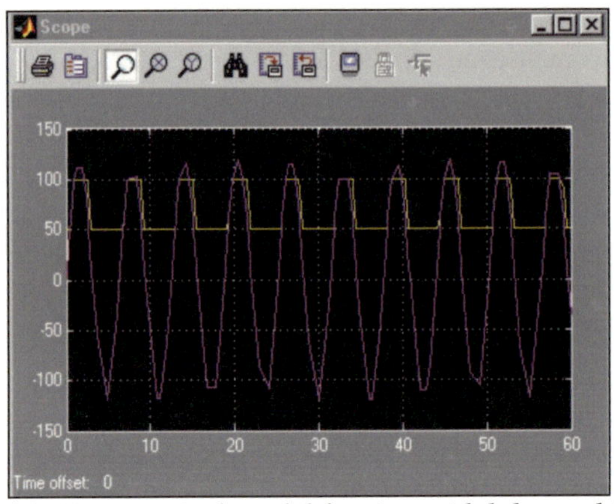

Fig. 11. The response of SMA numerical model for sinusoidal thermal input

The electrical activation of SMA actuator imposes the following relations for the current, temperature and response time:

For heating

$$T_{Max} = 4,928\frac{I}{d} + 1,632\frac{I^2}{d^2} = K_{T1}\frac{I}{\sqrt{F_L}} + K_{T2}\frac{I^2}{F_L} \tag{7}$$

T_{Max} - maximum temperature, d – wire diameter, I – electrical current, F_L - required force

$$t_{Heating} = J_h \ln\frac{T_{Max} - T_{medium}}{T_{Max} - T} \tag{8}$$

$t_{Heating}$ - heating time, Jh - heating coefficient, Tmedium - medium temperature, TA - ambient temperature, T - aim temperature upon heating

$$J_h = 6,72 + 3,922d^2 = K_{J_1} + K_{J_2}F_L \tag{9}$$

For cooling

$$t_c = J_c \ln\frac{T_H - T_A}{T_c - T_A} + \frac{28,88}{M_S - T_A} \quad (10)$$

$$J_c = 4,88 + 6,116d^2 = K_{J_{c1}} + K_{J_{c2}}F_L \quad (11)$$

tc - cooling time, T_H - initial temperature upon cooling, T_A - ambient temperature, d -wire diameter, T_C - aim cooling temperature, J_C - time constant for cooling, M - martensitic start temperature.

One can observe the time dependence of required force and required stroke.

The electrical calculations for direct current heating determine:

- The amount of current needed for actuation in the required time
- The resistance of the nickel titanium actuation element
- The voltage required to drive the current through element
- The power dissipated by the actuation element.

The first requirement can be establish using the material description tables (Waram, 1993).

The resistance is determined using the following expression:

$$\text{Resistance}/\text{mm} = \frac{1,019x10^{-3}}{d^3}\Omega/\text{mm} \quad (12)$$

The voltage and power requirements results from:

$$V = IR\,; \quad \text{Power} = I^2R \quad (13)$$

I - current in amps, V voltage in volts, R resistance in Ω.

In case of using pulse width modulation heating the following relation can be used:

$$\text{duty cycle}(\%) = \frac{t_1}{t_2}x100 \quad (14)$$

t_1 - the width of constant current pulse, t_2 the total cycle time.

$$\text{duty cycle}(\%) = \frac{100}{I_i}\sqrt{\frac{P_{avr}}{R}} \quad (15)$$

$$\text{duty cycle}(\%) = \frac{100}{V_i}\sqrt{P_{avr}R} \quad (16)$$

P_{avg} - average pulsed power (effective DC power), I_i applied pulse current, V_i applied pulsed voltage, R electric resistance.

4. Biomimetics design of mechatronics structure

4.1 Modular adaptive implant

Bionics or Biomechatronics is a fusion science which implies medicine, mechanics, electronics, control and computers. The results of this science are implants and prosthesis for human and animals. The roll of the implants and prosthesis is to interact with muscle, skeleton, and nervous systems to assist or enhance motor control lost by trauma, disease, or defect. Prostheses/implants are typically used to replace parts lost by injury (traumatic) or missing from birth (congenital) or to supplement defective body parts. In addition to the standard artificial limb for every-day use, many amputees have special limbs and devices to aid in the participation of sports and recreational activities.

4.1.1 The parametric 3D model of the bones

To obtain the bone cross sections of the bones, a PHILIPS AURA CT tomograph installed in the Emergency Hospital from Craiova was used –Fig. 12.

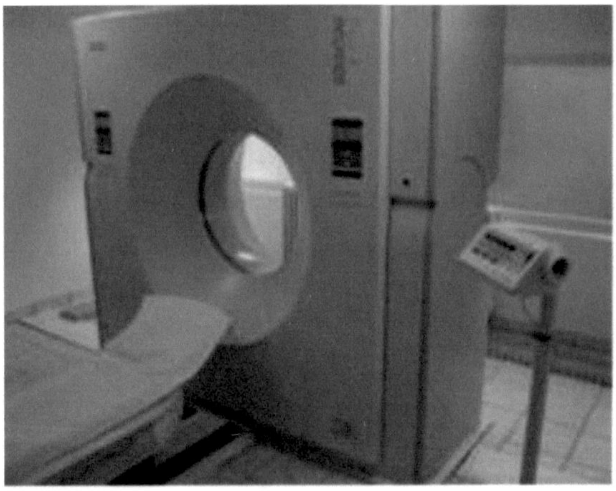

Fig. 12. The PHILIPS AURA CT tomography

To obtain the tomography of the two bones (tibia and femur) were used two scanning schemes presented in Fig. 13 ,Fig. 14 . For the ends of the bones the scanning operation was made at the distances of 1 mm and for the medial areas at the distances of 3 mm.

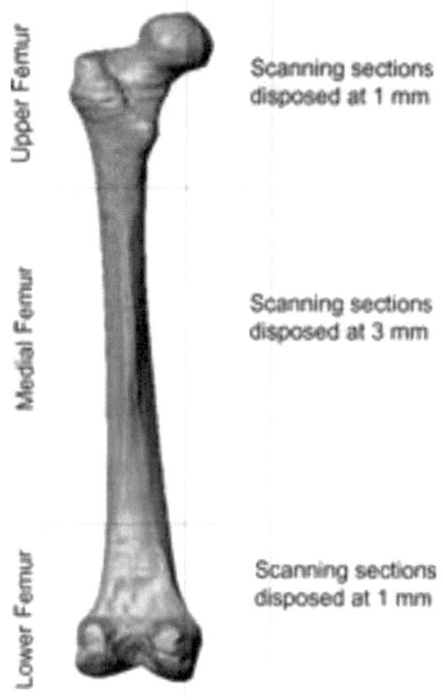

Fig. 13. Scanning schemes applied to the femur

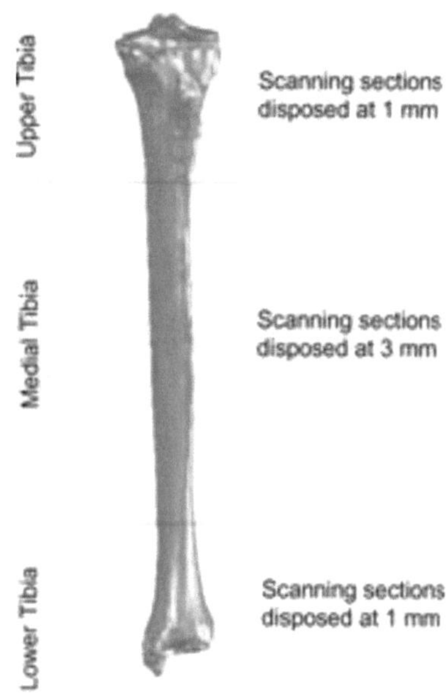

Fig. 14. Scanning schemes applied to the tibia

Were obtained 16 images folders and, after a strict selection, were used only 6, 3 for each bone component including the upper and lower areas (scanned at 1 mm) and the medial areas (scanned at 3 mm).

In Fig. 15 were presented two important images of the upper femur in the area of the femoral head.

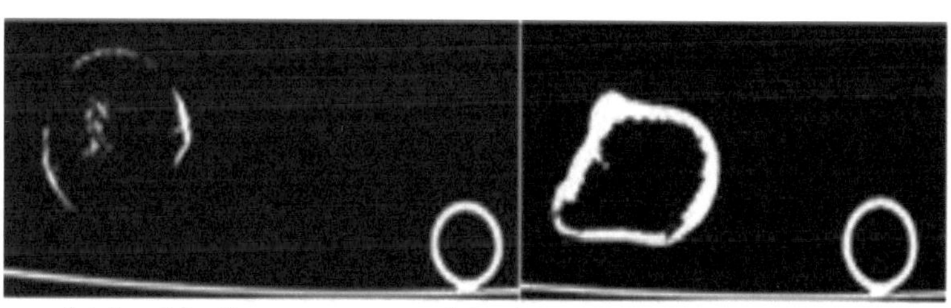

Fig. 15. Images obtained in the upper area of the femur

In Fig. 16 and Fig. 17 are presented main images of the medial and lower femur, which shown the changes of the shape of the bone.

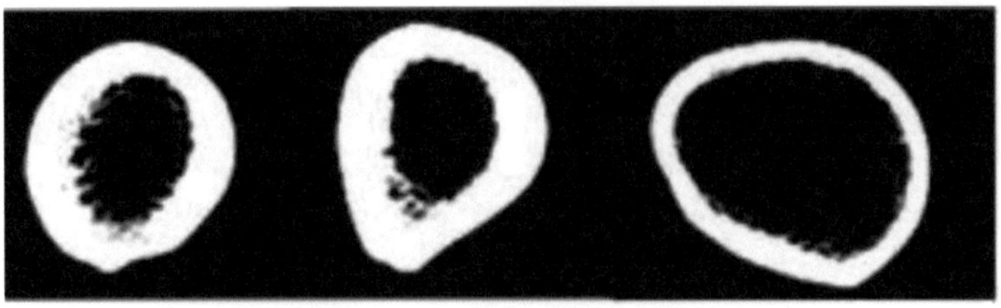

Fig. 16. Images obtained in the medial area of the femur

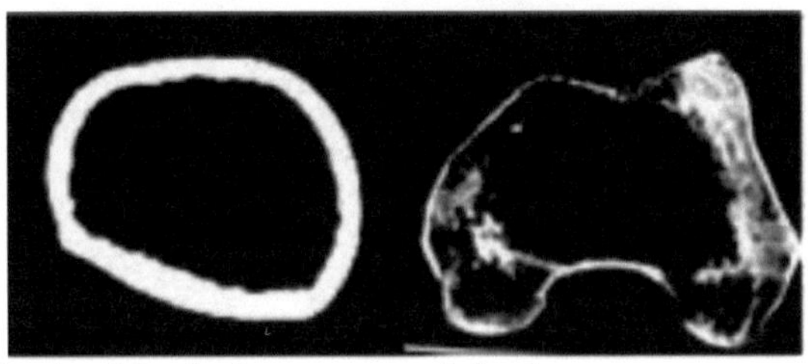

Fig. 17. Two images obtained in the lower area of the femur

In Fig. 18 , Fig. 19 and Fig. 20 important images of the upper tibia, the medial tibia and the lower tibia, which show the shape changes of the bone, are presented.

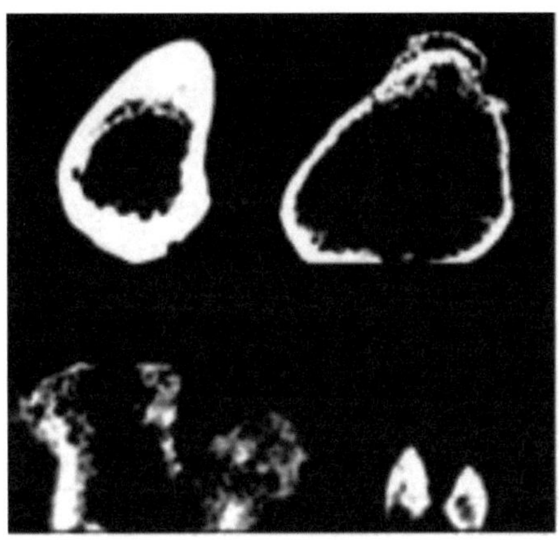

Fig. 18. Four main images of the upper tibia

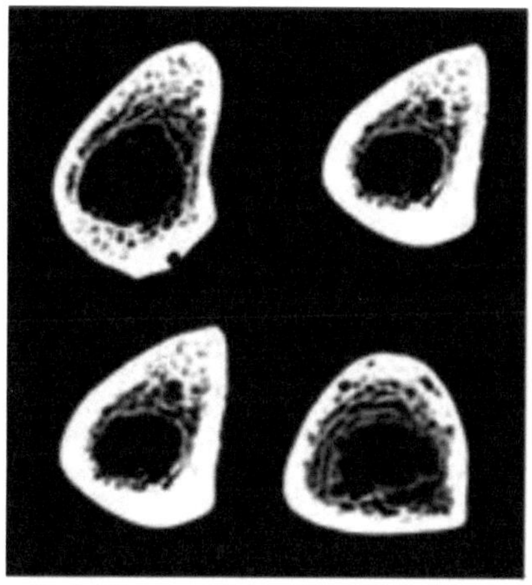

Fig. 19. Four main images made in the medial tibia area

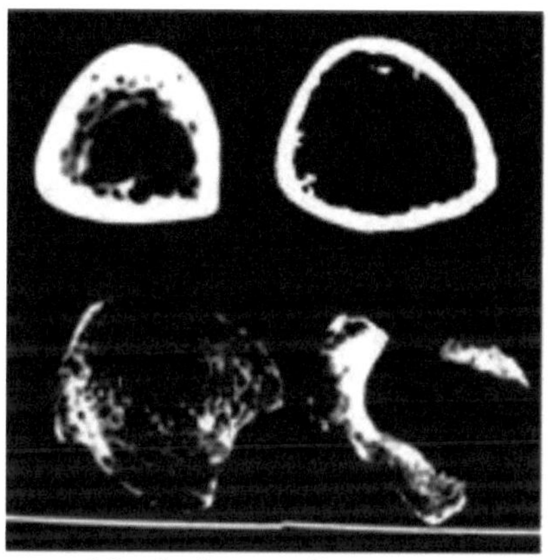

Fig. 20. Four main images scanned in the lower tibia area

The obtained images were re-drawn in AutoCad over the real tomographies and the drawings were imported in SolidWorks (a parametrical CAD software), section by section, in parallel planes. The sketches made in the upper and lower areas for the femur bone and for the tibia bone are presented Fig. 21, Fig. 22 and Fig. 23, Fig. 24.

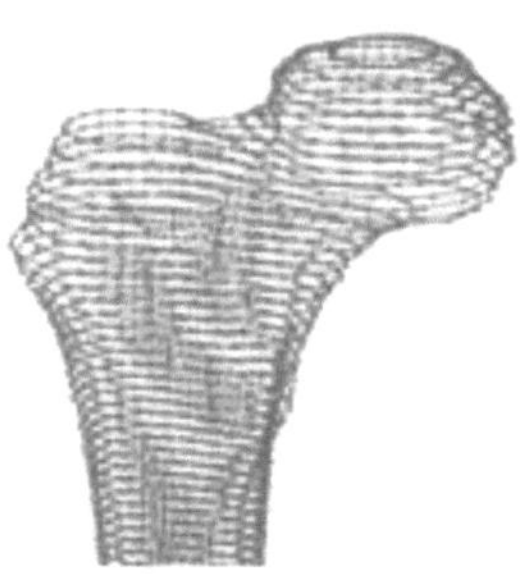

Fig. 21. Sections for the femur bone - upper femur

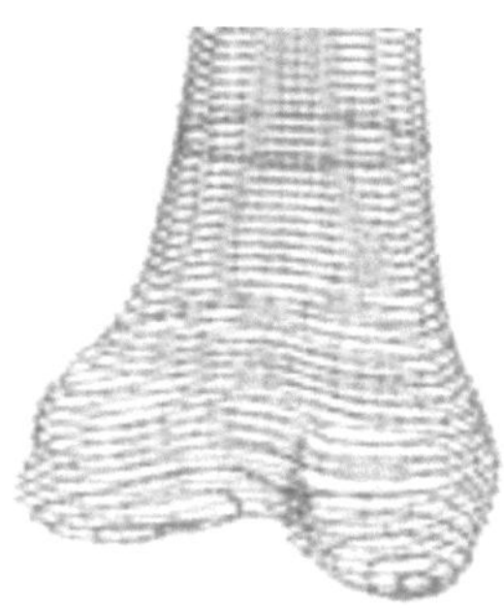

Fig. 22. Sections for the femur bone - lower femur

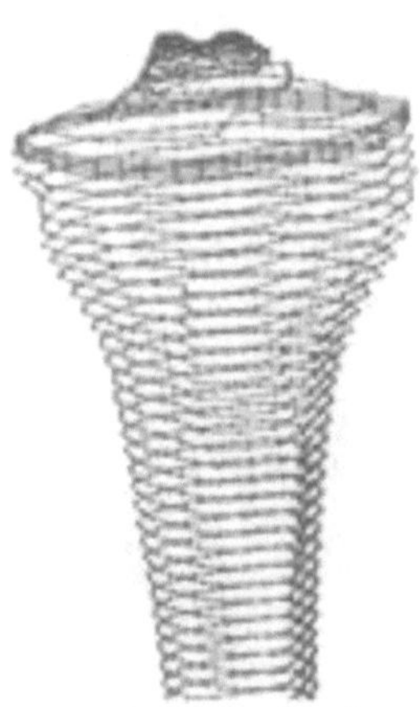

Fig. 23. Sections for the tibia bone - upper tibia

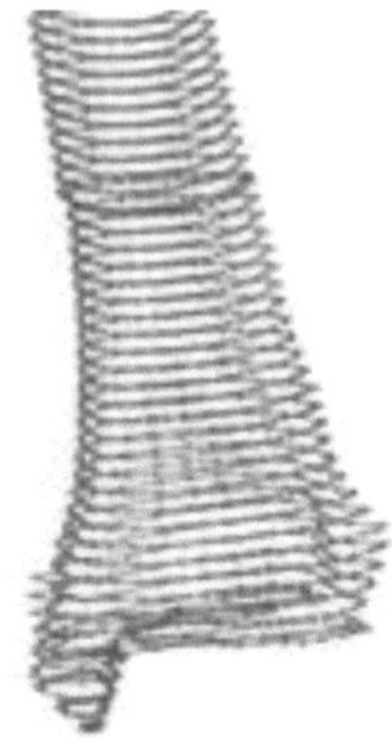

Fig. 24. Sections for the tibia bone - lower tibia

Solidworks permits to obtain a solid by "unifying" the sections drawn in parallel planes. The shape which solidifies these sections is the Loft Shape and it defines the solid starting with the sections and a Guide Curve defined automatically by the software. In Fig. 25, Fig. 26 is presented the definition scheme for the femur bone and for the tibia bone.

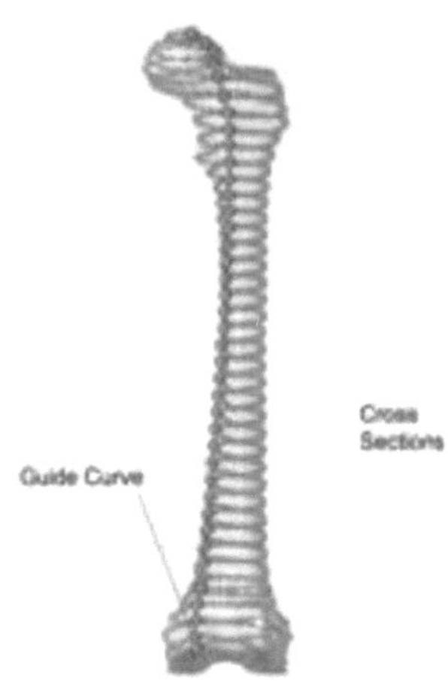

Fig. 25. The definition scheme of the virtual femur

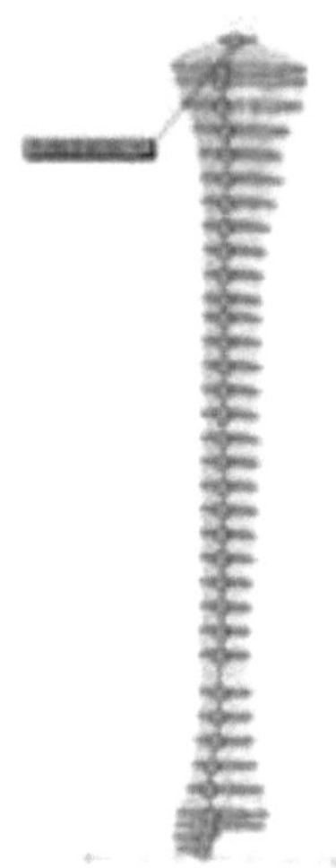

Fig. 26. The definition scheme of the virtual tibia

For defining the virtual models of the bones, the software SolidWorks are used. The virtual models can be transferred into kinematic simulation programs or into finite element analysis programs.

The structure of the femur bone **Fig. 27** has the following characteristics: the finite element size is 10 mm, the structure factor is 0.0132. A number of 13399 nods and 7178 finite elements are obtained for femur.

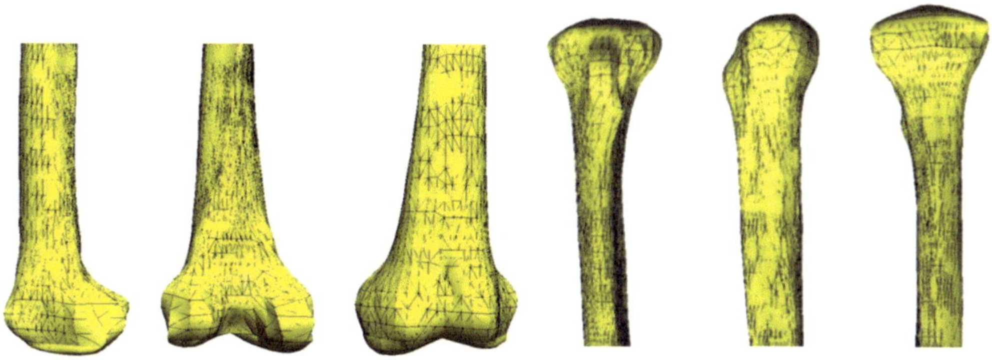

Fig. 27. The mesh structure of the virtual model of the femur bone (first 3) and tibia bone

4.1.2 Studies of normal regime bones mechanical loads

In order to identify the target shape of modular implants, a study for walking regime was developed (Tarnita et al. , 2006). The studies identify the tension maps developed by tibia and femur for different moments. In Fig. **28**, Fig. 29, Fig. **30** is exemplified only few simulation results, important for identifying tension distribution:

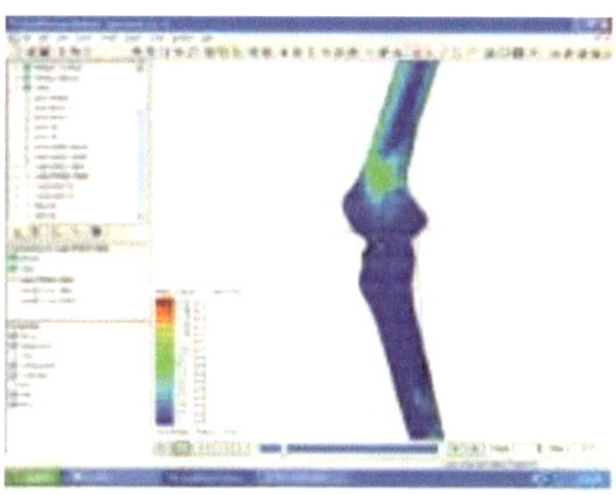

Fig. 28. Tension map for t=0.12 sec., angle tibia-femur α=30°

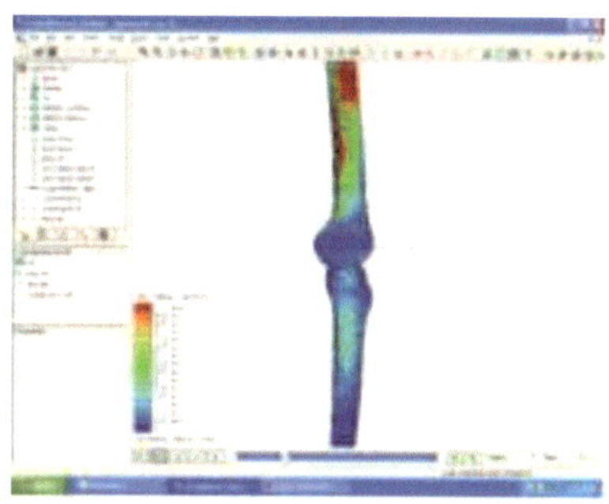

Fig. 29. Tension map for t=0.2 sec

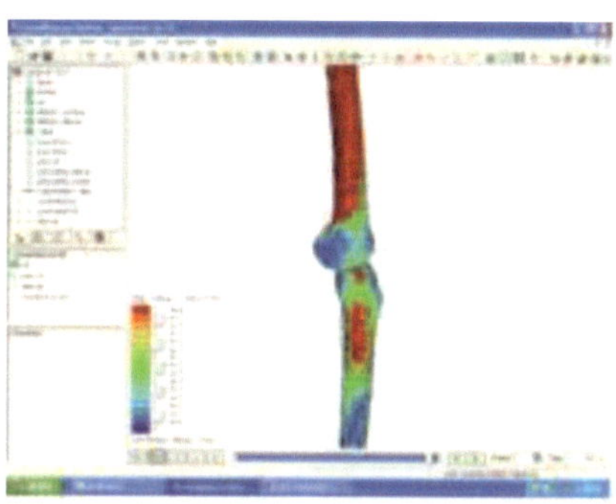

Fig. 30. Tension map for t=0.8 sec

4.1.3 Modular adaptive implant

The design idea of modular adaptive implants results from the following observations:

- the doctors have a limited degree of freedom in selecting the proper dimensional apparatus for bones fractures;

- the current mechanical devices used in orthopedics lose some of their mechanical characteristics after some time (especially elasticity, which should ensure a constant tension that is mandatory for the correct anatomical healing of the fractured bones) (Bizdoaca et al., 2006);

- the process of fracture healing has a particular dynamic, which imposes the necessity of particular progressive tension or discharge to improve the recovery time, depending on the normal structure and function of the bone (Crossley et al., 2006);

- to improve the healing process, the fractured parts have to be in permanent contact in order to ensure the proper conditions to develop bone calluses (Giladi et al., 1987). The actual or external fixator - Ilizarov apparatus - Fig. 31 and Taylor Spatial Frame Fixator - Fig. 32, has to be manually adjusted with respect to the main axis of the bone Fig. 33. Unfortunately, the degrees of freedom of current devices are limited to 3 or 4 vertical screws;

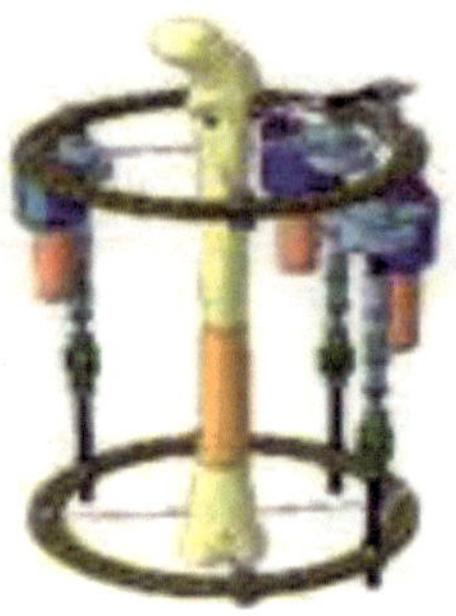

Fig. 31. Iliazov external apparatus

Fig. 32. Taylor Spatial Frame Fixate

Fig. 33. Manual adjust for external apparatus

- a minimally invasive surgery ensures protection from blood edema and improves bone recovery and vascularization of the region.

The solution to these problems is the **M**odular **A**daptive **I**mplant - MAI.

In order to identify the optimal design different implants was developed and experimented using numerical simulation and prototypes:

- Internal implants

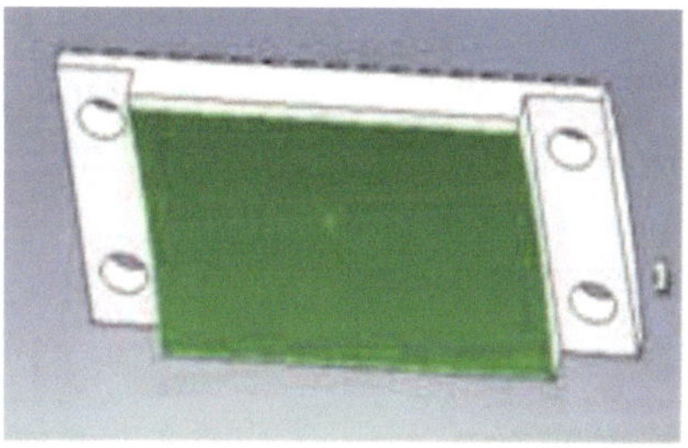

Fig. 34. Internal modular implants - design 1

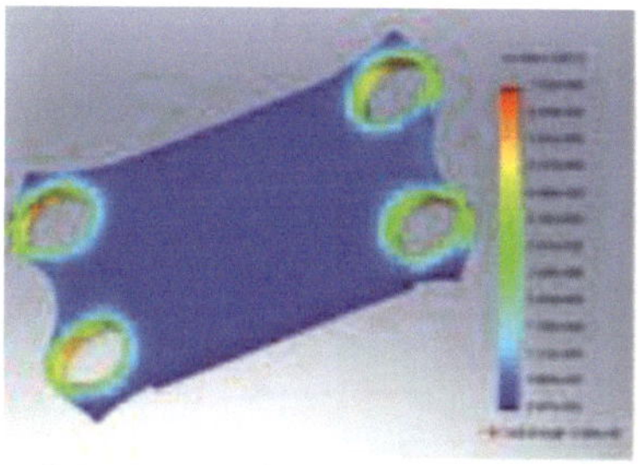

Fig. 35. Internal modular implants - design 1 - accidental tension and forces

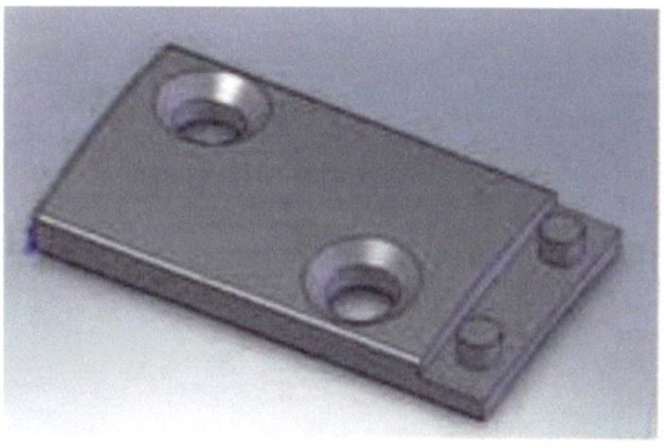

Fig. 36. Internal modular implants - design 2

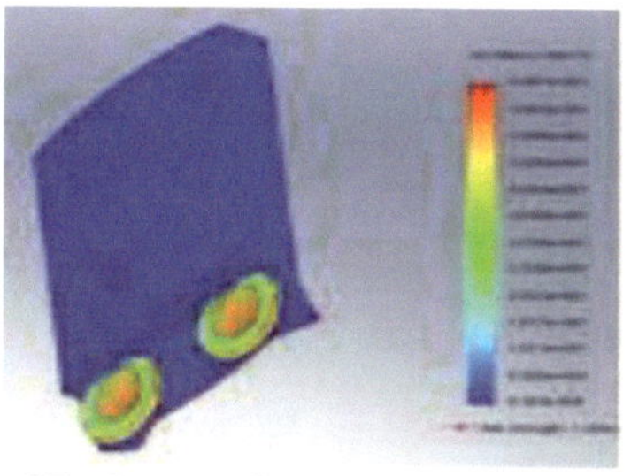

Fig. 37. Internal modular implants - design 2 - accidental tension and forces

Fig. 38. Internal modular implants - design 3

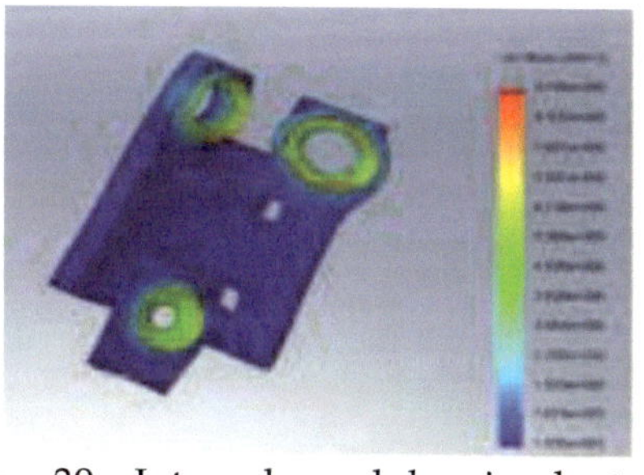

Fig. 39. Internal modular implants - design 3 - accidental tension and forces

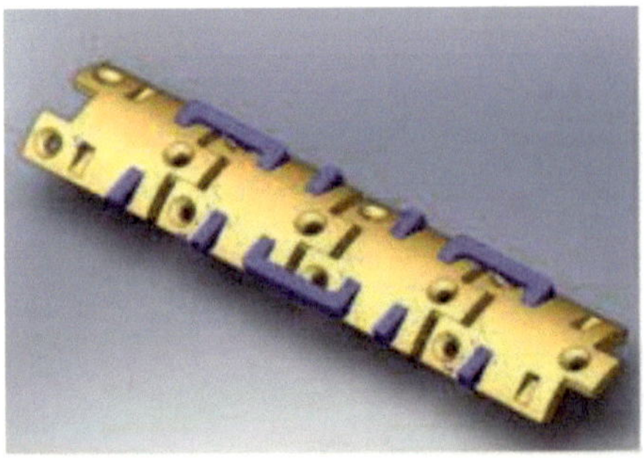

Fig. 40. Internal network implants - design 3

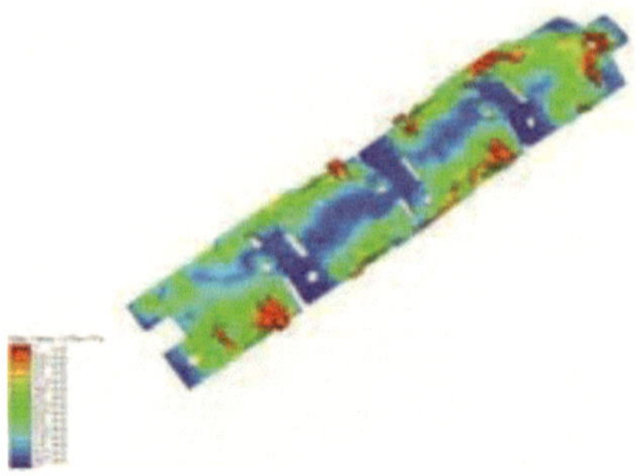

Fig. 41. Internal network implants - design 3 - accidental tension and force

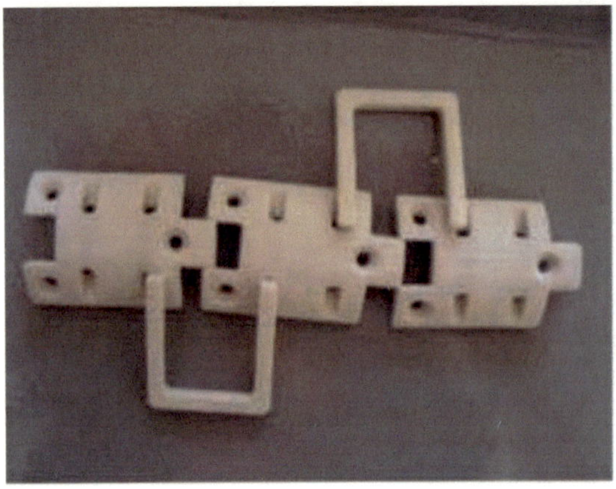

Fig. 42. Internal network implants - design 3 – prototype

External implants and network

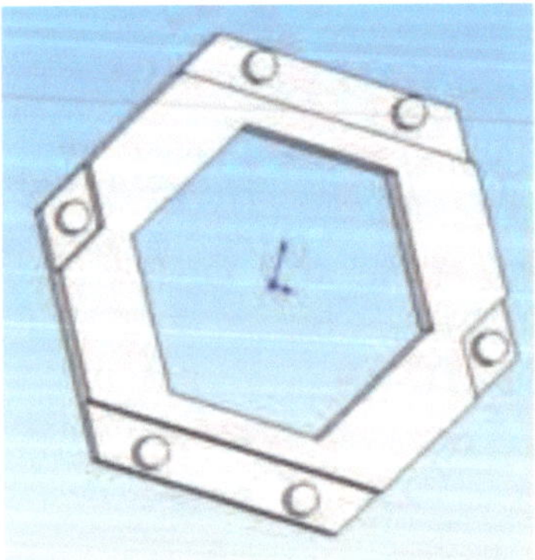

Fig. 43. External modular implants - design 1 – male

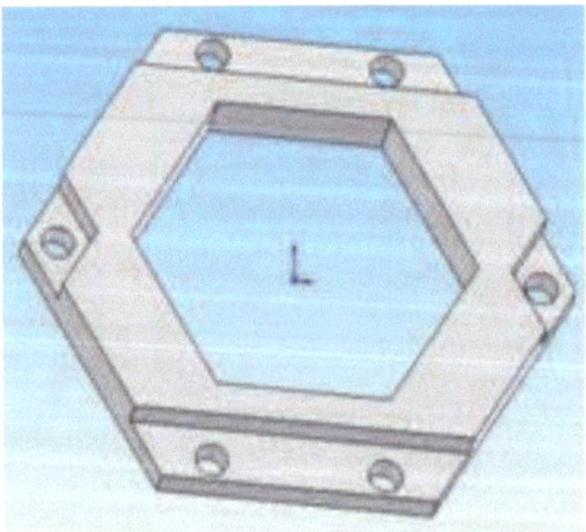

Fig. 44. External modular implants - design 1 – female

Fig. 45. External modular implants - design 1 – prototype

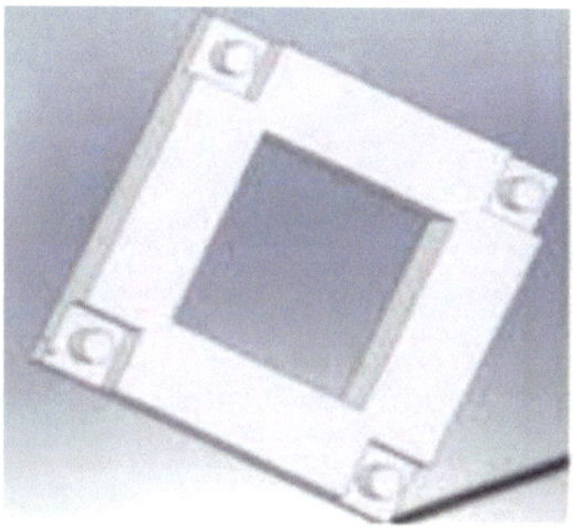

Fig. 46. External modular implants - design 2 – male

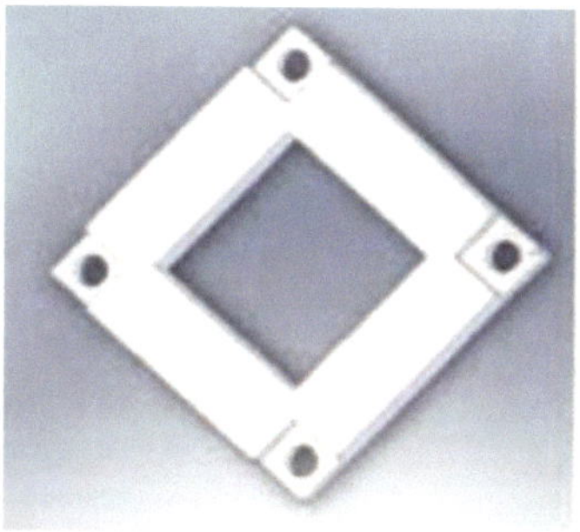

Fig. 47. External modular implants - design 2 – female

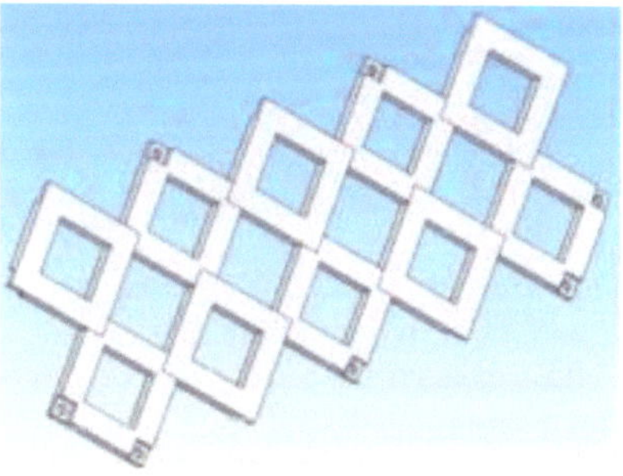

Fig. 48. External modular network implants - design 2

Fig. 49. External modular network implants - design 2 – prototype

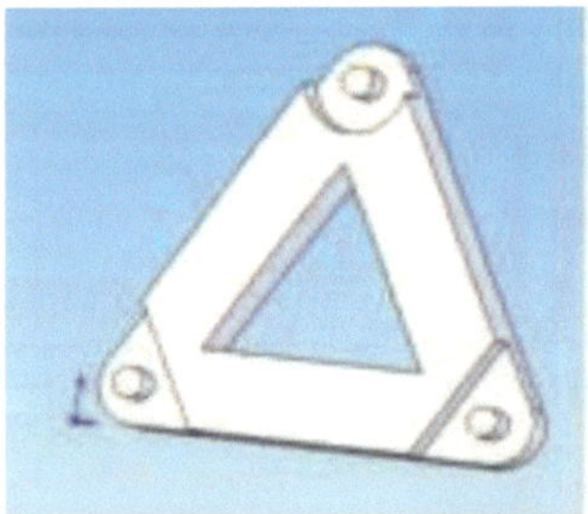

Fig. 50. External modular implants - design 3 - male

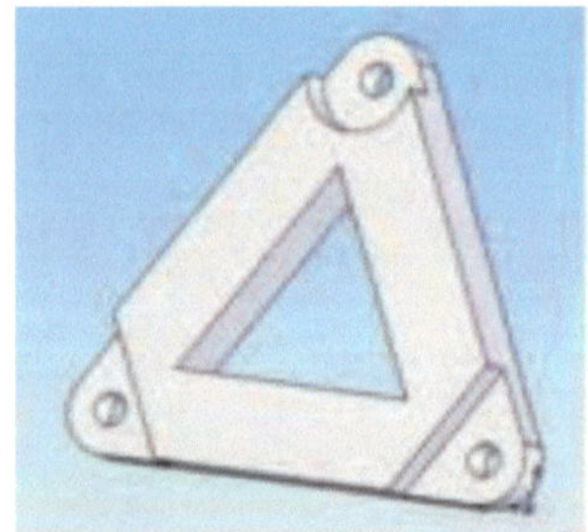

Fig. 51. External modular implants - design 3 - female

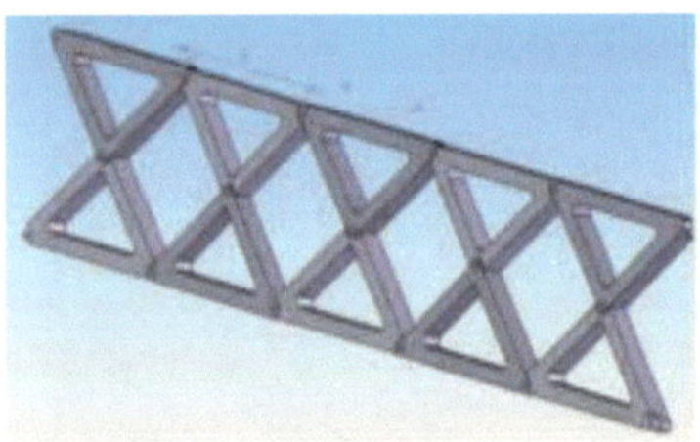

Fig. 52. External modular network implants - design 3

Fig. 53. External modular network implants - design 3 - prototype

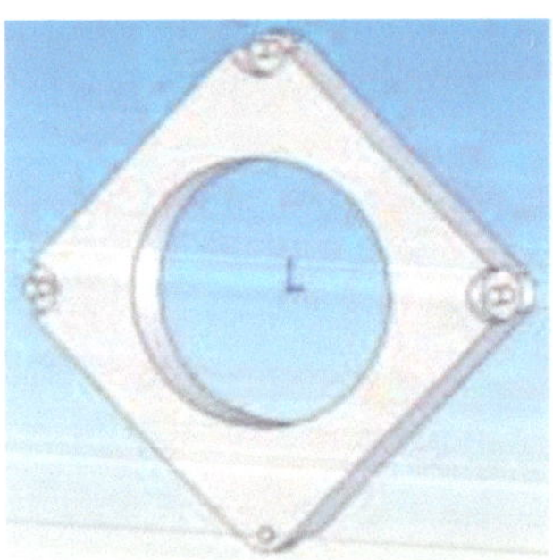

Fig. 54. External modular implants - design 4 - view 1

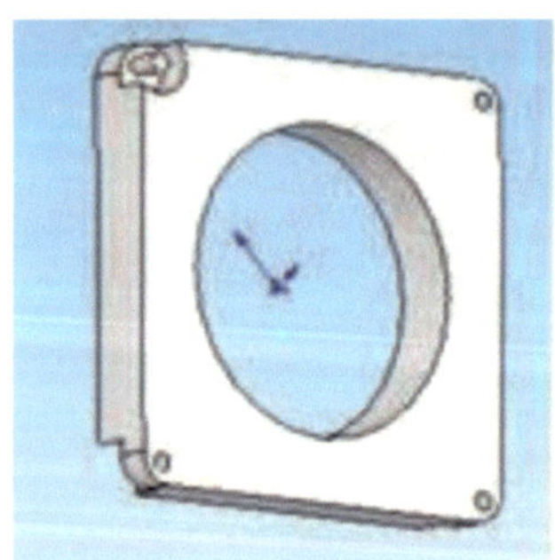

Fig. 55. External modular implants - design 4 - view 2

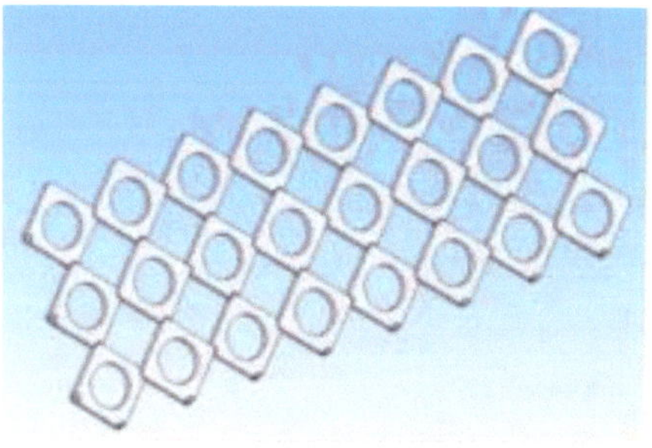

Fig. 56. External modular network implants - design 3

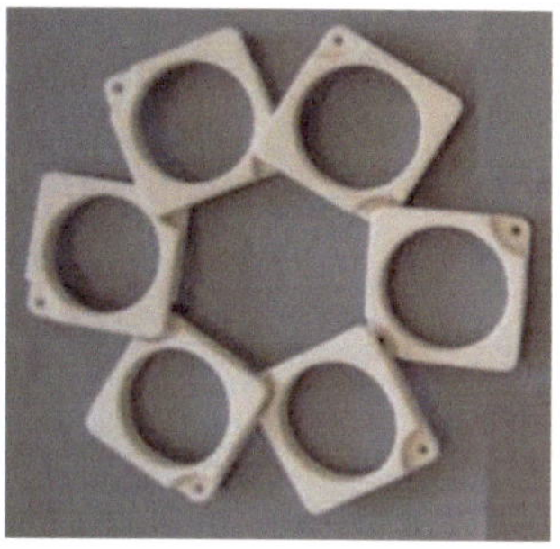

Fig. 57. External modular network implants - design 3 – prototype

The proper shape of MAI is related to the bones microscopic structure and to the numerical simulation presented in the previous chapter. As one can observe, comparing the structure of a healthy bone Fig. 58 with that of an osteoporotic bone Fig. 59, the internal architecture of the healthy bone has a regular modular structure (Burstein et al., 1976) .

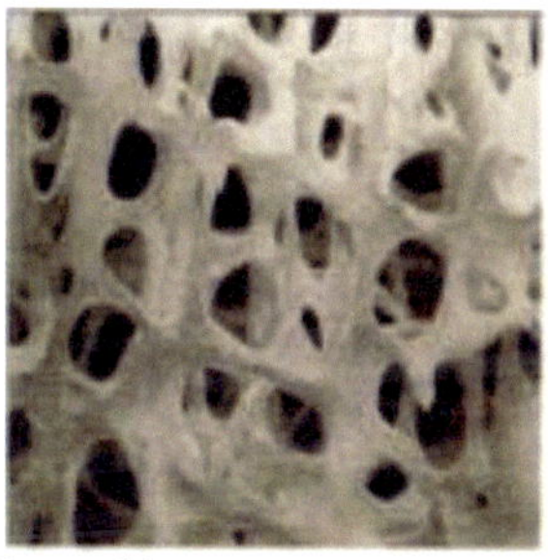

Fig. 58. Normal bone

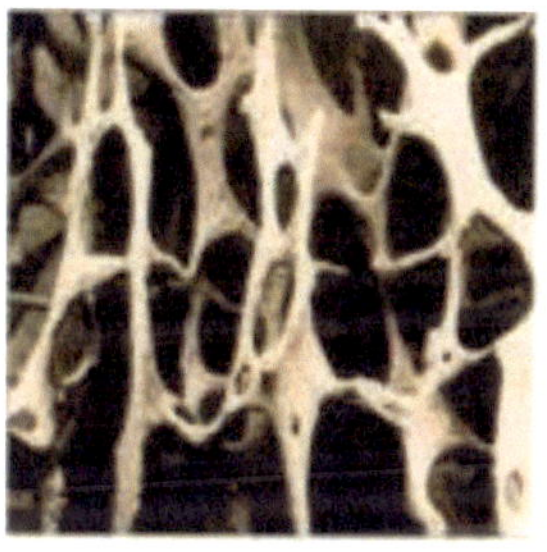

Fig. 59. Osteoporosis affected bone

Osteoporosis affect the bones net and leads to additional load. These problems increase the risks for bones fracture or they can limit the relative bones mobility (Evans, 1976).
A modular net, identical in structure with the bone and locally configurable in terms of tension and release, is best design solution in terms of biocompatibility. The identification of the mechanical solicitation of the particular bone structure, using finite element method, leads to the concept of the practical implementation of a feasible device able to undertake the functionality of normal bones. This device will partially discharge the tensions in the fractured bones (the fractured parts still need to be tensioned to allow the formation of the callus) improving the recovery time and the healing conditions.
The proposed intelligent device has a network structure, with modules made out of Nitinol, especially designed in order to ensure a rapid connection and/or extraction of one or more MAI modules. The binding of the SMA modules ensures the same function as other immobilization devices, but also respects additional conditions concerning variable tension and its discharge. Moreover, these modules allow little movement in the alignment of the fractured parts, reducing the risks of wrong orientation or additional bones callus.

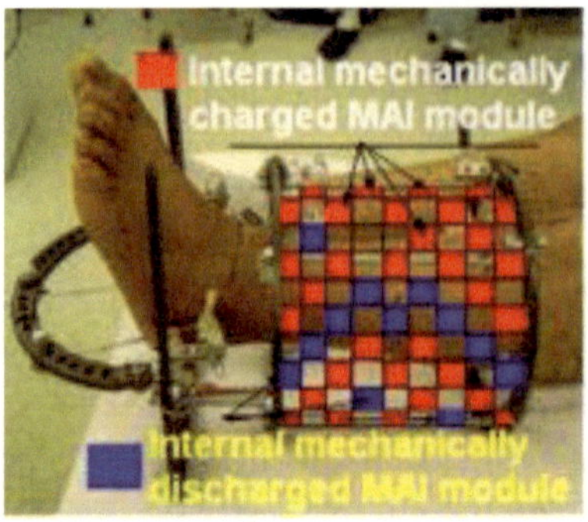

Fig. 60. MAI conceptual connection network

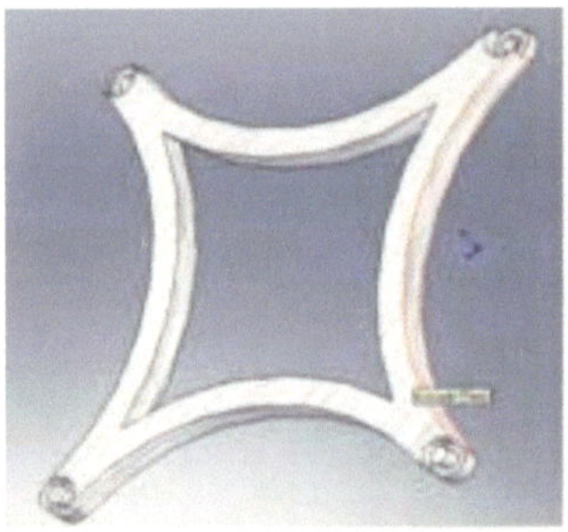

Fig. 61. SMA modul - design 1

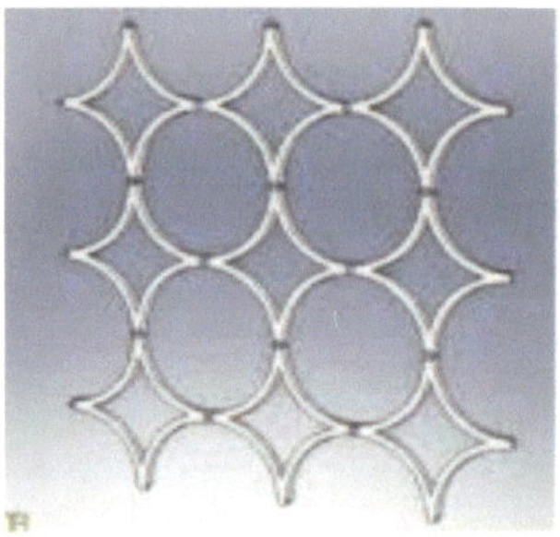

Fig. 62. MAI - SMA network - design 1

We suggest the design shown in Fig. 61 for the unitary SMA module structure, a design which ensures not only the stability of the super-elastic network Fig. 62 and constant force requirements, but also a rapid coupling/decoupling procedure.
Doctors can use SMA modules with different internal reaction tension, but all the modules will have same shape and dimension.
The connection with affected bones and the support for this net are similar to those of a classic external fixator, but allowing for the advantages of minimal invasive techniques.
Using Solid Works package and COSMOS software (Solidworks 98) we proceed to various numerical simulation of SMA module, in order to test the proper mechanical design. First design relives that applying high force -30N and torques to the MAI terminals, the coupling connectors will conduct to spike mechanical deformation Fig. 63, Fig. 64 , potential dangerous for the patients.

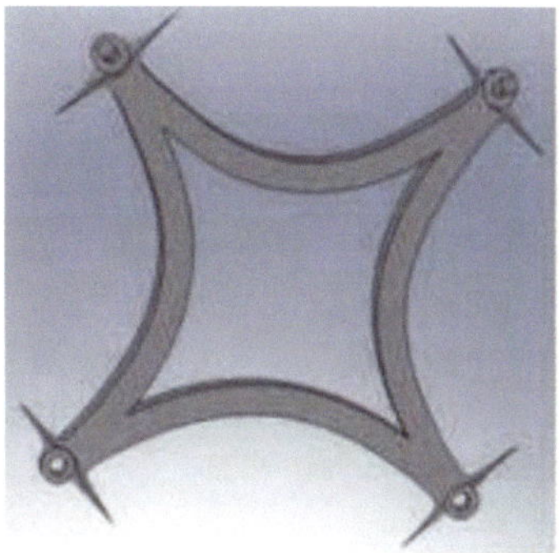

Fig. 63. MAI - design 1 - deformation for accidental tension and forces.

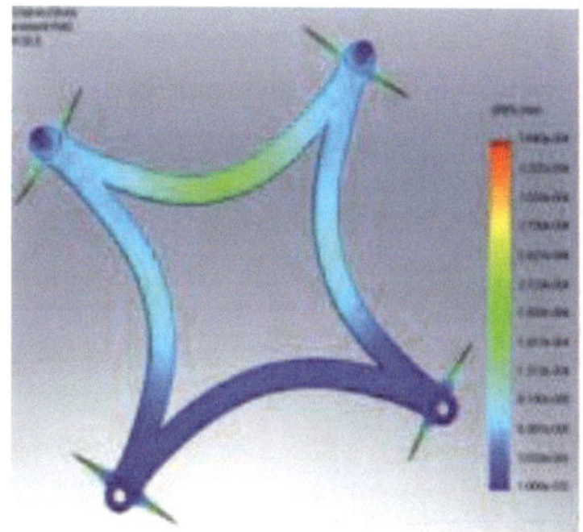

Fig. 64. MAI - design 1 - tension distribution for accidental tension and forces.

The shape of the MAI from Fig. 65 is an improved solution based on previous conclusions. This solution Fig. 66 respects the protection of the patients for accidental unusual mechanical tension.

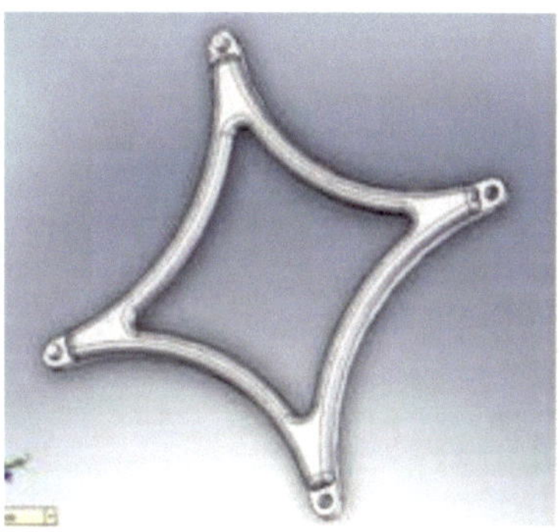

Fig. 65. MAI - SMA module - optimal design

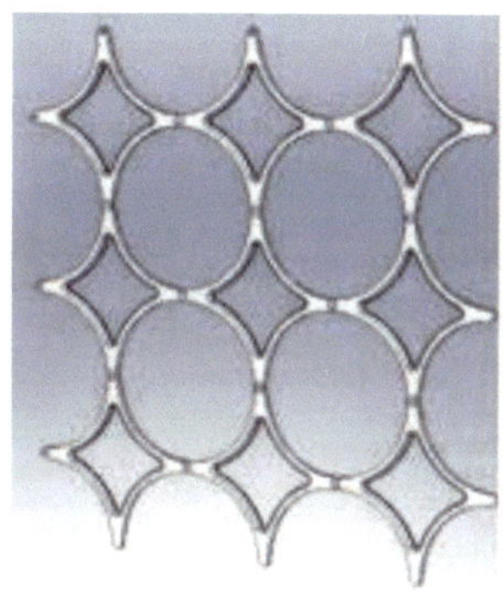

Fig. 66. MAI - SMA network optimal design

In Fig. 67 and Fig. 68, the MAI response to destructive tension and forces, which can appear in accidental cases, is shown.

Fig. 67. Deformations response of MAI - optimal design - for accidental tension and forces

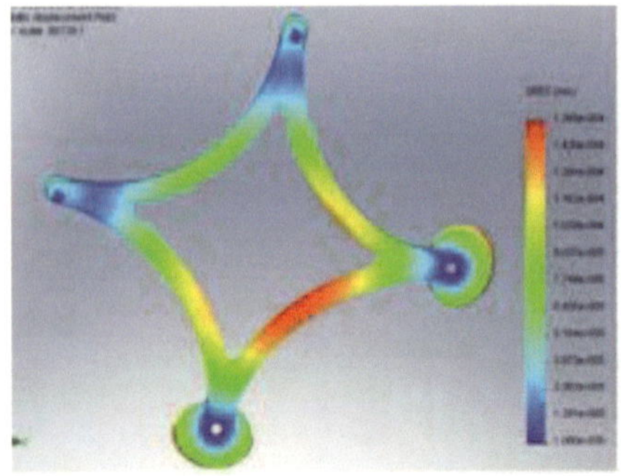

Fig. 68. Tension distribution of MAI - optimal design - for accidental tension and forces

The implant prototype and the experimental MAI network were obtained using a rapid prototyping device - 3D Printer Z Corp -Fig. 71.

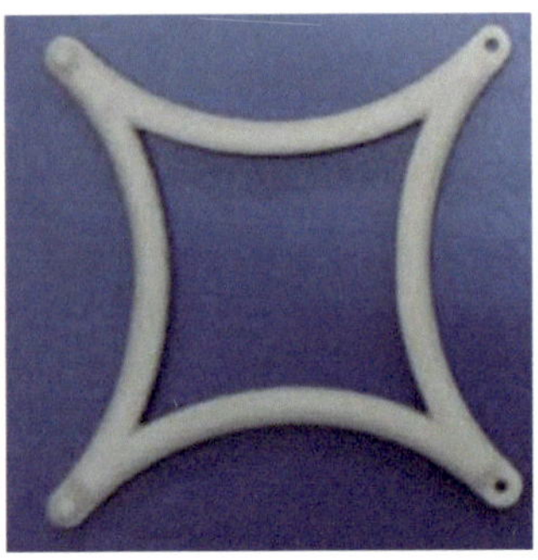

Fig. 69. Implant prototypes MAI network - optimal design

Fig. 70. Implant prototypes MAI network - optimal design

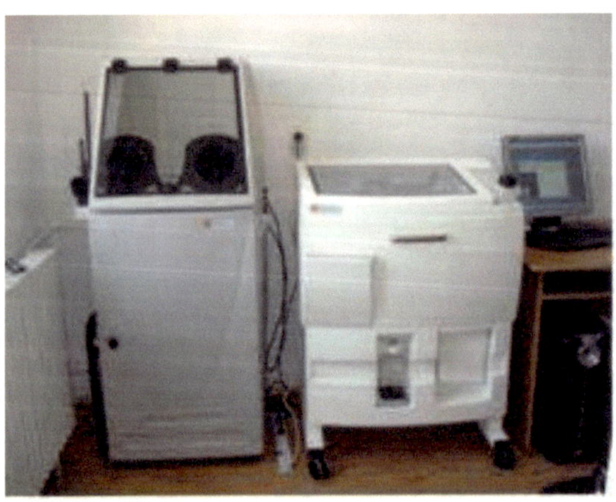

Fig. 71. Rapid prototyping device - 3D printer

The new device leads to a simple post-operatory training program of the patient. The relative advanced movement independence of patient with MAI network apparatus can lead to possibility of short distance walking. Actual devices (Fig. 72) are quite expensive and implies, for implementing, 3 persons: the patient, the current doctor and a kinetoterapeut.

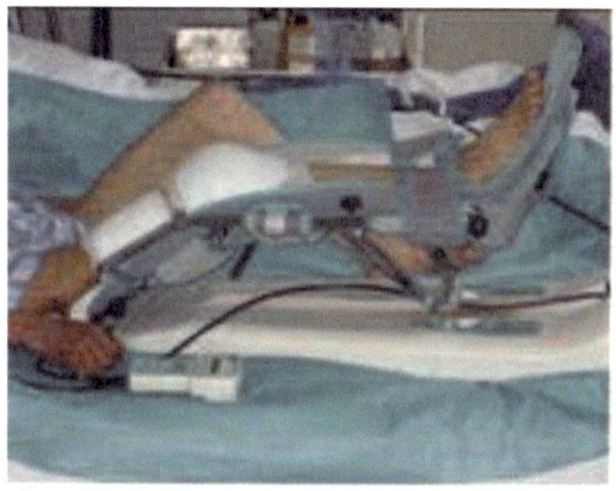

Fig. 72. KINETEK device for functional

4.2 Two-link tendon-driven finger

4.2.1 Dynamics of two-link finger

There are many methods for generating the dynamic equations of mechanical system. All methods generate equivalent sets of equations, but different forms of the equations may be better suited for computation different forms of the equations may be better suited for computation or analysis. The Lagrange analysis will be used for the present analysis, a method which relies on the energy proprieties of mechanical system to compute the equations of motion. We consider that each link is a homogeneous rectangular bar with mass *m_i* and moment of inertia tensor.

$$I_i = \begin{bmatrix} I_{xi} & 0 & 0 \\ 0 & I_{yi} & 0 \\ 0 & 0 & I_{zi} \end{bmatrix} \tag{17}$$

Letting $v_i \in R^3$ be the translational velocity of the center of mass for the ith link and $\omega_i \in R^3$ be angular velocity, the kinetic energy of the manipulator is:

$$T(\theta,\dot{\theta}) = \frac{1}{2}m_1\|v_1\|^2 + \frac{1}{2}m_1\omega_1^T I_1\omega_1 + \frac{1}{2}m_1\|v_2\|^2 + \frac{1}{2}m_1\omega_2^T I_2\omega_2 \tag{18}$$

Since the motion of the manipulator is restricted to xy plane, $\|v_i\|$ is the magnitude of xy velocity of the centre of mass and ω_i is a vector in the direction of the y axis, with $\|\omega_1\| = \dot{\theta}_1$ and $\|\omega_2\| = \dot{\theta}_1 + \dot{\theta}_2$ .We solve for kinetic energy in terms of the generalized coordinates by using the kinematics of the mechanism. Let $p_i = (x_i, y_i, 0)$ denote the position of the ith centre of mass.

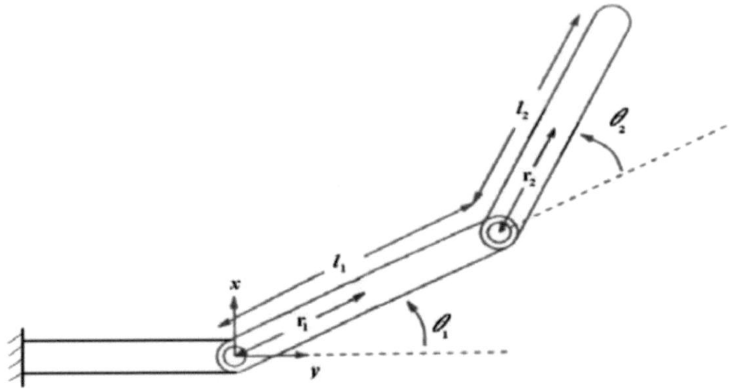

Fig. 73. Two link finger architecture

Letting r_1 and r_2 be the distance from the joints to the centre of mass for each link, results

$$\begin{aligned} x_1 &= r_1\cos(\theta_1) & \dot{x}_1 &= -r_1\dot{\theta}_1\sin(\theta_1) \\ y_1 &= r_1\sin(\theta_1) & \dot{y}_1 &= r_1\dot{\theta}_1\cos(\theta_1) \\ x_2 &= l_1\cos(\theta_1)+r_2\cos(\theta_1+\theta_2) & \dot{x}_2 &= -(l_1\sin(\theta_1)+r_2\sin(\theta_1+\theta_2))\dot{\theta}_1 - r_2\dot{\theta}_2\sin(\theta_1+\theta_2) \\ y_2 &= l_1\sin(\theta_1)+r_2\sin(\theta_1+\theta_2) & \dot{y}_2 &= -(l_1\cos(\theta_1)+r_2\cos(\theta_1+\theta_2))\dot{\theta}_1 - r_2\dot{\theta}_2\cos(\theta_1+\theta_2) \end{aligned} \tag{19}$$

Using the kinetic energy and Lagranage methods results:

$$\begin{bmatrix} \alpha+\beta c_2 & \delta+\frac{1}{2}\beta c_2 \\ \delta+\frac{1}{2}\beta c_2 & \delta \end{bmatrix}\begin{bmatrix}\ddot{\theta}_1 \\ \ddot{\theta}_2\end{bmatrix} + \begin{bmatrix} -\frac{1}{2}\beta\ s_2\dot{\theta}_2 & -\frac{1}{2}\beta\ s_2(\dot{\theta}_2+\dot{\theta}_1) \\ \frac{1}{2}\beta\ s_2\dot{\theta}_1 & 0 \end{bmatrix}\begin{bmatrix}\dot{\theta}_1 \\ \dot{\theta}_2\end{bmatrix} = \begin{bmatrix}\tau_1 \\ \tau_2\end{bmatrix} \tag{20}$$

where

$$\alpha = \frac{m_1}{12}\left(l_1^2+w_1^2\right)+\frac{m_2}{12}\left(l_2^2+w_2^2\right)+m_1r_1^2+m_2\left(l_1^2+r_2^2\right)$$

$$\beta = m_2l_1l_2$$

$$\delta = \frac{m_2}{12}\left(l_2^2+w_2^2\right)+m_2r_2^2;$$

with w_1, w_2, l_1,l_2 the width and respectively the length of link 1 and link 2.

4.2.2 Tendon actuated fingers.

Consider a finger which is actuated by a set of tendons such as the one shown in Fig. 73. Each tendon consists of a cable connected to a force generator. For simplicity we assume that each tendon pair is connected between the base of the hand and a link of the finger. Interconnections between tendons are not allowed. The routing of each tendon is modelled by an extension function $h_i : Q \to R$. The extension function measures the displacement of the end of the tendon as a function of the joint angles of the finger. The tendon extension is a linear function of the joint angles $h_i(\theta) = l_i \pm r_{i_1}\theta_i \pm \cdots \pm r_{i_n}\theta_n$ with l_i - nominal extension at $\theta = 0$ and r_{i_j} is the radius of the pulley at the jth joint. The sign depends on whether the tendon path gets longer or shorter when the angle is changed in a positive sense. The tendon connection, proposed is a classical one, as is exemplified in Fig. 74.

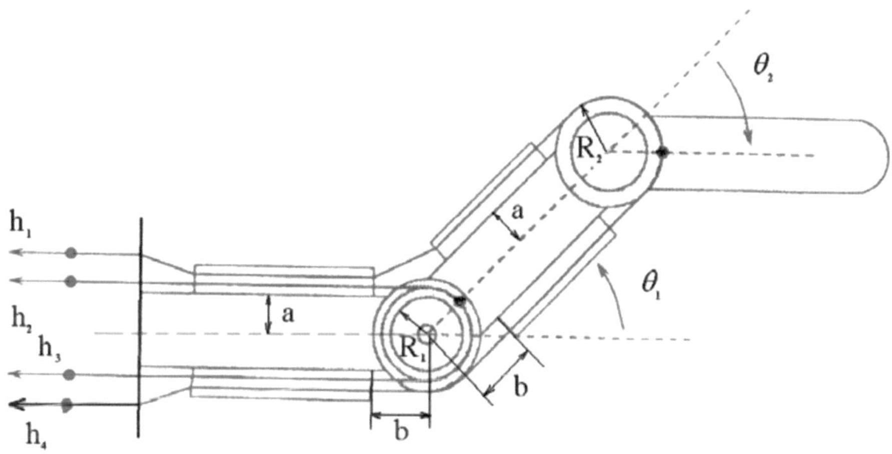

Fig. 74. Geometrical description of tendon driven finger

The extension function of the form is:

$$h_1(\theta) = l_1 + 2\sqrt{a^2 + b^2}\cos\left(\tan^{-1}\left(\frac{a}{b}\right) + \frac{\theta}{2}\right) - 2b \qquad (21)$$

$$\theta > 0 \ ,$$

while the bottom tendon satisfies: $h_2(\theta) = l_2 + R\ \theta \quad \theta > 0$, when $\theta < 0$ these relations are reversed.

Once the tendon extension functions have been computed, we can determine the relationships between the tendon forces and the joint torques by applying conservation of energy. Let $e = h(\theta) \in R^p$ represent the vector of tendon extensions for a system with p tendons and define the matrix $P(\theta) \in R^{nxp}$ as $P(\theta) = \frac{\partial h^T}{\partial \theta}(\theta)$. Then $\dot{e} = \frac{\partial h}{\partial \theta}(\theta)\dot{\theta} = P^T(\theta)\dot{\theta}$.

Since the work done by the tendons must equal that done by the fingers, we can use conservation of energy to conclude $\tau = P(\theta)f$ where $f \in R^p$ is the vector of forces applied to the ends of the tendons. The matrix $P(\theta)$ is called the coupling matrix.

The extension functions for the tendon network are calculated by adding the contribution from each joint. The two tendons attached to the first joint are routed across a pulley of radius R_1, and hence

$h_2 = l_2 - R_1\theta_1$

$h_3 = l_3 + R_1\theta_1$

The tendons for the outer link have more complicated kinematics due to the routing through the tendon sheaths. Their extension functions are

$$h_1 = l_1 + 2\sqrt{a^2+b^2}\cos\left(\tan^{-1}\left(\frac{a}{b}\right)+\frac{\theta_1}{2}\right) - 2b - R_2\theta_2 \tag{22}$$

$$h_4 = l_4 + R_1\theta_1 + R_2\theta_2 \tag{23}$$

The coupling matrix for the finger is computed directly from extension functions:

$$P(\theta) = \frac{\partial h^T}{\partial \theta} = \begin{bmatrix} -2\sqrt{a^2+b^2}\sin\left(\tan^{-1}\left(\frac{a}{b}\right)+\frac{\theta_1}{2}\right) & -R_1 & R_1 & R_1 \\ 0 & 0 & -R_2 & R_2 \end{bmatrix} \tag{24}$$

The pulling on the tendons routed to the outer joints (tendons 1 and 4) generates torques on the first joint as well as the second joint.

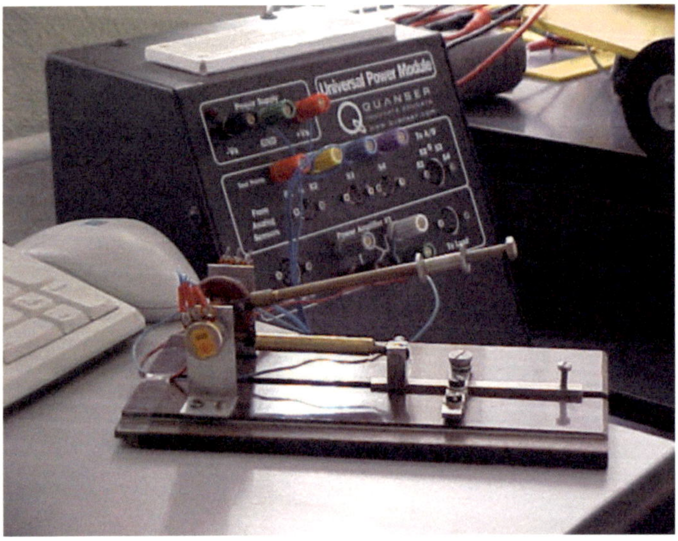

Fig. 75. Experimental model for a single link robotic structure

4.2.3 Numerical simulations

Based on the theoretical background presented, numerical simulations are required in order to evaluate the efficiency of real mechanism. For flexible studies all the elements are developed as configurable Simulink blocks:

- Shape memory alloy block wire - Fig. 9
- Dynamics of two link fingers block - based on equation 20 :

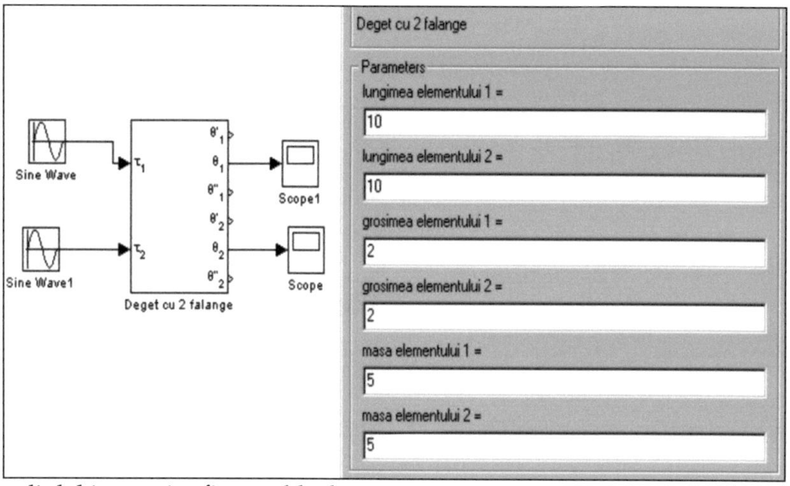

Fig. 76. Two link kinematics fingers block

Coupling block - based on equation 24

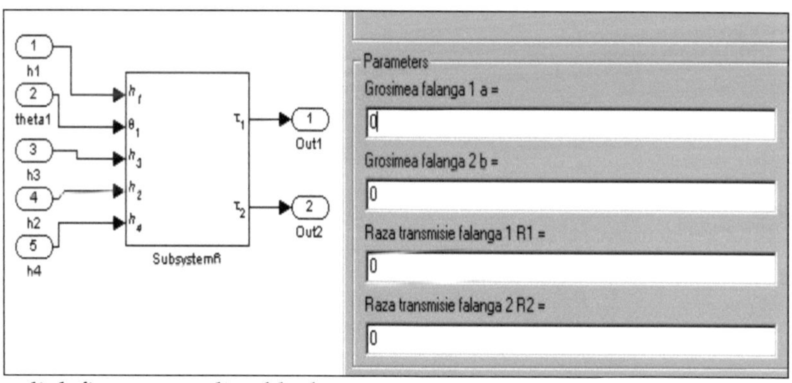

Fig. 77. Two link fingers coupling block

Connecting all this blocks , for numerical simulations the following parameters are used:

- for the elements of the finger:
  with = 2 cm
  length = 10 cm
  mass = 5 g
  radius pulley = 2cm
  height = 2 cm

- shape memory alloy parameters:
  start temperature of martensitic state = 60^{0}C
  final temperature of martensitic state = 10^{0}C
  start temperature of austenite state = 50^{0}C
  final temperature of austenite state = 100^{0}C
  lower force M_1 = .5 N
  higher force developed by the SMA M_2 = 5 N

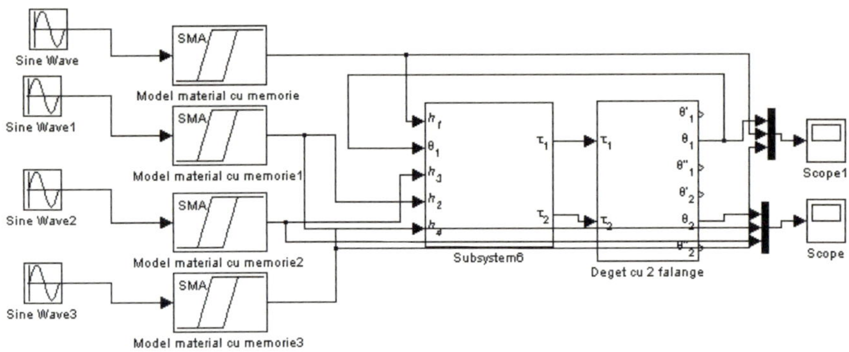

Fig. 78. Complete structure of two link fingers block

The results of numerical simulations are:

- For the element 1 of the finger:

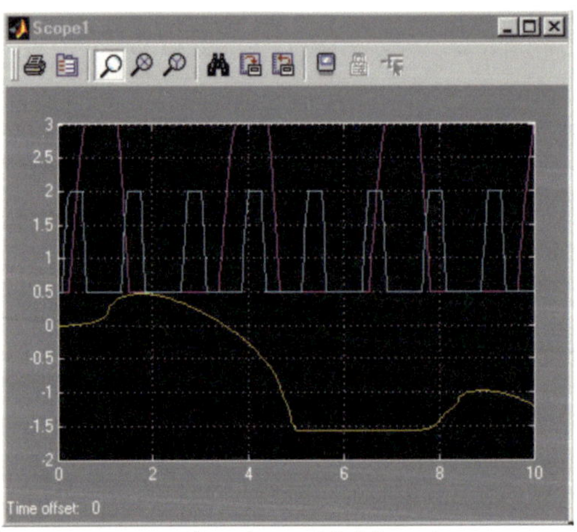

Fig. 79. Finger first phalange response for sinusoidal input

➢ For the second element of the finger:

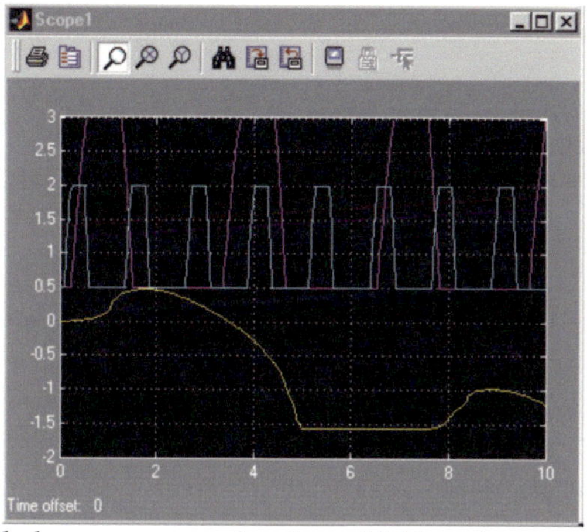

Fig. 80. Finger second phalange response for sinusoidal input

In the upper part of the figure is modelled the evolution of the SMA wire and in the lower part of the figure is the evolution of the angle θ_i of the joint. In the mathematical model the architectural limitation are implemented by imposing the angle evolution as $\theta_i \in \left[-\frac{\pi}{2}, \frac{\pi}{2}\right]$

4.3 Hopping robot

In this section is presented a model for hopping robot, based on the physical laws of mechanics. The variables in the model are positions and velocities, and the dynamic equations come from Newton's laws of motion. When humans walk, feet never lose contact with the ground and alternate between having both feet on the ground and a swing phase in which one foot is on the ground and the second leg swings like a pendulum. When run, we alternate between a flight phase in which both feet are off the ground and a stance phase in which one foot is on the ground. Kangaroos hop with a flight phase alternating with a stance phase in which both feet are on the ground simultaneously. A pogo stick with a single leg provides a simple model for running and hopping Fig. 81.

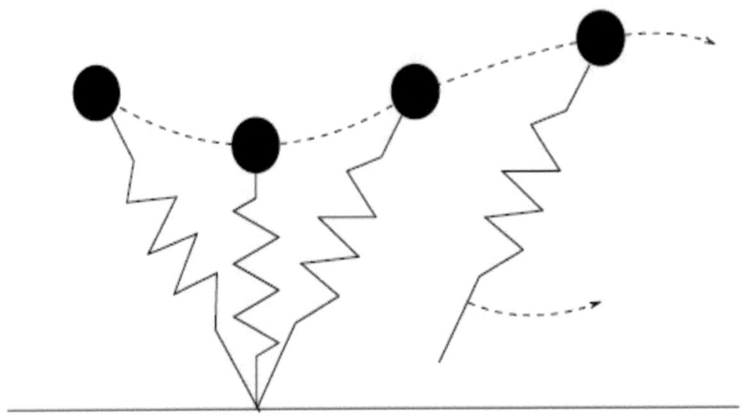

Fig. 81. A simple model for running and hopping

The robot is regarded as a "point mass" with a springy leg attached to the mass. As a first approximation, one can think of the spring as the tendons in the leg. By contracting muscles, the animal changes the force of the leg spring, enabling it to bounce off the ground. When in flight, we assume that the animal is able to swing the leg so that it will point in a new direction when the animal lands on the ground. At landing, the leg shortens, compressing the spring. The compressed spring exerts a vertical upward force that together with additional force exerted by the muscles propels the animal into its next flight phase. Simplifying, the hopping movement can be divided in hopping in place and hopping forward or backward.
Hopping in place and hopping forward/backward can both be divided into two different phases: an aerial phase (where the mass is airborne) and a ground phase (where the mass and spring are on the ground).

4.3.1 Hopping In Place

When the leg hops in place, the model is one of a spring, where the entire force of the spring is directed in the +y direction. The motion results from gravity trying to pull the mass toward the earth and the spring trying to push the mass away from the earth. In this case there is only one variable (y) because motion is only in one direction. As mentioned earlier, hopping in place has an aerial phase and a ground phase, with a different differential equation describing each.

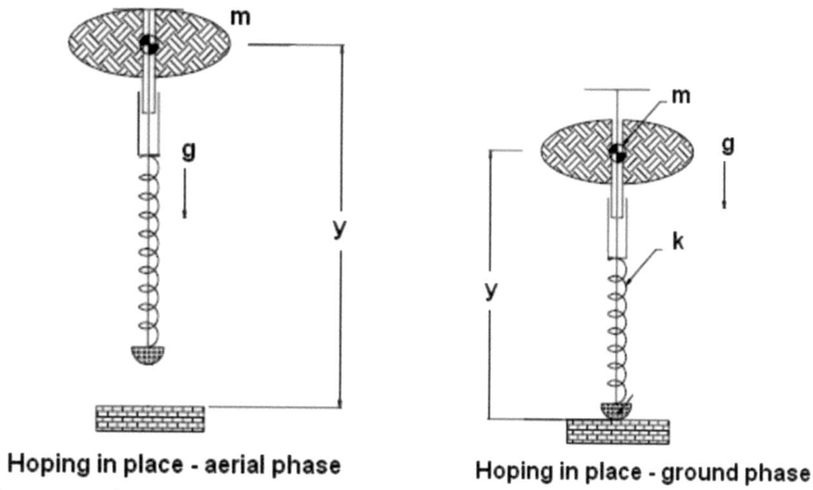

Fig. 82. Hoping in place pozitions

For hopping in place, the equations are:

$$m\frac{d^2y}{dt^2} = -mg \tag{25}$$

$$m\frac{d^2y}{dt^2} + ky = mg \tag{26}$$

For equation 25, while the mass is in the air, the acceleration of the object is equal to the acceleration due to gravity (9.8 m/s²) in the -*y* direction, and therefore as the mass goes up, it starts to decelerate to zero m/s and then begins to accelerate as it falls down.
The evolution of this simplified hopping robot have no influence regarding the stiffness of the spring, or actuator nature.
Equation 26 shows that force of the spring subtracted from the force of gravity equals the mass multiplied by its acceleration. This equation describe the touch down moment. For this simple model, last equations, if we choose to experiment the influence of temperature, in case of using a SMA spring, we obtain the results exemplified in Fig. 83.
From this simulation the advantage of using SMA spring is clear: a smart spring improve by reducing the touch-down time, and the energy losses.

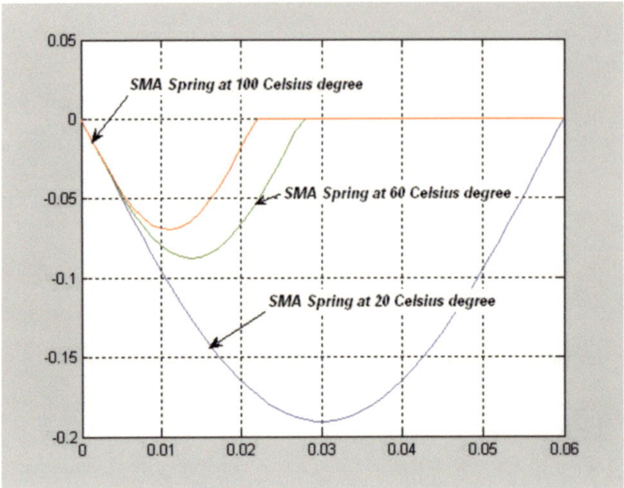

Fig. 83. Hopping in place for a SMA based robot

As one can easily observe, from the numerical simulations, the deformation of the contact terrain is up to 4 times reduced in case of a 100^0 C energized SMA spring.

4.3.2 Hopping Forward

Modeling a leg hopping forward is more complicated than hopping in place and can be thought of as running. It can be assumed that the mass spring takes off at an angle (theta) against the ground, which means that it is very important to keep the vectors straight. Instead of one direction in which to move, there are two directions: vertical (x) and horizontal (y). This means that there must be two equations for both the ground and aerial phases. The equations for the aerial phase are:

$$m\frac{d^2y}{dt^2} = -mg \tag{27}$$

$$m\frac{d^2x}{dt^2} = 0 \tag{28}$$

The equation 28, means that there is no acceleration in the x direction, and equations 27 means the same thing it did in the example of hopping in place.

The ground phase is such that the leg lands at an angle ($-\theta$) to the ground, pivots on the ground while being compressed, and then decompresses as it pivots to angle θ. At angle θ, the mass takes off into the aerial phase. The process repeats at the end of the aerial phase. The whole calculation assumes constant forward velocity (u) which is related to θ. The equations for the ground phases are given by the following equations:

$$\frac{d^2y}{dt^2} = \frac{ky}{m}\left(\frac{1}{\sqrt{x^2+y^2}} - 1\right) - g \tag{29}$$

$$\frac{d^2x}{dt^2} = \frac{kx}{m}\left(\frac{1}{\sqrt{x^2+y^2}} - 1\right) \quad (30)$$

$$l = \sqrt{x_0^2 + y_0^2} \quad (31)$$

$$w = \sqrt{\frac{k}{m}} \quad (32)$$

$$l\sin(\theta) = u\frac{t_c}{2} \Rightarrow \theta = \arcsin\left(\frac{u \cdot t_c}{2 \cdot l}\right) \quad (33)$$

$$\begin{aligned} \ddot{x} &= 0 \\ \ddot{z} &= -g \\ \ddot{\theta} &= \frac{k_{hip}\left(\theta_{hip} - \theta + \varphi\right)}{J_{leg}} \\ \ddot{\varphi} &= \frac{-k_{hip}\left(\theta_{hip} - \theta + \varphi\right)}{J_{body}} \end{aligned} \quad (34)$$

and the degree of freedom for stance stage are:

$$\begin{aligned} \ddot{r} &= -g\cos\theta + r\dot{\theta}^2 + \frac{k_{leg}\left(r_0 - r + p_{leg}\right)}{m} \\ \ddot{\theta} &= \frac{k_{hip}\left(p_{hip} - \theta + \varphi\right) - 2mr\dot{r}\dot{\theta} + rmg\sin\theta}{J_{leg} + mr^2} \\ \ddot{\varphi} &= -k_{hip}\frac{p_{hip} - \theta - \varphi}{J_{body}} \end{aligned} \quad (35)$$

For equations 31, l is the initial, unperturbed length of the spring: x_o and y_o are the coordinates of the mass at that length.
The coordinates (x_0,y_0) are most likely given by:

$$y_0 = l\sin(\theta); \qquad x_0 = l\cos(\theta) \quad (36)$$

Equations 29, 30, 31 and 32 separate the ground phase into horizontal (x) and vertical (y) directions, which relates the accelerations in the x and y directions to positions x and y, gravity, θ and the equilibrium position of the mass. In equations 33, t_c is the time the system is in contact with the ground.

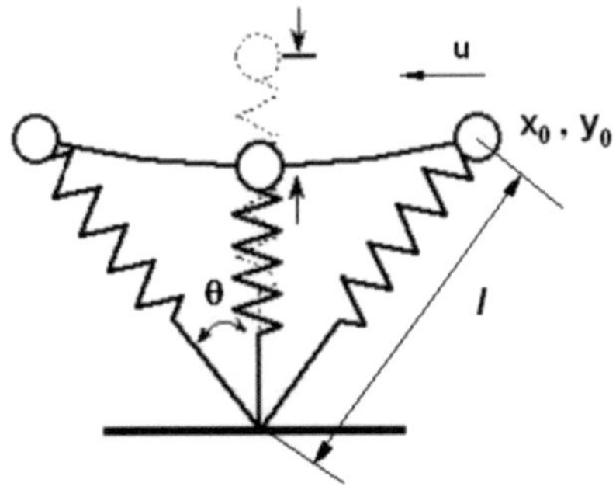

Fig. 84. Hopping forward schematically trajectory

If the scope of the experiment is to obtain the same forward velocity (u) the relation for ground time contact is:

$$t_c = \frac{2mg}{ku} \tag{37}$$

4.3.3 Shape memory alloy spring influences

Studying the influence of SMA spring in robotic structure we can observe, first, the evolution of time contact, considering constant forward velocity – u.

For the Nitionol, for spring temperature of 20⁰C, 60⁰C and 120⁰C the reduction of time contact is:

$$t_{cSMA20^0C} = \frac{C_1}{2750} \qquad t_{cSMA60^0C} = \frac{C_1}{13000} \qquad t_{cSMA120^0C} = \frac{C_1}{20700} \quad \text{with} \quad C_1 = \frac{2mg}{u} \tag{38}$$

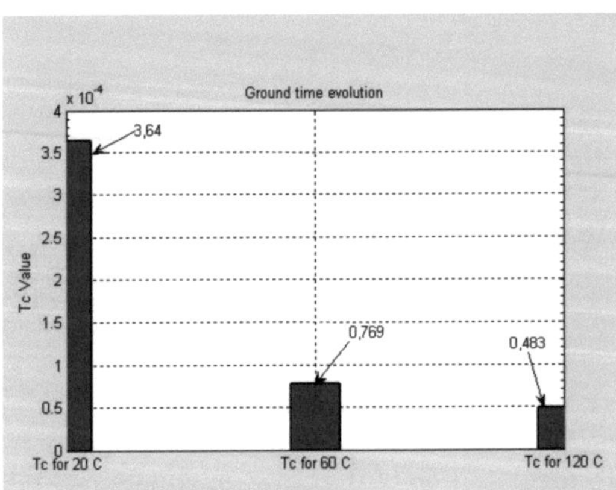

Fig. 85. Evolution of time contact, as energizing SMA spring temperature, considering constant forward velocity

Considering the contact time constant, as 8 msec, using equation 29- 35, the temperature influences for SMA spring can arise. A Matlab Simulink temperature comparative models was developed, based on the equations 29-38.

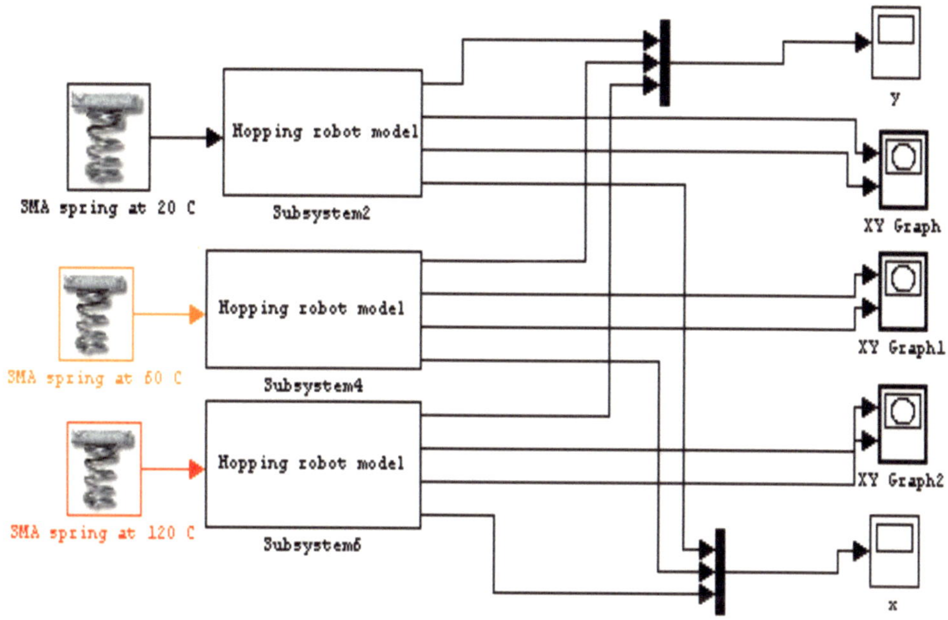

Fig. 86. Matlab Simulink temperature comparative models

The evolution of hopping robot, for case of 20ºC, 60ºC and 120ºC SMA spring energising temperature for x and y trajectory is shown in the Fig. 87, Fig. 88, Fig. 89.

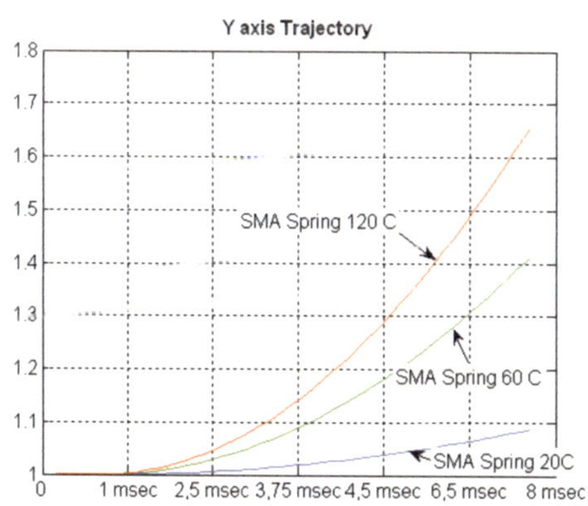

Fig. 87. Y Axis Trajectories for different energising SMA spring temperature

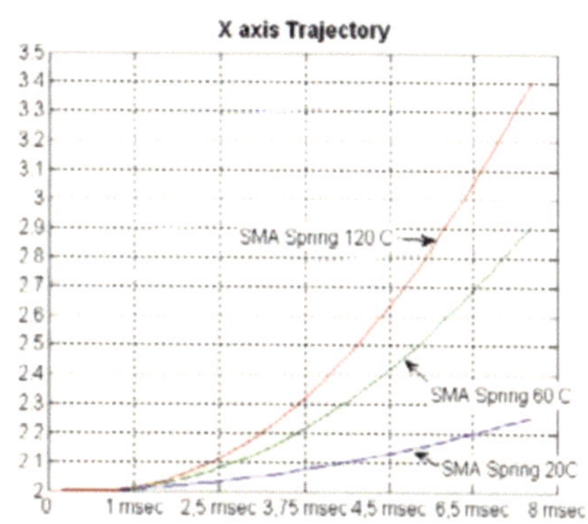

Fig. 88. X Axis Trajectories for different energising SMA spring temperature

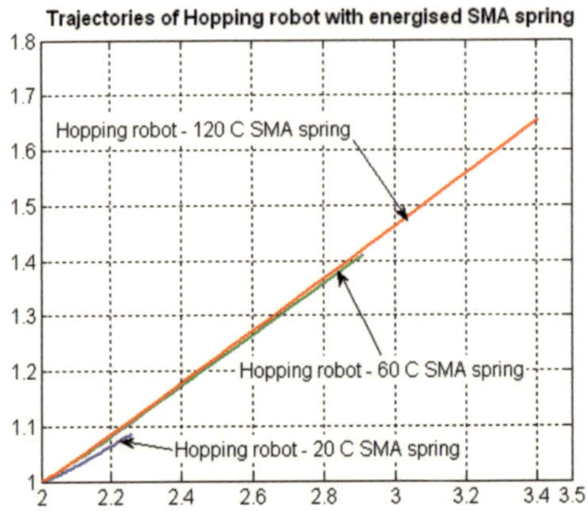

Fig. 89. Trajectories for different energising SMA spring temperature

The numerical simulations uncover the strong influence of energising temperature in case of SMA spring based hopping robot. The results of 20^0 C energising temperature of SMA spring correspond to usual steel spring. As the energising SMA spring temperature is increased to 60^0C and 120^0C, the total displacement is up to 5 times increased compare with the usual spring.

4.4 Tentacle SMA Robot

The tentacle robot structure contains units with SMA actuators. The first unit (starting from gripper) has the load equal with the weight of gripper and the work load. The unit contains 4 SMA actuators and a spherical articulation.
The actuator used is a SMA spring based actuator, structural presented in the Figure 1.
The superior units are obligatory to a highest load, because of the weight of the inferior unit. This is the reason that the superior unit use 2, 3 and 4 spring in parallel connection. The particular robotic structure demand special constructive requirements:
The load applied to a particular unit, is increased with the weight of inferior units, reason for every units has a individual parameters as shape memory alloy spring diameters wire, in and out spring diameter, number of turns, length, applied load, reset load. The relations used for unraveling these parameters are:

$$d = \sqrt{\frac{8W_C}{\pi T_c}}\sqrt{F_L} = K_1\sqrt{F_L} \tag{39}$$

d - wire diameter, W_C - Whal correction factor, TC - maximum shear stress, c - spring index, F_L force exerted by spring (required force)

$$n = \frac{1}{\pi c^3 \Delta\gamma\ K_1}\frac{S}{\sqrt{F_L}} = K_2\frac{S}{\sqrt{F_L}} \tag{40}$$

n - the number of coils, $\Delta\gamma$ difference between the low and high temperature shear strains, S - required stroke for the spring

$$F_{reset} = G_l\left(\frac{F_L}{G_h} + \frac{S^2}{F_L}\frac{T_c c}{\Delta\gamma W_c}\right) \tag{41}$$

F_{reset} - recovery force, G_h shear modulus at high temperature, G_l shear modulus at low temperature.

Numerical computation is simple and can be easy integrated in a SMA calculus tools.

The requirements applied to robotic structure conduct to a robotic tentacle structure with different elements. An important result expresses the angle dependence from the variation of SMA spring:

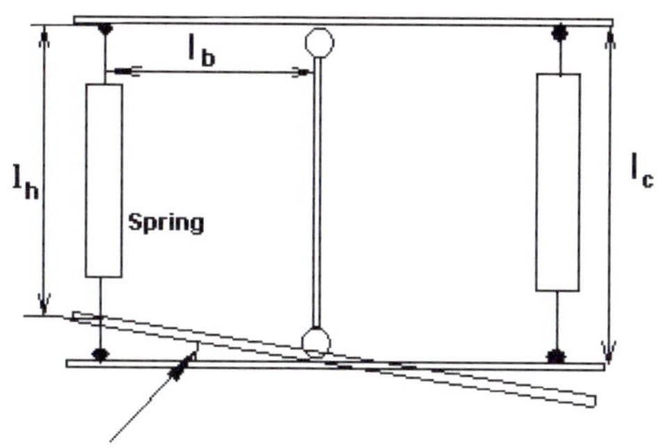

Fig. 90. Schematically representation of SMA robotic unit

$$\alpha = \arcsin\left[\cos(\mathrm{arctg}(l_h))\frac{2l_b^2 + l_c^2}{2l_b}\right] + \mathrm{arctg}(l_h) \tag{42}$$

l_h - the length after the heating process, l_c - the spring length after cooling, l_b - the base length

Fig. 91. The SMA robotic unit

The dependence is highly nonlinear but the graphical form of this dependence for real variation (between 100 and 108) results:

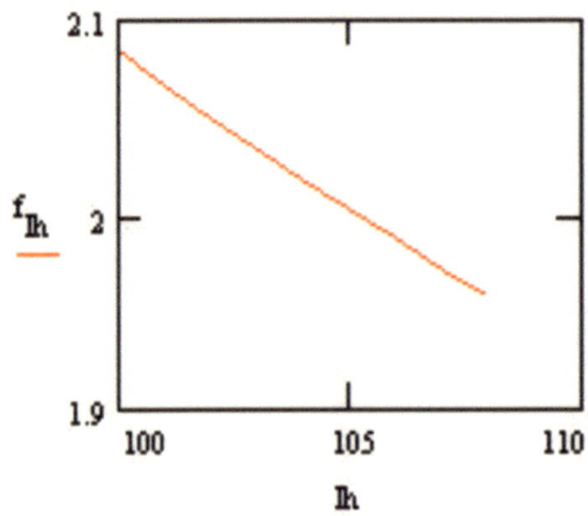

Fig. 92. The dependence of angle α as function of l_h

5. Control for biomimetic SMA robotic tentacle structure

5.1 Control architecture for SMA robotic tentacle unit

In order to investigate the SMA unit comportment a Quanser modified platform was used for experiments. The basic control structure uses a configurable PID controller and a Quanser Power Module Unit for energizing the SMA actuators.

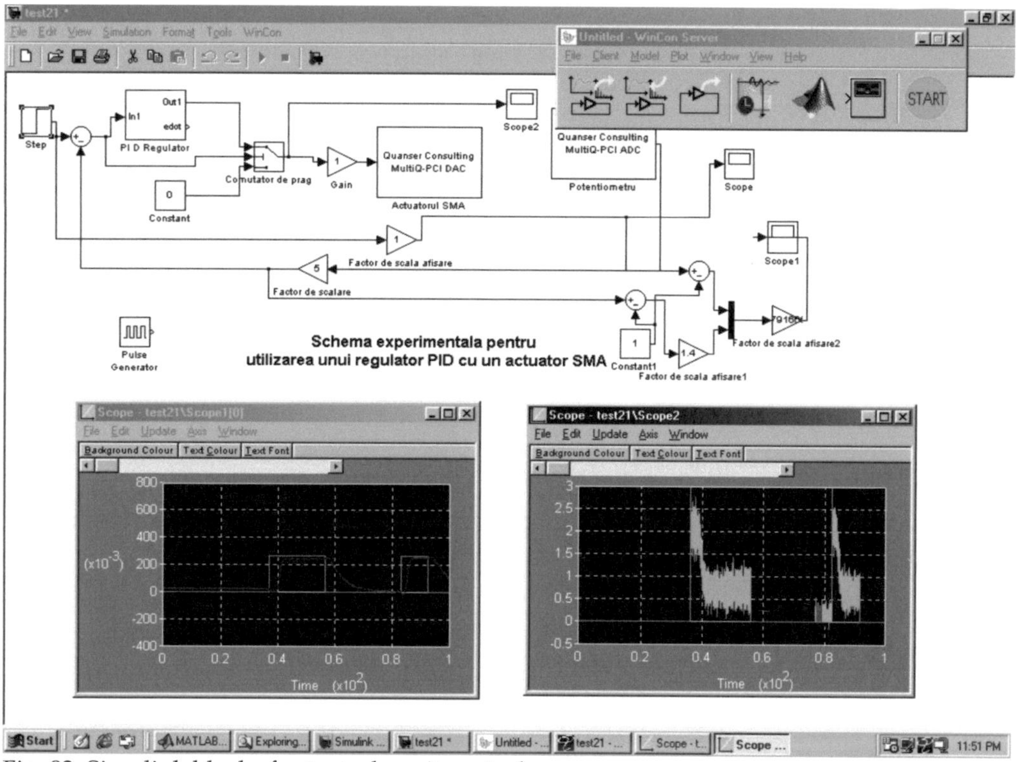

Fig. 93. Simulink blocks for tentacle unit control

PID controller was changed, in order to adapt to the particularities of the SMA actuator. A negative command for SMA actuator corresponds to a cooling source. The actual structure use for cooling only the ambient temperature. The best results arise when a PI controller is used. The PI experimented controller parameters are: the proportional parameter K_R = 10 and the integration parameter is K_I =0, 05. The input step is equivalently with 30^0 angle base variation and the evolution of this reference is represented with the response of real system in Fig. 94. The control signal variation is presented in Fig. 95.

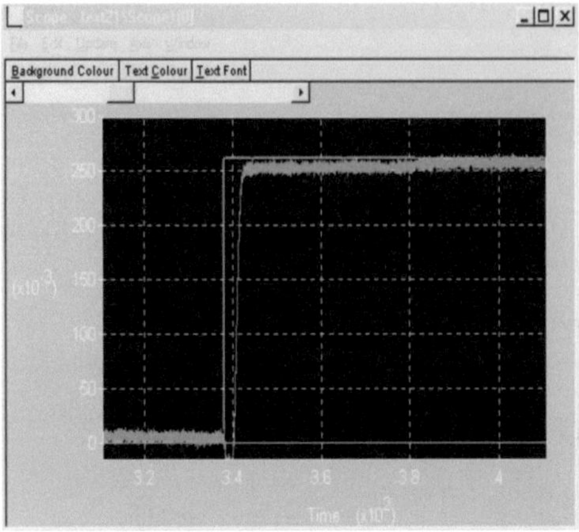

Fig. 94. System response, for step input

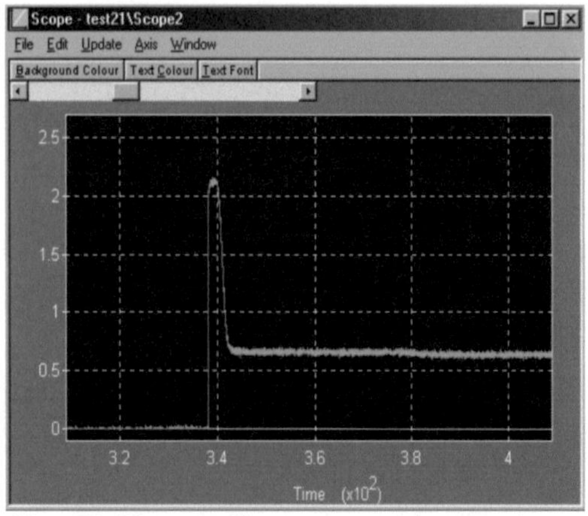

Fig. 95. PI controller response, for step input

For negative step, the evolution of the system and the control variable evolution are presented in Fig. 96 and Fig. 97.

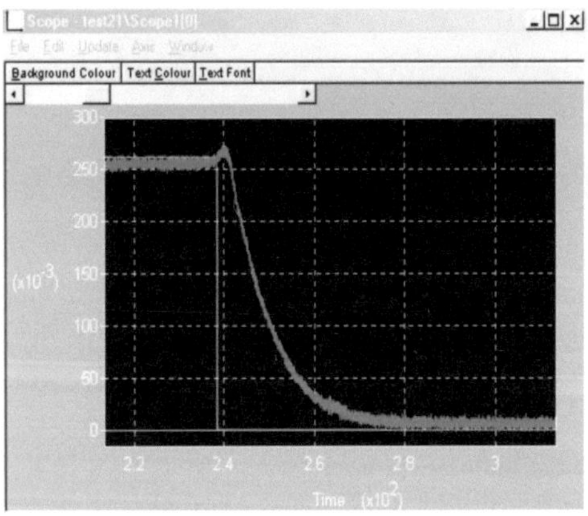

Fig. 96. System response, negative step input

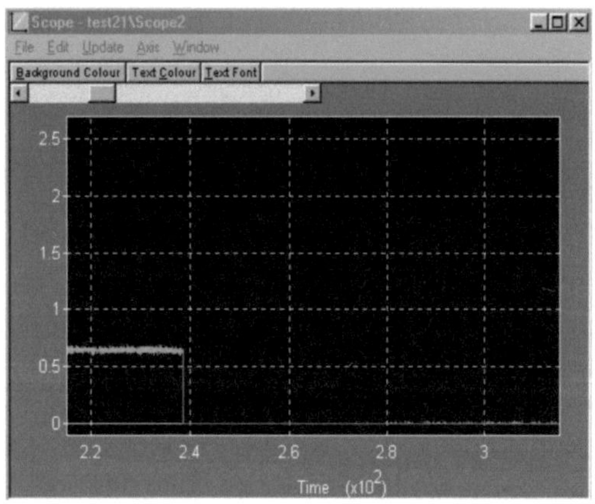

Fig. 97. PI controller response, negative step

The PD experimented controller parameters are: the proportional parameter $K_R = 10$ and the derivative parameter is $K_D = 2$.

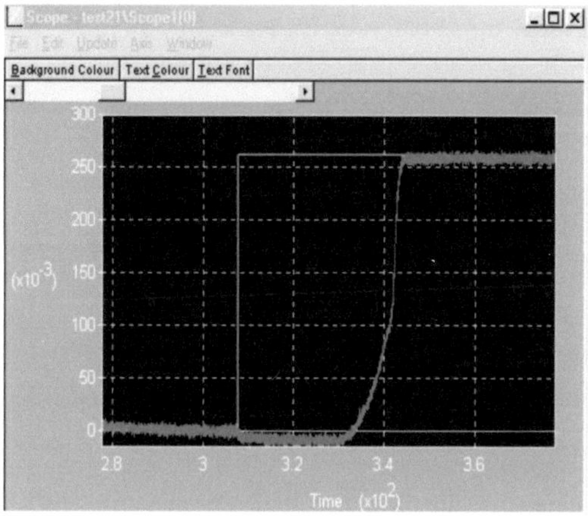

Fig. 98. System response, step input

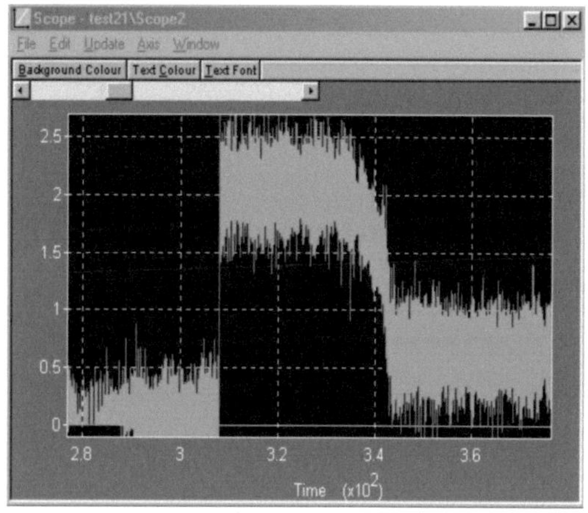

Fig. 99. PD controller response, step input

One can observe the high dynamics of the control variable - Fig. 99. The time response is longer then the case of PI controller, but the stationary error is less then the anterior case - Fig. 98.
The PID experimented controller parameters are: the proportional parameter K_R=10 and the integration component is K_I=0,005 and the derivative component is K_D=2.
The experimental results are illustrated in the figure, and the control signal variation is presented in Fig. 100 and Fig. 101.
Unfortunately, even with the complication of the controller, the time response is inferior to the case of the PI controller and the stationary error is near zero.

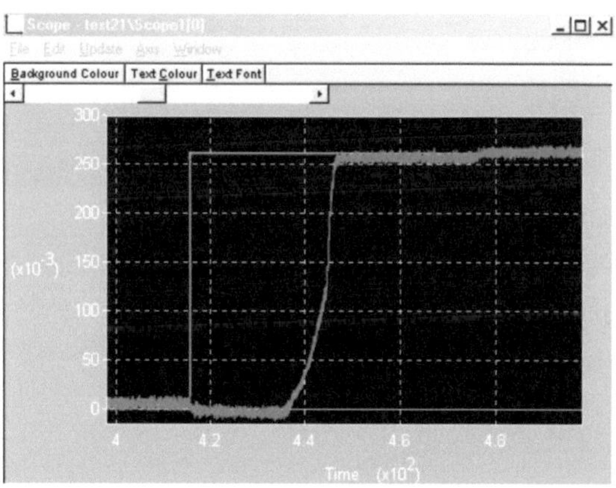

Fig. 100. System response, step input

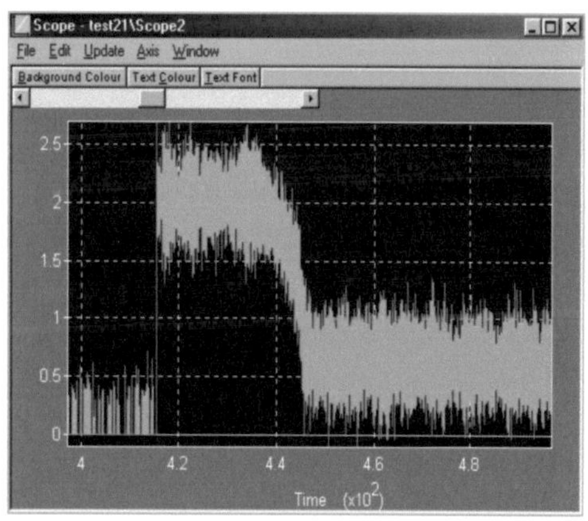

Fig. 101. PID controller response, step input

As a conclusion: the best results arise when a PI controller is used. The PI experimented controller parameters are: the proportional parameter K_R = 10 and the integration parameter is K_I =0, 05.
Using PID, PD controller the experiments conduct to less convenient results from the point of view of time response or controller dynamics.

Using heat in order to activate SMA wire, a human operator will increase or decrease the amount of heat in order to assure a desired position to robotic link.
Because of medium temperature influence, can not be establish, apriority, a clear control law, available for all the points of the robotic structure workspace.

5.2 A SMA Tentacle Manipulator model

A tentacle manipulator is a manipulator with a great flexibility, with a distributed mass and torque that can take any arbitrary shape. Technologically, such systems can be obtained by using a cellular structure for each element of the arm. The mathematical model developed (Ivănescu 1986) and (Ivănescu 1984) conduct to the following dynamics of bi-dimensional tentacle structure:

$$\begin{aligned} &\rho A \int_0^s \left(\sin(q - q(r))\dot{q}^2(r) + \cos(q - q(r))\ddot{q}(r) \right) dr + \\ &+M \int_0^L \left(\sin(q_L - q(r))\dot{q}^2(r) + \cos(q_L - q(r))\ddot{q}(r) \right) dr + \\ &+\rho \ A \int_0^L \cos(s) ds + Mg\cos(q) = U \end{aligned} \tag{43}$$

with initial conditions:

$$\begin{aligned} q(s,0) &= q^0(s), \\ \dot{q}(s,0) &= \dot{q}_1^0(s), \\ q_L &= q(L,t), \end{aligned} \tag{44}$$

with U is a generalised input, q generalised coordinates, ρ density of , M - generalised mass. Using a spatial discretisation $s_1, s_2, s_3, \ldots. s_n$ with $s_i - s_{i-1} = \Delta$

$$\begin{aligned} &\left| q(s) - q(s(r)) \right| < \varepsilon_1 \\ &\left| q_L(s) - q(s(r)) \right| < \varepsilon_2 \\ &s, s(r) \in [0, L], \ \text{and} \ \ \varepsilon_1, \varepsilon_2 \to 0 \end{aligned} \tag{45}$$

An integro-differential form of discretisation Δ and integration to the curve L conduct to the following expression:

$$\int_0^L\left(\int_0^s\left(\frac{\partial^2 q}{\partial t^2}\cos(q(r,t)-q(s,t))-\left(\frac{\partial q}{\partial t}\right)^2\sin(q(r,t)-q(s,t))-\frac{\partial q(s,t)}{\partial t}\frac{\partial q(r,t)}{\partial t}\cos(q(r,t)-q(s,t))\right)dr-g(\sin q(s,t)+\cos q(s,t))\right)ds=\frac{1}{\rho}\int_0^L u(s,t)ds \quad (46)$$

Using a discretisated form of the equations results:

$$\sum_{i=0}^{m}\frac{\partial^2 q_i}{\partial t^2}\cos(q_m-q_i)-\left(\frac{\partial q_i}{\partial t}\right)^2\sin(q_m-q_i)-\frac{\partial q_m}{\partial t}\frac{\partial q_i}{\partial t}\cos(q_m-q_i)-g(\sin(q_m)+\cos(q_m))=\frac{1}{\rho\Delta}u_m \quad (47)$$

$m=0,1\cdots(N-1)$. If $\varepsilon\to 0$ and with a suitable choosing of N and Δ for $|q_i(t)-q_{i-1}(t)|<\varepsilon$, $i=0,1\cdots(N-1)$, $t\in[0,t_F]$, results: $\frac{\partial^2 q_i}{\partial t^2}-\left(\frac{\partial q_i}{\partial t}\right)^2=\frac{1}{\rho\Delta}(u_i-u_{i-1})$

for $i=1\ldots(N-1)$, ui şi ui-1 are two control sequence for unit i and unit i-1.

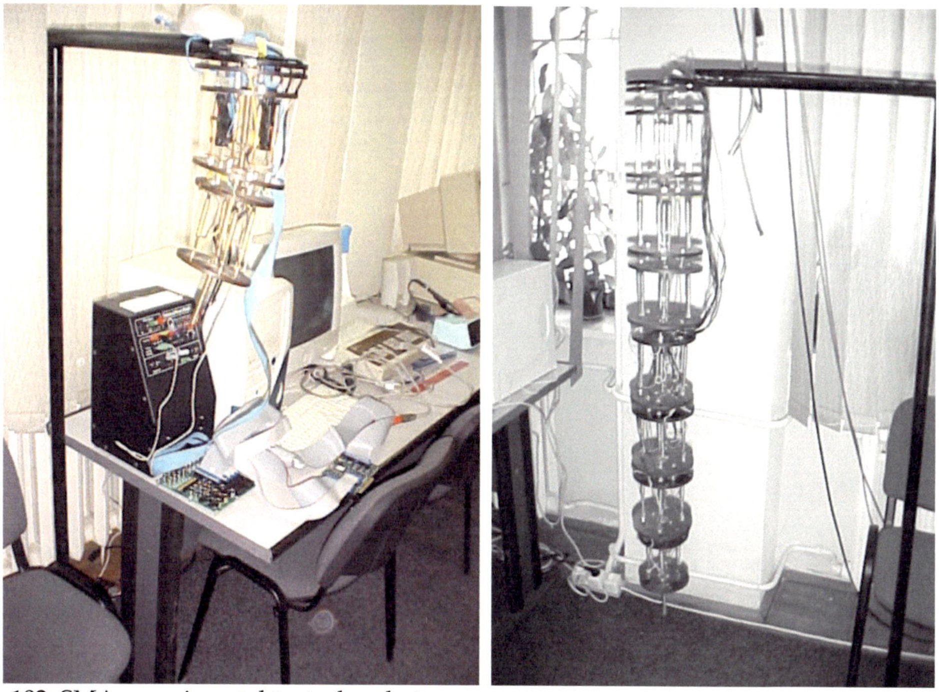

Fig. 102. SMA experimental tentacle robot

5.3 Tentacle's dynamics

It is considered the dynamics of the object defined by the equation (47). For a prescribed trajectory r_d, we define the object (load) motion error as:

$$e_1 = r_d - r \quad (48)$$

$$e_2 = \dot{r}_d - v$$

and the object error dynamic can be rewritten as:

$$\dot{e}_L = A_L e_L + b_L(r_d, \dot{r}_d) + c_L \tau \quad (49)$$

$$\text{where } e_L = \text{col}(e_1, e_2)$$

A is a constant matrix, b,c are vectors and τ is the resultant force which assures the object motion on the trajectory.
The dynamic of the manipulator can be approximated from (46),(47) by using a spatial discretization $s_1, s_2, \ldots\ldots s_N$. If this approximation verifies the condition

$$|q(s_i) - q(s_k)| < \varepsilon, \; i,k = 1,2,\ldots n^J \quad (50)$$

where ε is a constant sufficiently small, from (Ivanescu,1984) we obtain

$$M\ddot{q} + C\dot{q} + D + \tau = T \quad (51)$$

where M, C are (nxn) constant matrices, D is a nonlinear vector and

$$q = \text{col}(q_1, q_2 \cdots q_{n^J}) \quad (52)$$

$$T = \text{col}(T_1, T_2 \cdots T_{n^J})$$

$$\tau = \text{col}(\tau_1, \tau_2 \cdots \tau_{n^J})$$

$$q_i = q(s_i);\; T_i = T(s_i);\; \tau_i = \tau(s_i)$$

We define the system error as

$$e = q_d - q \quad (53)$$

where q_d represents the desired trajectory of the tentacle manipulator.
We assume the case in which is difficult to know exactly the influence of the load and we estimate as τ^* the value of the force which assures the desired motion on the trajectory.

Theorem (Ivanescu, Stoian, 1995)

The closed-loop system is stable if the control is defined by

$$T = +Ks + \tau^* - u_F - H(e, q_d) \tag{54}$$

$$K - M\sigma + C \text{ positive definite matrix,} \tag{55}$$

$$u_{F_i} = -k_{F_i} \operatorname{sgn} s_i \tag{56}$$

$$k_{F_i} \geq \left|\tau^* - \tau\right| \tag{57}$$

where $H(e,q_d)$ represents the nonlinear part of the dynamic model and τ^* is the estimated values of the forces required at the terminal points of the manipulators.
The Theorem 1 determines the conditions which assure the motion control in the neighborhood of the switching line. In order to accelerate the motion on the switching line we can use the DSMC (Direct Sliding Mode Control) method. This motion differs by the conventional Sliding Mode by the evolution on the switching line. When the trajectory of the system penetrates the switching line, the damping coefficient is increased and the motion is forced on the switching line toward the origin without the high frequency oscillations. In Fig. 103 is presented the conventional Sliding Mode Control and in Fig. 104, the Direct Sliding Mode Control.
This motion is assured by the following Proposition:

Proposition
The DSMC control is assured if the coefficients k_i verify the condition

$$(c_i + k_i)^2 \geq 4m_i\left(\sigma_i\left(-m_i\sigma_i + c_i + k_i\right) + h_i\right) \tag{58}$$

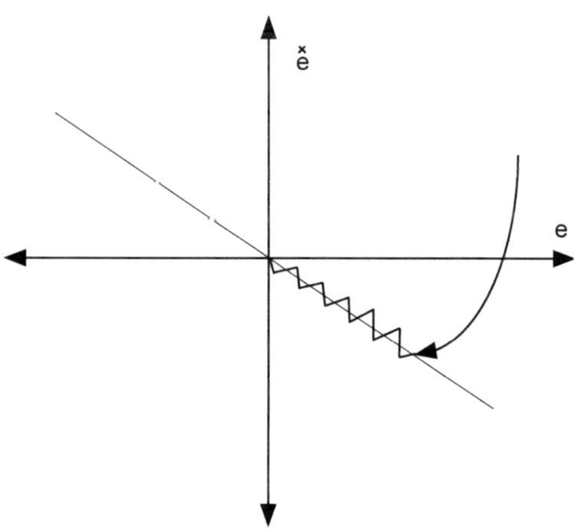

Fig. 103. Conventional Sliding Mode Control

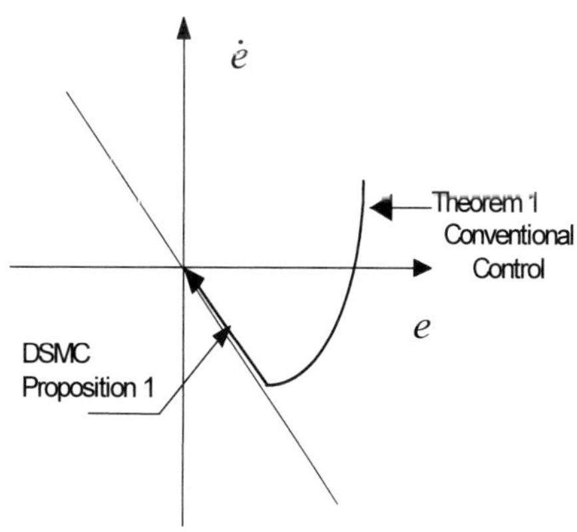

Fig. 104. Control trajectory

5.4 Control system

The control problem asks for determining the manipulatable torques T such that the trajectory of the overall system (object and manipulator) will correspond as closely as possible to the desired behavior.

The control system contains two parts: the first component is a conventional controller which implements a classic strategy of the motion control based on the Lyapunov stability and the second is a Fuzzy Controller Fig. 105.

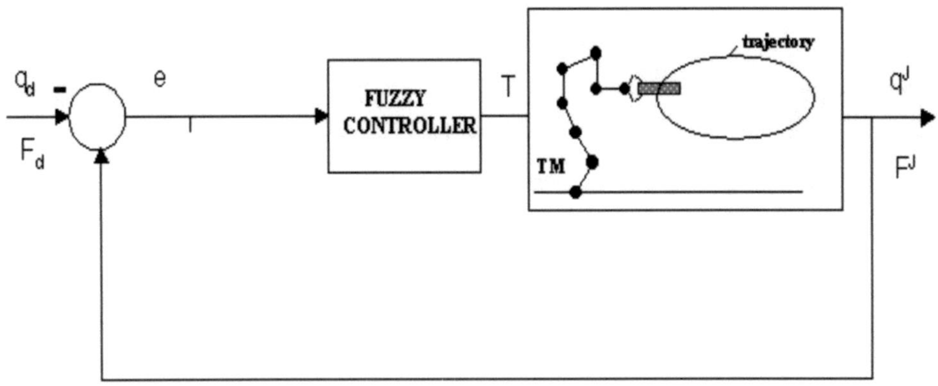

Fig. 105. Control system blocks

The controller receives the error and the change of the error components, $e^i, \dot{e}^i$ for each units of the tentacle manipulator and depending on the values of forces τ_{F_i}, generates the fuzzy control torques T_F^i. The control rules are determined by the motion in the neighborhood of the switching line as a variable structure controller. We adopted here a special class of SMC named DSMC (Direct Sliding Mode Control) (Tao, 1998). The physical meaning of the rules is as follows: the output is zero near the switching line, the output is negative above the switching line, the output is positive below the diagonal line, the magnitude of the output tends to increase in accordance with magnitude of the distance between the switching line and the state. The DSMC method introduces new rules. This method operates in two steps. First step assures the motion towards the switching line s, by the general stability conditions. The second step starts when the trajectory penetrates the switching line. In this case, the damping coefficients of the motion are changed and system is moving toward the origin, directly on the switching line. The procedure for the design of the Fuzzy Controller is the following:

We consider that all input/output fuzzy sets are assured to be designed on the normalized space. The basic membership of the input variables are proposed as in Fig. 106. The universes of the input variable $e_i, \dot{e}_i$ are initially partition on three fuzzy sets: negative (N), zero (Z) and positive (P) with trapezoid membership function.

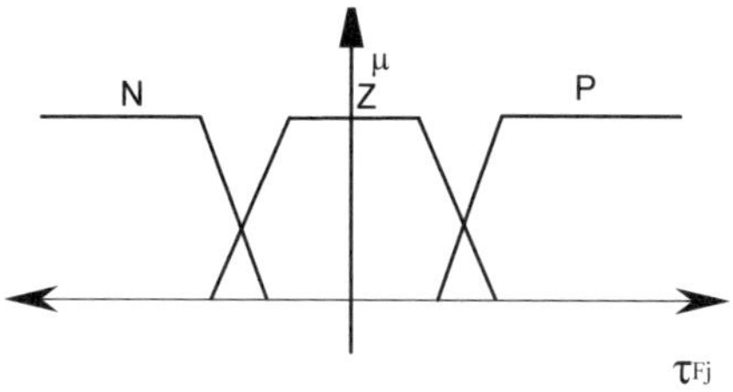

Fig. 106. Input fuzzy sets - 3 members

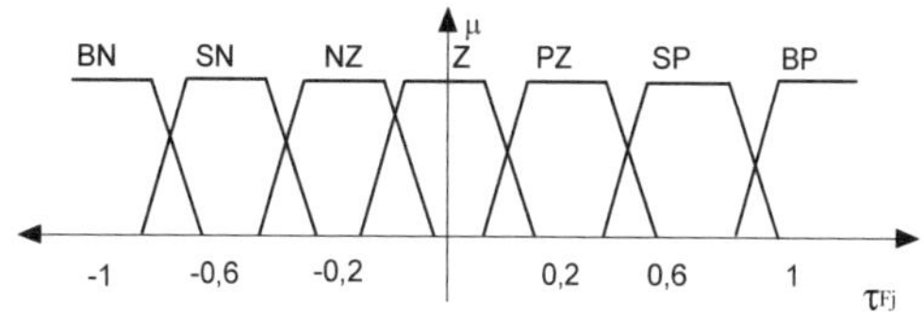

Fig. 107. Input fuzzy sets - 7 members

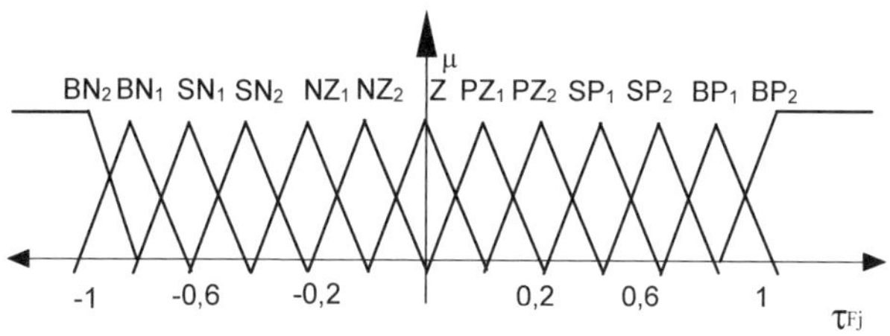

Fig. 108. Input fuzzy sets - 13 members

The state space of $e_i, \dot{e}_i$ will be partitioned into nine fuzzy regions. The fuzzy if-then rules for these fuzzy regions are presented in:

$\dot{e}^J \backslash e^J$	N	Z	P
P	Z	N	N
Z	P	Z	N
N	P	P	Z

Table 1. The initial fuzzy if-then rules

where the output membership are defined as singletons Fig. 109.

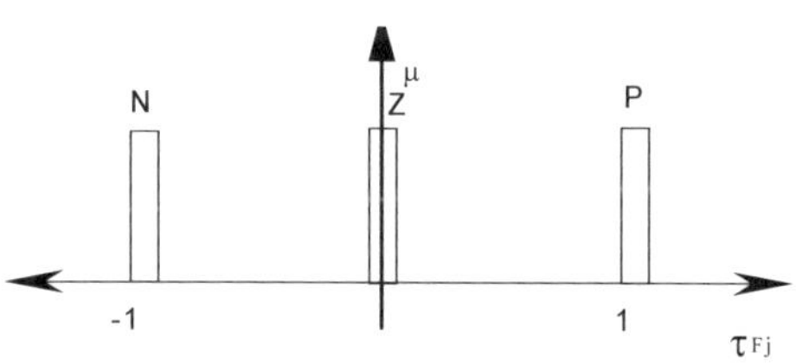

Fig. 109. Output singleton - 3 members

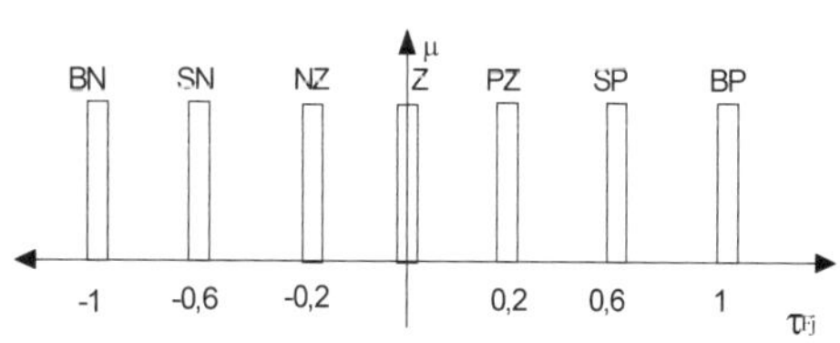

Fig. 110. Output singleton - 9 members

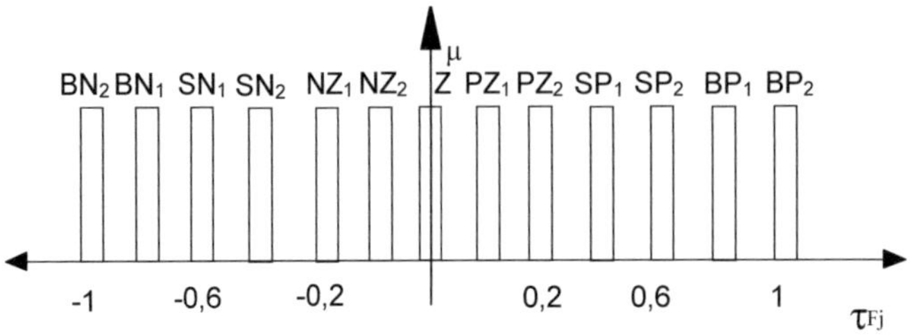

Fig. 111. Output singleton - 13 members

The output u_F of the fuzzy controller is derived to be (Soo, 1997)

$$u_F = \frac{\sum_{i=1}^{9} \rho_i f_{e_i}(e_i) f_{\dot{e}_i}(\dot{e}_i)}{\sum_{i=1}^{9} f_{e_i}(e_i) f_{\dot{e}_i}(\dot{e}_i)} \tag{59}$$

where ρ_i is one of the centers of the output singletons and $f_{e_i}(e_i), f_{\dot{e}_i}(\dot{e}_i)$ are the membership of the input variable $e_i, \dot{e}_i$ respectively.

The control 59 assure the motion of the system on the first part of the trajectory Fig. 115 with k_i SMALL $\left(k_i < K_i^S\right)$, when the trajectory penetrates the switching line the DSMC control is applied by the control of the coefficient k_i (Proposition 1) k_i is BIG $\left(k_i \in K_i^B\right)$.

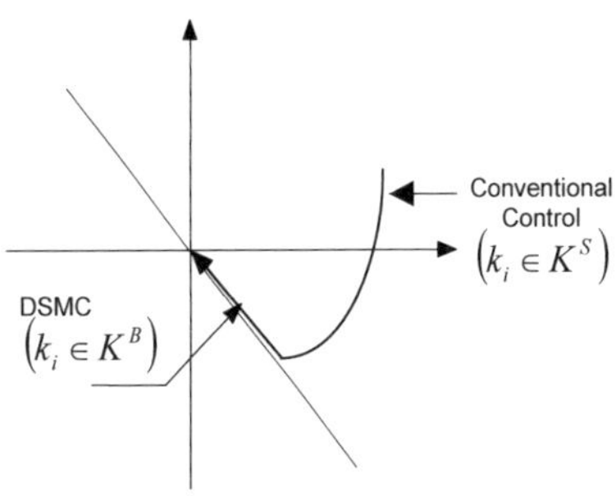

Fig. 112. Control trajectory strategy

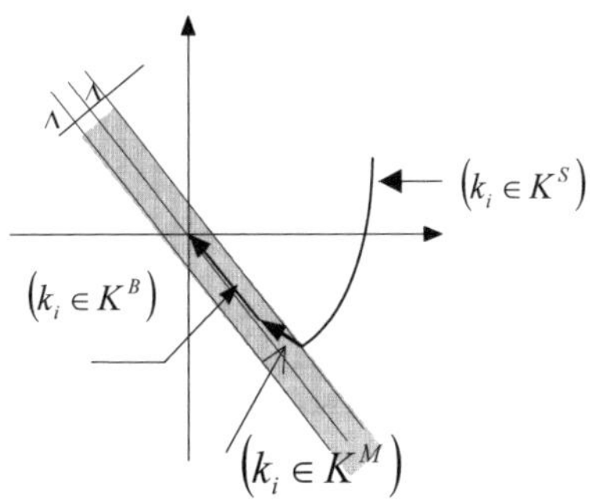

Fig. 113. DSMC Control trajectory

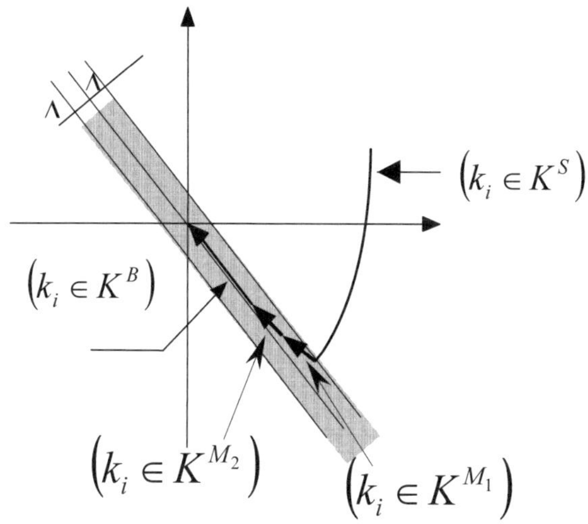

Fig. 114. DSMC Control trajectory increased control variable

The size of k_i is defined as singleton function Fig. 115.

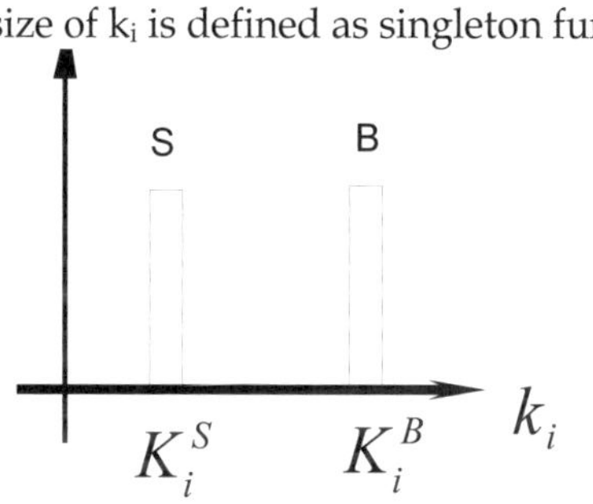

Fig. 115. Two member singleton k_i coefficients

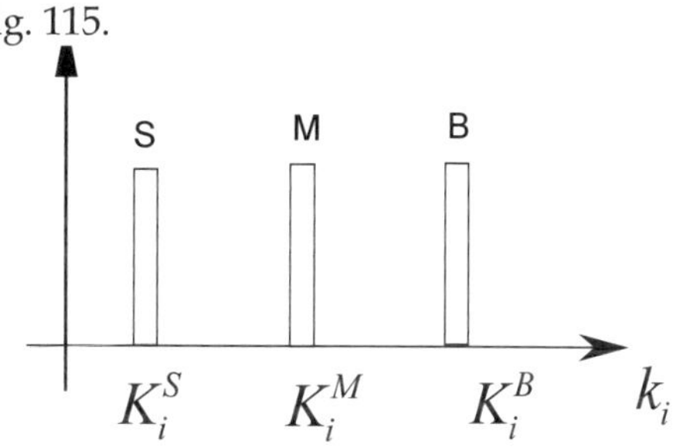

Fig. 116. Three member singleton k_i coefficients

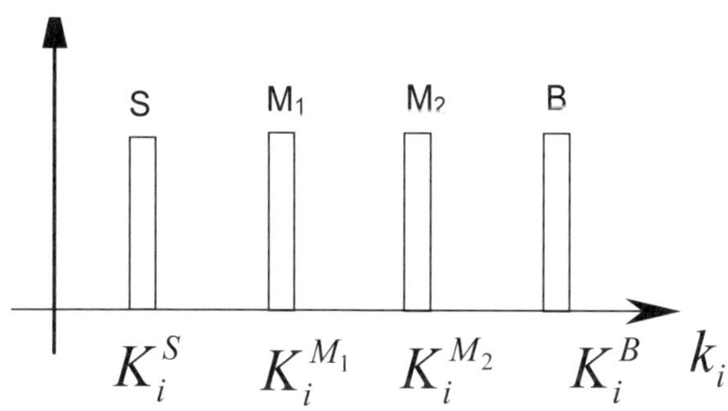

Fig. 117. Four member singleton k_i coefficients

If the evolution described in Fig. 113 is not satisfactory, a new control strategy is adopted. The finer fuzzy domains are introduced (Figure 6b) and new fuzzy partitions are used: big negative (BN), small negative (SN), negative zero (NZ), zero (Z), positive zero (PZ), small positive (SP), big positive (BP). The fuzzy if-then rules for these fuzzy regions are presented in the Table 2, where the outputs are the singletons defined in Fig. 110.

$\dot{e}_i^J, \dot{e}_L$ / e_i^J, e_L	BN	SN	NZ	Z	PZ	SP	BP
BP	Z	NZ	SN	SN	BN	BN	BN
SP	PZ	Z	SN	SN	SN	BN	BN
PZ	SP	SP	Z	NZ	SN	SN	BN
Z	SP	SP	PZ	Z	SN	SN	SN
NZ	BP	SP	SP	PZ	Z	NZ	SN
SN	BP	BP	SP	SP	PZ	Z	NZ
BN	BP	BP	BP	SP	SP	PZ	Z

Table 2. The fuzzy if-then rules

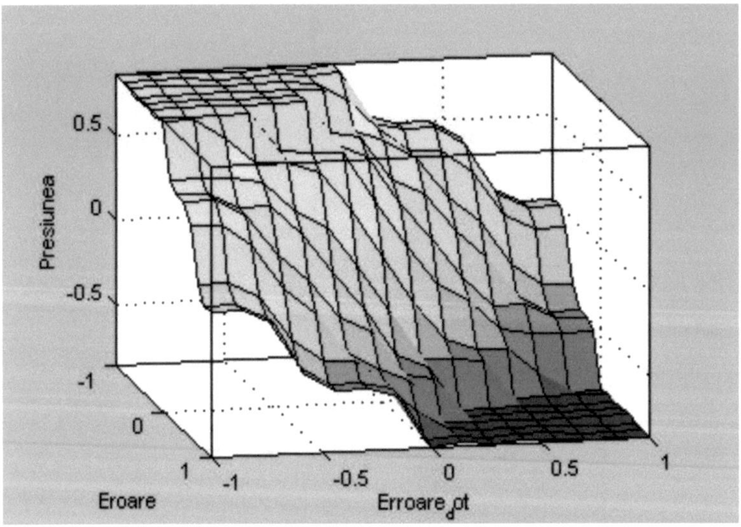

Fig. 118. The rules surface

The new membership of the inputs μ^* verify the inequality:

$$\mu^*(x_i) \leq \mu(x_i) \tag{60}$$

for every input x_i. Then, new control u_F from 60 will satisfy the condition 58, Theorem 1. Also the new finer distribution of the control allows a new trajectory Fig. 113 determined by the new values of the ki, small (S), medium (M), big (B).

The result of this strategy is evaluated with the performance indexes

The procedure of the modification of the fuzzy rule base will be repeated several times Fig. 111 , Fig. 111 , Fig. 114 until the performance requirements are satisfied.

5.5 Numerical results

The purpose of this section is to demonstrate the effectiveness of the method. This is illustrated by solving a fuzzy control problem for a tentacle manipulator system, which operates in XOZ plane (Figure 10). An approximate model (50) with Δ=0.36 m and n=7 is used. Also, the length and the mass of the object are 0.2 m and 1 kg, respectively. The initial positions of the arms expressed in the inertial coordinate frame are presented in Table 3.

TM	$q_1(0)$	$q_2(0)$	$q_3(0)$	$q_4(0)$	$q_5(0)$	$q_6(0)$	$q_7(0)$
TM	π/6	π/3	7π/12	2π/3	π/15	15π/8	0

Table 3. Initial positions of the arms

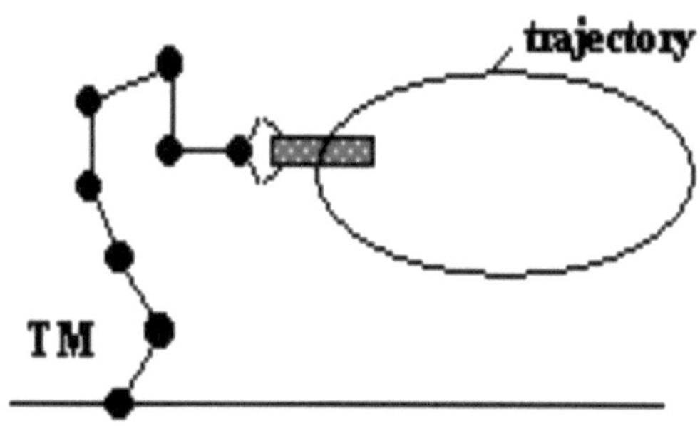

Fig. 119. Numerical simulation for tentacle biomimetic robotic structure

The desired trajectory of the terminal points is defined by: $x = x_0 + a\sin\omega t \quad z = z_0 + b\cos\omega t$

with x_0=0.2 m, z_0=0.1 m, a=0.3m, b=0.1 m, ω= 0.8 rad/s.

The trajectory lies the work envelope and does not go through any workspace singularities. The maximum force constraints are defined by:

$$F_X \le F_{MAX} = 50\text{N} \qquad F_Z \le F_{MAX} = 50\text{N}$$

and the optimal index $\min\left(\sum_J F_X^2\right), \min\left(\sum_J F_Z^2\right)$ are used. The uncertainty domain of the mass is defined as $0.8\text{kg} \le m \le 1.4\text{kg}$. The solution of the desired trajectory for the elements of the arms is given by solving the nonlinear differential equation:

$$\dot{q}_d(t) = \left[J^T(q) \quad J(q)\right]^{-1} J^T(q)\dot{w}(t)$$

where $w=(x,z)^T$ and $J(q)$ is the Jacobian matrix of the arms .

A conventional controller with k_i=0.5 (i =1,..7) is determined. A FLC is used with the scale factors selected as $G_{e_i} = G_{\dot{e}_i} = 10$, : i=1,..7.

The conventional and DSMC procedures are used and new switching line is computed. The condition 59 is verified and the new switching line is defined for p_i=1.03 : i=1,..7.

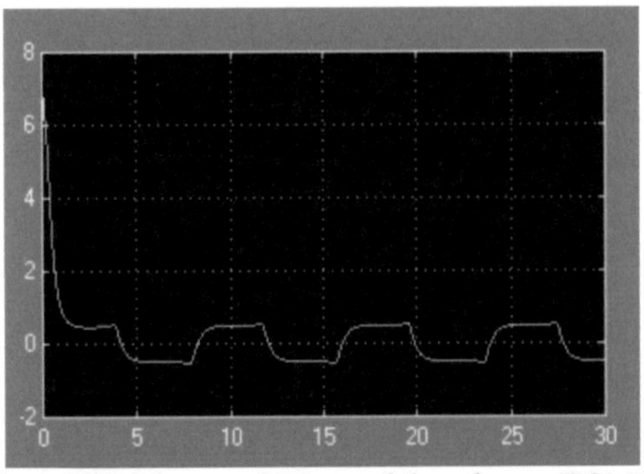

Fig. 120. The evolution of k_{51} for a DSMC procedure

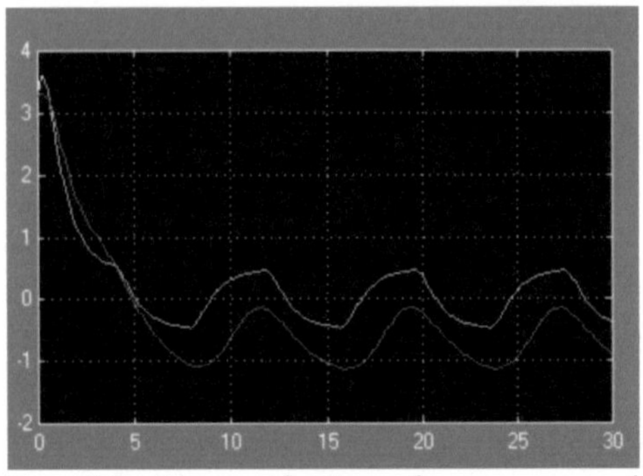

Fig. 121. The evolution of error e_{51} and error

In Fig. 120 is presented the evolution of k_5^1 for a DSMC procedure and the evolution of the position error e_5^1 and the position error rate $\dot{e}_5^1$ are presented in Fig. 121.

Fig. 122 represents the trajectory in the plane $\left(e_5^1, \dot{e}_5^1\right)$ for conventional procedure and Fig. 123 the same trajectory for a DSMC procedure for a new switching line.

Fig. 124 presents the final trajectory. We can remark the error during the 1th cycle and the convergence to the desired trajectory during the 2nd cycle.

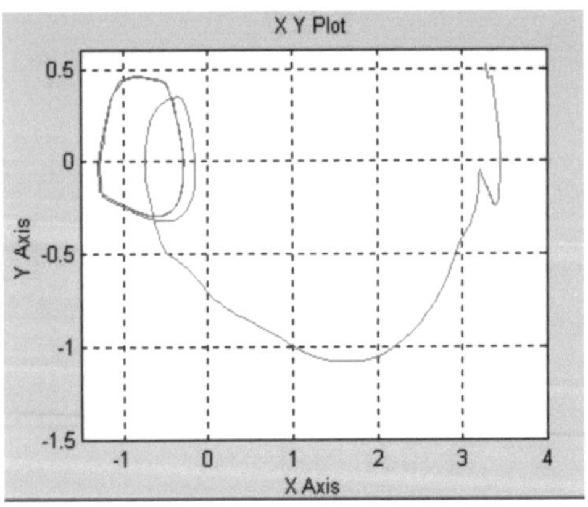

Fig. 122. Trajectory in the plane $\left(e_4, \dot{e}_4\right)$ for fuzzy SMC procedure

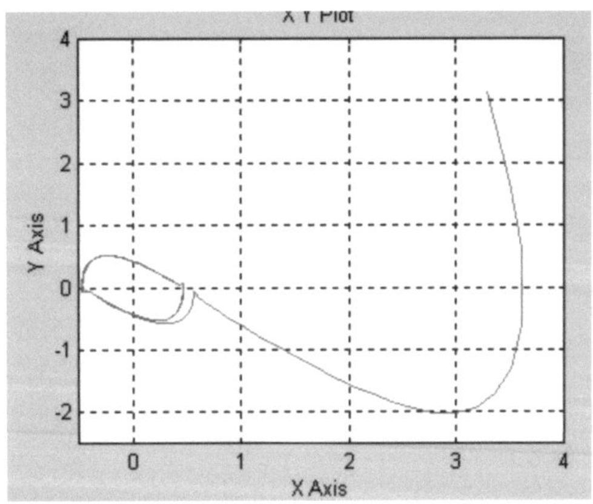

Fig. 123. Trajectory in the plane for fuzzy DSMC procedure

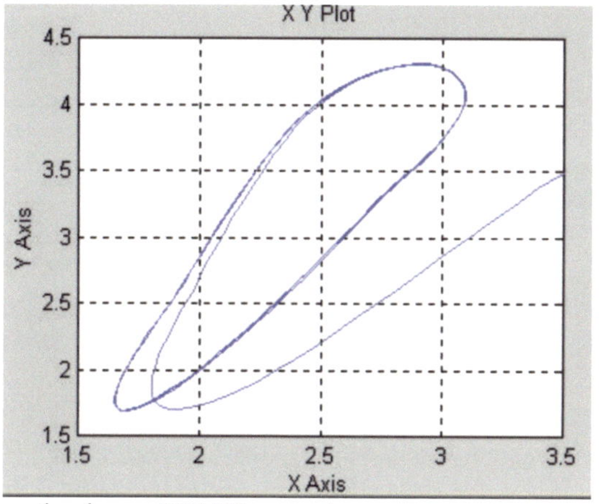

Fig. 124. Final trajectory for fuzzy DSMC procedure

6. Conclusion

The nickel titanium alloys, used in the present research, generally refereed to as Nitinol, have compositions of approximately 50 atomic % Ni/ 50 atomic % Ti, with small additions of copper, iron, cobalt or chromium. The alloys are four times the cost of Cu-Zn-Al alloys, but it possesses several advantages as greater ductility, more recoverable motion, excellent corrosion resistance, stable transformation temperatures, high biocompatibility and the ability to be electrically heated for shape recovery.

Shape memory actuators are considered to be low power actuators and such as compete with solenoids, bimetals and to some degree wax motors. It is estimated that shape memory springs can provide over 100 times the work output of thermal bimetals.

The use of shape memory alloy can sometimes simplify a mechanism or device, reducing the overall number of parts, increasing reliability and therefore reducing associated quality costs. Because of its high rezistivity of 80 - 89 micro ohm-cm, nickel titanium can be self heated by passing an electrical current through it. The basic rule for electrical actuation is that the temperature of complete transformation to martensite Mf, of the actuator, must be well above the maximum ambient temperature expected.

Scientists and engineers are increasingly turning to nature for inspiration. The solutions arrived at by natural selection are not only a good starting point in the search for answers to scientific and technical problems, but an optimal solution too. Equally, designing and building bio inspired devices or systems can tell us more about the original animal or plant model.

The connection between smart material and structures and biomimetics related with mechatronics offer o huge research domain. The present chapter explore using mathematical simulations and experiments only a modest part of the wonderful world of biomimetics.

Acknowledgment

This work was supported in part by a grant from **PNCDI-2 Idei 289/2008** - ***Reverse Engineering in Cognitive Modelling and Control of Biomimetics Structure.***

7. References

***, *Solidworks 98 Plus User's Guide*, SolidWorks Corporation, U.S.A..

A.F. Devonshire, Phil. Mag. 40, 1040 (1949) l, 42, 1065 (1951)

Agre, P. & Chapmam, D. (1990). What are plans for? *Designing Autonomous Agents*, pp 17-34, MIT Press Publisher

Burstein, A.H.; Reilly, D.T. & Martens, M. (1976). Aging of bone tissue: Mechanical properties, *J. Joint Surgery American*, No. 58, pp. 82–86, 1976

Arkin, R. (1989). Towards the Unication of Navigational Planning and Reactive Control, *Proceedings of American Association for Articial Intelligence Spring Symposium on Robot Navigation*, Palo Alto, CA, 1-5, 1989

Arkin, R. (1990). Integrating Behavioral, Perceptual and World Knowledge in Reactive Navigation, *Robotics and Autonomous Systems, Special Issue on Designing Autonomous Agents: Theory and Practice from Biology to Engineering and Back*, P. Maes, Vol. 6, No. 1-2, 1990, 105-122

Attanasio, M.; Faravelli, L. & Marioni, A. (1996). Exploiting SMA Bars in Energy Dissipators, *Proceedings of the 2nd International Workshop on Structural Control*, Hong Kong Hkust 41-50.

Bizdoaca, N. (2003). Robotic Finger Actuated with Shape Memory Alloy Tendon, *Proceedings of Soft computing, Optimization, Simulation & Manufacturing systems* (SOSM 2003), Malta, september 2003.

Bizdoaca, N.; Degeratu, S. & Diaconu, I. (2005). Behavior based control for robotics demining, *Proceedings of International Symposium on System Theory, SINTES 12*, pp. 249-254, ISBN 973-742-148-5, 973-742-152-3, Craiova, october 2005, Universitaria Publisher, Craiova

Bizdoaca, N. & Pana D. (2003). Strategy planning for mobile robot soccer, *Proceedings of 14th International Conference On Control Systems And Computer Science,* pp.238-243, Bucharest, July 2003, Politehnica Publisher, Bucharest

Bizdoaca, N. (2004). Shape Memory Alloy based robotic ankle, *Proceedings of ICCC,* pp. 709-715, ISBN 83-89772-00-0, Poland, 2004

Bizdoaca, N. & Degeratu, S. (2004). Shape memory alloy serial rotational link robotic structure, *Proceedings of the 5th International Carpathian Control Conference*, pp. 699-709, Poland, 2004

Bizdoaca, N.; Petrisor, A.; Diaconu, I. & Bizdoaca E. (2007). Sliding Mode Control of Shape Memory Alloy based Hopping Robot, *Proceedings of the 13th IEEE IFAC International Conference on Methods and Models in Automation and Robotics (MMAR),* pp.93-101, Szczecin, Poland, August, 2007

Bizzi, E.; Giszter, SF. & Mussa-Ivaldi, FA. (1991). Computations underlying the execution of movement: a biological perspective, *Science,* No.253, 1991, pp. 287-291

Breazeal, C. (1998). A motivational system for regulating human–robot interaction, *Proceedings of the Fifteenth National Conference on Arti.cial Intelligence, AAAI 98,* Madison, WI, 1998

Brooks, R. (1986). A robust layered control sustem for a mobile robot, *IEEE journal of Robotics and Automation*, Vol. RA-2, 1986, pp. 14-23

Brooks, R.(1991). Intelligence without reason, *Proceedings of International Joint Conference on Artificial Inteligence*, No. 47, 1991, pp. 139-159, MIT Press

Brooks, R. & Stein, L. (1994). Building brains for bodies, *Autonomous Robots*, No.1, 1994, pp. 7-25

Tao, C.W. (1998). Design of Fuzzy-Learning Fuzzy Controllers, *Proceedings of Fuzz IEEE'98*, pp. 416-421, 1998

Chun, M.; Wolfe, J. M. (2001). Visual Attention, In: *Blackwell Handbook of Perception*, pp. 272-310, Blackwell Publishers Ltd, Oxford, UK

Coman, D. & Petrisor A. (2006). Obstacle Avoidance of Robot Soccer Using the Fuzzy Univector Field Method, *Proceedings of MicroCAD International Scientific Conference*, pp. 13-18, ISBN 963-661-709-0, Hungary, 2006

Tarnita, D.; Popa, D.; Tarnita, D.N. & Bizdoaca, N.G. (2006). Considerations on the dynamic simulation of the 3D model of the human knee joint. *BIO Materialien Interdisciplinary Journal of Functional Materials, Biomechanics and Tissue Engineering*, No. 231, 2006, ISSN 1616-0177

Khatib, D.E. (1996). Coordination and Descentralisation of Multiple Cooperation of Multiple Mobile Manipulators, *Journal of Robotic Systems*, 13 (11) , 1996, pp. 755 - 764

Delay, L. & Chandrasekaran, M. (1987). *Les Editions Physique*, Les Ulis, 1987

Gat, E. (1998). On Three-layer architectures. In: *Artificial Intelligence and Mobile Robots: Case Studies of Successfid Robot Systems*, pp. 195-210, MIT Press, Cambridge MA

Cheng, F. T. & Orin, D. E. (1991). Efficient Formulation of the Force Distribution Equations for Simple Closed - Chain Robotic Mechanisms. *IEEE Trans on Sys. Man and Cyb.*, Vol. 21, 1991, pp. 25 -32

Cheng, F. T. & Orin, D. E . (1991). Optimal Force Distribution in Multiple-Chain Robotic Systems. *IEEE Trans. on Sys. Man and Cyb.*, Vol. 21, 1991, pp. 13 - 24

Cheng, F. T. (1995). Control and Simulation for a Closed Chain Dual Redundant Manipulator System. *Journal of Robotic Systems*, 1995, pp. 119 - 133

Evans, F.G. (1976). *Age changes in mechanical properties and histology of human compact bone*, pp.1361–1372, Phys Anthropol 20

Firby, R. J. (1987). An investigation into reactive planning in complex domains, *Proceedings of the Sixth National Conference on Artificial Intelligcncc*, pp. 202-206, 1987

Funakubo, H. (1987). *Shape Memory Alloys*, Gordon and Breach Science Publishers

Pioggia, G.; Ferro, M.; Sica, M.L.; Dalle Mura, G.; Casalini, S.; Ahluwalia, A.; De Rossi, D.; Igliozzi, R. & Muratori, F. (2006). Imitation and Learning of the Emotional Behaviour: Towards an android-based treatment for people with autism, *Proceedings of Sixth International Conference on Epigenetic Robotics: Modeling Cognitive Development in Robotic Systems Epirob06*, Parigi, September, 2006

Giralt, G.R. (1983). An integrated navigation and motion control system for autonomous mutisensory mobile robots, *Proceedings of the First International Symposium on Robotics Research*, pp. 191-214, 1983, MIT Press

Goetz, J.; Kiesler, S. & Powers, A. (2003). Matching Robot Appearance and Behavior to Tasks to Improve Human-Robot Cooperation, *Proceedings of IEEE Workshop on Robot and Human Interactive Communication*, 2003

Graesser, E.J. & Cozarelli, F.A. (1994). *Journal of Int. Material Systems and Structures*, 1994

http://ascon.net/blank.php?id=1634

http://ortho.smith-nephew.com/ca_en/Standard.asp?NodeId=2945

http://www0.sun.ac.za/ortho/webct-ortho/general/exfix/exfix.html

Ivanescu, M. & Bizdoaca, N. (2003). Dynamic control for a tentacle manipulator with SMA actuators, *Proceedings of IEEE International Conference on Robotics and Automation, ICRA 2003*, pp. 2079-2084,Taipei, Taiwan, September, 2003

Ivanescu, M. & Bizdoaca, N. (2003). A fuzzy controller for snake-type cooperative robots, *Proceedings of IFAC International Workshop on Intelligent Assembly and Disassembly*, pp. 102-107, October, 2003

Rosenblatt, J.; Willams, S. and & Durrant-Whyte, H. (2000). Behavior-based control for autonomous underwater exploration, *Proceedings of the 2000 IEEE International Conference on Robotics and Automation*, San Francisco, CA, April, 2000

Crossley, K.; Bennell, KL.; Wrigley, T. & Oakes, BW. (1999). *Ground reaction forces, bone characteristics, and tibial stress fracture in male runners.* Med Sci Sports Exerc 31, pp.1088–1093

Silverman, L. M. (1969). Inversion of Multivariable Linear Systems. *IEEE Trans. Aut. Contr.*, Vol AC - 14, 1969.

Wang, Li-Chun. (1996). Time-Optimal Control of Multiple Cooperating manipulators. *Jurnal of Robotic Systems*, 1996, pp. 229-241

Zadeh, L. (1965). Fuzzy sets. *Information and Control,* No.8, 1965, pp. 338-353

Giladi, M.; Milgrom, C.; Simkin, A. & M. Stein. (1987). Stress fractures and tibial bone width. A risk factor. *J Bone Joint Surg Br.*, No.69, 1987, pp.326–329

Mason, M. T. (1981). Compliance and Force Control. *IEEE Trans. Sys. Man Cyb.*, Nr. 6, 1981, pp. 418 - 432

Ivanescu, M. (1984). Dynamic Control for a Tentacle Manipulator, *Proceedings of International Conference*, Charlotte, USA, 1984

Ivanescu, M. (1986). A New Manipulator Arm: A Tentacle Model. *Recent Trends in Robotics*, Nov. 1986, pp. 51-57

Ivanescu, M. & Stoian, V. (1995). A Variable Structure Controller for a Tentacle Manipulator, *Proceedings of the 1995 IEEE International Conference on Robotics and Aut.*, Vol. 3, pp. 3155 - 3160, Nagoya, Japan, May,1995

Ivanescu, M. & Stoian, V. (1996). A Sequential Distributed Variable Structure Controller for a Tentacle Arm, *Proceedings of the 1996 IEEE International Conference on Robotics and Automation*, Vol. 4, pp. 3701 – 3706, Minneapolis, April 1996

Malcolm, C. & Smithers, T. (1990). Symbol grounding via a hybrid architecture in an autonomous assembly system. *Robotics and Autonomous Systems, Special Issue on Designing Autonomous Agents: Theory and Practice from Biology to Engineering and Back*, P. Maes, ed., Vol. 6, No. 1-2, 145-168, 1990

Mataric, M. (1990). Navigating With a Rat Brain: A Neurobiologically- Inspired Model for Robot Spatial Representation, *Proceedings, From Animals to Animats 1, First International Conference on Simulation of Adaptive Behavior,* pp. 169-175, 1990, MIT Press

Moravec, D.J. & Elfes, A. (1995). High resolutin maps from wide angle sonar, *Proccedings of IEEE*, 1985

Bizdoaca, N.G.; Tarnita, D.; Tarnita, D.N. (2006). Shape memory alloy programmable force medical staple. *BIO Materialien Interdisciplinary Journal of Functional Materials, Biomechanics and Tissue Engineering*, 2006, ISSN 1616-0177

Bizdoaca, N.G.; Degeratu, S.; Niculescu, M. & Pana, D. (2004). Shape Memory Alloy Based Robotic Ankle, Proceedings of 4th International Carpathian Control Conference, vol.I. Zakopane, Poland, 2004

Petrisor, A. & Bizdoaca, N. (2007). Mathematical model for walking robot with shape memory alloy ankle, *ISI Proceedings of the 4th International Conference on Informatics in Control, Automation and Robotics, ICINCO*, Angers, France, 9-12 may, 2007

Pirjanian, P.; Leger, C.; Mumm, E.; Kennedy, B.; Garrett, M.; Aghazarian, H.; Schenker, P. & Farritor, S. (2002). Behavior-Based Coordinated Control for Robotic Planetary Cliff Descent. *IEEE International Conference on Robotics and Automation*, May 2002.

Schilling, R.J. (1990). *Fundamentals Robotics*, Prent Hall

Brooks, R. (1990). The behavior language, user's guide, Tech.Rep. AIM-1227, MIT AI Lab, 1990

Rosenschein, S. & Kaebling, L.P. (1986). The synthesis of machines with provable epistemic proprieties. *Theoretical Aspects of Reasoning About Knowledge*, pp.83-98, Morgan Kaufmann Ed., Los Altos, CA, 1986

Han, S.S.; Choi, S.B.; Park, J.S.; Kim, J.H. & Choi, H.J. (2003). Robust sliding mode control of an electrorheological suspension system with parameter perturbations. *International Journal of Vehicle Design 2003*, Vol. 33, No.1/2/3, 2003, pp. 279 - 295

Schoppers, M. (1987). Universal plans for reactive robots in unpredictable domains, *Proceedings of IJCAI-87*, pp.1039-1046, Menlo Park, CA, 1987

Schroeder, B. Boller, Ch. (1998). Comparative Assessment of Models for Describing the Constitutive Behaviour of Shape Memory Alloy. *Smart Materials and Structures*, 1998

Yeong Yi, S. (1997). A robust Fuzzy Logic Controller for Robot Manipulators. *IEEE Transaction on Systems, Man and Cybernetics*, vol 27, No 4, 1997, pp. 706-713

Stalmans, R. (1993). *Doctorate Thesis*, Catholic University of Leuven, Dep. Of Metallurgy and Materials Science

Ross, T.J. (1995). Mc.Grow Hill, Inc.

Prescott, T.; Bryson, J. & Seth, A. (2007). Introduction. Modelling Natural Action Selection, *Proceedings of Philosophical Transactions of the Royal Society B 362(1485), special issue on Modelling Natural Action Selection*, pp. 1521–1529, Sept 2007

Utkin, V. I. (1993). Variable structure systems and sliding mode—State of the art assessment. *Variable Structure Control for Robotics and Aerospace Applications*, 1993, pp. 9-32, 1993, K. D. Young, Ed., New York: Elsevier

Utkin, V. I. (1977). Variable structure systems with sliding modes. *IEEE Transaction Automation Control*, Vol. AC-22, 1997, pp. 212-222

Waram, T. (1993). *Actuator Design Using Shape Memory Alloys*

Zheng, Y.F. & Luh, J.Y.S. (1988). Optimal Load Distribution for Two Industrial Robots Handling a Single Object, *Proceedings of IEEE International Conference Robotics Automation*, pp. 344 - 349, 1988